Ecophysiology and Biochemistry of Cyanobacteria

Rajesh Prasad Rastogi
Editor

Ecophysiology and Biochemistry of Cyanobacteria

Springer

Editor
Rajesh Prasad Rastogi
Ministry of Environment, Forest and
Climate Change
New Delhi, Delhi, India

ISBN 978-981-16-4872-4 ISBN 978-981-16-4873-1 (eBook)
https://doi.org/10.1007/978-981-16-4873-1

This Springer imprint is published by the registered company Springer Nature Singapore Pte Ltd.
The registered company address is: 152 Beach Road, #21-01/04 Gateway East, Singapore 189721, Singapore

Preface

Cyanobacteria are the most dominant prokaryotic floras on the Earth's surface and are of great importance in terms of ecological, economical and evolutionary perspectives. They are oldest groups of photosynthetic autotrophs, which create oxygenic atmosphere for the development and sustainability of ecosystems with different life forms. Recently, cyanobacteria have been employed in space research, bioremediation, as well as an efficient source of ecofriendly and alternative source of renewable energy in connection with photosynthesis—the most important life-supporting biological phenomena of the planet.

This book emphasizes and establishes the emerging information on ecophysiology and biochemistry of cyanobacteria with special emphasis on their biodiversity, molecular mechanisms of some important biological processes and survival mechanisms under myriad of environmental conditions as well as bioremediation. Also included is an integrative approach to their possible biotechnological application in the field of bioenergy and various aspects of biochemistry, biophysics and structural biology of photosynthesis.

This book has attempted to span the depth of cyanobacterial biology from the perspective of its basic ecophysiology and biochemistry starting with more general information about cyanobacteria such as evolution, distribution, taxonomy and photosynthesis in Chaps. 1–3. Chapters 4–7 focus on the impacts of environmental stress on physiology and biochemistry of cyanobacteria along with UV stress response and molecular mechanisms of stress tolerance in cyanobacteria. Chapter 8 presents an overview on stress proteins and signal transduction in cyanobacteria, whereas Chapter 9 focuses on molecular chaperones and their involvement in maintaining the cellular protein homeostasis under normal and stress conditions. Chapters 10 and 11 describe the chromatic acclimation in response to light quality and phenomenon of allelopathy in cyanobacteria, respectively. In Chapter 12, assembly of nitrogen-fixing machinery and role of key enzymes in nitrogen metabolism of cyanobacteria have been discussed. Chapters 13 and 14 address cyanobacteria-based phycoremediation for effective removal of numerous pollutants from waste effluents. Chapters 15–20 all deal with biochemistry of cyanobacteria uncovering their potential applications towards biotechnological values. Chapter 15 discusses antioxidant, anti-ageing and neuroprotective potential of various cyanobacterial biomolecules, while Chapter 16 describes the engineering challenges of carbon dioxide capture and sequestration by cyanobacteria to reach a better and greener world. Chapter 17 reviews the significant development and the recent progress in engineering cyanobacteria for photosynthetic production of sucrose and sucrose-synthesis mechanisms. Chapters 18–20 highlight the concept of cyanobacterial bio-refineries for future bio-energy/bio-fuel demand.

I believe that this book will be helpful to a great extent for the academicians and researchers in the field of cyanobacterial research. Certainly, the contents incorporated in this book can be used as a textbook by undergraduate and post-graduate students, teachers and researchers in the most interesting fields of physico-chemical ecology and biochemistry of cyanobacteria.

It is very sad to mention here that Mr. Mukesh Ghanshyam Chaubey, first author of the Book Chapter 15 passed away on 23rd November, 2020 due to Covid-19. We pray that his soul rests in peace and may God give enough strength to the bereaved family to bear the irreparable loss.

I am highly thankful to all the peer-reviewers for their thoughtful assistance in reviewing the manuscripts. I thank Dr. Madhurima Kahali, Editor (Book), Springer, India, for her assistance in seeing it through to completion. I am sincerely grateful to the entire team of Springer Nature for the coordination, support and implementation of this book project. Last but not least, I express my sincere gratitude to all the authors for their kind collaboration and scientific contributions towards completion of this book successfully.

New Delhi, Delhi, India
January 2021

Rajesh P. Rastogi

Contents

Editor and Contributors

About the Editor

Rajesh Prasad Rastogi is a scientist in the Ministry of Environment, Forest and Climate Change, New Delhi, India. He completed his Ph. D. in Botany at Banaras Hindu University, Varanasi, India. Dr. Rastogi is the recipient of some national and international awards and fellowships. He was a visiting scientist at Friedrich Alexander University, Nuremberg, Germany, and served as a visiting professor of biochemistry at Chulalongkorn University, Thailand. His major research areas are freshwater or marine algae and cyanobacterial ecology and biotechnology, plant physiology, photochemistry and photobiology and stress biology. He has published a number of research papers in journals of international repute and several book chapters and books, having more than 4000 citations with 32 h-index and 57 i10-index. He is serving as editorial board member of some national and international journals and is life member of several scientific organizations.

Contributors

Vessela Balabanova Department of Pharmacognosy, Faculty of Pharmacy, Medical University of Sofia, Sofia, Bulgaria

Deepika Bhatia Department of Biotechnology and Medical Sciences, Baba Farid College, Bathinda, Punjab, India

Gracjana Budzałek Division of Marine Ecosystems Functioning, Institute of Oceanography, University of Gdańsk, Gdynia, Poland

Mukesh Ghanshyam Chaubey Shri A. N. Patel P. G. Institute of Science and Research, Anand, Gujarat, India

Vivek Dalvi Applied Microbiology Laboratory, Center for Rural Development & Technology, Indian Institute of Technology, Delhi, Hauz Khas, New Delhi, India

Himanshu G. Dawda Ramniranjan Jhunjhunwala College, Mumbai, India

Daljeet Singh Dhanjal Department of Biotechnology, Lovely Professional University, Phagwara, Punjab, India

Reneta Gevrenova Department of Pharmacognosy, Faculty of Pharmacy, Medical University of Sofia, Sofia, Bulgaria

Amit Kumar Gupta Department of Botany, Mohanlal Sukhadia University, Udaipur, Rajasthan, India

Donat P. Häder Department of Biology, Emeritus of Friedrich-Alexander University, Möhrendorf, Germany

Harish Department of Botany, Mohanlal Sukhadia University, Udaipur, Rajasthan, India

Aran Incharoensakdi Laboratory of Cyanobacterial Biotechnology, Department of Biochemistry, Faculty of Science, Chulalongkorn University, Bangkok, Thailand
Academy of Science, Royal Society of Thailand, Bangkok, Thailand

Rahul Jain Applied Microbiology Laboratory, Center for Rural Development & Technology, Indian Institute of Technology, Delhi, Hauz Khas, New Delhi, India

Dhriti Kapoor Department of Botany, Lovely Professional University, Phagwara, Punjab, India

Kinga Kłodawska Faculty of Biochemistry, Biophysics and Biotechnology, Jagiellonian University in Krakow, Krakow, Poland

Zofia Konarzewska Division of Marine Ecosystems Functioning, Institute of Oceanography, University of Gdańsk, Gdynia, Poland

Alexander Dimitrov Kroumov Department of Biotechnology—Laboratory of Bioconversion and Biosynthesis of Microbial Metabolites, The Stephan Angeloff Institute of Microbiology—Bulgarian Academy of Sciences, Sofia, Bulgaria

Mukesh Kumar Department of Botany, Mohanlal Sukhadia University, Udaipur, Rajasthan, India

Vijay Kumar Regional Ayurveda Research Institute for Drug Development, Gwalior, Madhya Pradesh, India

Vinod Kumar Department of Botany, Centre of Advanced Study in Botany, Institute of Science, Banaras Hindu University, Varanasi, India

Geetanjali Kumawat Department of Botany, Mohanlal Sukhadia University, Udaipur, Rajasthan, India

Guodong Luan Key Laboratory of Biofuels, Qingdao Institute of Bioenergy and Bioprocess Technology, Chinese Academy of Sciences, Qingdao, China
Shandong Energy Institute, Qingdao, China
Qingdao New Energy Shandong Laboratory, Beijing, China
Dalian National Laboratory for Clean Energy, Dalian, China

Quan Luo Key Laboratory of Biofuels, Qingdao Institute of Bioenergy and Bioprocess Technology, Chinese Academy of Sciences, Qingdao, China
Shandong Energy Institute, Qingdao, China
Qingdao New Energy Shandong Laboratory, Beijing, China

Xuefeng Lu Key Laboratory of Biofuels, Qingdao Institute of Bioenergy and Bioprocess Technology, Chinese Academy of Sciences, Qingdao, China
Shandong Energy Institute, Qingdao, China
Qingdao New Energy Shandong Laboratory, Beijing, China
Dalian National Laboratory for Clean Energy, Dalian, China
Laboratory for Marine Biology and Biotechnology, Qingdao National Laboratory for Marine Science and Technology, Qingdao, China

Datta Madamwar P. D. Patel Institute of Applied Sciences, Charotar University of Science and Technology, Anand, Gujarat, India

Anushree Malik Applied Microbiology Laboratory, Center for Rural Development & Technology, Indian Institute of Technology, Delhi, Hauz Khas, New Delhi, India

Pankaj K. Maurya Department of Botany, Centre of Advanced Study in Botany, Institute of Science, Banaras Hindu University, Varanasi, India

Soumila Mondal Department of Botany, Centre of Advanced Study in Botany, Institute of Science, Banaras Hindu University, Varanasi, India

Hristo Najdenski Department of Infectious Microbiology, The Stephan Angeloff Institute of Microbiology—Bulgarian Academy of Sciences, Sofia, Bulgaria

Hitoshi Nakamoto Department of Biochemistry and Molecular Biology, Graduate School of Science and Engineering, Saitama University, Saitama, Japan

Harshita Nigam Applied Microbiology Laboratory, Center for Rural Development & Technology, Indian Institute of Technology, Delhi, Hauz Khas, New Delhi, India

Sunil Pabbi Division of Microbiology, Centre for Conservation and Utilisation of Blue Green Algae, ICAR—Indian Agricultural Research Institute, New Delhi, India

Aparna Pandey Ranjan Plant Physiology and Biochemistry Laboratory, Department of Botany, University of Allahabad, Allahabad, India

Neeraj Pandey Ranjan Plant Physiology and Biochemistry Laboratory, Department of Botany, University of Allahabad, Allahabad, India

Anuradha Patel Ranjan Plant Physiology and Biochemistry Laboratory, Department of Botany, University of Allahabad, Allahabad, India

Stuti Nareshkumar Patel Department of Biosciences, Sardar Patel University, Bakrol, Anand, Gujarat, India

Jainendra Pathak Department of Botany, Pt. Jawaharlal Nehru College (Affiliated to Bundelkhand University), Banda, India

Krutika Patil Division of Microbiology, Centre for Conservation and Utilisation of Blue Green Algae, ICAR—Indian Agricultural Research Institute, New Delhi, India

Sheo Mohan Prasad Ranjan Plant Physiology and Biochemistry Laboratory, Department of Botany, University of Allahabad, Allahabad, India

Krishna Kumar Rai Molecular Biology Section, Centre of Advanced Study in Botany, Institute of Science, Banaras Hindu University, Varanasi, India

L. C. Rai Molecular Biology Section, Centre of Advanced Study in Botany, Institute of Science, Banaras Hindu University, Varanasi, India

Ruchi Rai Molecular Biology Section, Centre of Advanced Study in Botany, Institute of Science, Banaras Hindu University, Varanasi, India

Alka Raj Molecular Biology Section, Centre of Advanced Study in Botany, Institute of Science, Banaras Hindu University, Varanasi, India

Praveen C. Ramamurthy Interdisciplinary Centre for Water Research (ICWaR), Indian Institute of Sciences, Bangalore, India

Rajesh P. Rastogi Ministry of Environment, Forest and Climate Change, New Delhi, Delhi, India

Nedeljka Rosic School of Health and Human Sciences, Southern Cross University, Gold Coast, QLD, Australia
Marine Ecology Research Centre, Southern Cross University, Lismore, NSW, Australia

Vishambhar Sangela Department of Botany, Mohanlal Sukhadia University, Udaipur, Rajasthan, India

Aniket Saraf Ramniranjan Jhunjhunwala College, Mumbai, India

Fabiano Bisinella Scheufele Graduation Program of Biotechnology and Bioprocess Engineering, Federal University of Technology—Paraná—UTFPR, Toledo, Paraná, Brazil

Kunal Seth Department of Botany, Government Science College, Valsad, Gujarat, India

Shweta Shekhar Department of Material Engineering, Indian Institute of Sciences, Bangalore, India

Garima Singh Ranjan Plant Physiology and Biochemistry Laboratory, Department of Botany, University of Allahabad, Allahabad, India

Joginder Singh Department of Microbiology, Lovely Professional University, Phagwara, Punjab, India

Niraj Kumar Singh Shri A. N. Patel P. G. Institute of Science and Research, Anand, Gujarat, India

Nitika Singh Department of Botany, Government College Bundi, Bundi, Rajasthan, India

Prashant R. Singh Laboratory of Photobiology and Molecular Microbiology, Centre of Advanced Study in Botany, Institute of Science, Banaras Hindu University, Varanasi, India

Prashant Singh Laboratory of Cyanobacterial Systematics and Stress Biology, Department of Botany, Institute of Science, Banaras Hindu University, Varanasi, India

Shailendra P. Singh Department of Botany, Centre of Advanced Study in Botany, Institute of Science, Banaras Hindu University, Varanasi, India

Shilpi Singh Molecular Biology Section, Centre of Advanced Study in Botany, Institute of Science, Banaras Hindu University, Varanasi, India

Simranjeet Singh Interdisciplinary Centre for Water Research (ICWaR), Indian Institute of Sciences, Bangalore, India

Rajeshwar P. Sinha Laboratory of Photobiology and Molecular Microbiology, Centre of Advanced Study in Botany, Institute of Science, Banaras Hindu University, Varanasi, India

Ramachandran Sivaramakrishnan Laboratory of Cyanobacterial Biotechnology, Department of Biochemistry, Faculty of Science, Chulalongkorn University, Bangkok, Thailand

Sylwia Śliwińska-Wilczewska Division of Marine Ecosystems Functioning, Institute of Oceanography, University of Gdańsk, Gdynia, Poland

Ravi R. Sonani Malopolska Centre of Biotechnology, Jagiellonian University, Krakow, Poland

Huili Sun Key Laboratory of Biofuels, Qingdao Institute of Bioenergy and Bioprocess Technology, Chinese Academy of Sciences, Qingdao, China
Shandong Energy Institute, Qingdao, China
Qingdao New Energy Shandong Laboratory, Beijing, China

Jiahui Sun Key Laboratory of Biofuels, Qingdao Institute of Bioenergy and Bioprocess Technology, Chinese Academy of Sciences, Qingdao, China
Shandong Energy Institute, Qingdao, China
Qingdao New Energy Shandong Laboratory, Beijing, China

Sanjesh Tiwari Ranjan Plant Physiology and Biochemistry Laboratory, Department of Botany, University of Allahabad, Allahabad, India

Kinga A. Wiśniewska Division of Marine Chemistry and Environmental Protection, Institute of Oceanography, University of Gdańsk, Gdynia, Poland

Maya Margaritova Zaharieva Department of Infectious Microbiology, The Stephan Angeloff Institute of Microbiology—Bulgarian Academy of Sciences, Sofia, Bulgaria

Shanshan Zhang Key Laboratory of Biofuels, Qingdao Institute of Bioenergy and Bioprocess Technology, Chinese Academy of Sciences, Qingdao, China
Shandong Energy Institute, Qingdao, China
Qingdao New Energy Shandong Laboratory, Beijing, China

Dimitrina Zheleva-Dimitrova Department of Pharmacognosy, Faculty of Pharmacy, Medical University of Sofia, Sofia, Bulgaria

Evolution and Distribution of Cyanobacteria

1

Jainendra Pathak, Prashant R. Singh, Rajeshwar P. Sinha, and Rajesh P. Rastogi

Abstract

Cyanobacteria are the ancient group of photosynthetic prokaryotes having pronounced variations in their physiological capacities, cellular differentiation strategies, and choice of habitats. They are the inventors of oxygenic photosynthesis on this planet and hence have played a crucial role in the evolution of biodiversity on Earth by gradually changing the atmospheric chemistry to be suitable for the evolution of eukaryotes. This conversion of atmosphere from anaerobic to an aerobic one was started by cyanobacteria through oxygenic photosynthesis, which finally supplied oxygen to the atmosphere for ~1.5 billion years leading to greater diversification of life on the Earth. Cyanobacteria inhabit a wide range of terrestrial and aquatic environments varying from the hot springs to polar region and other extreme environments. Their long-standing evolutionary history might be the reason for their success in acclimatization and sustenance in such diverse habitats. A high tolerance level of free sulfide and low oxygen, tolerance to lethal ultraviolet radiations, and the capacity to use H_2S in place of H_2O as a photoreductant are some of the various features of cyanobacteria that have aided in supporting their long history on this planet. Still, the picture

J. Pathak (✉)
Department of Botany, Pt. Jawaharlal Nehru College (Affiliated to Bundelkhand University, Jhansi), Banda, India

P. R. Singh · R. P. Sinha
Laboratory of Photobiology and Molecular Microbiology, Centre of Advanced Study in Botany, Institute of Science, Banaras Hindu University, Varanasi, India

R. P. Rastogi
Ministry of Environment, Forest and Climate Change, New Delhi, Delhi, India

R. P. Rastogi (ed.), *Ecophysiology and Biochemistry of Cyanobacteria*,
https://doi.org/10.1007/978-981-16-4873-1_1

regarding evolution and diversification of this ecologically and biotechnologically important group of photoautotrophs is not very clear. In this chapter, we present an overview of structural and genomic evolution of cyanobacteria and their distribution in diverse habitats on Earth.

Keywords

Cyanobacteria · Evolution · Phylogeny · Horizontal gene transfer · Cyanophages

1.1 Introduction

Cyanobacteria (blue-green algae) constitute a large and diverse group of photosynthetic bacteria, which range from single cells to large filamentous thallus and have tremendous potentials for applications in biotechnology, mariculture, agriculture, food and fuel, and biomedicals (Rastogi and Sinha 2009; Rajneesh et al. 2017; Singh 2017; Pathak et al. 2018). Being at the base of global carbon and nitrogen biogeochemical cycles, cyanobacteria have played crucial roles in the evolutionary past and in modern ecosystems (Kopp et al. 2005; Larsson et al. 2011). Global oxygenation of the atmosphere resulted in a radical transformation of the Earth, which occurred ~2.45–2.23 billion years ago, and this transformation was termed as the "Great Oxidation Event" (GOE), which changed the chemistry of atmosphere from a reduced state to an oxidized state, and this event was probably associated with the development of cyanobacteria-producing oxygen (Canfield 2005; Holland 2006; Shestakov and Karbysheva 2017; Sánchez-Baracaldo and Cardona 2020). Cyanobacteria might have emerged ~3 billion years ago and transformed the anoxygenic conditions of Earth to the oxygenic conditions through photosynthesis (Schirrmeister et al. 2011a). First oxygenic photosynthesis performing cyanobacteria could have arisen in the Archean time period in the local environments (Anbar et al. 2007; Lyons and Reinhard 2011), and this rise of oxygen on Earth facilitated the growth and development of complex multicellular life with aerobic respiration and profoundly altered the course of evolution on Earth (Soo et al. 2017). Surprisingly, the fossilized forms of cyanobacteria showed similarity to the cyanobacterial species of the present times, hence indicating the slow pace of evolutionary advancement in cyanobacteria (Henson et al. 2002). During their course of evolution, cyanobacteria became one of the most widely distributed and diverse prokaryotes, which occupy several niches within aquatic, benthic, and terrestrial habitats (Rastogi et al. 2012; Pathak et al. 2017; Walter et al. 2017; Gaysina et al. 2019).

Cyanobacteria have been named according to the Botanical Code as they share similar features with eukaryotic algae (Kauff and Büdel 2010; Walter et al. 2017). Currently, there is no consensus regarding taxa nomenclature of cyanobacteria and this has long been a topic of discussion (Hoffmann et al. 2005; Oren and Tindall 2005; Oren et al. 2009; Schirrmeister et al. 2011a; Oren and Ventura 2017; Singh 2017). Owing to their photosynthetic ability, the presence of chlorophyll *a* and distinct cell wall cyanobacteria have been grouped with plants and classified as

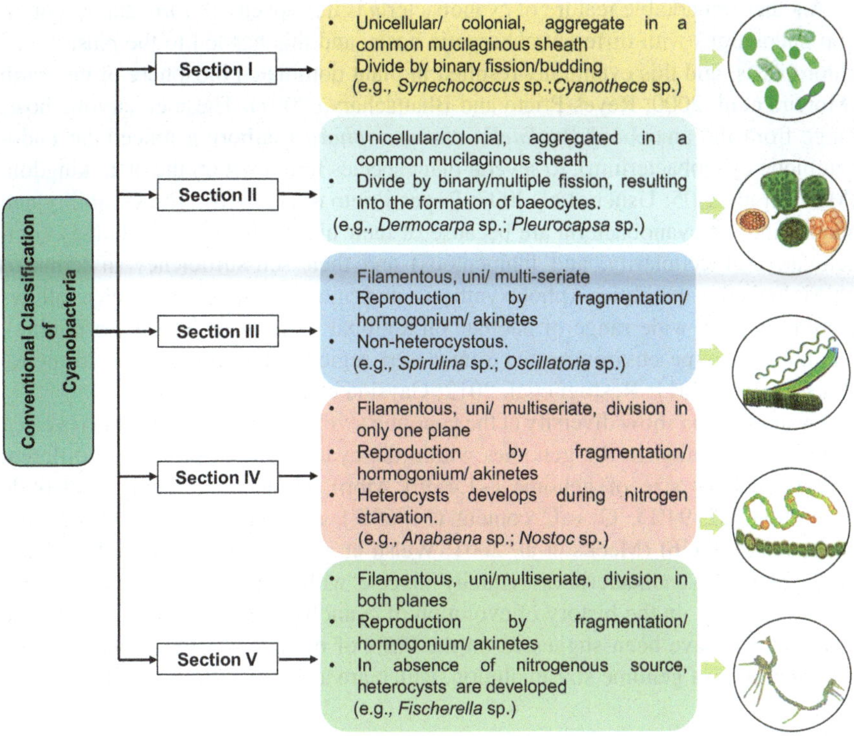

Fig. 1.1 Conventional classification of cyanobacteria. (For details, see the Refs. Rippka et al. 1979; Schirrmeister et al. 2016)

algae. The prime basis of this classification design was their morphological attributes and the developmental characteristics (Rippka et al. 1979; Rippka 1988; Rippka and Herdman 1992; Castenholz 2001). According to this classification design, cyanobacteria were formally recognized into five sections (Fig. 1.1).

Section I constitutes unicellular cyanobacteria in which division takes place by binary fission or budding, whereas in Section II division takes place by multiple fission, resulting in the development of baeocytes. Sections III, IV, and V constitute the filamentous forms of cyanobacteria. In Section III, members were filamentous but nonheterocystous that proliferated by trichome breakage. Sections IV and V represented the heterocystous cyanobacteria having ability to develop akinetes and heterocysts, and hormogonia formation was their main mode of reproduction. These two sections were further subdivided into two subsections, viz. Stigonematales and Nostocales on the basis of the plane of division. Section IV consisted of the nostocalean members, which comprised of the cyanobacterial strains that divided in only one plane, while the stigonematalean line represented by Section V constituted cyanobacteria, which have the ability to divide in more than one plane (Rippka et al. 1979; Rippka 1988).

Another remarkable feature of cyanobacteria is its capacity to form stable symbiotic interactions with different eukaryotic hosts, and this has led to the plastid, i.e., chloroplasts, and this eventually resulted in plant dominated biosphere of the Earth (Moreira et al. 2000; Reyes-Prieto and Bhattacharya 2007). These eukaryotic hosts range from the amoeboid *Paulinella chromatophora* (harboring unicellular endosymbiotic cyanobacterium), to several plant species found within the plant kingdom (Marin et al. 2005; Usher et al. 2007; Reyes-Prieto et al. 2010). The flexibility and adaptability of cyanobacteria are because of their highly diverse morphology (unicellular, multicellularity, and filamentous) and their self-sufficiency in terms of physiological capabilities (photosynthesis and nitrogen fixation), which allows them to occupy wide range of habitats on a global scale (soils/freshwater/marine), including extreme environments (from desert regions, hot springs to cold arctic) (Larsson et al. 2011; Rastogi et al. 2012; Gaysina et al. 2019).

Cyanobacteria show diversity at the genomic level also. Sequencing data revealed significant variation in the genomes within the cyanobacterial phylum in different aspects such as size of genome (~1.4–9.1 Mbp), number of coding nucleotide proportion (52–94%), G + C content (31–63%), and number of protein-coding genes (1214–8446) (Meeks et al. 2001; Welsh et al. 2008; Ran et al. 2010; Tripp et al. 2010). Cyanobacteria are equally diverse with respect to size and protein-coding capacity. In the history of evolution of cyanobacteria, two routes of genome development have been suggested on the basis of multicopy gene abundance and different rates of genome size evolution, which are as follows (Larsson et al. 2011):

1. The genome expansion
2. The genome streamlining

The genome expansion is achieved by gene family enlargement and develops a broad adaptive potential, whereas the genome streamlining imposes adaptations to highly specific niches and is also indicated in their different functional capacities (Larsson et al. 2011). Based on *16S rRNA* gene sequences, 170 genera of cyanobacteria have been proposed (Kozlov et al. 2016). Farrant et al. (2016) delineated 15 *Synechococcus* and 121 *Prochlorococcus* ecologically significant taxonomic units in the marine ecosystems utilizing single-copy petB sequences (encoding cytochrome b6) and different environmental cues. Cyanobacteria constitute a challenging group for the ecologists and microbiologists. Traditional taxonomy based on only morphologic traits does not completely reflect the results of phylogenetic analyses in cyanobacteria (Singh 2017). The *16S rRNA* gene sequences can be a useful data in characterizing and charting microbial communities, but it lack the sensitivity for evolutionary changes that take place in ecological dynamics, where physicochemical parameters determine the microbial diversity (Choudoir et al. 2012; Becraft et al. 2015; Kozlov et al. 2016). The long history of cyanobacteria provided them with a broad heterogeneity comprising multicellular and unicellular with genomes sizes ranging from 1 to 10 Mb, nonphotosynthetic (*Melainabacteria*) and photosynthetic, symbiotic, free-living, toxic, and predatory organisms (Schirrmeister et al. 2011b; Di Rienzi et al. 2013; Shih et al. 2013; Soo et al. 2014, 2015). The processes that shaped cyanobacterial communities over time and space are still not very clear, and this chapter endeavors to decipher the complex evolutionary pattern in this group of oxygenic photoautotrophs along with their mesmerizing diversity in a wide range of habitats.

1.2 Evolution of Cyanobacteria

It is believed that the first cyanobacteria could have appeared ~2.7–2.6 billion years ago in local warm shallow water bodies that formed small oxygen "oases" within the biosphere, which was anoxygenic (Buick 2008; Blank and Sanchez-Baracaldo 2010). In Archean sediments, microfossils morphologically similar to filamentous cyanobacteria were found (Schopf 1993; Buick 2008). Previously, carbonate stromatolites were considered as analogs of modern cyanobacterial mats, which were dated back ~3.5–3.2 billion years old. In later studies, it was found that both ancient mats and microfossils were probably formed by filamentous anaerobic bacteria and not by cyanobacteria (Brasier et al. 2006; Fedonkin 2006; Bosak et al. 2013). Among some eukaryotic clades, oxygenic photosynthesis spread at least 1.05 Ga ago, which resulted in diverse types of algae and plants (Fig. 1.2) (Demoulina et al. 2019).

This crucial evolutionary process was because of the primary endosymbiosis of a cyanobacterium within a unicellular eukaryote and subsequent endosymbiotic events of higher order (Sagan 1967; Delwiche 1999; Schirrmeister et al. 2011b). Despite the importance of cyanobacteria in the early evolution of life on Earth, basic questions remain about the origin of these ancient groups of photoautotrophs, origin of oxygenic photosynthesis, and pattern and timing of diversification of cyanobacteria, in the time range from the Archean to the GOE (Fischer et al. 2016). Discrepancy between the unambiguous record of cyanobacterial fossil (starting at 1.9 Ga, the GOE at 2.4 Ga), and several older geochemical data suggestive of oxygenic photosynthesis, is one crucial problem, which needs to be addressed carefully (Rosing and Frei 2004; Shen and Buick 2004). For reconstructing the fossil record of cyanobacteria, several types of evidence are used but all have their challenges and limitations (Demoulina et al. 2019). Cyanobacterial fossil stromatolites are usually associated with cyanobacterial activity; however, although conical stromatolites seem indicate for oxygenic photosynthesis, microbially induced sedimentary structures (MISSs) and other types of stromatolites may have been produced by noncyanobacterial lineages such as anoxygenic phototrophs and their association with methanotrophs (Noffke et al. 2001; Bosak et al. 2009, 2013; Heubeck 2009; Slotznick and Fischer 2016; Homann et al. 2018). These studies indicate that MISS and stromatolites do not reflect cyanobacterial activity and not even photosynthesis by cyanobacteria (Suosaari et al. 2016). Direct evidence for cyanobacteria may be provided by microfossils, but because of their ambiguous identification they are not very reliable (Demoulina et al. 2019). Presently, identity of only three cyanobacterial fossil taxa is not debated namely *Polybessurus, Eohyella,* and *Eoentophysalis*. The oldest cyanobacterial

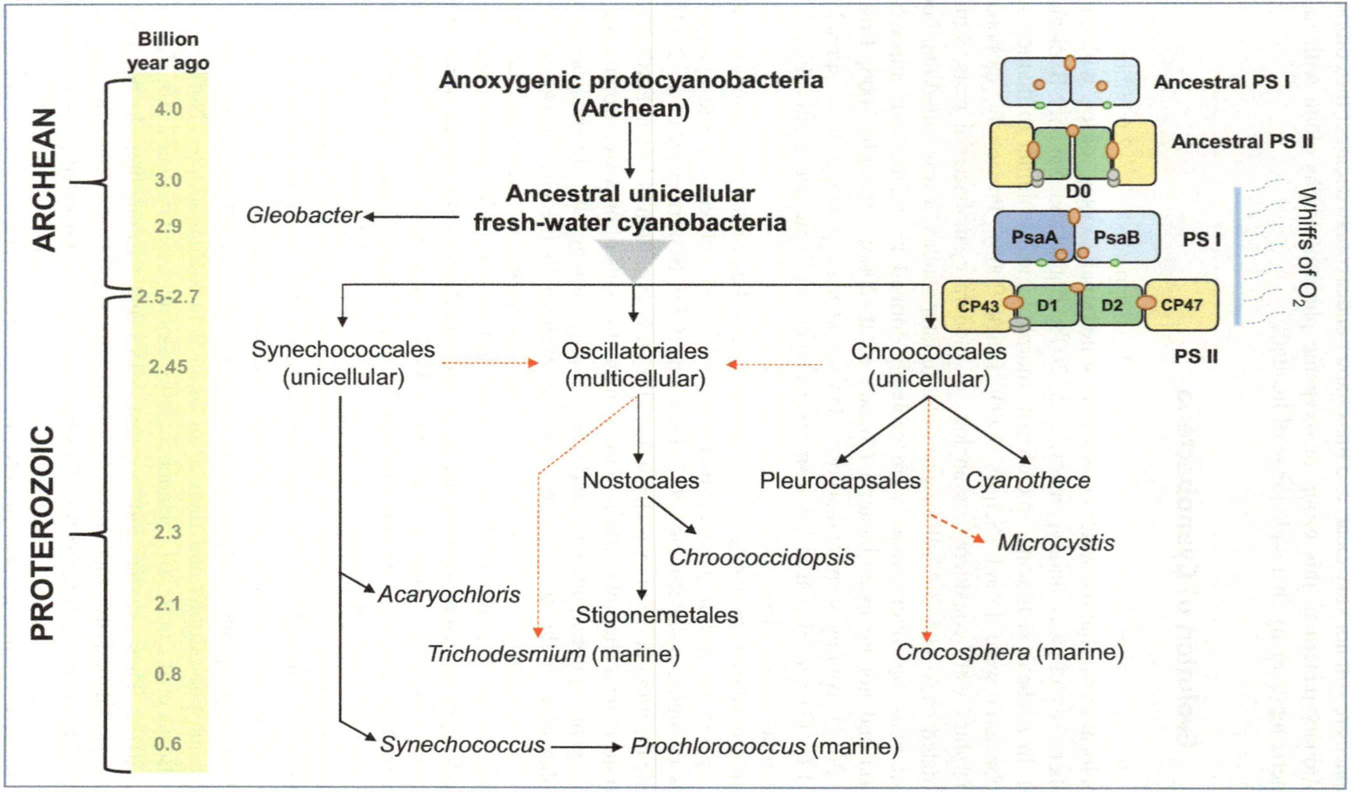

Fig. 1.2 Chronology of the evolution and origin of the cyanobacterial orders and some families in the geological scale of the Earth's history. Timeline of the appearance of photosystem (PS) I, PSII, and cyanobacterial lineages has been shown. D0 depicts an ancestral core subunit before the gene duplication that led to evolution of D1 and D2 proteins. (For details, see the Refs. Shestakov and Karbysheva 2017; Sánchez-Baracaldo and Cardona 2020)

microfossil interpreted with certainty as a cyanobacterium is *Eoentophysalis belcherensis,* which is silicified stromatolites of the Belcher supergroup found from Hudson Bay, Canada, and dates back to 1.89–1.84 Ga (Hofmann 1976).

Biomarkers (fossil molecules) can also serve as indicator of oxygenic photosynthesis, but these biomolecules are present only in the well-preserved unmetamorphosed rocks and their contamination is a big challenge to study these fossil molecules (Alleon and Summons 2019). Among such biomarkers, pigments such as porphyrins with N isotope composition and lipids such as 2-methyl-hopanes are produced by some cyanobacteria (Rashby et al. 2007; Schinteie and Brocks 2017; Gueneli et al. 2018). The fossilized porphyrins exhibit a specific fractionation of N isotope reflecting a cyanobacterial source and also indicate that in mid-proterozoic oceans cyanobacteria were the main primary producers (Gueneli et al. 2018). Ultraviolet-absorbing (sunscreen) pigments/compounds such as the mycosporine-like amino acids (MAAs), and two pigments specific to cyanobacteria, i.e., scytonemin and gloeocapsin, may be used as biosignature for bacterial life (Rastogi and Sinha 2009; Rastogi et al. 2013; Pathak et al. 2015; Demoulina et al. 2019).

Scytonemin, the novel multipurpose pigment, is a sun-screening molecule composed of phenolic and indolic subunits and is specific to cyanobacteria (Proteau et al. 1993; Rastogi et al. 2013; Pathak et al. 2020). It is biosynthesized in several species of cyanobacteria having exopolysaccharide sheaths (Rastogi et al. 2013; Pathak et al. 2017; Pandey et al. 2020) including benthic filaments of *Calothrix* sp. (Lepot et al. 2014), *Hyella* sp., and *Solentia* sp. (the endolithic cyanobacteria) from coastal carbonates (Storme et al. 2015). Scytonemin may be a promising biosignature of cyanobacterial presence given that it can be fossilized (Fulton et al. 2012; Lepot et al. 2014). Carotenoids and derivatives of scytonemin can be extracted from 125,000 years BP sediments in older deposits from Antarctica (Hodgson et al. 1997). However, there is scarce information about the preservation potential of scytonemin in older rocks. The recalcitrance of filamentous polysaccharide sheaths, possibly helped by the presence of pigments, was observed by artificial taphonomic experiments of decaying cyanobacterial cultures (Bartley 1996). However, both transparent (scytonemin-poor) and brown (scytonemin-rich) filamentous sheaths were found to be well preserved in lake sediments from Antarctica; hence, it was found that scytonemin probably was not the factor responsible for their preservation (Lepot et al. 2014).

For piecing together the events that occurred around the Earth's oxygenation, understanding of the origins of cyanobacteria and oxygenic photosynthesis is required (Sánchez-Baracaldo and Cardona 2020). It is believed that evolution of photosynthesis occurred within bacterial lineages, which are not extant; hence, studying the early history of photosynthesis becomes challenging. Our knowledge about the evolution of cyanobacteria and evolution of photosynthetic reaction centers has changed significantly due to recent findings on molecular and genomic and evolution (Sánchez-Baracaldo and Cardona 2020). In cyanobacteria, the photosynthetic apparatus was optimized and various strategies for protection against the lethal effects of the oxygen produced from cyanobacteria were developed during the slow evolution of these ancient photoautotrophs, which lasted for hundreds of millions of years (Garcia-Pichel 1998). After oxidation of the oceans which occurred for a long time cyanobacteria relatively and rapidly occupied the photic zone on surface of the oceans, whereas oxygenation of deeper layers of oceans occurred much later (Johnston et al. 2009). As mentioned previously, these ancient photoautotrophs changed the ecological and geochemical parameters of the planet *via* production of oxygen and played crucial role in the evolution of the aerobic atmosphere, which led to the formation of complex communities and eukaryotes (Shestakov and Karbysheva 2017). Combination of paleobiological and geological approaches aids in better understanding of microbiology of modern cyanobacteria (Demoulina et al. 2019). Estimation of the origin of the oxygenic photosynthesis and the origin of phylum cyanobacteria can be better understood through the increasing cyanobacterial genetic data, which allow molecular clock analyses and phylogenetic reconstructions (Demoulina et al. 2019). However, these estimates are quite variable because of the contamination of genetic sequences, lack of tree calibrations from the fossil record, chosen dataset, and differences/limitations in models (Schirrmeister et al. 2016). Thus, there are discrepancies between the fossil and geological records and molecular phylogenies and the origin and evolution of oxygenic photosynthesis, cyanobacteria, and chloroplast are still debated (Demoulina et al. 2019).

1.2.1 Structural Evolution

Phylogenetic relationship based on conservative housekeeping genes and *16S rRNA* gene sequences reveals the relationship between the cyanobacterial taxa but does not give a complete picture of the evolutionary relationships between different taxa reflecting the pathways of losses and acquisitions of ecologically crucial properties such as halophility, thermophility, production of toxins, and motility (Shestakov and Karbysheva 2017). These properties can be expressed independently in cyanobacteria, which are phylogenetically distant due to duplication, horizontal gene transfer (HGT), genome rearrangements, and neofunctionalization, which affect the regulation of cellular metabolism (Shestakov and Karbysheva 2017).

Certain tendencies of cyanobacteria such as their ability of aggregation and colonies formation, specialized cells, and multicellular filaments increase adaptation to varying environmental conditions due to enhancement in the reliability of metabolic cooperation and functions. Initially, it was believed that multicellular blue-green appeared from anoxygenic nitrogen-fixing heterotrophic bacteria during the earliest stages of evolution on Earth (Gupta 1982). These multicellular forms could have evolved simultaneously with the unicellular forms, which evolved from photoautotrophic bacterium. Molecular phylogenetics studies suggested that phylum cyanobacteria have monophyletic origin (Shestakov and Karbysheva 2017). Phylogenetic data indicate that multicellular cyanobacteria evolved from small unicellular coccoid cyanobacteria inhabiting aquatic (freshwater) habitats (Blank and Sanchez-Baracaldo 2010; Larsson et al. 2011). The phylogenetic trees constructed with different methods revealed that bacterium *Gloeobacter violaceus* (living fossil), which possessed a photosynthetic apparatus of primitive organization, occupies a root position distant from other cyanobacteria (Shi and Falkowski 2008; Gupta and Mathews 2010; Nguyen et al. 2012; Shih et al. 2013).

During early stages of evolution, different groups of both unicellular and multicellular families of cyanobacteria originated from ancestral unicellular form (related to protocyanobacterium) (Schirrmeister et al. 2013). Polyphasic analysis revealed representatives of phylogenetically related families Leptolyngbyaceae and Pseudoanabaenaceae forming linear filaments composed of identical cells belonged to the polyphyletic order Synechococcales, which earlier included unicellular species of *Acaryochloris, Prochlorococcus, Synechococcus,* and other cyanobacteria, which divide by binary fission (Komarek et al. 2014). Order Chroococcales consisted of other unicellular cyanobacteria (*Aphanothece*, *Microcystis,* and *Crocosphaera*) that are able to form colonies and cellular aggregates. Chroococcales is phylogenetically distant from Synechococcales and more related to Pleurocapsales, which is characterized by baeocytes formation and irregular multiple fission. Chroococcales is more close to the recently separated order Spirulinales, which consists of cyanobacteria having spiral filaments. *Spirulina platensis* was excluded from this group (Spirulinales) as molecular biological and phylogenetic tree studies revealed that it belonged to the genus *Arthrospira* of the order Oscillatoriales, which lies in between Chroococcales and Synechococcales (unicellular) (Komarek et al. 2014).

Families differing in thylakoid structure and cell division type such as multicellular filamentous Microcoleaceae and unicellular Cyanothecaceae belonged to the order Oscillatoriales (polyphyletic). *Oscillatoria limnetica* (filamentous cyanobacterium), which can use H_2S as instead of H_2O as electron donor, was previously considered as an evidence of phylogenetic relationship between green sulfur bacteria and cyanobacteria. However, it was found that under selective environmental conditions, cyanobacteria gained the capability for sulfide oxidation later through HGT as revealed by the data of comparative ecolological genomics (Sanchez-Baracaldo et al. 2005). *Trichodesmium erythraeum*, the marine nitrogen-fixing cyanobacterium, which is phylogenetically related to *Arthrospira* and *Lyngbya*, belonged to the family Microcoleaceae, but specificities of physiological, cytological, and biochemical properties of this cyanobacterium make the issue of its origin debatable. It differs from other nitrogen-fixing cyanbacteria of the monophyletic cluster Nostocales, which is capable of differentiation of specialized cells such as akinetes and heterocysts, which allow them to survive under unfavorable environmental conditions. Several members of order Nostocales inhabiting different environments live in symbiosis with different plants. The origin of symbiotic cyanobacteria imprinted on the genomic structure, which occurred during the late stages of evolution. For example, *Anabaena azollae* (obligate symbiont) is characterized by presence of a high number of pseudogenes and genome reduction indicating the incompleteness of the evolutionary optimization of the cyanobacterial species (Larsson et al. 2011).

Among prokaryotes, multicellular cyanobacteria of the family Stigonematales such as *Scytonema hofmanii*, *Mastigocladus laminosus*, *Fischerella thermalis,* and others forming branched trichomes possess the most complex morphology. These cyanobacteria having branched filamentous were previously kept under a separate morphotypes V, but the modern taxonomy included them in the order Nostocales, as they were found to have the same ancestors (Rippka et al. 1979; Komarek et al. 2014). Their proteomes contained only few specific proteins coding for the capability of "branching," and these findings suggest that functioning of certain regulatory genes encoding for intercellular communications and cell division determines branching in these cyanobacteria (Dagan et al. 2013; Nürnberg et al. 2014).

Phylogenetic analysis of large number of cyanobacterial species belonging to different orders suggested that cyanobacterial evolution was not in a unidirectional pathway from unicellular forms to multicellular taxa. The process of loss of multicellularity was going simultaneously along with complications of morphotypes in cyanobacteria (Schirrmeister et al. 2011b). Secondary transitions from unicellular forms to multicellular taxa occurred during the course of evolution as indicated by the polyphasic analysis, and this probably occurred during the appearance of the cyanobacterial genus *Spirulina* (Schirrmeister et al. 2011b). Thus, in order to evaluate phylogenetic relationships between different cyanobacterial taxa the morphological properties such as cell shape, size, and cell division type are taxonomically important but insufficient to reach at any reliable conclusion as morphologically taxa could be formed due to a convergence of phylogenetically different cyanobacteria for allowing them to adapt to the same environmental conditions/ecological habitats (Dvořák et al. 2015). Variability in cyanobacterial phylogeny is typical of certain crucial physiological property such as nitrogen fixation, which is specific to many multicellular and some unicellular cyanobacteria. During the course of cyanobacterial evolution, they were selected on the basis of the possession of the "nitrogenase gene cluster" and different strategies for protection of nitrogenase enzyme from oxygen, whereas selection of symbiotic forms was on the basis of their ability to interact with their host organism/partner. The enzyme complex "nitrogenases" appeared for the first time in archaea in anoxygenic bacteria and later on through HGT it could be transferred in cyanobacteria (Raymond et al. 2004).

1.2.2 Genome Evolution

The enormous biodiversity of phylum cyanobacteria is also reflected in the sizes of their genomes, which range from 1 to 13 Mb (Larsson et al. 2011). The combination of various genetic processes forms the evolutionary trajectories of cyanobacterial genomes (Fig. 1.3) as these trajectories are not constituted by simple bifurcation schemes (Zhaxybayeva et al. 2006; Shestakov 2007; Shi and Falkowski 2008).

Discovery of new cyanobacterial species and strains and increasing data of their genome sequencing have resulted in continuously change in the sizes of pangenomes of the phylum cyanobacteria along with its taxonomic groups. Recently, significant progress in this field of research has been observed owing to metagenomic studies and the advancements in the methods/techniques for the analysis of genomes of

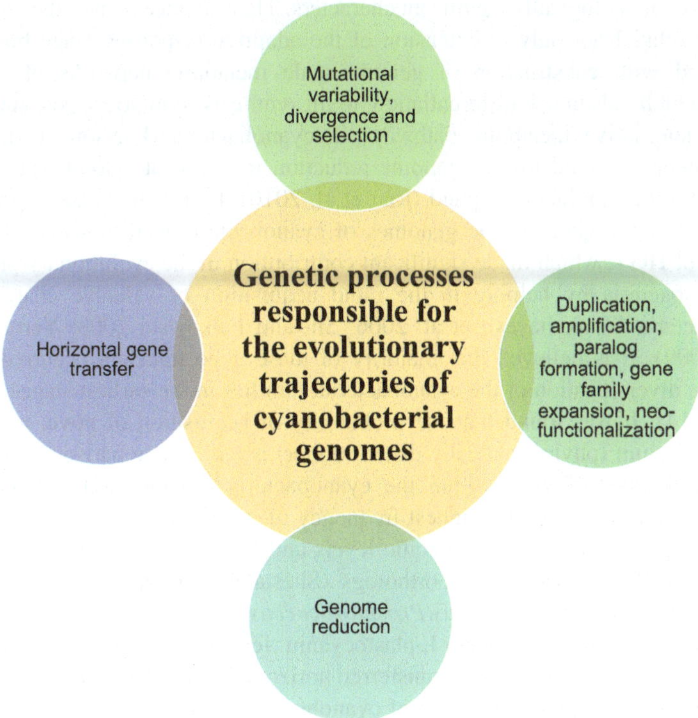

Fig. 1.3 The evolutionary trajectories of cyanobacterial genomes. (For details, see the Ref. Shestakov and Karbysheva 2017)

noncultivated cyanobacteria. Decrease in the size of the general cyanobacterial core genome has been observed as studies showed that the core set comprised 1044 genes in 2006, whereas it included only 559 genes in 2015 showing a significant decrease (Mulkidjanian et al. 2006; Simm et al. 2015). On the basis of genomic analyses of 60 cyanobacterial species and strains, two main trajectories of cyanobacterial genomic transformations have been suggested (Sun and Blanchard 2014):

1. Reduction in genome size
   It is achieved through deletion along the entire genome sequence and fixed by stabilizing selection.
2. Increase genome size
   It occurs *via* gene family expansion and the presence of repeated sequences, plasmids, and mobile elements.

Majority of cyanobacterial species having a large number of mobile elements show low gene polymorphism and their genomes evolution occurred primarily *via* genomic rearrangements through site-specific transposases and integrases responsible for movements, which altered the nature of regulation of genes responsible for expression of ecologically significant characters. This enhancement in the number of genes is related not only to extension of the adaptive responses range but is also associated with construction of genome-scale metabolic networks of complex nature, which aids in cell differentiation, toxin synthesis, symbiogenesis, and operation of alternative metabolic pathways in cyanobacteria (Larsson et al. 2011). Simultaneously, trend toward genome reduction may operate along with the tendency of gene families to expand (Ran et al. 2010). Different studies suggest that ~10–50% of the genes in the genomes of cyanobacteria were transferred *via* the process of HGT, which made significant contribution in the evolutionary processes of cyanobacteria by helping in the rapid acquisition of valuable characters in cyanobacteria (Zhaxybayeva et al. 2006; Shi and Falkowski 2008; Yerrapragada et al. 2009). It is believed that majority of these gene transfers occurred during intensive diversification of the cyanobacterial families in the earliest stages (Puigbò et al. 2014). It was found that the probability of acquisition of novel gene from another phylum (phylogenetically distant donor) is less in comparison with probability of transfer of gene within the cyanobacteria phylum itself. Homologous recombination results in the highest frequency of genetic exchange at intraspecific (between the strains) and interspecific levels and helps in selecting more valuable variants by the replacement of orthologs (Shestakov and Karbysheva 2015). In representatives of *Synechococcus*/*Prochlorococcus* group, *psbAD* genes of photosystem II, genes of photosystem I, plastocyanin, ferredoxin, and other components of energy metabolism have been transferred horizontally (Lindell et al. 2004; Millard et al. 2004). During the interaction of cyanobacteria with cyanophages, the factors which determine strain specificity (such as formation of light-proof Hli proteins and proteins contributing in the cell surface formation) were acquired by the process of HGT (Shestakov and Karbysheva 2015). Still, the mechanisms involved in the process of HGT in cyanobacteria are not very clear. In the evolution of cyanobacteria, cyanophages have thought to play a crucial role as they control the number of natural populations and providing preservation of their gene pool during unfavorable environmental conditions (Shestakov and Karbysheva 2015). Although the involvement of cyanophages in gene transfer between cyanobacteria is obvious, reproducible transduction systems have not yet been developed (Lindell et al. 2004; Dammeyer et al. 2008; Ignacio-Espinoza and Sullivan 2012; Shestakov and Karbysheva 2015). However, recent large-scale genome sequencing studies suggested that the viruses such as Chlorovirus, Coccolithovirus, Pandoravirus, Marseillevirus, and Tupanvirus have played crucial role in the evolution of microalgae (Nelson et al. 2021).

1.3 Diverse Habitats of Cyanobacteria

Cyanobacteria can be found in diverse and extreme habitats ranging from the very extreme hot springs to extremely cold deserts of the Arctic and Antarctic Zones and thus represent an interesting and diverse form of life in a variety of terrestrial and aquatic environments (Whitton and Potts 2000a, b; Mataloni and Komarek 2004;

Rastogi et al. 2012; Gaysina et al. 2019). A wide range of symbiotic relationships are formed by cyanobacteria (nitrogen fixing) with almost all groups of plant such as *Geosiphon pyriforme* (fungi) with *Nostoc*, *Hemiaulus hauckii* (algae) with *Richelia intracellularis*, *Anthoceros* (bryophyte) with *Nostoc*, *Azolla* (pteridophyte) with *Anabaena*, *Cycas* (gymnosperm) with *Nostoc*, and *Gunnera* (angiosperm) with *Nostoc* (Morot-Gaudry and Touraine 2001). Their enormous physiological flexibility and plasticity enable them to be present in almost all geographical regions of the earth (Castenholz 1973; Whitton 1973; Skulberg 1994; Laamanen 1996; Gaysina et al. 2019). Figure 1.4 depicts the wide distribution of cyanobacteria in different habitats on Earth.

1.3.1 Terrestrial Habitats

Cyanobacteria constitute the major microorganisms in the biological soil crusts (Büdel et al. 2009). In different regions of India, biological soil crusts constitute genera with sheath such as *Plectonema*, *Lyngbya*, and *Scytonema*, which were found to be dominant, whereas *Phormidium*, *Oscillatoria*, *Nostoc*, *Microcoleus*, *Aulosira*, *Calothrix*, *Westiellopsis*, *Hapalosiphon*, and *Fischerella* were also found frequently (Tirkey and Adhikary 2005). In Baja California Desert in Mexico, *Desmonostoc muscorum* (*Nostoc muscorum*) and *Schizothrix calcicola* were found to be the dominant taxa (Flechtner et al. 1998). Cyanobacteria *Chroococcidiopsis* sp., *Microcoleus paludosus*, *Phormidium* spp., *Pseudanabaena* spp., *Nostoc* spp., and *Leptolyngbya* spp. were detected frequently in biological soil crusts in four biomes in Africa (Büdel et al. 2009). *Microcoleus*, *Scytonema*, *Nostoc*, *Lyngbya*, and

Fig. 1.4 Distribution of cyanobacteria in different habitats on Earth such as marine water (**A**), rock surface (**B**), rice paddy field (**C**), fresh water (**D**), tree bark (**E**) and mudflatS (**F**)

Phormidium were also found frequently (Issa et al. 1999). In biological crusts around the world, *Microcoleus vaginatus* was found to be the most dominant and ecologically important cyanobacteria (Johansen and Shubert 2001). In soils of North American deserts together with the cyanobacterium, *Microcoleus vaginatus*, *Nostoc commune*, *Schizothrix calcicola*, *N. paludosum*, *N. punctiforme*, *N. muscorum*, *Leptolyngbya tenuis* (as *Phormidium tenue*), *Trichormus variabilis* (as *Anabaena variabilis*), *Phormidium minnesotense,* and *Tolypothrix tenuis* have been found in the biological crusts (Johansen 1993). *Microcoleus vaginatus*, *Scytonema* sp., and *Nostoc* spp. were the dominant cyanobacteria found in the desert crusts of Southeastern Utah (Garcia-Pichel and Belnap 1996). In steppes and semideserts in the territory of USSR, *Scytonema ocellatum*, *Nostoc commune*, and *Microcoleus vaginatus* formed *Nostoc–Scytonema* communities (Gollerbach and Shtina 1969). *Scytonema* sp., *Scytonema* cf. *ocellatum*, *Microcoleus* cf. *paludosus*, *M.* cf. *sociatus*, *Calothrix* cf. *marchica*, *Calothrix* cf. *elenkinii*, *Phormidium* sp., and *Nostoc* cf. *microscopicum* were detected in microbiotic crusts in eroded soils of a tropical dry forest in Mexico (Maya et al. 2002). *Xenococcus lyngbyae*, *Microcoleus paludosus*, and *M. vaginatus* were the most dominant cyanobacteria in the biological soil crusts in the Gurbantunggut Desert in Western China (Chen et al. 2007). Several cyanobacteria including *Microcoleus vaginatus* were detected in the microbiotic crusts on sand dunes (artificially stabilized) in Tengger Desert, China, during first stages of dune stabilization (after 0–8 years); however, in stylized dune after 24 years, these species were not found (Li et al. 2002). *Anabaena azotica*, *Jaaginema pseudogeminatum* (as *Oscillatoria pseudogeminata*), *Limnoraphis cryptovaginata* (as *Lyngbya cryptovaginata*), *Oscillatoria obscura*, *O. subbrevis*, *Leptolyngbya tenuis* (as *Phormidium tenue*), *Leptolyngbya lurida* (as *Phormidium luridum*), *Microcoleus autumnalis* (as *Phormidium autumnale*), *Schizothrix rupicola*, *Scytonema javanicum,* and *S. millei* were also found together with *Microcoleus vaginatus*. Filamentous cyanobacteria *Scytonema* sp. and *Sypmplocastrum purpurascens* were found to be the dominating cyanobacteria in the dry savanna ecosystem in Australia (Büdel et al. 2018).

In temperate forest soils, *Nostoc punctiforme*, *Desmonostoc muscorum* (*Nostoc muscorum*), *Leptolyngbya foveolarum* (*Phormidium foveolarum*), and *Microcoleus autumnalis* (*Phormidium autumnale*) were the dominant cyanobacterial species (Aleksakhina and Shtina 1984). *Microcoleus autumnalis* (*Phormidium autumnale*) and *Leptolyngbya foveolarum* were detected in the algal flora of unlimed and limed forest soils in the Ardennes (Belgium) (Hoffmann et al. 2007). *Symplocastrum friesii* was detected in the soils of the northern part of the Great Smoky Mountains National Park, USA (Khaybullina et al. 2010). Several cyanobacterial species such as *Aphanothece stagnina*, *Leptolyngbya* cf. *nostocorum*, *Leptolyngbya* cf. *hansgirgiana*, *Hormoscilla pringsheimii*, *Kamptonema laetevirens*, *Kamptonema animale*, *Oxynema* cf. *acuminatum*, *Phormidium* cf. *retzii*, *Phormidium aerugineo-caeruleum*, *Phormidium uncinatum*, *Phormidium tergestinum,* and *Nostoc* cf. *ellipsosporum* were reported only in the boreal forest zone. In the broad-leaved forest zone, cf. *Trichocoleus hospitus* was the widely distributed cyanobacteria.

Chroococcus varius and *Myxosarcina* cf. *tatrica* were found only in this type of environment (Gaysina et al. 2018). In flood plain forest having trees *Padus avium* Mill. and *Alnus glutinosa* (L.) Gaertn., a maximum number of cyanobacteria were found namely *Borzia trilocularis*, *Cylindrospermum* sp., *Cylindrospermum majus*, *Leptolyngbya voronichiniana*, *Leptolyngbya foveolarum*, *Microcoleus vaginatus*, *Nostoc* cf. *calcicola*, *N.* cf. *punctiforme*, *Phormidium ambiguum*, *P. breve*, *P. corium*, *P. dimorphum*, *Roholtiella bashkiriorum*, *Trichormus variabilis,* and cf. *Trichocoleus hospitus* (Gaysina et al. 2018). In the Yuraktau and Tratau Mounts in the forest steppe zone of Bashkiria, 56 species of cyanobacteria were reported, among which dominant species were *Phormidium jadinianum*, *Leptolyngbya foveolarum*, *Microcoleus autumnalis* (*Phormidium autumnale*), and *Nostoc punctiforme* (Bakieva et al. 2012). Unique cenoses in the arid regions were created by *Nostoc commune*, *Microcoleus vaginatus,* and *Scytonema ocellatum* (Gollerbach and Shtina 1969). In a forb–grass steppe near Sibay town and a sand savanna of Northwestern Ohio, *Cyanothece aeruginosa* was found in the biological soil crusts (Neher et al. 2003; Gaysina et al. 2018).

Filamentous cyanobacteria like *Anabaena* and *Tolypothrix* were dominant in the restoration of soils damaged by volcano eruption (Treub 1888). Cyanobacteria were dominant only near lava flows after volcanic activity in Surtsey Island (Schwabe 1972). Several *Nostoc* Vausher species and *Anabaena variabilis* Küzting were reported (Henriksson et al. 1972). On the volcanic ash of Kuril–Kamchatka arcs, inside the edge of the crater nine cyanobacterial taxa were found namely *Aphanocapsa muscicola* (*Microcystis muscicola*), *Synechocystis aquatilis*, *Desmonostoc muscorum* (*Nostoc muscorum*), *Mastigocladus laminosus*, *Aphanothece castagnei*, *Nostoc gelatinosum*, *N. humifusum*, *Oscillatoria geminata* f. *sulphurea*, *Leptolyngbya* (*Plectonema nostocorum*), and *Leptolyngbya gracillima* (*Plectonema gracillimum*) (Shtina et al. 1992). *Mastigocladus laminosus* Cohn is usually found in the hot springs (Shtina et al. 1992). Phylogenetic analysis of cyanobacterial strains through *16S rRNA* gene sequencing was done for the cyanobacteria isolated from hot springs in Rajgir, India. These cyanobacteria were identified as *Cyanothece* sp. strain HKAR-1, *Nostoc* sp. strain HKAR-2, *Scytonema* sp. strain HKAR-3, and *Rivularia* sp. strain HKAR-4 (Rastogi et al. 2012).

Reclamation of the highly alkaline "usar" soil in India by blue-green algae with the dominance of *Nostoc commune* was detected by Singh (1950). In deserts of USSR, cyanobacteria were extensively grown in the wet period on "takyr" soils having pH 9–10 and *Nostoc commune* (*Desmonostoc commune*), *Microcoleus,* and *Phormidium* were the dominant species (Gollerbach et al. 1956). It was found that *Microcoleus vaginatus* crusts started to grow in liquid media after cultivation in salt solutions (Bolyshev et al. 1965). In halophytic solonchaks (salted soils) of the Sahara–Gobi desert area, cyanobacteria *Anabaena*, *Anabaenopsis*, *Aulosira*, *Calothrix*, *Nostoc*, and *Tolypothrix* were found to be widely distributed. In various types of salted soils and vegetation true solonchak, saline steppes, meadow halophilous 49 cyanobacterial species were reported and the dominant genera were *Calothrix*, *Leptolyngbya*, *Lyngbya*, *Phormidium*, *Anabaena*, *Jaaginema,* and *Nostoc*. *Nostoc linckia*, *Leptolyngbya fragilis,* and *L. tenuis* were the most dominant species. *Phormidium paulsenianum, Leptolyngbya fragilis,* and *Nostoc linckia* were reported to grow on soils covered by meadow halophilous vegetation (Vinogradova and Darienko 2008).

Different cyanobacteria such as *Phormidium paulsenianum, P. jadinianum, P. breve* (*Oscillatoria brevis*), and *Leptolyngbya foveolarum* (*Phormidium foveolarum*) were found to be grown in all types of solonchaks, and *Microcoleus autumnalis* (*Phormidium autumnale*) was typically found in the meadow solonetz (Khaibullina and Gaisina 2008). Recently, for examining the cyanobacterial community structure, pooled mat sample was studied from the Rann of Kachchh, India, which is desert area on the western part of India and is exposed to dynamic environmental changes such as temperature, salinity, and nutrients (Patel et al. 2019). Taxonomic profiling revealed that the mats predominately contained the members of Pseudanabaenales and Oscillatoriales. Other abundant cyanobacterial orders were Nostocales, Chroococcales, and unclassified cyanobacteria (Patel et al. 2019).

Cyanobacteria also play an important role in the restoration of disturbed ecological areas by colonizing the lifeless substrates left after anthropogenic degradation such as mine spoils, heavy metals, and contaminated soils. Such degraded habitats are characterized by lack of water high concentrations of heavy metals, deficient nutrient contents, and high levels of isolation (Trzcińska and Pawlik-Skowrońska 2008). Cyanobacterial species such as *Lyngbya, Microcoleus, Nostoc edaphicum, Nostoc* sp., *Oscillatoria* sp., and *Phormidium* sp. were present in the soils polluted with heavy metal contaminations (García-Meza et al. 2006; Trzcińska and Pawlik-Skowrońska 2008; Cabala et al. 2011). In reclaimed soils in brown coal and lignite postmining area of Czech Republic and Germany, *Microcoleus vaginatus, M. autumnalis, Nostoc muscorum, N.* cf. *calcicola,* and representatives of the genera *Phormidium, Leptolyngbya, Pseudophormidium,* and *Schizothrix* were found (Lukešová 2001). The cyanobacterial genera *Microcoleus, Oscillatoria,* and *Phormidium* were reported as dominant taxa in the spoils of age 1–2 years of coal deposits of Russia where as on the spoils of age 5–9 years, *Pseudophormidium, Phormidium,* and *Oscillatoria* were reported as dominant genera (Kabirov 1997).

The polar region of Earth comprises the Antarctic and Arctic regions and constitutes about 14% of the Earth's biosphere (Rampelotto 2014). In these ecosystems, cyanobacteria have been reported as dominant phototrophs because of their ability to tolerate the abiotic stresses such as low temperature and ultraviolet radiation in these regions of Earth (Vincent 2007). Cyanobacterial species such as *Aphanocapsa fusco-lutea, A. grevillei, Chroococcus cohaerens, C. spelaeus, Desmonostoc muscorum, Gloeocapsa ralfsii, G. sanguinea, G. violacea, Kamptonema animale, Leptolyngbya boryana, L. foveolarum, Microcoleus autumnalis, Nostoc commune, N. punctiforme,* and *Phormidium ambiguum* were reported from aerophytic habitats in Hypoarctic and Arctic regions, and these were on the soil surface and inside the soil layer (Davydov and Patova 2018). Cyanobacterial diversity in the Arctic was found to be higher as compared to the Antarctic regions (dry valleys) (Zakhia et al. 2008). *Chroococcus* and *Gloeocapsa* were found to be dominant in the crust in the Arctic conditions, whereas *Stigonema ocellatum*, *S. minutum*, and *S. informe* with associated *Gloeocapsopsis magma* and *Gloeocapsa violascea* were found to be most frequent species in crusts in hypoarctic regions (Davydov and Patova 2018). *Gloeocapsopsis magma*, *Leptolyngbya foveolarum*, *Nostoc commune*, *Scytonema hofmannii*, *Stigonema minutum*, and *S. ocellatum* were reported as permanent species of BSC in the mountain tundras of the Polar and Subpolar Urals (Patova et al. 2018). Several cyanobacterial taxa such as *Microcoleus autumnalis*, *Merismopedia tenuissima*, *Nostoc punctiforme*, *N. commune*, *Pseudanabaena frigida*, and *Schizothrix* cf. *calcicola* were identified in the Hornsund area, Spitsbergen (Matuła et al. 2007). On wet soils in Antarctica, wide distribution of filamentous cyanobacteria from the order Oscillatoriales, especially *Microcoleus autumnalis,* was found (Strunecký et al. 2012).

1.3.2 Aquatic Habitats

Cyanobacteria inhabiting aquatic habitats can be divided into two broad ecological groups (Fogg et al. 1973):

1. Planktonic cyanobacteria (float freely in the water column)
2. Benthic cyanobacteria (adhere to submerged solid surfaces)

In many ocean regions, cyanobacteria genera such as *Cyanobium, Prochlorococcus, Synechococcus,* and *Synechocystis* are widely distributed as marine planktonic communities (Flombaum et al. 2013; Costa et al. 2014). Some filamentous genus such as *Romeria* also inhabits oceans as marine plankton (Komárek 2001). During favorable environmental conditions, cyanobacteria form blooms as a result of their rapid growth (Sellner 1997; De Figueiredo et al. 2006; Sciuto and Moro 2015). The colonial filamentous cyanobacteria *Trichodesmium* is one of the most abundant bloom-forming genus in the marine pelagic zone and is distributed panglobally in subtropical and tropical oceans having oligotrophic environments (Capone et al. 1997; LaRoche and Breitbarth 2005). Cyanobacterium *Crocosphaera watsonii* contributes significantly to oceanic nitrogen fixation, and *Crocosphaera* also inhabit regions having low iron content due to its ability to reduce its iron metalloenzyme inventory (Zehr et al. 2001; Montoya et al. 2004; Moisander et al. 2010; Saito et al. 2011). In the Baltic Sea, cyanobacterial genera *Anabaena, Aphanizomenon,* and *Nodularia* are found as the most important bloom-forming cyanobacteria (O'Neil et al. 2012). Worldwide, filamentous cyanobacteria *Lyngbya* are commonly found as benthic communities (Paul et al. 2005; Jones et al. 2011; O'Neil et al. 2012). The cyanobacterial genus *Lyngbya majuscule* belongs to the benthic zones forming dense mats and is widely distributed in tropics in reef and lagoons (Whitton and Potts 1982, 2000a, b; Hoffmann 1994; Thacker and Paul 2004). Another filamentous genus *Moorea* belongs to a cosmopolitan pantropical ecological group, which is abundant in the marine benthos. In intertidal flats of the German Wadden Sea, the cyanobacterial genera *Coleofasciculus, Hydrocoleum,* and *Lyngbya* are dominant in all the sediment types in cyanobacterial populations (Vogt et al. 2018). Common cyanobacterial species in marine littoral and intertidal habitats are constituted by *Microcoleus ethnoplasts* and representatives of the genera *Oscillatoria* sp. and *Spirulina* (Kulasooriya 2011). In the Portugal coast, the filamentous cyanobacterial genus *Leptolyngbya*, *Nodosilinea*, *Pseudanabaena*, and *Romeria* constitute a large group of the marine cyanobacterial strains (Costa et al. 2014). Among the most widely distributed cyanobacterial mangrove dwellers worldwide, *Aphanocapsa*, *Calothrix*, *Chroococcus*, *Coleofasciculus*, *Lyngbya*, *Oscillatoria*, and *Schizothrix* constitute the most important genera (Alvarenga et al. 2015).

In the oceans and large transparent lakes, the autotrophic picoplanktons constitute the major primary producers (Callieri and Stockner 2002; Ting et al. 2002). The phycoerythrin-rich freshwater cyanobacteria *Synechococcus* is the dominant genus among the autotrophic picoplanktons in oligotrophic lakes (Fahnenstiel and Carrick 1992; Ting et al. 2002). The cyanobacterial genera *Cyanobium* and *Synechocystis* are also very important plankton in freshwater ecosystems (Stockner 1988; Albertano et al. 1997; Komárek 2003). In freshwater bodies, large populations are formed by the genus *Aphanothece* (Mur et al. 1999). In freshwater ecosystems, common cyanobacterial genera are *Chroococcus*, *Coelosphaerium*, *Coelomoron*, *Cyanodictyon*, *Gomphosphaeria*, *Rhabdoderma*, *Merismopedia,* and *Snowella* (Komárek and Anagnostidis 1999; Komárek 2003). Ecostrategists focusing on scum formation constitute large colonies of filaments or coccoid cells and genera *Anabaena*, *Aphanizomenon,* and *Microcystis* belong to such ecological group. In freshwater habitats, the genus *Microcystis* is one of the most widely distributed microcystin-producing cyanobacteria, which forms blooms in eutrophic lakes and springs of the temperate zone (Reynolds et al. 1981; Kurmayer et al. 2002; Rastogi et al. 2014, 2015). Filamentous cyanobacterial species such as *Limnothrix redekei* and *Planktothrix agardhii* inhabit eutrophic and hypertrophic shallow (<3 m depth) lakes (Mur et al. 1999). *Aphanothece*, *Oscillatoria*, and *Phormidium* constitute benthic mats, which usually grow on the sediments of ponds and lakes (Komárek 2003). Among epilithic cyanobacteria, *Aphanocapsa*, *Aphanothece*, *Chroococcus*, *Nostoc,* and *Leptolyngbya* are the most distributed cyanobacterial genera from freshwater streams of India (Saha et al. 2007). *Oscillatoria*, *Phormidium*, *Lyngbya*, *Leptolyngbya*, *Microcoleus*, *Tychonema*, and *Schizothrix* are usually found as benthic cyanobacteria (Steppe et al. 1996; Mez et al. 1997, 1998; Hitzfeld et al. 2000; Aboal et al. 2005; Gugger et al. 2005). In freshwater habitats, *Aphanothece* and *Synechococcus* along with nitrogen-fixing cyanobacteria *Anabaena* and *Scytonema* are usually found as toxic cyanobacteria (Krienitz et al. 2003; Dasey et al. 2005; Mohamed et al. 2006; Mohamed 2008; Smith et al. 2011). Macroscopic colonies forming cyanobacteria of order Nostocales namely *Nostoc caeruleum*, *N. commune*, *N. microscopicum*, *N. parmelioides*, *N. pruniforme*, *N. verrucosum*, and *N. zetterstedtii* have been found from inland aquatic habitats (Mollenhauer et al. 1999).

1.3.3 Symbiotic Associations

Corals, diatoms, dinoflagellates, seagrass, and sponges are the common marine organisms, which form associations with cyanobacteria. Colonies of the coral *Montastraea cavernosa* form endosymbiotic association with cyanobacteria, which express nitrogenase and thus also provide fixed nitrogen to the host coral (Lesser et al. 2007). *Calothrix rhizosoleniae* and *Richelia intracellularis* (heterocystous cyanobacteria) form symbiotic relationship with diatoms such as *Chaetoceros, Hemiaulus,* and *Rhizosolenia* (Foster et al. 2011). A unicellular nitrogen-fixing cyanobacterium is present as endosymbiont in diatoms belonging to the family *Epithemiaceae* (DeYoe et al. 1992). In sponges, *Synechococcus* sp. is commonly found in symbiotic association, and *Oscillatoria spongeliae* has also been reported to form association with sponges over a wide geographic range in oceans (Usher 2008). The leaves of the seagrass *Cymodocea rotundata* bear cyanobionts as small attached patches of thin biofilms having pigmented microbial aggregates. The cyanobacterium *Nostoc* is a prolific symbiotic partner, which forms association with several eukaryotic organisms such as protists, fungi, plants, and animals (Rai et al. 2002). Nostocacean cyanobacteria form the symbiotic association with members of the plant kingdom ranging from bryophyta to pteridophyta (*Azolla*) and from gymnosperms (family Cycadaceae) to angiosperms (family Gunneraceae). High strain diversity has been observed both among and within different host species as revealed by most of the studies on identification and diversity of the cyanobionts from the individual hosts except *Azolla* (West and Adams 1997; Rasmussen and Svenning 1998; Nilsson et al. 2000; Costa et al. 2001; Guevara et al. 2002; Rasmussen and Nilsson 2002). *Nostoc muscorum* and *N. punctiforme* have been identified as cyanobionts, which form symbiotic relationship with *Cycas* (Costa et al. 1999). Approximations of these cyanobionts have been assigned to the genera *Anabaena*, *Nostoc*, and *Trichormus*, or all of these symbionts have been shifted to a new separate genus, but all of these cyanobionts certainly belong to the order Nostocales (Komárek and Anagnostidis 1989; Plazinski et al. 1990; Gebhardt and Nierzwicki-Bauer 1991; Caudales et al. 1995; Baker et al. 2003; Pabby et al. 2003; Svenning et al. 2005).

1.4 Perspective and Conclusion

Undoubtedly, the ancient photoautotrophs cyanobacteria have played crucial role in the evolution of early Earth and its biosphere and are also responsible for the oxygenation of the oceans and atmosphere. Diversity of cyanobacteria is expressed by their morphological, physiological, and biochemical properties, which enable them to survive and sustain in diverse range of ecological niches ranging from the polar regions to the hot springs, thus representing life in almost every possible environments on Earth. Their success in acclimatizing such wide range of diverse habitats can be attributed to their long course of evolutionary process. Despite the important role of cyanobacteria in the early evolution of life and Earth, fundamental questions still remain unanswered about the origin, timing, and pattern of diversification of cyanobacteria. Hence, it is required to define new biosignatures, which could serve as indicator of cyanobacteria in order to reassess their fossil record and could aid in providing new calibration points for molecular clocks. These biosignatures will help in combining analyses of the ultrastructure, morphology, and ecology of cyanobacterial microfossils with their biomolecular (pigments and lipids), metal, and isotopic composition. Identification of these promising fossils, not only as cyanobacteria, but of specific clades within this ancient group of photoautotrophs will improve the understanding of the diversification record of cyanobacteria.

Acknowledgement Prashant R. Singh (09/013(0795)/2018-EMR-I) is thankful to Council of Scientific and Industrial Research, New Delhi, for the financial support in the form of junior research fellowship.

References

Aboal M, Puig MA, Asencio AD (2005) Production of microcystins in calcareous Mediterranean streams: the Alharabe River, Segura River basin in south-east Spain. J Appl Phycol 17:231–243

Albertano P, Di Somma D, Capucci E (1997) Cyanobacterial picoplankton from the Central Baltic Sea: cell size classification by image-analyzed fluorescence microscopy. J Plankton Res 19:1405–1416

Aleksakhina TI, Shtina EA (1984) Terrestrial algae of forest biogeocoenoses (Pochvennye vodorosli lesnych biogeotsenozov). Nauka, Moskow

Alleon J, Summons RE (2019) Organic geochemical approaches to understanding early life. Free Radic Biol Med 140:103–112

Alvarenga DO, Rigonato J, Branco LHZ, Fiore MF (2015) Cyanobacteria in mangrove ecosystems. Biodivers Conserv 24:799–817

Anbar AD, Duan Y, Lyons TW, Arnold GL, Kendall B, Creaser RA, Kaufman AJ, Gordon GW, Scott C, Garvin J, Buick R (2007) A whiff of oxygen before the great oxidation event? Science 317:1903–1906

Baker JA, Entsch B, McKay DB (2003) The cyanobiont in an *Azolla* fern is neither *Anabaena* nor *Nostoc*. FEMS Microbiol Lett 229:43–47

Bakieva GR, Khaibullina LS, Gaisina LA, Kabirov RR (2012) Ecological-floristic analysis of soil algae and cyanobacteria on the Tra-Tau and Yurak-Tau mounts, Bashkiria. Eurasian Soil Sci 45(9):873–881

Bartley JK (1996) Actualistic taphonomy of cyanobacteria; implications for the Precambrian fossil record. PALAIOS 11:71–586

Becraft ED, Wood JM, Rusch DB, Kühl M, Jensen SI, Bryant DA, Roberts DW, Cohan FM, Ward DM (2015) The molecular dimension of microbial species: 1. Ecological distinctions among, and homogeneity within, putative ecotypes of *Synechococcus* inhabiting the cyanobacterial mat of Mushroom Spring, Yellowstone National Park. Front Microbiol 6:590

Blank CE, Sanchez-Baracaldo P (2010) Timing of morphological and ecological innovations in the cyanobacteria-a key to understanding the rise in atmospheric oxygen. Geology 8:1–23

Bolyshev NN, Shtina EA, Konnova EN (1965) The influence of different salts and their concentrations on the species composition of algae. In: Moscow University Sciences bulletin. Series VI, Biology, pedology, vol 2. Moscow State University, Moscow, pp 72–80

Bosak T, Liang B, Sim MS, Petroff AP (2009) Morphological record of oxygenic photosynthesis in conical stromatolites. Proc Natl Acad Sci U S A 106:10939–10943

Bosak T, Knoll AH, Petroff AP (2013) The meaning of stromatolites. Annu Rev Earth Planet Sci 41:21–44
Brasier M, McLoughlin N, Green O, Wacey D (2006) A fresh look at the fossil evidence for early Archean cellular life. Philos Trans R Soc Lond B 361:887–902
Büdel B, Darienko T, Deutschewitz K, Dojani S, Friedl T, Mohr KI, Salisch M, Reisser W, Weber B (2009) Southern African biological soil crusts are ubiquitous and highly diverse in drylands, being restricted by rainfall frequency. Microb Ecol 57:229–247
Büdel B, Williams WJ, Reichenberger H (2018) Annual net primary productivity of a cyanobacteria dominated biological soil crust in the Gulf savanna, Queensland, Australia. Biogeosciences 15: 491–505
Buick R (2008) When did oxygenic photosynthesis evolve? Philos Trans R Soc B 263:2731–2743
Cabala J, Rahmonov O, Jablonska M, Teper E (2011) Soil algal colonization and its ecological role in an environment polluted by past Zn-Pb mining and smelting activity. Water Air Soil Pollut 215:339–348
Callieri C, Stockner JS (2002) Freshwater autotrophic picoplankton: a review. J Limnol 61(1):14
Canfield DE (2005) The early history of atmospheric oxygen. Annu Rev Earth Planet Sci 33:1–36
Capone DG, Zehr JP, Paerl HW, Bergman B, Carpenter EJ (1997) *Trichodesmium*, a globally significant marine cyanobacterium. Science 276:1221–1229
Castenholz RW (1973) Ecology of blue-green algae in hot springs. In: Carr NG, Whitton BA (eds) The biology of blue-green algae. Blackwell Scientific Publications, Oxford, pp 379–414
Castenholz RW (2001) Phylum BX. Cyanobacteria, oxygenic photosynthetic bacteria. In: Boone DR, Castenholz RW (eds) Bergey's manual of systematic bacteriology, vol 1. Springer, New York, p 721
Caudales R, Wells JM, Antoine AD, Butterfield JE (1995) Fatty acid composition of symbiotic cyanobacteria from different host plant (*Azolla*) species: evidence for co-evolution of host and symbiont. Int J Syst Bacteriol 45:364–370
Chen YN, Wang Q, Li WH, Ruan X (2007) Microbiotic crusts and their interrelations with environmental factors in the Gurbantonggut desert, western China. Environ Geol 52:691–700
Choudoir ML, Campbell AN, Buckley DH (2012) Grappling with Proteus: population-level approaches to understanding microbial diversity. Front Microbiol 3:336
Costa J-L, Paulsrud P, Lindblad P (1999) Cyanobiont diversity within coralloid roots of selected cycad species. FEMS Microbiol Ecol 28:85–91
Costa J-L, Paulsrud P, Lindblad P (2001) Genetic diversity of *Nostoc* symbionts endophytically associated with two bryophyte species. Appl Environ Microbiol 67:4393–4396
Costa M, Garcia M, Costa-Rodrigues J, Costa MS, Ribeiro MJ, Fernandes MH, Barros P, Barreiro A, Vasconcelos V, Martins R (2014) Exploring bioactive properties of marine cyanobacteria isolated from the Portuguese coast: high potential as a source of anticancer compounds. Mar Drugs 12:98–114
Dagan T, Roettger M, Stucken K, Landan G, Koch R, Major P, Gould SB, Goremykin VV, Rippka R, Tandeau de Marsac N, Gugger M (2013) Genomes of Stigonematalean cyanobacteria (subsection V) and the evolution of oxygenic photosynthesis from prokaryotes to plastids. Genome Biol Evol 5:31–44
Dammeyer T, Bagby SC, Sullivan MB, Chisholm SW, Frankenberg-Dinkel N (2008) Efficient phage-mediated pigment biosynthesis in oceanic cyanobacteria. Curr Biol 18:442–448
Dasey M, Ryan N, Wilson J, McGregor G, Fabbro L, Neilan BA, Burns BP, Kankaanpaa H, Morrison LF, Codd GA, Rissik D, Bowling L (2005) Investigations into the taxonomy, toxicity and ecology of benthic cyanobacterial accumulations in Myall Lake, Australia. Mar Freshw Res 56:45–55
Davydov D, Patova E (2018) The diversity of cyanoprokaryota from freshwater and terrestrial habitats in the Eurasian Arctic and Hypoarctic. Hydrobiologia 811:119–137
De Figueiredo DR, Reboleira ASSP, Antunes SC, Abrantes N, Azeiteiro U, Gonçalves F, Pereira MJ (2006) The effect of environmental parameters and cyanobacterial blooms on phytoplankton dynamics of a Portuguese temperate lake. Hydrobiologia 568:145–157

Delwiche CF (1999) Tracing the thread of plastid diversity through the tapestry of life. Am Nat 154: S164–S177

Demoulina CF, Laraa YJ, Corneta L, Francoisa C, Baurainb D, Wilmottec A, Javauxa EJ (2019) Cyanobacteria evolution: insight from the fossil record. Free Radic Biol Med 140:206–223

DeYoe HR, Lowe RL, Marks JC (1992) Effects of nitrogen and phosphorus on the endosymbiont load of *Rhopalodia gibba* and *Epithemia turgida* (Bacillariophyceae). J Phycol 28(6):773–777

Di Rienzi SC, Sharon I, Wrighton KC, Koren O, Hug LA, Thomas BC, Goodrich JK, Bell JT, Spector TD, Banfield JF, Ley RE (2013) The human gut and groundwater harbor non-photosynthetic bacteria belonging to a new candidate phylum sibling to cyanobacteria. eLife 2:e01102

Dvořák P, Poulíčková A, Hašler P, Belli M, Casamatta DA, Papini A (2015) Species concepts and speciation factors in cyanobacteria, with connection to the problems of diversity and classification. Biodivers Conserv 24:739–757

Fahnenstiel GL, Carrick HJ (1992) Phototrophic picoplankton in Lakes Huron and Michigan: abundance, distribution, composition, and contribution to biomass and production. Can J Fish Aquat Sci 49:379–388

Farrant GK, Doré H, Cornejo-Castillo FM, Partensky F, Ratin M, Ostrowski M, Pitt FD, Wincker P, Scanlan DJ, Iudicone D, Acinas SG (2016) Delineating ecologically significant taxonomic units from global patterns of marine picocyanobacteria. Proc Natl Acad Sci U S A 113:E3365–E3374

Fedonkin MA (2006) Two chronicles of life: comparison experience (paleobiology and genomics about early stages of evolution of biosphere). In: Problemy geologii i mineralogii (Problems of geology and mineralogy). Geoprint, Syktyvkar, pp 331–350

Fischer WW, Hemp J, Johnson JE (2016) Evolution of oxygenic photosynthesis. Annu Rev Earth Planet Sci 44:647–683

Flechtner VR, Johansen JR, Clark WH (1998) Algal composition of microbiotic crusts from the central desert of Baja California, Mexico. Great Basin Naturalist 58:295–311

Flombaum P, Gallegos JL, Gordillo RA, Rincon J, Zabala LL, Jiao N, Karl DM, Li WK, Lomas MW, Veneziano D, Vera CS, Vrugt JA, Martiny AC (2013) Present and future global distributions of the marine cyanobacteria *Prochlorococcus* and *Synechococcus*. Proc Natl Acad Sci U S A 110:9824–9829

Fogg GE, Stewart WDP, Fay P, Walsby AE (1973) The blue-green algae. Academic, New York

Foster RA, Kuypers MMM, Vagner T, Paerl RW, Musat N, Zehr JP (2011) Nitrogen fixation and transfer in open ocean diatom-cyanobacterial symbioses. ISME J 5:1484–1493

Fulton JM, Arthur MA, Freeman KH (2012) Subboreal aridity and scytonemin in the holocene black sea. Org Geochem 49:47–55

García-Meza JV, Carrillo-Chávez A, Morton-Bermea O (2006) Sequential extractions on mine tailings samples after and before bioassays: implications on the speciation of metals during microbial re-colonization. Environ Geol 49(3):437–448

Garcia-Pichel F (1998) Solar ultraviolet and the evolutionary history of cyanobacteria. Origins Life Evol Biospheres 28:321–347

Garcia-Pichel F, Belnap J (1996) Microenvironments and microscale productivity of cyanobacterial desert crusts. J Phycol 32:774–782

Gaysina LA, Bohunická M, Hazuková V, Johansen JR (2018) Biodiversity of terrestrial cyanobacteria of the South Ural region. Cryptogamie Algol 39(2):1–32

Gaysina LA, Saraf A, Singh P (2019) Cyanobacteria in diverse habitats. In: Mishra AK, Tiwari DN, Rai AN (eds) Cyanobacteria: from basic science to applications. Academic, London, pp 145–171

Gebhardt JS, Nierzwicki-Bauer SA (1991) Identification of a common cyanobacterial symbiont associated with *Azolla* spp. through molecular and morphological characterization of free-living and symbiotic cyanobacteria. Appl Environ Microbiol 57:2141–2146

Gollerbach MM, Shtina EA (1969) Pochvennyje vodorosli (Soil algae). Nauka, Leningrad, p 228

Gollerbach MM, Novichkova LN, Sdobnikova NV (1956) The algae of takyrs. In: Takyrs of Western Turkmenia and ways of their agricultural development. Izd AN SSSR, Moscow, pp 38–54

Gueneli N, Mckenna AM, Ohkouchi N, Boreham CJ, Beghin J, Javaux EJ, Brocks JJ (2018) 1.1-Billion-year-old porphyrins establish a marine ecosystem dominated by bacterial primary producers. Proc Natl Acad Sci U S A 115:E6978–E6986

Guevara R, Armesto JJ, Caru M (2002) Genetic diversity of *Nostoc* microsymbionts from *Gunnera tinctoria* revealed by PCRSTRR fingerprinting. Microb Ecol 44:127–136

Gugger M, Lenoir S, Berger C, Ledreux A, Druart JC, Humbert J-F, Guette C, Bernard C (2005) First report in a river in France of the benthic cyanobacterium *Phormidium favosum* producing anatoxin-a associated with dog neurotoxicosis. Toxicon 45:919–928

Gupta RS (1982) Evolution of blue-green algae. Bionature 2:47–51

Gupta RS, Mathews DW (2010) Signature proteins for the major clades of cyanobacteria. BMC Evol Biol 10:24

Henriksson LE, Enekell PH, Henriksson E (1972) Determination of the nitrogen-fixing capacity of algae in soil. Oikos 23:420–423

Henson BJ, Watson LE, Barnum SR (2002) Molecular differentiation of the heterocystous cyanobacteria, *Nostoc* and *Anabaena*, based on complete NifD sequences. Curr Microbiol 45: 161–164

Heubeck C (2009) An early ecosystem of Archean tidal microbial mats (Moodies Group, South Africa, ca. 3.2 Ga). Geology 37:931–934

Hitzfeld BC, Lampert CS, Spaeth N, Mountfort D, Kaspar H, Dietrich DR (2000) Toxin production in cyanobacterial mats from ponds on the McMurdo Ice Shelf, Antarctica. Toxicon 38:1731–1748

Hodgson DA, Wright SW, Davies N (1997) Mass spectrometry and reverse phase HPLC techniques for the identification of degraded fossil pigments in lake sediments and their application in palaeolimnology. J Paleolimnol 18:335–350

Hoffmann L (1994) Marine Cyanophyceae of Papua New Guinea. VI. The genus *Lyngbya* S.L. Belg J Bot 127:79–86

Hoffmann L, Komárek J, Kaštovský J (2005) System of cyanoprokaryotes (cyanobacteria)-state in 2004. Algol Stud 117:95–115

Hoffmann L, Ector RL, Kostikov I (2007) Algal flora from limed and unlimed forest soils in the Ardenne (Belgium). Syst Geogr Plants 77:15–90

Hofmann HJ (1976) Precambrian microflora, belcher islands, Canada: significance and systematics. J Paleontol 50:1040–1073

Holland HD (2006) The oxygenation of the atmosphere and oceans. Philos Trans R Soc B 361:903–915

Homann M, Sansjofre P, Van Zuilen M, Heubeck C, Gong J, Killingsworth B, Foster IS, Airo A, Van Kranendonk MJ, Ader M, Lalonde SV (2018) Microbial life and biogeochemical cycling on land 3,220 million years ago. Nat Geosci 11:665–671

Ignacio-Espinoza JC, Sullivan MB (2012) Phylogenomic of T4 cyanophages: lateral gene transfer in the "core" and origins of host genes. Environ Microbiol 14:2113–2126

Issa OM, Trichet J, Défarge C, Couté A, Valentin C (1999) Morphology and microstructure of microbiotic soil crusts on a tiger bush sequence (Niger, Sahel). Catena 37:175–196

Johansen JR (1993) Cryptogamic crusts of semiarid and arid lands of North America. J Phycol 29: 140–147

Johansen JR, Shubert LE (2001) Algae in soil. Nova Hedwig Beih 123:297–306

Johnston DT, Wolfe-Simon F, Pearson A, Knoll AH (2009) Anoxygenic photosynthesis modulated Proterozoic oxygen and sustained Earth's middle age. Proc Natl Acad Sci U S A 106:16925–16929

Jones AC, Monroe EA, Podell S, Hess WR, Klages S, Esquenazi E, Niessen S, Hoover H, Rothmann M, Lasken RS, Yates JR III, Reinhardt R, Kubed M, Burkart MD, Eric E, Allen EE, Dorrestein PC, William H, Gerwick WH, Gerwick L (2011) Genomic insights into the physiology and ecology of the marine filamentous cyanobacterium *Lyngbya majuscula*. Proc Natl Acad Sci U S A 108:8815–8820

Kabirov RR (1997) Soil algae involved in the formation of plant cover on the dumping grounds of the Kan-Achinskcoal field. Russ J Ecol 28(3):188–190

Kauff F, Büdel B (2010) Phylogeny of cyanobacteria: an overview. In: Lüttge U, Beyschlag W, Büdel B, Francis D (eds) Progress in botany (Genetics-physiology-systematics-ecology), vol 72. Springer, Berlin, pp 209–224

Khaibullina LS, Gaisina LA (2008) Effect of salinization on the species composition and morphological features of soil algae. Eurasian Soil Sci 41(2):215–221

Khaybullina LA, Gaysina LA, Johansen JR, Krautova M (2010) Examination of the terrestrial algae of the Great Smoky National Park, USA. Fottea 10(2):201–215

Komárek J (2001) Review of cyanoprokaryotic genus *Romeria*. Czech Phycol Bull Phycol Sec Czech Bot Soc 1:5–19

Komárek J (2003) Coccoid and colonial cyanobacteria. In: Wehr JD, Sheath RG (eds) Freshwater algae of North America. Ecology and classification. Academic, Amsterdam, pp 59–116

Komárek J, Anagnostidis K (1989) Modern approach to the classification system of the cyanophytes 4-Nostocales. Archiv für Hydrobiologie 56:247–345

Komárek J, Anagnostidis K (1999) Cyanoprokaryota. 1. Teil: Chroococcales. In: Ettl H, Gärtner G, Heynig H, Mollenheuer D (eds) Süßwasserflora von Mitteleuropa, Bd. 19/1. Spektrum Akademische Verlag GmbH, Berlin, p 548

Komarek J, Kastovski J, Mares J, Johansen J (2014) Taxonomic classification of cyanoprokaryotes (cyanobacterial genera) 2014, using a polyphasic approach. Preslia 86:295–335

Kopp RE, Kirschvink JL, Hilburn IA, Nash CZ (2005) The Paleoproterozoic snowball Earth: a climate disaster triggered by the evolution of oxygenic photosynthesis. Proc Natl Acad Sci U S A 102:11131–11136

Kozlov AM, Zhang J, Yilmaz P, Glöckner FO, Stamatakis A (2016) Phylogeny-aware identification and correction of taxonomically mislabelled sequences. Nucleic Acids Res 44:5022–5033

Krienitz L, Ballot A, Kotut K, Wiegand C, Puetz S, Metcalf JS, Codd GA, Pflugmacher S (2003) Contribution of hot spring cyanobacteria to the mysterious deaths of Lesser Flamingos at Lake Bogoria, Kenya. FEMS Microbiol Ecol 43:141–148

Kulasooriya SA (2011) Cyanobacteria: pioneers of planet Earth. Ceylon J Sci 40(2):71–88

Kurmayer R, Dittman E, Fastner J, Chorus I (2002) Diversity of microcystin genes within a population of the toxic cyanobacterium *Microcystis* spp in Lake Wannsee (Berlin, Germany). Microb Ecol 43:107–118

Laamanen M (1996) Cyanoprokaryotes in the Baltic Sea ice and winter plankton. Arch Hydrobiol/ Algol Stud 83:423–433

LaRoche J, Breitbarth E (2005) Importance of the diazotrophs as a source of new nitrogen in the ocean. J Sea Res 53:67–91

Larsson J, Nylander JAA, Bergman B (2011) Genome fluctuations in cyanobacteria reflect evolutionary, developmental and adaptive traits. BMC Evol Biol 11:187

Lepot K, Compere P, Gerard E, Namsaraev Z, Verleyen E, Tavernier I, Hodgson DA, Vyverman W, Gilbert B, Wilmotte A, Javaux EJ (2014) Organic and mineral imprints in fossil photosynthetic mats of an East Antarctic lake. Geobiology 12:424–450

Lesser MP, Falcon LI, Rodriguez-Roman A, Enriquez S, Hoegh-Guldberg O, Iglesias-Prieto R (2007) Nitrogen fixation by symbiotic cyanobacteria provides a source of nitrogen for the scleractinian coral *Montastraea cavernosa*. Mar Ecol Prog Ser 346:143–152

Li XR, Wang XP, Li T, Zhang JG (2002) Microbiotic soil crust and its effect on vegetation and habitat on artificial stabilized desert dunes in Tengger Desert, North China. Biol Fertil Soils 35: 147–154

Lindell D, Sullivan MB, Johnson ZI, Tolonen AC, Rohwer F, Chisholm SW (2004) Transfer of photosynthesis genes to and from *Prochlorococcus* viruses. Proc Natl Acad Sci U S A 101: 11013–11018

Lukešová A (2001) Soil algae in brown coal and lignite post-mining areas in central Europe (Czech Republic and Germany). Restor Ecol 9:341–350
Lyons TW, Reinhard CT (2011) Sea charge for the rise of oxygen. Nature 478:194–195
Marin B, Nowack ECM, Melkonian M (2005) A plastid in the making: evidence for a second primary endosymbiosis. Protist 156:425–432
Mataloni G, Komarek J (2004) *Gloeocapsopsis aurea*, a new subaerophytic cyanobacterium from maritime Antarctica. Polar Biol 27:623–628
Matuła J, Pietryka M, Richter D, Wojtuń B (2007) Cyanoprokaryota and algae of Arctic terrestrial ecosystems in the Hornsund area, Spitsbergen. Pol Polar Res 28(4):283–315
Maya Y, Lopéz-Cortés A, Soeldner A (2002) Cyanobacterial microbiotic crusts in eroded soils of a tropical dry forest in the Baja California peninsula. Mexico Geomicrobiol J 19:505–518
Meeks JC, Elhai J, Thiel T, Potts M, Larimer F, Lamerdin J, Predki P, Atlas R (2001) An overview of the genome of *Nostoc punctiforme*, a multicellular, symbiotic cyanobacterium. Photosynth Res 70:85–106
Mez K, Beattie KA, Codd GA, Hanselmann K, Hauser B, Preisig HR (1997) Identification of a microcystin in benthic cyanobacteria linked to cattle deaths on alpine pastures in Switzerland. Eur J Phycol 32:111–117
Mez K, Hanselmann K, Preisig HR (1998) Environmental conditions in high mountain lakes containing toxic benthic cyanobacteria. Hydrobiologia 368:1–15
Millard A, Clokie MRJ, Shub DA, Mann NH (2004) Genetic organization of the psbAD region in phages infecting marine *Synechococcus* strains. Proc Natl Acad Sci U S A 101(30):11007–11012
Mohamed ZA (2008) Toxic cyanobacteria and cyanotoxins in public hot springs in Saudi Arabia. Toxicon 51:17–27
Mohamed ZA, El-Sharouny HM, Ali WSM (2006) Microcystin production in benthic mats of cyanobacteria in the Nile River and irrigation canals, Egypt. Toxicon 47:584–590
Moisander PH, Beinart RA, Hewson I, White AE, Johnson KS, Carlson CA, Montoya JP, Zehr JP (2010) Unicellular cyanobacterial distributions broaden the oceanic N_2 fixation domain. Science 327:1512–1514
Mollenhauer D, Bengtsson R, Lindrstøm E (1999) Macroscopic cyanobacteria of the genus *Nostoc*: a neglected and endangered constituent of European inland aquatic biodiversity. Eur J Phycol 34:349–360
Montoya JP, Holl CM, Zehr JP, Hansen A, Villareal TA, Capone DG (2004) High rates of N_2 fixation by unicellular diazotrophs in the oligotrophic Pacific Ocean. Nature 430:1027–1032
Moreira D, Guyader HL, Philippe H (2000) The origin of red algae and the evolution of chloroplasts. Nature 405:69–72
Morot-Gaudry JF, Touraine B (2001) Sources of nitrogen, nitrogen cycle, root structure and nitrogen assimilation. In: Morot-Gaudry JF (ed) Nitrogen assimilation by plants: physiological, biochemical and molecular aspects. Science Publishers Inc., Boca Raton, FL, pp 5–14
Mulkidjanian AY, Koonin EV, Makarova KS, Mekhedov SL, Sorokin A, Wolf YI, Dufresne A, Partensky F, Burd H, Kaznadzey D, Haselkorn R (2006) The cyanobacterial genome core and the origin of photosynthesis. Proc Natl Acad Sci U S A 103:13126–13131
Mur LR, Skulberg OM, Utkilen H (1999) Cyanobacteria in the environment. In: Chorus I, Bartram JE (eds) Toxic cyanobacteria in the water. E & F.N. Spon, London, pp 15–40
Neher DA, Walters TL, Tramer E, Weicht TR, Veluci RM, Saiya-Cork K, Will-Wolf S, Toppin J, Traub J, Johansen JR (2003) Biological soil crust and vascular plant communities in a sand savanna of northwestern Ohio. J Torrey Bot Soc 130:244–252
Nelson DR, Hazzouri KM, Lauersen KJ, Jaiswal A, Chaiboonchoe A, Mystikou A, Fu W, Daakour S, Dohai B, Alzahmi A, Nobles D, Hurd M, Sexton J, Preston MJ, Blanchette J, Lomas MW, Amiri KMA, Salehi-Ashtiani K (2021) Large-scale genome sequencing reveals the driving forces of viruses in microalgal evolution. Cell Host Microbe. https://doi.org/10.1016/j.chom.2020.12.005

Nguyen TA, Brescic J, Vinyard DJ, Chandrasekar T, Dismukes GC (2012) Identification of an oxygenic reaction center psbADC operon in the cyanobacterium *Gloeobacter violaceus* PCC 7421. Mol Biol Evol 29:35–38

Nilsson M, Bergman B, Rasmussen U (2000) Cyanobacterial diversity in geographically related and distant host plants of the genus *Gunnera*. Arch Microbiol 173:97–102

Noffke N, Gerdes G, Klenke T, Krumbein WE (2001) Microbially induced sedimentary structures-a new category within the classification of primary sedimentary structures-discussion. J Sediment Res 72:587–588

Nürnberg DJ, Mariscal V, Parker J, Mastroianni G, Flores E, Mullineaux CW (2014) Branching and intercellular communication in the Section V cyanobacterium *Mastigocladus laminosus*, a complex multicellular prokaryote. Mol Microbiol 91:935–949

O'Neil JM, Davis TW, Burford MA, Gobler CJ (2012) The rise of harmful cyanobacteria blooms: the potential roles of eutrophication and climate change. Harmful Algae 14:313–334

Oren A, Tindall BJ (2005) Nomenclature of the cyanophyta/cyanobacteria/cyanoprokaryotes under the international code of nomenclature of prokaryotes. Algol Stud 117:39–52

Oren A, Ventura S (2017) The current status of cyanobacterial nomenclature under the "prokaryotic" and the "botanical" code. Antonie Van Leeuwenhoek 110:1257–1269

Oren A, Komárek J, Hoffmann L (2009) Nomenclature of the Cyanophyta/Cyanobacteria/ Cyanoprokaryotes-what has happened since IAC Luxembourg? Algol Stud 130:17–26

Pabby A, Prasanna R, Nayak S, Singh PK (2003) Physiological characterization of the cultured and freshly isolated endosymbionts from different species of *Azolla*. Plant Physiol Biochem 41:73–79

Pandey A, Pathak J, Singh DK, Ahmed H, Singh V, Kumar D (2020) Photoprotective role of UV-screening pigment scytonemin against UV-B-induced damages in the heterocyst-forming cyanobacterium *Nostoc* sp. strain HKAR-2. Braz J Bot 43:67–80

Patel HM, Rastogi RP, Trivedi U, Madamwar D (2019) Cyanobacterial diversity in mat sample obtained from hypersaline desert, Rann of Kachchh. 3 Biotech 9:304

Pathak J, Rajneesh, Richa, Kannaujiya VK, Sonker AS, Sinha RP (2015) Diverse functions and applications of novel and unique cyanobacterial sheath pigment, scytonemin. In: Sinha RP, Richa, Rastogi RP (eds) Biological sciences: innovations and dynamics. New India Publishing Agency, New Delhi, India, pp 237–262

Pathak J, Sonker AS, Richa, Rajneesh, Kannaujiya VK, Singh V, Ahmed H, Sinha RP (2017) Screening and partial purification of photoprotective pigment scytonemin from cyanobacterial crusts dwelling on the historical monuments in and around Varanasi, India. Microbiol Res 8(1):4–12

Pathak J, Rajneesh, Maurya PK, Singh SP, Häder D-P, Sinha RP (2018) Cyanobacterial farming for environment friendly sustainable agriculture practices: innovations and perspectives. Front Environ Sci 6:7

Pathak J, Pandey A, Maurya PK, Rajneesh, Sinha RP, Singh SP (2020) Cyanobacterial secondary metabolite scytonemin: a potential photoprotective and pharmaceutical compound. Proc Natl Acad Sci India Sect B Biol Sci 90:467–481

Patova EN, Novakovskaya IV, Deneva SV (2018) The influence of edaphic and orographic factors on algal diversity in biological soil crusts on bare spots in the Polar and Subpolar Urals. Eurasian Soil Sci 51(3):309–320

Paul VJ, Thacker RW, Banks K, Golubic S (2005) Benthic cyanobacterial bloom impacts the reefs of South Florida (Broward County, USA). Coral Reefs 24:693–697

Plazinski J, Zheng Q, Taylor R, Croft L, Rolfe BG, Gunning BES (1990) DNA probes show genetic variation in cyanobacterial symbionts of the *Azolla* fern and a closer relationship to free-living *Nostoc* strains than to free-living *Anabaena* strains. Appl Environ Microbiol 56:1263–1270

Proteau PJ, Gerwick WH, Garcia-Pichel F, Castenholz R (1993) The structure of scytonemin, an ultraviolet sunscreen pigment from the sheaths of cyanobacteria. Experientia 49:825–829

Puigbò P, Lobkovsky AE, Kristensen DM, Wolf YI, Koonin EV (2014) Genomes in turmoil: quantification of genome dynamics in prokaryote supergenomes. BMC Biol 12:66

Rai H, Bergman B, Rasmussen U (2002) Cyanobacteria in symbiosis. Kluwer Academic, Dordrecht
Rajneesh, Singh SP, Pathak J, Sinha RP (2017) Cyanobacterial factories for the production of green energy and value-added products: an integrated approach for economic viability. Renew Sust Energ Rev 69:578–595
Rampelotto PH (2014) Polar microbiology: recent advances and future perspectives. Biology 3:81–84
Ran L, Larsson J, Vigil-Stenman T, Nylander JA, Ininbergs K, Zheng WW, Lapidus A, Lowry S, Haselkorn R, Bergman B (2010) Genome erosion in a nitrogen-fixing vertically transmitted endosymbiotic multicellular cyanobacterium. PLoS One 5(7):11486
Rashby SE, Sessions AL, Summons RE, Newman DK (2007) Biosynthesis of 2-methylbacteriohopanepolyols by an anoxygenic phototroph. Proc Natl Acad Sci U S A 104: 15099–15104
Rasmussen U, Nilsson M (2002) Cyanobacterial diversity and specificity in plant symbioses. In: Rai AN, Bergman B, Rasmussen U (eds) Cyanobacteria in symbiosis. Kluwer Academic, Dordrecht, pp 313–328
Rasmussen U, Svenning MM (1998) Fingerprinting of cyanobacteria based on PCR with primers derived from short and long tandemly repeated repetitive sequences. Appl Environ Microbiol 64:265–272
Rastogi RP, Sinha RP (2009) Biotechnological and industrial significance of cyanobacterial secondary metabolites. Biotechnol Adv 27:521–539
Rastogi RP, Kumari S, Richa, Han T, Sinha RP (2012) Molecular characterization of hot spring cyanobacteria and evaluation of their photoprotective compounds. Can J Microbiol 58(6):719–727
Rastogi RP, Sinha RP, Incharoensakdi A (2013) Partial characterization, UV-induction and photoprotective function of sunscreen pigment, scytonemin from *Rivularia* sp. HKAR-4. Chemosphere 93(9):1874–1878
Rastogi RP, Sinha RP, Incharoensakdi A (2014) The cyanotoxin-microcystins: current overview. Rev Environ Sci Biotechnol 13:215–249
Rastogi RP, Madamwar D, Incharoensakdi A (2015) Bloom dynamics of cyanobacteria and their toxins: environmental health impacts and mitigation strategies. Front Microbiol 6:1254
Raymond J, Siefert JL, Staples CR, Blankenship RE (2004) The natural history of nitrogen fixation. Mol Biol Evol 21:541–554
Reyes Prieto A, Bhattacharya D (2007) Phylogeny of Calvin cycle enzymes supports Plantae monophyly. Mol Phylogenet Evol 45:384–391
Reyes Prieto A, Yoon HS, Moustafa A, Yang EC, Andersen RA, Boo SM, Nakayama T, Ishida K-i, Bhattacharya D (2010) Differential gene retention in plastids of common recent origin. Mol Biol Evol 27:1530–1537
Reynolds CS, Jaworski GHM, Cmieche HA, Leedale GF (1981) On the annual cycle of the blue-green alga *M. aeruginosa* Kütz. Emend. Elenkin. Philos Trans R Soc Lond B 293:419–477
Rippka R (1988) Recognition and identification of cyanobacteria. Methods Enzymol 167:28–67
Rippka R, Herdman M (1992) Pasteur culture collection of cyanobacterial strains in axenic culture, catalogue and taxonomic handbook. Institut Pasteur, Paris
Rippka R, Deruelles J, Waterbury JB, Herdman M, Stanier RY (1979) Generic assignments, strain histories and properties of pure cultures of cyanobacteria. Microbiology 111:1–61
Rosing MT, Frei R (2004) U-rich Archaean sea-floor sediments from Greenland-indications of >3700 Ma oxygenic photosynthesis. Earth Planet Sci Lett 217:237–244
Sagan L (1967) On the origin of mitosing cells. J Theor Biol 14:225–274
Saha KS, Das R, Bora KN, Uma L (2007) Biodiversity of epilithic cyanobacteria from freshwater streams of Kakoijana reserve forest, Assam, India. Indian J Microbiol 47:219–232
Saito MA, Bertrand EM, Dutkiewicz S, Bulygin VV, Moran DM, Monteiro FM, Follows MJ, Valois FW, Waterbury JB (2011) Iron conservation by reduction of metalloenzyme inventories in the marine diazotroph *Crocosphaera watsonii*. Proc Natl Acad Sci U S A 108:2184–2189

Sánchez-Baracaldo P, Cardona T (2020) On the origin of oxygenic photosynthesis and cyanobacteria. New Phytol 225(4):1440–1446

Sanchez-Baracaldo P, Hayes PK, Blank CE (2005) Morphological and habitat evolution in the cyanobacteria using a compartmentalization approach. Geobiology 3:145–165

Schinteie R, Brocks JJ (2017) Paleoecology of Neoproterozoic hypersaline environments: biomarker evidence for haloarchaea, methanogens, and cyanobacteria. Geobiology 15:641–663

Schirrmeister BE, Anisimova M, Antonelli A, Bagheri HC (2011a) Evolution of cyanobacterial morphotypes: taxa required for improved phylogenomic approaches. Commun Integr Biol 4: 424–427

Schirrmeister BE, Antonelli A, Bagheri HC (2011b) The origin of multicellularity in cyanobacteria. Evol Biol 11:45

Schirrmeister BE, Vos JM, Antonelli A, Bagheri HC (2013) Evolution of multicellularity coincided with increased diversification of cyanobacteria and the Great Oxidation Event. Proc Natl Acad Sci U S A 110:1791–1796

Schirrmeister BE, Sanchez-Baracaldo P, Wacey D (2016) Cyanobacterial evolution during the Precambrian. Int J Astrobiol 15:187–204

Schopf JW (1993) Microfossils of the early Archaean apex chert: new evidence of the antiquity of life. Science 260:640–646

Schwabe GH (1972) Blue-green algae as pioneers on postvolcanic substrate (Surtsey/Iceland). In: Desikachary TV (ed) Proceedings of the 1st international symposium of taxonomy and biology of blue-green algae. University of Madras, Madras, pp 419–424

Sciuto K, Moro I (2015) Cyanobacteria: the bright and dark sides of a charming group. Biodivers Conserv 24:711–738

Sellner KG (1997) Physiology, ecology, and toxic properties of marine cyanobacteria blooms. Limnol Oceanogr 42:1089–1104

Shen Y, Buick R (2004) The antiquity of microbial sulfate reduction. Earth Sci Rev 64:243–272

Shestakov SV (2007) Horizontal transfer of genes in bacteria: process and limitations. Ekol 5:12–24

Shestakov SV, Karbysheva EA (2015) The role of viruses in the evolution of cyanobacteria. Biol Bull Rev 5:527–537

Shestakov SV, Karbysheva EA (2017) The origin and evolution of cyanobacteria. Biol Bull Rev 7: 259–272

Shi T, Falkowski PG (2008) Genome evolution in cyanobacteria: the stable core and the variable shell. Proc Natl Acad Sci U S A 105:2510–2515

Shih PM, Wu D, Latifi A, Axen SD, Fewer DP, Talla E, Calteau A, Cai F, De Marsac NT, Rippka R, Herdman M (2013) Improving the coverage of the cyanobacterial phylum using diversity-driven genome sequencing. Proc Natl Acad Sci U S A 110:1053–1058

Shtina EA, Andreyeva VM, Kuzyakina TI (1992) Algae settlement of volcanic substrates (Zaseleniye vodoroslyami vulkanicheskikh substratov). Bot Zhurn 8:33–42

Simm S, Keller M, Selymesi M, Schleiff E (2015) The composition of the global and feature specific cyanobacterial core-genome. Front Microbiol 6:219

Singh RN (1950) Reclamation of "usar" lands of India through blue-green algae. Nature 165:325–326

Singh P (2017) Cyanobacterial taxonomy and systematics: a brief review. In: Singh SS (ed) Plants and microbes in ever changing environment. Nova Science Publishers Inc., Hauppauge, NY, pp 1–29

Skulberg OM (1994) Oscillatoialean cyanoprokaryotes and their application for algal culture technology. Algol Stud/Archiv für Hydrobiologie 75:265–1278

Slotznick SP, Fischer WW (2016) Examining Archean methanotrophy. Earth Planet Sci Lett 441: 52–59

Smith FMJ, Wood SA, van Ginkel R, Broady PA, Gaw S (2011) First report of saxitoxin production by a species of the freshwater benthic cyanobacterium, *Scytonema* Agardh. Toxicon 57:566–573

Soo RM, Skennerton CT, Sekiguchi Y, Imelfort M, Paech SJ, Dennis PG, Steen JA, Parks DH, Tyson GW, Hugenholtz P (2014) An expanded genomic representation of the phylum cyanobacteria. Genome Biol Evol 6:1031–1045

Soo RM, Woodcroft BJ, Parks DH, Tyson GW, Hugenholtz P (2015) Back from the dead; the curious tale of the predatory cyanobacterium *Vampirovibrio chlorellavorus*. Peer J 3:e968

Soo RM, Hemp J, Parks DH, Fischer WW, Hugenholtz P (2017) On the origins of oxygenic photosynthesis and aerobic respiration in cyanobacteria. Science 355:1436–1440

Steppe TF, Olson JB, Paerl HW, Belnap J (1996) Consortial N_2 fixation: a strategy for meeting nitrogen requirements of marine and terrestrial cyanobacterial mats. FEMS Microbiol Ecol 21: 149–156

Stockner JG (1988) Phototrophic picoplankton: an overview from marine and freshwater ecosystems. Limnol Oceanogr 33:765–775

Storme J-Y, Golubic S, Wilmotte A, Kleinteich J, Velazquez D, Javaux EJ (2015) Raman characterization of the UV-protective pigment gloeocapsin and its role in the survival of cyanobacteria. Astrobiology 15:843–857

Strunecký O, Elster J, Komárek J (2012) Molecular clock evidence for survival of Antarctic cyanobacteria (Oscillatoriales, *Phormidium autumnale*) from Paleozoic times. FEMS Microbiol Ecol 82:482–490

Sun Z, Blanchard JL (2014) Strong genome-wide selection early in the evolution of *Prochlorococcus* resulted in reduced genome through the loss of large number of small effect genes. PLoS One 9:e88837

Suosaari EP, Reid RP, Playford PE, Foster JS, Stolz JF, Casaburi G, Hagan PD, Chirayath V, Macintyre IG, Planavsky NJ, Eberli GP (2016) New multiscale perspectives on the stromatolites of Shark Bay, western Australia. Sci Rep 6:1–13

Svenning MM, Eriksson T, Rasmussen U (2005) Phylogeny of symbiotic cyanobacteria within the genus *Nostoc* based on 16S rDNA sequence analyses. Arch Microbiol 183:19–26

Thacker RW, Paul VJ (2004) Morphological, chemical, and genetic diversity of tropical marine cyanobacteria, *Lyngbya* spp. and *Symploca* spp. (Oscillatoriales). Appl Environ Microbiol 70: 3305–3312

Ting CS, Rocap G, King J, Chisholm SW (2002) Cyanobacterial photosynthesis in the oceans: the origins and significance of divergent light harvesting strategies. Trends Microbiol 10:134–142

Tirkey J, Adhikary SP (2005) Cyanobacteria in biological soil crusts of India. Curr Sci 89:515–521

Treub M (1888) Notice sur la nouvelle flore de Krakatau. Ann Jard Bot Buiten 7:213–223

Tripp HJ, Bench SR, Turk KA, Foster RA, Desany BA, Niazi F, Affourtit JP, Zehr JP (2010) Metabolic streamlining in an open-ocean nitrogen-fixing cyanobacterium. Nature 464:90–94

Trzcińska M, Pawlik-Skowrońska B (2008) Soil algal communities inhabiting zinc and lead mine spoils. J Appl Phycol 20:341–348

Usher KM (2008) The ecology and phylogeny of cyanobacterial symbionts in sponges. Mar Ecol 29(2):178–192

Usher KM, Bergman B, Raven JA (2007) Exploring cyanobacterial mutualisms. Annu Rev Ecol Evol Syst 38:255–273

Vincent WF (2007) Cold tolerance in cyanobacteria and life in the cryosphere. In: Seckbach J (ed) Algae and cyanobacteria in extreme environments. Springer, Heidelberg, pp 289–304

Vinogradova ON, Darienko TM (2008) Algae of Azovo-Syvashsky National Nature Park (Ukraine). Int J Algae 8:163–178

Vogt JC, Albach DC, Palinska KA (2018) Cyanobacteria of the Wadden Sea: seasonality and sediment influence on community composition. Hydrobiologia 811(1):103–117

Walter JM, Coutinho FH, Dutilh BE, Swings J, Thompson FL, Thompson CC (2017) Ecogenomics and taxonomy of cyanobacteria phylum. Front Microbiol 8:2132

Welsh EA, Liberton M, Stöckel J, Loh T, Elvitigala T, Wang C, Wollam A, Fulton RS, Clifton SW, Jacobs JM, Aurora R, Ghosh BK, Sherman LA, Smith RD, Wilson RK, Pakrasi HB (2008) The genome of *Cyanothece* 51142, a unicellular diazotrophic cyanobacterium important in the marine nitrogen cycle. Proc Natl Acad Sci U S A 105:15094–15099

West NJ, Adams DG (1997) Phenotypic and genotypic comparison of symbiotic and free-living cyanobacteria. Appl Environ Microbiol 63:4479–4484

Whitton BA (1973) Freshwater plankton. In: Fogg GE, Stewart WDP, Fay P, Walsby AE (eds) The blue-green algae. Academic, London, pp 353–367

Whitton BA, Potts M (1982) Marine littoral. In: Carr NG, Whitton BA (eds) The biology of cyanobacteria. Blackwell Science, Oxford, pp 515–542

Whitton BA, Potts M (2000a) The ecology of cyanobacteria. Their diversity in time and space. Kluwer Academic, Dordrecht

Whitton BA, Potts M (2000b) Introduction of cyanobacteria. In: Whitton BA, Potts M (eds) The ecology of cyanobacteria. Their diversity in time and space. Kluwer Academic, Dordrecht, pp 1–11

Yerrapragada S, Siefert JL, Fox GE (2009) Horizontal gene transfer in cyanobacterial signature genes. Methods Mol Biol 532:339–366

Zakhia F, Jungblut AD, Taton A, Vincent WF, Wilmotte A (2008) Cyanobacteria in cold ecosystems. In: Margesin R, Schinner JC, Gerday C (eds) Psychrophiles: from biodiversity to biotechnology. Springer, Heidelberg, pp 121–135

Zehr JP, Waterbury JB, Turner PJ, Montoya JP, Omoregie E, Steward GF, Hansen A, Karl DM (2001) Unicellular cyanobacteria fix N_2 in the subtropical North Pacific Ocean. Nature 412:635–638

Zhaxybayeva O, Gogarten JP, Charlebois RL, Doolittle WF, Papke RT (2006) Phylogenetic analyses of cyanobacterial genomes: quantification of horizontal gene transfer events. Genome Res 16:1099–1108

Polyphasic Approach and Cyanobacterial Taxonomy: Some Perspectives and Case Studies

2

Aniket Saraf, Himanshu G. Dawda, and Prashant Singh

Abstract

Cyanobacteria are oxygen-producing, photosynthesizing, gram-negative prokaryotes, which played a major role in the development of the atmosphere of the present earth. Despite being so old and omnipresent, it is surprising that proper and correct identification of cyanobacteria is still a challenge and has often created confusing patterns. The primary reason for all this confusion is the morphological plasticity of these organisms, which eventually creates confusion during long-term studies. This fact makes the study of cyanobacteria both challenging and interesting too. The taxonomy of cyanobacteria for a long time was based only on the morphological criterion, which, in the modern times, has raised many questions, which need to be answered by adopting an approach that respects both the classical morphology and the modern methods based on genetic information and phylogeny. The amalgamation of both the classical and the modern methods has led to the development of the polyphasic approach. Unfortunately, at the present moment what all constitutes a polyphasic approach is still under scrutiny, thus making the taxonomy of cyanobacteria complicated and challenging. Modern taxonomists must solve all the abovementioned problems by adopting an approach that reflects in a considerate way the morphology, ecology, and the molecular phylogeny. Unequal, biased preference or convenience-based methods are posing hindrances and thus lead to ambiguities that are tough to resolve. The future of cyanobacterial taxonomy may lie in the

A. Saraf · H. G. Dawda
Ramniranjan Jhunjhunwala College, Mumbai, India

P. Singh (✉)
Laboratory of Cyanobacterial Systematics and Stress Biology, Department of Botany, Institute of Science, Banaras Hindu University, Varanasi, India

R. P. Rastogi (ed.), *Ecophysiology and Biochemistry of Cyanobacteria*,
https://doi.org/10.1007/978-981-16-4873-1_2

proper usage of the polyphasic approach with more revisionary works being anticipated.

Keywords

Cyanobacteria · Taxonomy · Systematics · Polyphasic approach · *Nostoc* · *Calothrix*

2.1 Introduction

Cyanobacteria are gram-negative, photosynthetic, and oxygen-evolving prokaryotes that are found in diverse habitats varying from freshwater to marine, acidic soils to alkaline marshes, and cold deserts to thermal springs (Brock 1969, 1978; Whitton and Potts 2002; Whitton 2012; Seckbach and Oren 2007; Gaysina et al. 2019). Cyanobacteria have been inhabiting the earth since the early Precambrian (Seckbach and Oren 2007), and their long-lasting evolutionary history may be attributed to their successful life strategies. The presence of exopolysaccharides, which reduces the water loss and provides protection in extreme hot and cold environmental conditions, is one of the important features of the phylum cyanobacteria. Some of the other features include high level of tolerance to ultraviolet radiations, low oxygen level, and free sulfide. Moreover, the ability to utilize H_2S as a reducing agent in the absence of H_2O and production of toxins helps them to survive in competitive ecosystems (Whitton and Potts 2002; Whitton 2012; Gaysina et al. 2019). Cyanobacteria are also considered to be a crucial link in the course of evolution because it is now known that the endosymbiosis of cyanobacteria by the eukaryotic heterotrophs has led to the development of chloroplast (Klein and Cronquist 1967; Margulis 1981; Seckbach and Oren 2007). Due to the evolutionary significance and ecological importance of cyanobacteria, it is important to correctly identify and perform complete characterization of cyanobacterial strains isolated from different habitats.

2.2 The Development of Cyanobacterial Taxonomy

Cyanobacteria were traditionally classified as algae and were governed under the botanical code, even though their relationship with bacteria was recognized by Ferdinand Cohn in 1875 (Oren 2011). Different taxonomists over the years used different nomenclatures to represent the oxygen-evolving photosynthetic microorganisms. Cyanophyceae (Sachs 1874) along with blue-green algae were the most widely used names, whereas Myxophyceae (Stitzenberger 1860) and Schizophyceae (Cohn 1875) were the lesser used terms. Cyanobacterial taxonomy has undergone multiple revisions over the years in order to achieve a stable and a consistent taxonomic system. Traditionally, the taxonomic classification of blue-green algae was based on morphological traits and the blue-green algae were grouped under the tribes Coccogoneae (with unicellular reproductive bodies) and Hormogoneae (reproduction through hormogonia) (Thuret 1875). Filamentous cyanobacteria are morphologically more diverse, and the morphological characters such as presence or absence of heterocytes and akinetes, branching pattern, morphology of filaments, and other characters formed the basis of classification. The earliest taxonomic monographs for blue-green algae were given by Thuret (1875), Bornet and Flahault (1885, 1886a, b, c, d, 1888), Gomont (1892), and Borzì (1878, 1879, 1882, 1895, 1914, 1916, 1917). The taxonomic framework proposed by Bornet and Flahault (1885, 1886a, b, c, d, 1888) was widely accepted and is considered as the starting point for modern cyanobacterial taxonomy. Later, Geitler (1925) proposed a new classification system, which grouped blue-green algae into seven orders: Chroococcales, Entophysalidales, Pleurocapsales, Dermocarpales, Siphononematales, Nostocales, and Stigonematales. On the contrary, Frémy (1929) recognized only three orders Chroococcales, Chamaesiphonales, and Hormogonales in his work, which was also adopted by Geitler (1932). A decade later again, Geitler retained Chroococcales and Hormogonales; however, Chamaesiphonales was replaced by Dermocarpales and Pleurocapsales (Geitler 1942). Further, Fritsch (1942, 1944, 1945) broadly accepted Geitler's classification and recognized Chroococcales, Chamaesiphonales, Pleurocapsales, Nostocales, and Stigonematales. Desikachary (1959) also adopted Fritsch's classification in recognizing five orders; however, there were few differences at the family level among both the classification systems. Prescott (1962) adopted Frémy's classification system and Bourrelly (1970) followed Desikachary's' classification system.

From the latter half of the nineteenth century till the beginning of the seventh decade into the twentieth century, the nomenclature of oxygen-evolving photosynthetic microorganisms was governed under the botanical code. It was Roger Stanier, who introduced the name cyanobacteria and further proposed that the nomenclature of cyanobacteria should be governed under the provisions of the bacteriological code because of their prokaryotic nature (Stanier 1977; Stanier et al. 1978). The proposal of naming cyanobacteria under the provisions of the bacteriological code was opposed by phycologists (Bourrelly 1970; Golubic 1979). This contrasting opinion gave rise to a long-lasting debate with two different school of thoughts, which has been documented in different publications (Oren 2004, 2011; Hoffmann 2005; Oren and Tindall 2005; Oren et al. 2009; Oren and Komárek 2015; Oren and Garrity 2014). At present, the cyanobacterial classification is governed under both botanical and bacteriological codes (Oren and Ventura 2017); however, recent taxonomic descriptions indicate that the botanical code is preferred by the taxonomists over the bacteriological code (Singh et al. 2016; Singh et al. 2017a, b; Suradkar et al. 2017; Becerra-Absalón et al. 2018; Kabirnatj et al. 2018; Saraf et al. 2018; Shalygin et al. 2018; Mareš et al. 2019; Pietrasiak et al. 2019; Casamatta et al. 2020; Mishra et al. 2020; Saraf et al. 2020). After the classification of Bourrelly (1970), Rippka et al. (1979) proposed another classification system in which the cyanobacteria were divided into five sections (I to V) rather than the orders as done in earlier classification systems. Further, an important aspect of cyanobacteria which is the morphological plasticity with change in environmental and laboratory conditions was also addressed by Rippka et al (1979). Moving ahead of Rippka et al. (1979), Anagnostidis and Komárek (1985) put forward another classification system in which the authors proposed to recognize small genera as compared to the fewer genera comprising of many species.

The path-breaking research by Carl Woese toward the end of the twentieth century revolutionized the approach of taxonomists across the globe (Woese and Fox 1977). Taxonomic studies thereafter considered genetic markers as an important tool to differentiate taxa. This new approach that includes morphological, ecological, physiological, and genetic markers is termed as polyphasic approach (Hofmann et al. 2005). Initial phylogenetic analysis demonstrated certain genera to be heterogeneous, indicating the need for taxonomic revisions (Lyra et al. 2001; Gugger et al. 2002a, b; Rajaniemi et al. 2005). Moreover, the unicellular and simple filamentous forms (without heterocytes and akinetes) were also reported to be heterogeneous (Wilmotte and Golubić 1991; Turner et al. 1999; Honda et al. 1999; Castenholz 2001; Ishida et al. 2001; Wilmotte and Herdman 2001), which disregarded earlier classification systems wherein all filamentous forms (with or without heterocytes and akinetes) were grouped under Hormogonales. Another important observation was made by Gugger and Hoffmann (2004) with respect to the orders Nostocales and Stigonematales. The authors reported that the members of both the orders were intermixed and their phylogenetic analysis did not support two different orders. Based on the abovementioned results, Hofmann et al. (2005) proposed a revised classification system, which was based on the polyphasic approach that reflected the evolutionary relationships in cyanobacteria more accurately as compared to earlier systems, which were based mainly on morphological characters. The highlight of this classification system was the unification of Nostocales and Stigonematales into a single order and grouping of unicellular and the simple filamentous cyanobacteria according to their phylogenetic positioning. Komárek (2013) also followed Hofmann et al. (2005) and grouped all the heterocystous genera under the order Nostocales.

With the advancement of molecular taxonomy, many studies reported the polyphyletic nature of traditional genera like *Nostoc*, *Calothrix*, *Scytonema*, *Chroococcus*, *Phormidium*, and *Leptolynbya* (Gugger and Hoffmann 2004; Rajaniemi et al. 2005; Casamatta et al. 2005; Berrendero et al. 2008; Komárková et al. 2009; Komárek et al. 2013; Saraf et al. 2018). Furthermore, several new genera were described using polyphasic approach in order to achieve monophyly. From the year 2000 to 2013, 50 novel cyanobacterial genera were described and 16 new putative genera were proposed in the 19th IAC symposium held in Cleveland (Komárek et al. 2014). Later, Komárek et al. (2014) proposed a new classification system, which was based on the polyphasic approach, and several changes at the higher taxonomic level were made. Komárek et al. (2014) described two new orders along with six new families; moreover, the status of four subfamilies was elevated to family level. The ultimate aim of the classification scheme of Komárek et al. was to achieve monophyly. Furthermore, the authors also emphasized the importance of polyphasic approach in the taxonomic and systematic studies of cyanobacteria. With more clarity and increased understanding in the usage of polyphasic approach, the last 6 years has seen a surge in the description of new families and genera along with taxonomic revisions at the family level (Hentschke et al. 2016; Rigonato et al. 2016; Hašler et al. 2017; Shalygin et al. 2017; Kilgore et al. 2018; Sendall and McGregor 2018; Saraf et al. 2019c are some of the examples). More than 80 new genera and nine new families have been described after the classification system proposed by Komárek et al.

The taxonomic studies from the end of twentieth century have indicated that the polyphasic approach is indeed the best choice for characterizing cyanobacteria. In this chapter, we discuss the importance of polyphasic approach in addressing the taxonomic complexities with a special focus on the hugely diverse, heterogeneous, and polyphyletic genera *Nostoc* and *Calothrix*.

2.2.1 The Taxonomy of *Nostoc* and *Nostoc*-Like Genera

The genus *Nostoc* (Agardh ex Bornet and Flahault) (type species; *Nostoc commune*) is the type genus of the family Nostocaceae and is characterized by the presence of thick gelatinous colonies and complex life cycle. The filaments of *Nostoc* are irregularly coiled, loosely or densely agglomerated, and the trichome is surrounded by a thin mucilaginous sheath. The extracellular matrix may be fragile and diffluent or it may be present as a thick peridermal layer enclosing a mass of filaments. Trichomes are uniseriate and isopolar and of the same width along the whole length (Komárek 2013). The filaments of *Nostoc* are usually unbranched; however, true branching has been reported in some studies (Mollenhaur et al. 1999; Dodds et al. 1995; Singh et al. 2020). Recently, a new species, *Nostoc neudorfense,* has been described from Czech Republic, which showed true branching in the natural samples (Singh et al. 2020). The vegetative cells vary from cylindrical to barrel shaped to spherical; however, the apical cells within a single trichome are morphologically similar to the other cells. Heterocytes are solitary and develop terminally and/or intercalary. Akinetes are observed and develop apoheterocytically (Komárek 2013). Approximately 300 species of *Nostoc* (and *Nostoc*-like genera) have been described till date, and these species have been reported from diverse habitats (Komárek 2013). The members of the genus *Nostoc* display complex life cycle, and the younger filaments are formed by either hormogenic cycle or sporogenic cycle (Lazaroff 1966). In the hormogenic cycle, differentiation of hormogonia starts with the fragmentation of old filaments at site adjacent to heterocytes. Further, the vegetative filament develops from the newly formed hormogonia. In sporogenic cycle, the new vegetative filament is formed by the germination of akinetes (Lazaroff 1966). Based on its life cycle, the genus *Nostoc* is reported to be heterogeneous (Hrouzek et al. 2013). Even though the members of the genus *Nostoc* exhibit a range of interesting morphological characters; however, lack of synapomorphic character makes *Nostoc* a morphologically complex genus to study taxonomically (Řeháková et al. 2007).

The heterogeneity within *Nostoc* was first established by Rajaniemi et al. (2005) while evaluating the phylogenetic and morphological characters of the Nostocalean genera. The authors reported that the percentage similarity among the *Nostoc* strains included in their study was below the threshold limit recommended for separation of different genera. Moreover, in the 16S rRNA gene tree the bootstrap support for the clade consisting of *Nostoc* strains was weak. Furthermore, the *Nostoc* strains did not form a monophyletic cluster in *rbc*LX and *rpo*B gene trees. Based on the above observations, Rajaniemi et al. (2005) concluded that the *Nostoc* strains included in their study probably represented two different genera. In the same year, Hrouzek et al. (2005) compared the morphological characters and 16S rRNA gene sequences of their soil *Nostoc* strains with the other original isolates and PCC strains available at that time. The authors reported considerable morphological and phylogenetic variability among the strains included in their study. The observations of Rajaniemi et al. (2005) and Hrouzek et al. (2005) were in coherence with each other and indicated the need for taxonomic revisions within the genus *Nostoc*. The study of Řeháková et al. (2007) is a landmark study in resolving the phylogenetic complexities related to the genus *Nostoc* because of two reasons: First, the authors established the *Nostoc* sensu stricto; and second, they described the first *Nostoc*-like genus (*Mojavia*) based on the polyphasic approach. The genus *Mojavia* morphologically resembled the genus *Nostoc* but was phylogenetically distant from *Nostoc* sensu stricto and also had different secondary structures for the ITS region. Moreover, the locality from which the strain of *Mojavia* was isolated was also different from the *Nostoc* species originally described from Europe (Řeháková et al. 2007).

Further, Hrouzek et al. (2013) in continuation of their initial study gathered additional morphological and phylogenetic evidence to prove that the *Nostoc muscorum* strains included in their study represent a novel evolutionary lineage. Based on the results obtained in their study, the authors described a new *Nostoc*-like genus *Desmonostoc* (Hrouzek et al. 2013) to accommodate the *Nostoc muscorum* strains. Although the genus *Desmonostoc* was morphologically similar to the genus *Nostoc,* certain noteworthy morphological traits were reported. The filaments of the genus *Desmonostoc* were comparatively longer and densely coiled; moreover, the akinete and the hormogonia formed by the members were different from true *Nostoc*. Thereafter, Genuário et al. (2015) while studying strains isolated from Antarctica and Brazilian mangroves observed that the morphological characters of their strains resembled *Nostoc*, *Mojavia,* and *Desmonostoc*. However, all the newly isolated strains could be differentiated on the basis of ecological and physiological aspects. Moreover, the Antarctic and Brazilian strains clustered distantly from *Nostoc* sensu stricto, *Mojavia,* and *Desmonostoc* in the phylogenetic tree inferred using 16S rRNA gene. Based on the polyphasic approach, Genuário et al. (2015) described *Halotia* as the third *Nostoc*-like genus. Further, Genuário et al. in another study isolated *Nostoc*-like strains from Brazil and the phylogenetic analysis indicated the strains to be distant from *Nostoc* sensu stricto (Genuário et al. 2017). However, this time the authors did not establish a new genus. Later, Bagchi et al. (2017) also isolated a *Nostoc*-like strain from India that clustered with the Brazilian strains. Further, based on polyphasic studies, Bagchi et al. (2017) described *Aliinostoc* as the new *Nostoc*-like genus. At present, the members of the genus *Aliinostoc* have been reported from habitats that are rich in dissolved salts and ions. Morphologically, it is difficult to distinguish the members of the genus *Aliinostoc* from the other *Nostoc*-like genera;

however, formation of motile hormogonia with gas vesicles is considered to be the characteristic feature of the genus *Aliinostoc* (Bagchi et al. 2017).

Komarekiella is another *Nostoc*-like genus described by Hentschke et al. (2017) based on polyphasic approach. In their study, the authors isolated six cyanobacterial strains from Brazilian Atlantic rainforest and one strain from Kauai, Hawaii Islands, and observed that all the strains were morphologically similar to *Nostoc*, *Desmonostoc*, *Halotia,* and *Mojavia* and indistinguishable from the genus *Chlorogloeopsis*. However, the phylogenetic analysis based on 16S rRNA gene indicated their strains to be different from *Nostoc*, *Desmonostoc*, *Halotia*, *Mojavia,* and *Chlorogloeopsis*. The special type of germination of akinetes and the absence of macroscopic mucilaginous colonies are considered as the diacritical features of the genus *Komarekiella.* In another study, Saraf et al. (2019a, b) also described a new genus—*Desikacharya* based on the polyphasic approach. The members of the genus *Desikacharya* are morphologically similar to the genus *Nostoc;* however, notable differences such as prominent coiling and well-defined constrictions separated the members of the genus *Desikacharya* from the other related genera. Moreover, in the phylogenetic analysis inferred through 16S rRNA gene, the genus *Desikacharya* clustered distantly from *Nostoc* sensu stricto and other *Nostoc*-like genera. Two well-established strains, *Nostoc piscinale* CENA21 and *Trichormus azollae* KomBAI/1983, were also transferred to the genus *Desikacharya* based on their phylogenetic positioning (Saraf et al. 2019a, b). Furthermore, in the last 2 years, four new *Nostoc*-like genera—*Minunostoc* (Cai et al. 2019a), *Compactonostoc* (Cai et al. 2019b), *Violetonostoc* (Cai et al. 2020a), and *Purpurea* (Cai et al. 2020b)—were described based on polyphasic approach. Among the four new genera, the taxonomic status of *Minunostoc* has been questioned in a recent study by Singh et al. (2020), because the sequences corresponding to the genus *Minunostoc* clustered within the *Desikacharya* clade. The phylogenetic positioning of all the newly described *Nostoc*-like genera is illustrated in Fig. 2.1.

The introduction of polyphasic approach has significantly contributed in determining the correct taxonomic status of *Nostoc*-like taxa; however, recent studies have indicated a lack of similar approach within *Nostoc* sensu stricto. This is evident from the fact that till date only four species of true *Nostoc* have been described using polyphasic approach (Řeháková et al. 2007; Mesfin et al. 2020; Singh et al. 2020). Singh et al. (2020) in their study reported the polyphyletic nature of *Nostoc commune* (type species of *Nostoc*) and *Nostoc punctiforme,* and the authors further recommended the need to perform polyphasic revisions within *Nostoc* sensu stricto. The phylogenetic inconsistencies within *Nostoc* sensu stricto is illustrated in Fig. 2.2. From the above studies, it is evident that identification of strains through polyphasic approach is the only way to resolve the taxonomic complexities hovering around *Nostoc* and *Nostoc*-like taxa.

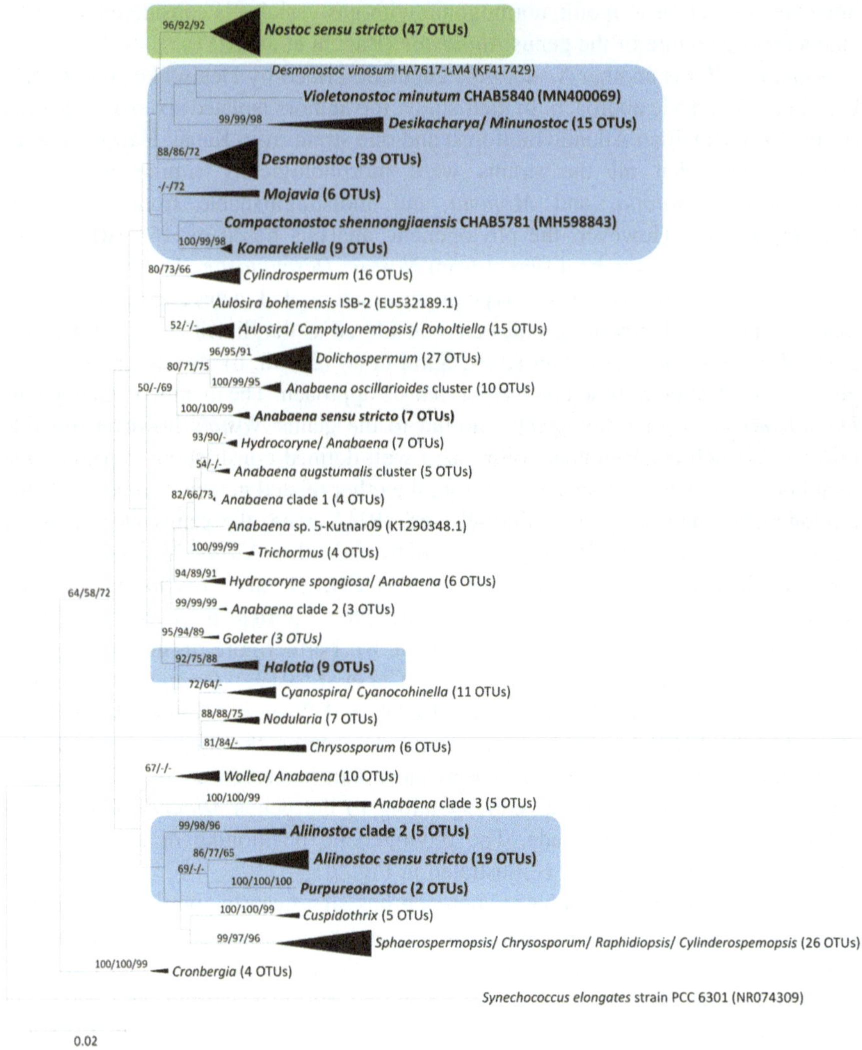

Fig. 2.1 Phylogenetic positioning of *Nostoc* sensu stricto and *Nostoc*-like genera inferred using 16S rRNA gene with the bootstrap values representing neighbor joining/maximum likelihood/ maximum parsimony. Bar, 0.02 changes per nucleotide position. Bootstrap values >50 are shown

2.2.2 The Taxonomy of *Calothrix* and Related Genera

The family Rivulariaceae is one of the most studied cyanobacterial family and has undergone multiple taxonomic revisions over the years. Traditionally, tapering cyanobacteria with the ability to produce heterocytes were grouped under Rivulariaceae and Mastichotricheae (Bornet and Flahault 1885, 1886a, b, c, d, 1888). Rivulariaceae included *Rivularia*, *Isactis*, *Gloeotrichia,* and *Brachytrichia,*

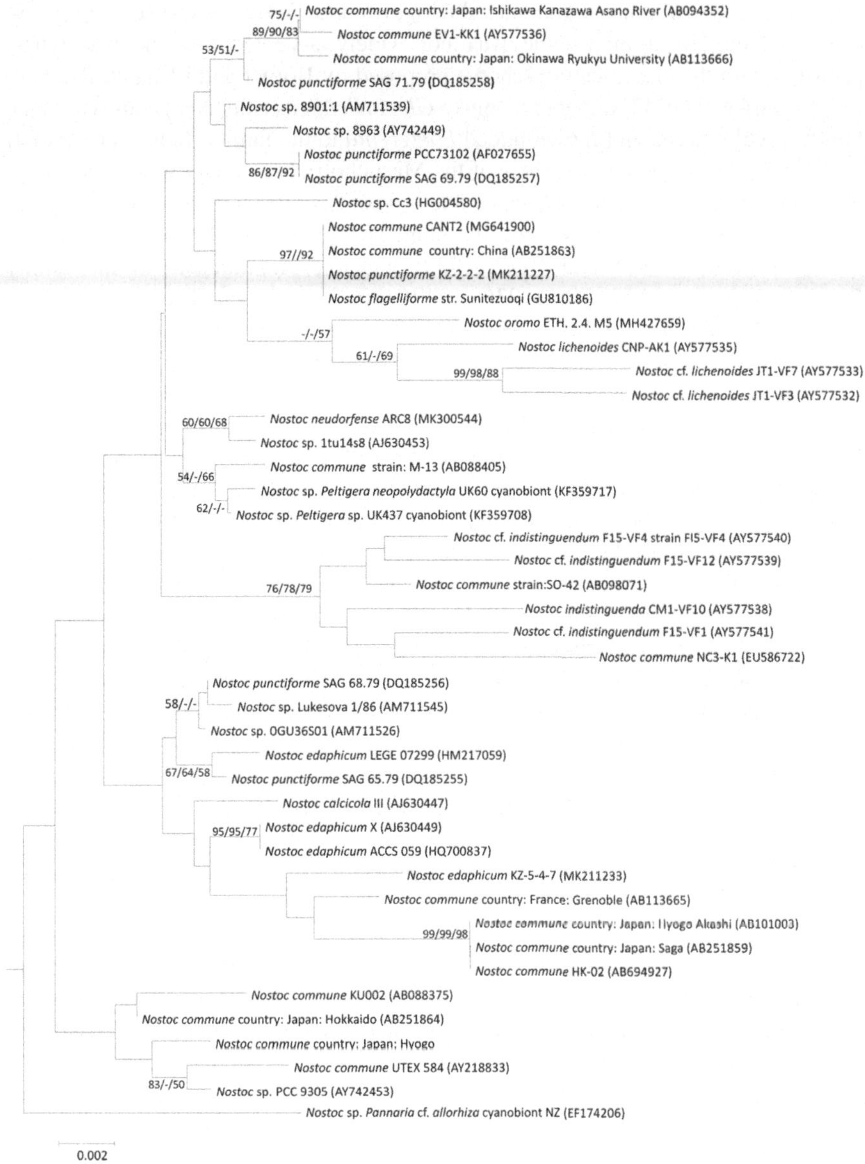

Fig. 2.2 Phylogenetic tree representing *Nostoc* sensu stricto clade, inferred using 16S rRNA gene with the bootstrap values representing neighbor joining/maximum likelihood/maximum parsimony. Bar, 0.002 changes per nucleotide position. Bootstrap values >50 are shown

whereas the genera *Calothrix*, *Dichothrix*, *Polythrix* (= *Gardnerula*), and *Sacconema* were included under Mastichotricheae. Alternatively, Bennet and Murray (1889) mentioned two different families—Rivulariaceae and

Calotrichaceae—which accommodated *Rivularia* and *Calothrix*, respectively. However, this classification scheme was not widely accepted and the researchers persisted with the classification scheme proposed by Bornet and Flahault. Kirchner (1898) and Forti (1907) did not recognize *Gloeotrichia* as a distinct genus within the family Rivulariaceae and further added *Loefgrenia* to the family. In the early part of the twentieth century, the members of the Mastichotricheae along with some tapering nonheterocystous genera (*Leptochaete*, *Amphithrix*, *Homoeothrix*, *Hominoidea*, and *Tapinothrix*) were merged into the family Rivulariaceae (Frémy 1929; Geitler 1932). Further, Fritsch and Rich (1929) included *Raphidopsis* to the family Rivulariaceae. Later, Elenkin (1936, 1949) transferred *Leptochaete*, *Amphithrix*, and *Homoeothrix* to the family Homoeotrichaceae and *Loefgrenia* to the family Loefgreniaceae. Geitler accepted the transfer of *Loefgrenia* to the family Loefgreniaceae, whereas *Leptochaete*, *Amphithrix*, and *Homoeothrix* were classified under the family Rivulariaceae. Moreover, Geitler transferred *Brachytrichia* to the family Mastigocladaceae. Desikachary (1959) followed Geitler's classification and included *Homoeothrix*, *Calothrix*, *Dichothrix*, *Rivularia*, *Gloeotrichia*, and *Leptochaete* under the family Rivulariaceae. Later, the nonheterocystous genera were removed from the family Rivulariaceae by Komárek and Anagnostidis (1989) and this classification scheme was also adopted by Komárek (2013) and Komárek et al. (2014). Furthermore, Komárek et al. (2014) included *Microchaete* under the family Rivulariaceae and transferred *Gloeotrichia* to a new family Gloeotrichaceae. Creation of Gloeotrichaceae was the first instance wherein a new cyanobacterial family was established from the Rivulariaceae based on the polyphasic approach. The family Rivulariaceae has remarkable morphological information available in the literature; however, the molecular data are very limited. According to the most recent classification system, the family Rivulariaceae consists of seven genera and the 16S rRNA gene sequence is available for *Calothrix*, *Rivularia*, *Dichothrix*, and *Microchaete*. It must be noted that there is only a single sequence (393 bp) for *Dichothrix*. Furthermore, three new genera—*Phyllonema*, *Macrochaete*, and *Nunduva*—have been described within the family Rivulariaceae based on polyphasic studies (Alvarenga et al. 2016; Berrendero et al. 2016; González-Resendiz et al. 2018). Moreover, *Kyrtuthrix*, which was initially classified under the family Scytonemataceae, was transferred to Rivulariaceae based on polyphasic study (León-Tejera et al. 2016).

The family Rivulariaceae is one of the morphologically complex family of cyanobacteria, and the genus *Calothrix* is considered as the most problematic genus of the family (Whitton 1987; Berrendero et al. 2016). The genus *Calothrix* is characterized by the presence of heteropolar filaments with heterocytes developed usually at the basal end, and the apical cells are usually hyaline, long, and narrow, creating a thin hair-like appearance. The members of the genus *Calothrix* appear as solitary filaments or small bundles of filaments, whereas the other genera within the family Rivulariaceae show colonial nature (Komárek 2013). Due to the simplistic definition of the genus *Calothrix*, researchers isolating tapered filaments, which occurred as solitary or in small bundles, were classified as *Calothrix* (Berrendero 2016). Furthermore, the phylogenetic analysis based on 16S rRNA gene, *cpc*BA-

IGS, and *nif*D indicated *Calothrix* to be polyphyletic (Henson et al. 2004; Sihvonen et al. 2007; Berrendero et al. 2011; Domínguez-Escobar et al. 2011; Komárek et al. 2012; Whitton and Mateo 2012). Moreover, the absence of the sequence corresponding to the type species of *Calothrix* has further contributed to the taxonomic complexity hovering around the genus (Saraf et al. 2019c). Unlike the genus *Nostoc*, there have been comparatively fewer polyphasic studies on *Calothrix*-like taxa and this is evident from the fact that till date only two *Calothrix*-like genera have been described (Berrendero et al. 2016; Saraf et al. 2019c). Berrendero et al. in their study on *Calothrix*-like strains isolated from different ecological habitats observed that all the strains included in their study formed a separate monophyletic clade, and therefore, the authors characterized their strains as species of novel genus *Macrochaete* (Berrendero et al. 2016). Through this study, Berrendero et al. described the first *Calothrix*-like genus. The genus *Macrochaete* showed morphological similarities with *Calothrix* but differed by the ability to produce a pair of heteromorphic basal heterocytes. Berrendero et al. also reported *Calothrix* to be polyphyletic and emphasized the need for further taxonomic revisions. The authors reported two distinct clusters of *Calothrix* strains that could be differentiated on the basis of their ecological preferences and terminal hair forming ability. Moreover, based on the phylogenetic analysis, the authors observed that *Calothrix* and *Macrochaete* clustered distantly from *Rivularia*, the type genus of the family Rivulariaceae. Further, similar observation was reported in different studies (Shalygin et al. 2017; Saraf et al. 2018; Villanueva et al. 2019). Later, Saraf et al. (2019c), in their study, created a new genus *Dulcicalothrix* to accommodate the freshwater/terrestrial *Calothrix* strains that did not form terminal hairs. The creation of genus *Dulcicalothrix* was also supported by the phylogenetic analysis in which the genus *Calothrix* and *Dulcicalothrix* were separated by the genus *Macrochaete*. The phylogenetic positioning of the genus *Calothrix*, *Dulcicalothrix,* and *Macrochaete* is illustrated in Fig. 2.3. Furthermore, based on the strong phylogenetic evidence, Saraf et al. (2019c) re-erected the family Calotrichaceae to accommodate *Calothrix*, *Dulcicalothrix,* and *Macrochaete*. The family Rivulariaceae at present consists of *Rivularia*, *Microchaete*, *Dichothrix*, *Sacconema*, *Isactis*, *Gardnerula*, *Phyllonema*, *Kyrtuthrix,* and *Nunduva* (Saraf et al. 2019c). The phylogenetic positioning of Calotrichaceae and Rivulariaceae is illustrated in Fig. 2.3. From the above studies, it is evident that the polyphasic approach has indeed contributed in resolving the taxonomic complexities surrounding the genus *Calothrix* to a certain extent. However, similar approach is required to understand the correct taxonomic identities of the *Calothrix* strains that cluster in close proximity of the members of the family Nostocaceae.

2.3 Conclusions

The above case studies have proven that identifying cyanobacteria through the polyphasic approach is the way forward to resolve the taxonomic complexities within the phylum. Traditionally, the taxonomy of cyanobacteria was primarily

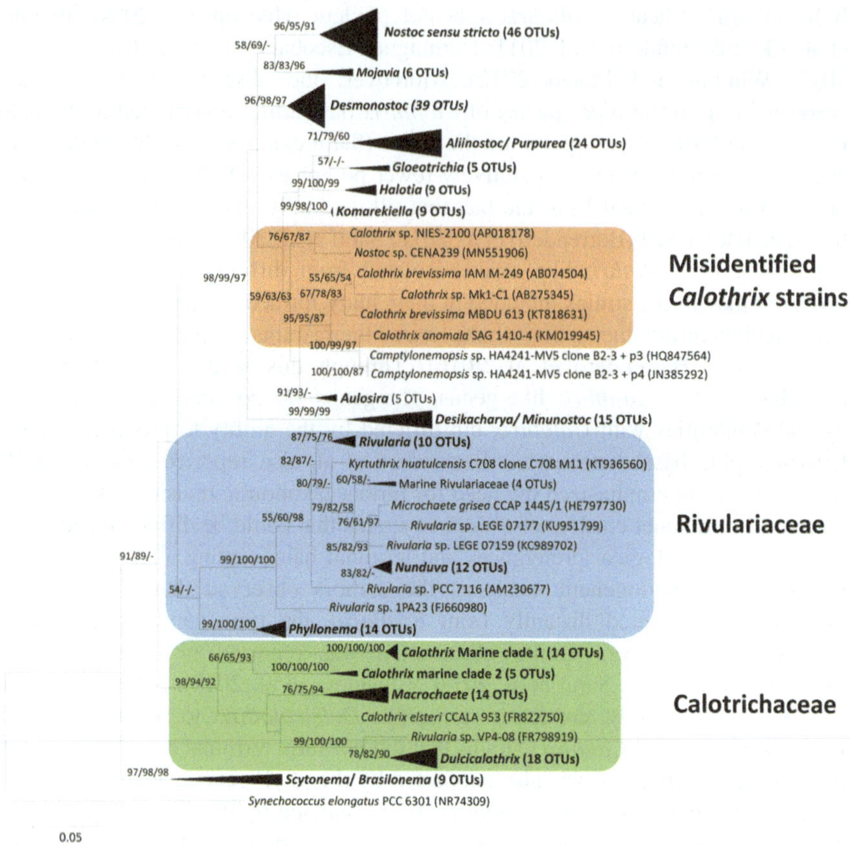

Fig. 2.3 Phylogenetic positioning of the members of the family Rivulariaceae and Calotrichaceae inferred using 16S rRNA gene with the bootstrap values representing neighbor joining/maximum likelihood/maximum parsimony. Bar, 0.05 changes per nucleotide position. Bootstrap values >50 are shown

driven by the differences in the morphological characters, while traits such as false branching and tapering of filaments were thought to have evolved only once in the evolution. However, the phylogenetic studies in recent times have indicated that these characters may have evolved multiple times. With the advancement in the phylogenetic studies, the primary criterion for delineating cyanobacterial taxa has shifted to genetic markers; however, the morphological, ecological, and physiological characters are still important in certain cyanobacterial groups. At present, the phylogenetic studies also have limitations due to the lack of sequences as compared to the total number of cyanobacterial species described and in certain cases the absences of sequence corresponding to the type species further complicate the scenario. In such cases, the morphological and ecological data may prove to be valuable. Therefore, it is important to cautiously apply polyphasic approach depending on the taxa under investigation.

Acknowledgements We thank the Head, Department of Botany, BHU, for providing the necessary facilities and encouragement. We thank the principal, Ramniranjan Jhunjhunwala College, Mumbai, for support. Aniket Saraf is thankful to the Council of Scientific and Industrial Research (CSIR-HRDG) for providing financial assistance in the form of Senior Research Fellowship.

Funding: The work was supported by the Department of Science and Technology (DST-SERB), India, through the Core Research Grant Project CRG/2018/004111.

Conflict of Interest The authors declare that they have no conflict of interest.

References

Alvarenga DO, Rigonato J, Branco LH, Melo IS, Fiore MF (2016) *Phyllonema aviceniicola* gen. nov., sp. nov. and *Foliisarcina bertiogensis* gen. nov., sp. nov., epiphyllic cyanobacteria associated with *Avicennia schaueriana* leaves. Int J Syst Evol Microbiol 66(2):689–700

Anagnostidis K, Komárek J (1985) Modern approach to the classification system of the cyanophytes 1: Introduction. Algol Stud 38(39):291–302

Bagchi SN, Dubey N, Singh P (2017) Phylogenetically distant clade of *Nostoc*-like taxa with the description of *Aliinostoc* gen. nov. and *Aliinostoc morphoplasticum* sp. nov. Int J Syst Evol Microbiol 67:3329–3338

Becerra-Absalón I, Johansen JR, Muñoz-Martín A, Montejano G (2018) *Chroakolemma* gen. nov. (Leptolyngbyaceae, Cyanobacteria) from soil biocrusts in the semi-desert central region of Mexico. Phytotaxa 367(3):201–218

Bennet A, Murray G (1889) A handbook of cryptogamic botany. Longmans, Green and Co., London

Berrendero E, Perona E, Mateo P (2008) Genetic and morphological characterization of *Rivularia* and *Calothrix* (Nostocales, Cyanobacteria) from running water. Int J Syst Evol Microbiol 58:447–460

Berrendero E, Perona E, Mateo P (2011) Phenotypic variability and phylogenetic relationships of the genera *Tolypothrix* and *Calothrix* (Nostocales, Cyanobacteria) from running water. Int J Syst Evol Microbiol 61:3039–3051

Berrendero E, Johansen JR, Kaštovský J, Bohunická M, Capková K, (2016) *Macrochaete* gen. nov. (Nostocales, Cyanobacteria), a taxon morphologically and molecularly distinct from *Calothrix*. J Phycol 52:638–655

Bornet É, Flahault C (1885) Tableau synoptique des Nostochacées filamenteuses hétérocystées. Mémoires de la Soc nat des Sci nat de Cherboug 25:195–223

Bornet É, Flahault C (1886a) Revision des Nostocacées hétérocystées contenues dans les principaux herbiers de France. Ann Sci Nat Bot Septième série 3:323–381

Bornet É, Flahault C (1886b) Revision des Nostocacées hétérocystées contenues dans les principaux herbiers de France (deuxième fragment). Ann Sci Nat Bot Septième Série 4:343–373

Bornet É, Flahault C (1886c) Revision des Nostocacées hétérocystées contenues dans les principaux herbiers de France (Troisième fragment). Ann Sci Nat Bot Septième Série 5:51–129

Bornet É, Flahault C (1886d) Revision des Nostocacées hétérocystées contenues dans les principaux herbiers de France (quatrième et dernier fragment). Ann Sci Nat Bot Septième Série 7:177–262

Bornet É, Flahault C (1888) Note sur deux nouveaux genres d'algues perforantes. J Bot 2:161–165

Borzì A (1878) Nachtrage zur Morphologie und Biologie der Nostocaccen. Flora 61:465–471

Borzì A (1879) Note alla morfologia e biologia della Alghe Ficocromacee. Fam. IIa. Scytonemaceae. Nuovo Giornale Botanico Italiano 11:347–388

Borzì A (1882) Note alla morfologia e biologia delle Alghe Ficocromacee. Fam. IIIa. Rivulariacee. Nuovo Giornale Botanico Italiano 14:272–315

Borzì A (1895) Studi algologici: saggio di richerche sulla biologia delle alghe. Fasc. II. Alberto Reber Libreria, Carlo Clausen, Palermo, pp 121–378

Borzì A (1914) Studi sulle Mixoficee. I. Cenni generali - Systema Myxophycearum. Nuovo Giornale Botanico Italiano 2(21):307–360

Borzì A (1916) Studi sulle Mixoficee. II. Stigonemaceae. Nuovo Giornale Botanico Italiano 2 (23):559–588

Borzì A (1917) Studi sulle Mixoficee. II (continuaz.). Nuovo Giornale Botanico Italiano 24:17–25

Bourrelly P (1970) Les algues d'eau douce III. N. Boubée & Cie, Paris

Brock TD (1969) Microbial growth under extreme conditions. Symp Soc Gen Microbiol 19:15–41

Brock TD (1978) Thermophilic microorganisms and life at high temperatures. Springer, New York

Cai F, Li X, Geng R, Peng X, Li R (2019a) Phylogenetically distant clade of *Nostoc*-like taxa with the description of *Minunostoc* gen. nov. and *Minunostoc cylindricum* sp. nov. Fottea 19 (1):13–24

Cai F, Li X, Yang Y, Jia N, Huo D, Li R (2019b) *Compactonostoc shennongjiaensis* gen. & sp. nov. (Nostocales, Cyanobacteria) from a wet rocky wall in China. Phycologia 58(2):200–210

Cai F, Peng X, Li R (2020a) Violetonostoc minutum gen. et sp. nov. (Nostocales, Cyanobacteria) from a rocky substrate in China. Algae 35(1):1–16

Cai F, Wang Y, Yu G, Wamg J, Peng X, Li R (2020b) Proposal of *Purpurea* gen. nov. (Nostocales, Cyanobacteria), a novel cyanobacterial genus from wet soil samples in Tibet, China. Fottea 20 (1):86–97

Casamatta DA, Johansen JR, Morgan VL, Broadwater ST (2005) Molecular and morphological characterization of ten polar and near-polar strains within the Oscillatoriales (Cyanobacteria). J Phycol 41(2):421–438

Casamatta DA, Villanueva CD, Garvey AD, Stocks HS, Vaccarino M, Dvorák P, Hasler P, Johansen JR (2020) Reptodigitus chapmanii (Nostocales, Hapalosiphonaceae) gen. nov.: a unique nostocalean (Cyanobacteria) genus based on a polyphasic approach. J Phycol 56 (2):425–436

Castenholz RW (2001) Oxygenic photosynthetic bacteria. In: Boone DR, Castenholz RW (eds) Bergey's manual of systematic bacteriology, vol 1, 2nd edn. Springer, New York, pp 473–600

Cohn F (1875) Die Entwicklungsgeschichte der Gattung *Volvox*. Beitr Biol Pflanz 1:93–115

Desikachary TV (1959) Cyanophyta, I.C.A.R. monographs on algae. I.C.A.R, New Delhi

Dodds WK, Gudder DA, Mollenhauer D (1995) The ecology of *Nostoc*. J Phycol 31:2–18

Domínguez-Escobar J, Beltrán Y, Bergman B, Díez B, Ininbergs K, Souza V, Falcón LI (2011) Phylogenetic and molecular clock inferences of cyanobacterial strains within Rivulariaceae from distant environments. FEMS Microbiol Lett 316(2):90–99

Elenki AA (1949) Monographie algarum Cyanophycearum aquidulcium et terrestrium infinibus URSS inventatrum, Pars specialis (Systematica) Fasc I, vol 1. Acad Nauk URSS Moscow, Leningrad, p 675

Elenkin AA (1936) Monographie algarum Cyanophycearum aquidulcium et terrestrium infinibus URSS inventatrum, vol 1. Acad Nauk URSS Moscow, Leningrad, p 675

Forti A (1907) Sylloge Myxophycearum. In: Toni D (ed) Sylloge Algarum 5. Patavii, Padova, p 761

Frémy P (1929) Les Nostocacées de la Normandie. Not Mem Doc Soc Agric Archeol Hist nat Manche 41:197–228

Fritsch FE (1942) The interrelations and classification of the Myxophyceae (Cyanophyceae). New Phytol 41(2):134–148

Fritsch FE (1944) Present-day classification of algae. Bot Rev 10:233–277

Fritsch FE (1945) The structure and reproduction of the algae, Foreword, Phaeophyceae, Rhodophyceae, Myxophyceae, vol II. Cambridge University Press, Cambridge, pp 1–939

Fritsch FE, Rich F (1929) Freshwater algae from Griqualand West. Trans R Soc S Afr 18:1–123

Gaysina L, Saraf A, Singh P (2019) Cyanobacteria in diverse habitats. In: Mishra AK, Tiwari DN, Rai AN (eds) Cyanobacteria. From basic science to applications. Academic, London, pp 1–28

Geitler L (1925) Cyanophyceae. In: Pascher A (ed) Süswasserflora 12. Gustav Fischer Verl, Jena, p 48
Geitler L (1932) Cyanophyceae. In: Rabenhorst's Kryptogamen Flora von Deutschland, vol 14. Akad Verlagsges, Leipzig, p 1196
Geitler L (1942) Schizophyta (Klasse Schizophyceae). In: Engler A, Prantl K (eds) Natürliche Pflanzenfamilien 1b. Duncker & Humblot, Berlin, pp 1–232
Genuário DB, Vaz MGMV, Hentschke GS, Sant'Anna CL, Fiore MF (2015) *Halotia* gen. nov., a phylogenetically and physiologically coherent cyanobacterial genus isolated from marine coastal environments. Int J Syst Evol Microbiol 65:663–675
Genuário DB, Andreote APD, Vaz MGMV, Fiore MF (2017) Heterocyte forming cyanobacteria from Brazilian saline-alkaline lakes. Mol Phylogenet Evol 109:105–112
Golubic S (1979) Taxonomy of stromatolite building cyanophytes. In: Walter MR (ed) Stromatolites, Development in sedimentology, vol 20. Elsevier, New York, pp 127–140
Gomont M (1892) Monographie des Oscillatoriées (Nostocacées homocystées). Ann Sci Nat Bot 7 (15):263–368, 16: 91–264
González-Resendiz L, Johansen JR, Alba-Lois L, Segal-Kischinevzky C, Escobar-Sámche V, Jiménez-García LF, Huaer T, León-Tejera H (2018) *Nunduva*, a new marine genus of Rivulariaceae (Nostocales, Cyanobacteria) from marine rocky shores. Fottea 18(1):86–105
Gugger M, Hoffmann L (2004) Polyphyly of true branching cyanobacteria (Stigonematales). Int J Syst Evol Microbiol 54:349–357
Gugger M, Lyra C, Henriksen P, Couté A, Humbert JF, Sivonen K (2002a) Phylogenetic comparison of the cyanobacterial genera *Anabaena* and *Aphanizomenon*. Int J Syst Evol Microbiol 52:1–4
Gugger M, Lyra C, Suominen I, Tsitko I, Humbert JF, Salkinoja-Salonen M, Sivonen K (2002b) Cellular fatty acids as chemotaxonomic markers of the genera *Anabaena*, *Aphanizomenon*, *Microcystis*, *Nostoc* and *Planktothrix* (Cyanobacteria). Int J Syst Evol Microbiol 52:1007–1015
Hašler P, Casamatta DA, Dvořák P, Poulíčková A (2017) *Jacksonvillea apiculata* (Oscillatoriales, Cyanobacteria) gen. & sp. nov.: a new genus of filamentous, epipsamic cyanobacteria from North Florida. Phycologia 56(3):284–295
Henson BJ, Hesselbrock SM, Watson LE, Barnum SR (2004) Molecular phylogeny of the heterocystous cyanobacteria (subsections IV and V) based on *nifD*. Int J Syst Evol Microbiol 54 (2):493–497
Hentschke GS, Johansen JR, Pietrasiak N, Fiore MF, Rigonato J, Sant'Anna CL, Komárek J (2016) Phylogenetic placement of *Dapisostemon* gen. nov. and *Streptostemon*, two tropical heterocytous genera (Cyanobacteria). Phytotaxa 245(2):129–143
Hentschke GS, Johansen JR, Pietrasiak N, Rigonato J, Fiore MF, Sant'Anna CL (2017) *Komarekiella atlantica* gen. et sp. nov. (Nostocaceae, Cyanobacteria): a new subaerial taxon from the Atlantic Rainforest and Kauai, Hawaii. Fottea 17(2):178–190
Hoffmann L (2005) Nomenclature of Cyanophyta/Cyanobacteria: roundtable on the unification of the nomenclature under the Botanical and Bacteriological codes. Algol Stud 117(6):13–29
Hofmann L, Komárek J, Kaštovský J (2005) System of cyanoprokaryotes (cyanobacteria) – state in 2004. Algol Stud 117:95–115
Honda D, Yokota A, Sugiyama J (1999) Detection of seven major evolutionary lineages in Cyanobacteria based on the 16S rRNA gene sequence analysis with new sequences of five marine *Synechococcus* strains. J Mol Evol 48:723–739
Hrouzek P, Ventura S, Lukešová A, Mugnai MA, Turicchia S, Komárek J (2005) Diversity of soil *Nostoc* strain: phylogenetic and phenotypic variability. Algol Stud 117:1–14
Hrouzek P, Lukešová A, Mareš J, Ventura S (2013) Description of the cyanobacterial genus *Desmonostoc* gen. nov. including *D. muscorum* comb. nov. as a distinct, phylogenetically coherent taxon related to the genus *Nostoc*. Fottea 13(2):201–213
Ishida T, Watanabe MM, Sugiyama J, Yokota A (2001) Evidence for polyphyletic origin of the members of the orders of Oscillatoriales and Pleurocapsales as determined by 16S rDNA analysis. FEMS Microbiol Lett 201:79–82

Kabirnatj S, Nematzadeh GA, Talebi AF, Tabatabaei M, Singh P (2018) *Neowestiellopsis* gen. nov, a new genus of true branched cyanobacteria with the description of *Neowestiellopsis persica* sp. nov. and *Neowestiellopsis bilateralis* sp. nov., isolated from Iran. Plant Syst Evol 301:501–510

Kilgore CH, Johansen JR, Mai T, Hauer T, Casamatta DA, Sheil CA (2018) Molecular characterization of *Geitleria appalachiana* sp. nov. (Nostocales, Cyanobacteria) and formation of Geitleriaceae fam. nov. Fottea 18(2):150–163

Kirchner O (1898) Schizophyceae. In: Engler A, Prantl K (eds) Die natürlichen Pflanzenfamilien. I. Teil, Abt. 1a. Wilhelm Engelmann, Leipzig, pp 45–92

Klein RM, Cronquist A (1967) A consideration of evolutionary and taxonomic significance of some biochemical, micromorphological, and physiological characters in the Thallophytes. Q Rev Biol 42:105–296

Komárek J, Anagnostidis K (1989) Modern approach to the classification system of Cyanophytes 4 – Nostocales. Algol Stud 56:247–345

Komárek J (2013) Cyanoprokaryota. 3. Heterocytous genera. In: Büdel B, Gärtner G, Krienitz L, Schagerl M (eds) Süswasserflora von Mitteleuropa/Freshwater flora of Central Europe. Springer Spektrum, Berlin

Komárek J, Nedbalová L, Hauer T (2012) Phylogenetic position and taxonomy of three heterocytous cyanobacteria dominating the littoral of deglaciated lakes, James Ross Island, Antarctica. Polar Biol 35:759–774

Komárek J, Sant'Anna CL, Bohunická M, Mareš J, Hentschke GS, Rigonato J, Fiore MF (2013) Phenotype diversity and phylogeny of selected *Scytonema*-species (Cyanoprokaryota) from SE Brazil. Fottea 13(2):173–200

Komárek J, Kaštovský J, Mareš J, Johansen JR (2014) Taxonomic classification of cyanoprokaryota (cyanobacterial genera) 2014, using a polyphasic approach. Preslia 86:295–335

Komárková J, Jezberová J, Komárek O, Zapomělová E (2009) Variability of *Chroococcus* (Cyanobacteria) morphospecies with regard to phylogenetic relationships. Hydrobiologia 639:69–83

Lazaroff N (1966) Photoinduction and photoreversal of the Nostocacean developmental cycle. J Phycol 2:7–17

León-Tejera H, González-Resendiz L, Johansen JR, Segal-Kischinevsky C, Escobar-Sánchez V, Alba-Lois L (2016) Phylogenetic position reevaluation of *Kyrtuthrix* and description of a new species *K. huatulcensis* from Mexico's Pacific coast. Phytotaxa 278(1):1–18

Lyra C, Suomalanien S, Gugger M, Vezie C, Sundman PM, Sivonen K (2001) Molecular characterization of planktic cyanobacteria of *Anabaena*, *Aphanizomenon*, *Microcystis* and *Planktothrix* genera. Int J Syst Evol Microbiol 51:513–526

Mareš J, Johansen JR, Hauer T, Ventura S, Cuzman O, Tiribilli B, Kaštovský J (2019) Taxonomic resolution of the genus *Cyanothece* (Chroococcales, Cyanobacteria), with a treatment on *Gloeothece* and three new genera, *Crocosphaera*, *Rippkaea*, *Zehria*. J Phycol 55(3):578–610

Margulis L (1981) Symbioses in cell evolution. Freeman and Co., San Francisco

Mesfin M, Johansen JR, Pietrasiak N, Baldarelli LM (2020) *Nostoc oromo* sp. nov. (Nostocales, Cyanobacteria) from Ethiopia: a new species based on morphological and molecular evidence. Phytotaxa 433(2):21–93

Mishra D, Suradkar A, Saraf A, Singh P (2020) Phylogenetic evaluation of the true-branched heterocytous cyanobacteria and description of soil dwelling *Westiellopsis akinetica* sp. nov. FEMS Microbiol Lett 367(5):fnaa046

Mollenhaur D, Bengtsson R, Lindstrøm E-A (1999) Macroscopic cyanobacteria of the genus *Nostoc*: a neglected and endangered constituent of European inland aquatic biodiversity. Eur J Phycol 34:349–360

Oren A (2004) A proposal for further integration of the Cyanobacteria under the Bacteriological Code. Int J Syst Evol Microbiol 54:1895–1902

Oren A (2011) Naming Cyanophyta/Cyanobacteria – a bacteriologist's view. Fottea 11(1):9–16

Oren A, Garrity GM (2014) Proposal to change General Consideration 5 and Principle 2 of the International Code of Nomenclature of Prokaryotes. Int J Syst Evol Microbiol 64:309–310

Oren A, Komárek J (2015) Nomenclature of the Cyanobacteria/Cyanophyta: current problems and proposed solutions. Notes based on a roundtable discussion held on August 16, 2010 during the 18th Symposium of the International Association for Cyanophyte Research, České Budějovice, Czech Republic. Bull BISMiS 1:25–33

Oren A, Tindall BJ (2005) Nomenclature of the cyanophyta/cyanobacteria/cyanoprokaryotes under the International Code of Nomenclature of Prokaryotes. Algol Stud 117(6):39–52

Oren A, Ventura S (2017) The current status of cyanobacterial nomenclature under the "prokaryotic" and the "botanical" code. Antonie Van Leeuwenhoek 110:1257–1269

Oren A, Komárek J, Hoffmann L (2009) Nomenclature of the Cyanophyta/Cyanobacteria/Cyanoprokaryotes: what has happened since IAC Luxembourg? Algol Stud 130:17–26

Pietrasiak N, Osorio-Santos K, Shalygin S, Martin MP, Johansen JR (2019) When is a lineage a species? A case study in *Myxacorys* gen. nov. (Synechococcales: Cyanobacteria) with the description of two new species from the Americas. J Phycol 55(5):976–996

Prescott GW (1962) Algae of the western Great Lakes area. W.C. Brown Co, Dubuque, IA

Rajaniemi P, Hrouzek P, Kaštovská K, Willame R, Rantala A, Hoffmann L, Komárek J, Sivonen K (2005) Phylogenetic and morphological evaluation of the genera *Anabaena*, *Aphanizomenon*, *Trichormus* and *Nostoc* (Nostocales, Cyanobacteria). Int J Syst Evol Microbiol 55(1):11–26

Řeháková K, Johansen JR, Casamatta DA, Xuesong L, Vincent J (2007) Morphological and molecular characterization of selected desert soil cyanobacteria: three species new to science including *Mojavia pulchra* gen. et sp. nov. Phycologia 46:481–502

Rigonato J, Gama WA, Alvarenga DA, Branco LHZ, Brandini FP, Genuário DB, Fiore MF (2016) *Aliterella atlantica* gen. nov., sp. nov., and *Aliterella antarctica* sp. nov., novel members of coccoid Cyanobacteria. Int J Syst Evol Microbiol 66:2853–2861

Rippka R, Stanier RY, Deruelles J, Herdman M, Waterbury JB (1979) Generic assignments, strain histories and properties of pure cultures of Cyanobacteria. Microbiology 111:1–61

Sachs J (1874) Lehrbuch der Botanik, vol XVI, 4th edn. W. Engelmann, Leipzig, p 928

Saraf A, Dawda HG, Suradkar A, Batule P, Behere I, Kotulkar M, Kumat A, Singh P (2018) Insights into the phylogeny of false-branching heterocytous cyanobacteria with the description of *Scytonema pachmarhiense* sp. nov. isolated from Pachmarhi Biosphere Reserve, India. FEMS Microbiol Lett 365:fny160

Saraf A, Dawda HG, Singh P (2019a) *Desikacharya* gen. nov., a phylogenetically distinct genus of Cyanobacteria along with the description of two new species, *Desikacharya nostocoides* sp. nov. and *Desikacharya soli* sp. nov., and reclassification of *Nostoc thermotolerans* to *Desikacharya thermotolerans* comb. nov. Int J Syst Evol Microbiol 69(2):307–315

Saraf A, Dawda HG, Singh P (2019b) Validation of the genus *Desikacharya* gen. nov. (Nostocaceae, Cyanobacteria) and three included species. Notulae Algarum 107:1–3

Saraf A, Suradkar A, Dawda HG, Gaysina LA, Gabidullin Y, Kumat A, Behere I, Kotulkar M, Batule P, Singh P (2019c) Phylogenetic complexities of the members of Rivulariaceae with the re-creation of the family Calotrichaceae and description of *Dulcicalothrix necridiiformans* gen nov., sp nov., and reclassification of *Calothrix desertica*. FEMS Microbiol Lett 366:fnz219

Saraf A, Dawda HG, Suradkar A, Agre V, Singh P (2020) *Fortiea necridiiformans* sp. nov., a soil-dwelling cyanobacterium from Pachmarhi Biosphere Reserve, India. Int J Syst Evol. https://doi.org/10.1099/ijsem.0.004337

Seckbach J, Oren A (2007) Oxygenic photosynthetic microorganisms in extreme environments: possibilities and limitations. In: Seckbach J (ed) Algae and cyanobacteria in extreme environments. Springer, Netherlands, pp 5–25

Sendall BC, McGregor GB (2018) Cryptic diversity within the *Scytonema* complex: characterization of the paralytic shellfish toxin producer *Heterosyctonema crispum*, and the establishment of the family Heteroscytonemataceae (Cyanobacteria/Nostocales). Harmful Algae 80:158–170

Shalygin S, Shalygina R, Johansen JR, Pietrasiak N, Berrendero E, Bohunická M, Mareš J, Sheil CA (2017) *Cyanomargarita* gen. nov. (Nostocales, Cyanobacteria): convergent evolution resulting in a cryptic genus. J Phycol 53:762–777

Shalygin S, Pietrasiak N, Gomez F, Mlewski C, Gerard E, Johansen JR (2018) Riuvlaria halophila sp. nov. (Nostocales, Cyanobacteria): the first species of Rivularia described with modern polyphasic approach. Eur J Phycol 53(4):537–548

Sihvonen LM, Lyra C, Fewer DP, Rajaniemi-Wacklin P, Lehtimäki JM, Wahlsten M, Sivonen K (2007) Strains of the cyanobacterial genera *Calothrix* and *Rivularia* isolated from the Baltic Sea display cryptic diversity and are distantly related to *Gloeotrichia* and *Tolypothrix*. FEMS Microbiol Ecol 61(1):74–84

Singh P, Minj RA, Kunui K, Shaikh ZM, Suradkar A, Shouche YS, Mishra AK, Singh SS (2016) A new species of *Scytonema* isolated from Bilaspur, Chhattisgarh, India. J Syst Evol 54 (5):519–527

Singh P, Dubey N, Bagchi SN (2017a) *Westiellopsis ramosa* sp. nov., intensely branched species of *Westiellopsis* (cyanobacteria) from a freshwater habitat of Jabalpur, Madhya Pradesh, India. Plant Syst Evol 303:1239–1249

Singh P, Minj RA, Kunui K, Shaikh ZM, Suradkar A, Shouche YS, Mishra AK, Singh SS (2017b) A new species of *Scytonema* isolated from Bilaspur, Chattisgarh, India using the polyphasic approach. Plant Syst Evol 303:249–258

Singh P, Šnokhousová J, Saraf A, Suradkar A, Elster J (2020) Phylogenetic evaluation of the genus *Nostoc* and description of *Nostoc neudorfense* sp. nov., from the Czech Republic. Int J Syst Evol Microbiol 70(4):2740–2749

Stanier RY (1977) The position of cyanobacteria in the world of phototrophs. Carlsberg Res Commun 42:77

Stanier RY, Sistrom WR, Hansen TA, Whitton BA, Castenholz RW, Pfennig N, Gorlenko VN, Kondratieva EN, Eimhjellen KE, Whittenburry R, Gherna RL, Trüper HG (1978) Proposal to place the Nomenclature of the Cyanobacteria (Blue-Green Algae) under the rules of the International Code of Nomenclature of Bacteria. Int J Syst Evol Microbiol 28:335–336

Stitzenberger E (1860) Dr. L Rabenhoosts Algew Sachens, manual of phycology. Chronica Botanica, Waltham, MA, pp 243–262

Suradkar A, Villanueva C, Gaysina L, Casamatta D, Saraf A, Dighe G, Mergu R, Singh P (2017) *Nostoc thermotolerans* sp. nov., a soil-dwelling species of *Nostoc* (Cyanobacteria). Int J Syst Evol Microbiol 67(5):1296–1305

Thuret G (1875) Essai de classification des Nostochinées. Ann Sci Nat Bot 6(1):372–382

Turner S, Pryer KM, Miao VPW, Palmer JD (1999) Investigating deep phylogenetic relationships among cyanobacteria and plastids by small subunit rRNA sequence analysis. J Eukaryot Microbiol 46:327–338

Villanueva CD, Garvey AD, Hašler P, Dvořák P, Poulícková A, Norwich AR, Casamatta DA (2019) Descriptions of *Brasilonema geniculatum* and *Calothrix dumus* (Nostocales, Cyanobacteria): two new taxa isolated from cemetery tombstones. Phytotaxa 387(1):1–20

Whitton BA (1987) The biology of Rivulariaceae. In: Fay P, Van Maalen C (eds) The cyanobacteria a comprehensive review. Elsevier, Amsterdam, pp 513–534

Whitton BA (ed) (2012) The ecology of cyanobacteria II. Their diversity in time and space. Springer, Netherlands

Whitton BA, Mateo P (2012) Rivulariaceae. In: Whitton BA (ed) The ecology of cyanobacteria II. Their diversity in time and space. Springer, Netherlands, pp 561–591

Whitton BA, Potts M (eds) (2002) The ecology of cyanobacteria. Their diversity in time and space. Springer, Netherlands

Wilmotte A, Golubić S (1991) Morphological and genetic criteria in the taxonomy of Cyanophyta/ Cyanobacteria. Algol Stud 64:1–24

Wilmotte A, Herdman M (2001) Phylogenetic relationships among the cyanobacteria based on 16S rRNA sequences. In: Boone DR, Castenholz RW (eds) Bergey's manual of systematic bacteriology, 2nd edn. Springer, New York, pp 487–493

Woese CR, Fox GE (1977) Phylogenetic structure of the prokaryotic domain: the primary kingdoms. Proc Natl Acad Sci U S A 74(11):5088–5090

Photosynthesis and Energy Flow in Cyanobacteria

3

Sanjesh Tiwari, Anuradha Patel, Neeraj Pandey, Garima Singh, Aparna Pandey, and Sheo Mohan Prasad

Abstract

Sunlight is the most important source of energy for life on the earth. Photosynthetic organisms (lower and higher green plants) have capability to capture this light energy and convert it into chemical energy. This chemical energy is incorporated for biomass production. During the course of evolution, oxygen is released into the environment primarily by cyanobacteria due to the presence of two photosystems linked in a series. The two photosystems cause splitting of water, which behaves as an electron donor. Harvesting of light in cyanobacteria is accomplished by chlorophyll-containing photosystems and photosystem-associated light-harvesting antennae (phycobiliproteins; PBPs) embedded on thylakoid membrane that captures the light energy. This light energy can be used for the formation of reduced NADPH, which is supplied for anabolic processes such as the Calvin–Benson cycle for CO_2 fixation. Cyanobacteria also possess accessory pigment complex such as carotenoids (Cars) that protect the photosynthetic apparatus from photoinhibition by the process of nonphotochemical quenching (NPQ). Cyanobacterial thylakoid membranes are densely packed with membrane-integral proteins that restrict the mobility of membrane-integral proteins. Electrons that are released from splitting of water are moved from photosystem II (PS II) to photosystem I (PS I) acquire two types of pathways: One is mediated by the NADPH: plastoquinone oxidoreductase (NDH) complexes, and second is an NADH-independent but ferredoxin-dependent pathway. Current methods such as PS II fluorescence give detailed mechanism of electron transport flow in cyanobacteria. Further, these organisms have the potential to use solar energy for the production and isolation of important

S. Tiwari · A. Patel · N. Pandey · G. Singh · A. Pandey · S. M. Prasad (✉)
Ranjan Plant Physiology and Biochemistry Laboratory, Department of Botany, University of Allahabad, Allahabad, India

R. P. Rastogi (ed.), *Ecophysiology and Biochemistry of Cyanobacteria*,
https://doi.org/10.1007/978-981-16-4873-1_3

bioproducts including biofuels. This article summarizes the process of photosynthesis and mechanism of electron transport in cyanobacteria.

Keywords

Cyanobacteria · Electron transport chain · Fluorescence · Nonphotochemical quenching · Photosynthetic pigments

3.1 Introduction

Sun is the ultimate source of energy for all living organisms ranging from lower to higher plants and also for humans. Solar energy is converted into chemical energy, which is further used in various metabolic processes of plants and considered as a major environmental factor that regulates the growth and development of plants. This solar energy is captured by plants that possess light-harvesting complexes (LHCs) or photosynthetic pigments and the organisms, thus known as photosynthetic organism. In this series, about 2.8 billion years ago cyanobacteria or blue-green algae (BGA) came into existence. They are considered as the first photosynthetic prokaryotes performing oxygenic photosynthesis. They release molecular oxygen due to the presence of two photosystems linked in series and water molecule behaves as electron donor that makes environment reducing (Dismukes et al. 2001; Nowicka and Kruk 2016). Excitation of both photosystems (PS II and PS I) is balanced to maximize the quantum yield of photosynthetic light reactions, but any change in the light quality/quantity decreases the ability of photosynthetic apparatus along with reduced photosynthetic efficiency (Dietzel et al. 2008).

Cyanobacteria are the ancient photosynthetic, Gram-negative prokaryotes that functions as "dual-edge sword"; on one hand, they have peculiar role in global CO_2 fixation and N_2 fixation thus actively participate in carbon cycle (Stock et al. 2014), and on other hand, cyanobacteria fix atmospheric nitrogen as they have unique thick-walled, barrel-shaped structure known as heterocyst that subsequently increase the fertility of agricultural lands particularly paddy fields and increase the productivity of rice by 30% (Zehr 2011). Besides this, blue-green algae are considered as the precursor of eukaryotic chloroplast and primary producers of aquatic ecosystem (Lane 2017). They perform both photosynthesis and respiration, and therefore have evolved a diverse range of electron transport pathways. Cyanobacteria easily grow under ambient sunlight and water. Important elements such as C, K, P, S, N, and Fe are ubiquitous and abundantly found in water bodies like lagoon, lakes, ponds, rivers, and different stagnant water bodies. Besides its role as biofertilizer, they are potent source of carbohydrates, lipids, phenolics, vitamins, amino acids, sugars, and plant growth regulators that directly or indirectly enhance the crop yield. Cyanobacteria also enhance the physiochemical properties of soil and increase the water-holding capacity. In cyanobacteria, three types of pigments are present: First is major pigment content, i.e., chlorophyll *a*, second is accessory pigment, i.e., carotenoids, and third is light-harvesting pigment that constitutes the phycobiliproteins (PBPs) that include phycocyanin (PC), allophycocyanin (APC), and phycoerythrin (PE) (Page et al. 2012; Kumar et al. 2015).

The majority of cyanobacterial species have two membrane systems: the cytoplasmic membrane and internal thylakoid membranes. The light reactions occur in the thylakoid membranes (Vermaas 2001) except *Gloeobacter violaceus* that lacks thylakoid membranes and localizes electron transport pathways to specific domains in the cytoplasmic membrane (Rexroth et al. 2011). The cytoplasmic membrane has also been reported to contain incompletely assembled nonfunctional PSI and PSII complexes, perhaps as part of the photosystem biogenesis pathway (Smith and Howe 1993; Keren et al. 2005). Similar to other photosynthetic membranes, cyanobacterial thylakoid membranes packed densely with membrane-integral proteins that restrict the mobility of membrane-integral proteins (Folea et al. 2008). Further, to pinpoint the photochemistry of PS II and flow of electrons, the chlorophyll fluorescence is considered as best method and extensively studied by several authors in cyanobacteria (Patel et al. 2018; Tiwari et al. 2018). Hence, in this article we summarized the basic concept of photosynthesis and energy flow with special reference to cyanobacteria.

3.2 Component of Photosynthetic Apparatus

Photosynthetic process totally depends on interactive role of major and accessory photosynthetic pigments present on thylakoid membranes. In cyanobacteria, three types of photosynthetic pigment content are present: First is chlorophyll *a* (major pigments); second is phycobiliproteins (PBPs: light-harvesting pigments) (Fig. 3.1); and third is the carotenoids (photoprotective nature) that protect the photosynthetic apparatus against adverse environmental factors (Kirilovsky 2010; Masojidek et al. 2013).

Among three photosynthetic pigments, chlorophyll is the leading light-harvesting pigment directly link with the healthiness of photosynthetic organisms and acts as reaction center and very sensitive pigment against adverse stress factors. Decrease in the chlorophyll is directly link with reduction in biomass of photoautotrophs (Kühl et al. 2005). Six different types of chlorophyll present in cyanobacteria such as Chls *a*, *b*, *d*, and *f*, divinyl-Chl *a* and *b*, but Chl *a* is most common and green in color (Gan and Bryant 2015). Chlorophyll has tetrapyrrole ring formed from protoporphyrin IX in which four pyrrole rings (porphyrin) are connected. The central metal ion in the porphyrin ring is magnesium (Mg^{2+}) attached by N molecules through coordinate and covalent bonding and four subunits named A, B, C, and D present in the pyrrole ring. Molecule cyclopentanone has attachment to "C" ring, and a phytol chain known as tail attached to "D" ring with the help of esterification of propionic acid. The phytol chain is basically a hydrocarbon showed similarity with isoprenoid compounds, which have C-C bonding (Chakdar and Pabbi 2017). Chlorophyll is an incredible photoreceptor because of the presence of interconnecting single and double bonds, which allow delocalization of electron in their structure (Song et al.

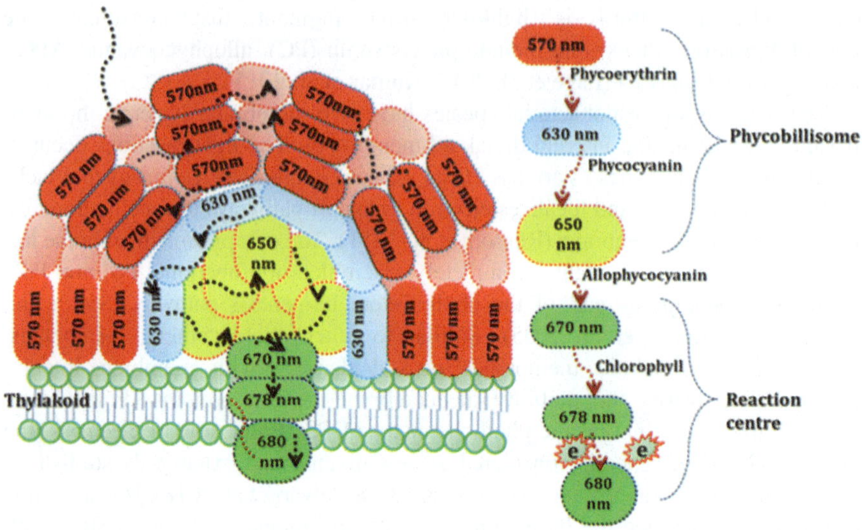

Fig. 3.1 Structure of phycobilisome organization in cyanobacteria

2015). This delocalization of electron that permits polyene structures mediates absorption of light from different bands of the visible spectrum of sunlight and initiates electron transport chain reaction. That is why the chlorophyll pigment is present in the center of the redox reaction of water photolysis and CO_2 fixation (Huang et al. 2016).

Biomass production of cyanobacteria depends upon light harvesting mediated by brilliantly colored, water-soluble accessory antenna complex called phycobiliproteins (PBPs) (Fig. 3.1). Besides occurrence of PBPs in cyanobacteria, they are also present in some eukaryotic algae such as red algae and cryptomonads (Grossman et al. 1993). PBPs play prominent role in absorption of the light energy and transfer it to photosystems for initiating the photochemistry. Further, the color of cyanobacteria, which is bluish-green, is only because of PBPs because these proteins can constitute up to 40–50% of the total proteins in the cell when cyanobacteria are cultured under low-light conditions. PBPs possess light-harvesting property due to the presence of covalently attached chromophores called bilins and the prosthetic groups mediate the absorption of light in the visible region where chlorophyll *a* has minimal absorption (Manirafasha et al. 2016). PBPs are classified into three major categories: allophycocyanin (APC), phycocyanin (PC), and phycoerythrin (PE) with characteristics spectrum referred as blue-green (650–655 nm), blue (610–620 nm), and pink (540–570 nm), respectively, in both cyanobacteria and red algae (Moraes and Kalil 2009; Sonani et al. 2014a, b). It is estimated that around 50% of absorbed light by PBPs is transferred to photosystems that involve in CO_2 fixation during the process of photosynthesis (Tavanandi et al. 2018). Individual subunits of PBPs composed of two polypeptide chains (α- and β-subunits) and assembled to form monomeric, trimeric, and hexameric structure, but majority of PBPs are in trimers (αβ)3 or hexamers (αβ)6. Among three types of PBPs, APC is core pigment, while PE and PC located on exterior (periphery) side of thylakoid membranes (Kannaujiya et al. 2017). Final structure of phycobiliproteins formed by attachment of phycobilins posttransnationally by enzyme lyases and then asparagine residue on β-subunits undergo methylation with the help of enzyme methyltransferase (Schluchter et al. 2010). Phycocyanin is most common and composed of two α- and β-subunits with a hexameric conformation (αβ)6 at pH 5.0–6.0 having single chromophore named phycocyanobilin (Dumay and Morancais 2016). Phycoerythrin is hexameric structure (αβ)6 with four chromophores named phycoerythrobilin (two bind with α-subunit and two with β-subunits) (Dumay and Morancais 2016; Glazer 1994). Allophycocyanin is trimeric in structure (α3β3) acts as mediator of energy transfer between photosystems (PSII and PSI) (Glazer 1994; Lundell and Glazer 1981). Apart from their major role photochemistry, PBPs function as high-valued natural products with great potential for biotechnological applications not only for nutraceutical and pharmaceutical, but also in cosmetic, feed, and food industries (Pandey et al. 2013; Manirafasha et al. 2016). They have immense potential to serve as antioxidants to scavenge free radicals (Zhou et al. 2005; Patel et al. 2018), anticancer activity (Ravi et al. 2015; Jiang et al. 2017), and antidiabetic agent demonstrated by oral administration in mice (Ou et al. 2012; Zheng et al. 2013), used in the production of beauty care products and due to low molecular weight they are also used in flow cytometry (Telford et al. 2001).

To protect the photosynthetic apparatus, photosynthetic organisms have photoprotective pigment named carotenoids. In general, carotenoids perform dual role as it behaves as accessory light-harvesting pigment and also involved in photochemical quenching. On the other hand, they have antioxidative property and participated in minimizing the oxidative stress by quenching singlet oxygen (Melnicki et al. 2016). In cyanobacteria, the orange carotenoid protein (OCP) is unique in nature, which is water-soluble and carry out both photosensory and photoprotective functions. In the blue-green light spectrum, absorption of light interchanged the stable form of OCP^{O} into OCP^{R} that binds to the primary light-harvesting antenna (Wilson et al. 2008; Gwizdala et al. 2011). First evidence of presence of OCP in cyanobacteria described by Holt and Krogmann (1981) and its structure and functions is studied after isolation and crystallization of the OCP from *Arthrospira maxima* (Kerfeld et al. 2003). OCP is composed of two structural domains that have different functions: One is α/β C-terminal domain (CTD) that has regulatory role and second is α-helical N-terminal domain (NTD) considered as effector domain (Kerfeld et al. 2003; Kerfeld 2004; Leverenz et al. 2014). In recent study of Leverenz et al. (2015), the isolated NTD is sometime known as red carotenoid protein (RCP) that actively participates in energy quenching by interacting with phycobilisome (PBS) antenna complexes. Molecule zeaxanthin lacks the carbonyl group of keto-carotenoids, and OCP binds with it to form zea-OCP, which is photo-inactive and yellow in color (Punginelli et al. 2009). Cyanobacteria when exposed with high light irradiance the activated OCP^{R} participated in thermal energy dissipation and protects photosynthetic apparatus against damage and after completion of energy dissipation the OCP^{R} interchange into OCP^O in the dark (Wilson et al. 2006). After analyzing fluorescence recovery kinetics, it is suggested that the OCP^R form is more stable in vivo than in vitro (Rakhimberdieva et al. 2007). Besides, OCP cyanobacteria also have different carotenoids such as myxoxanthophyll, β-carotene, and its derivatives, that is, echinenone and zeaxanthin. Decrease in the carotenoid contents under adverse environmental factors proves severe toxicity in cyanobacteria against particular stress conditions (Patel et al. 2018; Tiwari et al. 2018). Hence, basal level of carotenoid is necessary for maintaining the structure and functions of photosynthetic apparatus.

3.3 Structure of Photosystems (PS I and PS II)

Fixation of CO_2 in biomass in cyanobacteria is processed by three membrane-bound pigment–protein complexes: photosystems II (PS II), photosystem I (PS I), and the Cyt b6f complex (Nelson 2011). Along with these pigment–protein complexes, the mobile electron carriers such as plastoquinone (PQ) and plastocyanin (PC) form the photosynthetic electron transport chain (PET) and mediate the photophosphorylation (Figs. 3.2a, b and 3.3) (Rast et al. 2015). The PS I is located in the stroma lamella of thylakoid, while the PS II is in the stacked grana domain (Dekker and Boekema 2005). The process of photosynthesis is divided into two parts one is light reaction and second is dark reaction; former is involved in generating assimilating powers as ATP and NADPH, while later synthesized carbohydrate by using the assimilating powers. Reaction centers of pigment system are classified on the basis of their terminal electron acceptor as type I that has iron-sulfur cluster acceptor and type II that has quinone terminal acceptor (Mazor et al. 2014). After harvesting the light by photosynthetic pigments, solar energy get transferred to photosynthetic reaction center (RC) where the chlorophyll molecules get excited as P_{680} for PSII and P_{700} for PSI to generate proton motive force across the membrane (Nelson and Junge 2015).

Photosystem I (PS I) is an important component of noncyclic photophosphorylation mediate the transfer of electron from luminal side via plastocyanin (PC) to stromal side via ferredoxin (Fd) in cyanobacteria (Croce et al. 2018) (Fig. 3.2a). The structure and function of cyanobacterial PS I are identical to its counterpart present in the chloroplast of plants and higher algae. Photosystem I is the largest complex proteins that have crystallized, and high-resolution structure has been determined (Jordan et al. 2001). PS I exists in oligomeric forms, but in plants PS I is monomeric having light-harvesting complex I (LHCI), while in cyanobacteria it is trimeric. Detailed structure of trimeric PS I is first described in *Thermosynechococcus elongates* having a molecular weight of 1080 kDa with a diameter of 210 Å and height of 90 Å. Cyanobacterial PS I is spectrally and kinetically heterogeneous and absorbs longer wavelengths of light, i.e., 710–750 nm (Karapetyan et al. 2014). Single unit (monomeric) of trimeric PS I having 12 different proteins to which 127 cofactors are noncovalently bound and around 96 chlorophyll *a* molecules, 22 β-carotenes, 2 phylloquinones, 3 [4Fe-4S] clusters, and 4 lipids (Jordan et al.

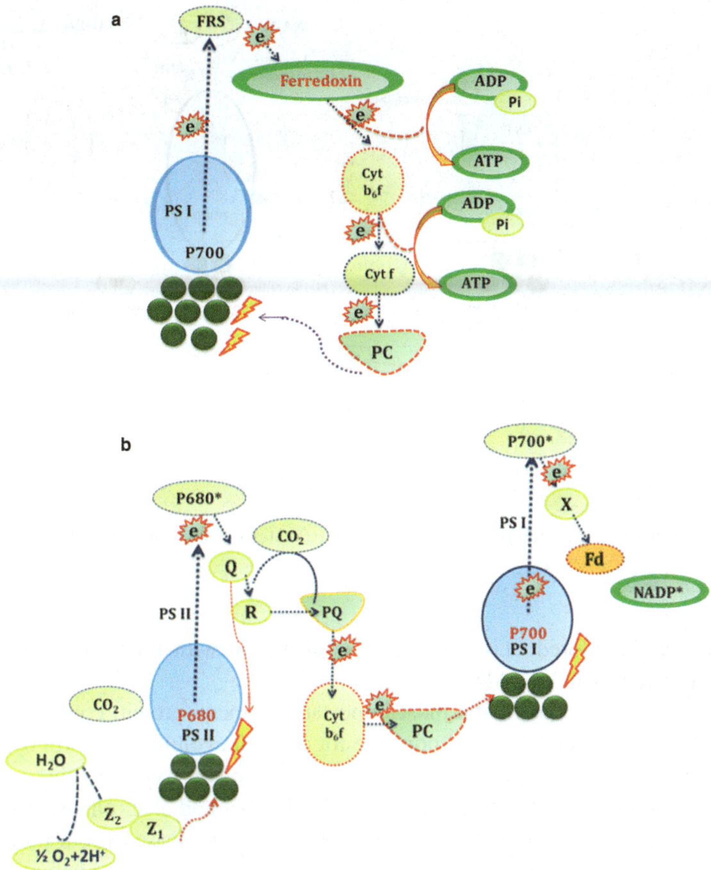

Fig. 3.2 Cyclic electron flow (**a**) and noncyclic electron flow (**b**) in cyanobacteria

2001). PS I has some major proteins that form reaction center, and core antenna complex of PS I complex is formed by major proteins such as PsaA and PsaB and the core of PsaA/B is further surrounded by small membrane-intrinsic proteins that are seven in numbers: PsaF, PsaI, PsaJ, PsaK, PsaL, PsaM, and PsaX. Further, the stromal hump of PS I is formed by combination of three stromal subunits as PsaC, PsaD, and PsaE (Jensen et al. 2003, 2004). Among them, the peripheral subunit PsaC binds with the two iron-sulfur rieske proteins (4Fe-4S) and with terminal electron acceptors F_A and F_B having redox potentials of 520 mV and 580 mV, respectively (Nelson and Yocum 2006). After receiving the solar energy, the reaction center of PS I, i.e., $P_{700,}$ gets excited and designated as P_{700}* and electrons are transported from A_0 to A_1 and finally electrons reduced the ferredoxin (Fd) (Mamedov et al. 2015) (Fig. 3.2a). The utmost notable property of PS I is its high efficacy; it operates with a quantum yield close to 1.0 (Nelson and Yocum 2006). PS I functions as a

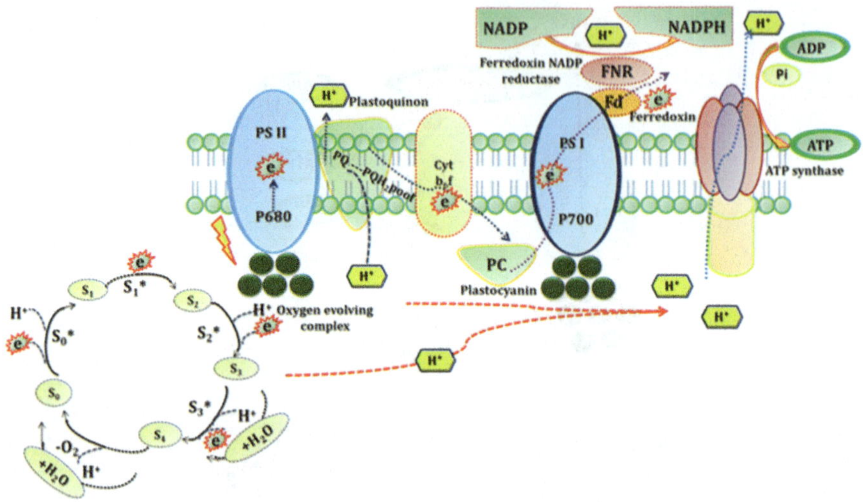

Fig. 3.3 Schematic representation of electron flow (transfer of electron and proton) in thylakoid membrane

PC/Fd (photo)-oxidoreductase, which, in cooperation with PS II, leads to a linear electron transfer H_2O to $NADP^+$.

Photosystem II is one of the most important pigment–protein complexes firstly appeared in cyanobacteria and responsible for oxygen evolution on earth by absorbing shorter wavelength light, i.e., 680 nm and changing an-oxygenic photosynthesis into oxygenic around 2.5 billion years ago. PS II liberated electrons by splitting of water and transferred it to plastoquinone (PQ) results in release of four protons into the lumen, and oxygen is evolved (Barber 2016). PS II is a multi-subunit protein complex having molecular mass of 350 kDa, and per monomer has dimensions of 105 Å in depth (45 Å in membrane), 205 Å in length, and 110 Å in width (Ferreira et al. 2004) and embedded in the thylakoid membrane. By X-ray crystallography, monomer of PS II from *Thermosynechococcus vulcanus* on resolution of 2.9 Å (Guskov et al. 2009) and 1.9 Å (Umena et al. 2011) has around 20 different polypeptides out of which 17 are on integral membrane, while 3 are extrinsic subunits on the lumen side and also possess 90 cofactors. Further, each monomer of PS II has 35 chlorophyll *a* molecules, 20–25 lipids, 12 β-carotenes, 2–3 plastoquinones, 2 pheophytins, 2 hemes, the WOC (Mn_4O_5Ca), 4 Ca^{2+} ions, 3 Cl^- ions, and a nonheme iron (Zouni et al. 2001; Shi et al. 2012). There are four important components of PS II monomer: (1) Active reaction centers proteins D1 and D2 (homologous of PsbA and PsbD) participate in photochemical events; (2) α- and β-subunits of Cytb559; (3) antenna subunits CP47 and CP43 where CP denotes chlorophyll protein complex (homologous of PsbC and PsbB); and (4) 13 membrane-intrinsic small subunits (PsbE, PsbF, PsbH-M, PsbN, PsbX, PsbY, PsbZ, and PsbYcf12) and 3 extrinsic subunits (PsbO, PsbU, and PsbV) attached with the lumenal surface. The major functions of these intrinsic and extrinsic proteins are to protect the oxygen-evolving complex (OEC) (Fig. 3.3) (Bricker et al. 2012). Furthermore, these proteins are also responsible for making the channels for water to come to the Mn_4CaO_5 cluster and for molecular oxygen and protons to go out of the membrane (Vogt et al. 2015). Both D1 and D2 proteins have five helices, all tilted against the membrane planes, while there are six helices in CP43 and CP47 that surround the D1/D2 core complex (Zouni et al. 2001). D1/D2 heterodimer axis has six chlorophylls *a*; two pheophytins (PheoD1 and PheoD2); two quinones (Q_A on D2, and Q_B on D1); a nonheme iron between Q_A and Q_B; (5) two β-Cars; four Mn ions; three or four Ca^{2+} ions; three Cl^- ions; and one carbonate (CO_3^{2-}) or hydrogen carbonate (HCO_3^-) ion bound to the nonheme iron (Shevela et al. 2012). Splitting of water by oxygen-evolving complex (OEC) consists of Mn_4CaO_5 cluster, and this process fulfilled in five consecutive stages named S0 to S4 (Mukherjee et al. 2012). The structure of OEC Mn_4CaO_5 cluster has "distorted chair" conformation in which three Mn, one Ca, and four oxygen atoms form an asymmetric cubane like seat base and the fourth Mn (Mn_4) together with the fifth oxygen atom (O_4) forming the chair back (Umena et al. 2011). Further, it was reported that the chloride ion is essential for oxygen evolution.

Besides major role of PS I and PSII, cyanobacterial thylakoid membrane has key complex that actively participates in photosynthetic and respiratory electron transport chain is the Cyt b_6f complex. The Cyt b_6f complex is capable for generating proton motive force (PMF) by translocating the protons across the thylakoid membrane (DeRuyter and Fromme 2008) (Fig. 3.3). Structurally, Cyt b_6f is a homodimer and each monomer composed of eight subunits with four components: cytochrome b_6, the cytochrome f, Rieske-type 2Fe-2S protein, and subunit IV, which does not carry redox-active cofactors. The Cyt b_6f complex is in many respects analogous to the cytochrome bc1 complex in mitochondria and gram-negative bacteria. Important function of Cyt b_6f is to generate PMF for ATP synthesis via the redox loop Q-cycle to maintain the ATP/NAD(P)H ratio for CO_2 fixation. In Q-cycle, the PQ accepts two electrons and forms plastoquinol (PQH_2) and after then the Rieske-type 2Fe-2S protein of cytochrome f accepts one electron and in the end transferring one of the two electrons to PS I via PC; the other electron is donated to the lower potential cytochrome b hems (Cramer and Zhang 2006) (Fig. 3.4). Cyanobacterial Cyt b_6f complex has multiple copies of petC that code for Rieske 2Fe-2S centers, which is a remarkable characteristic and this multiplicity enables adaptation to stress conditions such as fluctuating light intensities (Tsunoyama et al. 2009) and low oxygen (Summerfield et al. 2008).

3.4 Electron Flow in Cyanobacteria

Thylakoid membrane of cyanobacteria has both an oxygen-evolving photosynthetic apparatus and a full complement of respiratory enzymes (Ohkawa et al. 2000; Cooley and Vermaas 2001). The electron transport chain of photosynthesis is a complex process depends on various components that includes PBPs, PS II, PS I, Cyt b_6f, and ATP synthase (ATPase) and also some electron carriers mainly

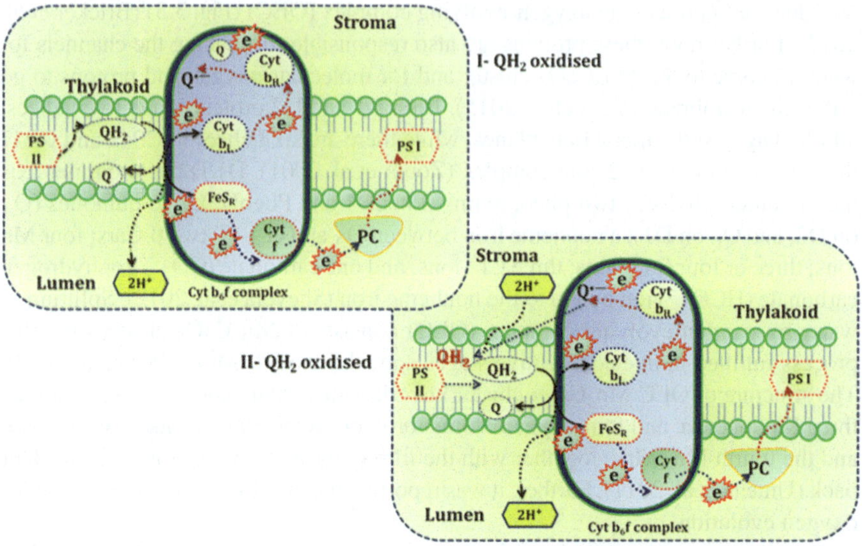

Fig. 3.4 Mechanism of electron and proton transfer in cytochrome b_6f complex via plastohydroquinone PQH_2 (Q-cycle)

plastoquinone (PQ), plastocyanin (PC), and cytochrome b6. All components work in sequence and convert solar energy into chemical energy in form of ATP and NADPH that utilized during carbon fixation (Liu 2016) (Fig. 3.3). Photosynthetic energy production in cyanobacteria is of two types: (1) the linear electron transport pathway in which electrons liberated from splitting of water and travel to $NADP^+$ and molecular oxygen evolved by involving both photosystems (PS II and PS I); (2) the second pathway is the cyclic electron transport in which electrons from PS I are returned to the PQ pool.

The linear or noncyclic electron transport chain is shown in Figs. 3.2 and 3.3 in which splitting or oxidation of water takes place on the lumenal (p) side of the membrane and reduction in the PQ on the stromal side by accepting the electron from water. Solar energy is captured by light-harvesting pigment–protein complex and is quickly transferred to PS II and PSI and excited them. Reaction center site of PS II, i.e., P680 after excitation through light, converts into P680* and initiates the light-induced electron transfer that leads the transition of electrochemical potential energy and water-splitting reaction. Splitting of water occurs on the (electron) donor side of PSII, and this is best explained by Kok et al. (1970) who developed a model of water oxidation in which oxygen-evolving complex interchanges into five oxidation states, labeled S0, S1, S2, S3, and S4 (Fig. 3.3). This model clearly explained that a single electron is transferred from the OEC after each photochemical reaction at reaction center of PSII and change the OEC to the next higher S-state. Thus, after four such reactions, two water molecules are oxidized to one molecule of oxygen (Mar and Govindjee 1972; Retegan et al. 2014). After excitation of P680*, the electrons are now accepted by two electron acceptors Q_A and Q_B in PSII located inside the thylakoid membrane and transferred it to PSI by cytochrome b_6f complex and plastocyanin (PC). The difference in the electron-accepting capacity between QA and QB is that former is a one-electron acceptor tightly bound with D_2 proteins, while later is a two electron acceptor and loosely attached with D1 protein. After accepting the one electron, QA reduced and forms $PheoQA^-$ and transfers one electron to QB and reduced it into QB^-. After accepting the two electrons and two protons, QB^- is fully reduced into QBH2 (sometime also known as PQH2). Formed PQH2 is released in the membrane and is substituted by a PQ molecule from the PQ pool. The electron flow from PQH2 to PS I via cytochrome b_6f complex is explained on the basis of new mechanism known as Q-cycle (Fig. 3.4). In this mechanism, plastohydroquinone (PQH2) is oxidized, and out of two electrons, one is accepted by blue-colored copper protein PC and passed toward photosystem I by following linear electron transport chain and reducing oxidized P700 of PS I. The second electron goes through a cyclic process that increases the number of protons pumped across the membrane and generates proton motive force. After reaching the reaction center of PS I, i.e., P700, the electrons again excite by absorbing the light and form excited P700* before reaching the final acceptors in the cytoplasm, i.e., oxygen or $NADP^+$. Electrons from PS I are transferred to additional electron acceptors that include Fe-S proteins, or bound ferredoxins (Fd), also known as Fe–S centers FeSX, FeSA, and FeSB. Electrons linearly transported via canters A and B to ferredoxin (Fd). In the last steps of noncyclic electron transport chains, there is need of two electrons to reduce ferredoxin-NADP+ oxidoreductase (FNR) to generate NADPH and electrons come from Fd. Thus, completing the sequence of noncyclic electron transport that begins with the oxidation of water and splitting of one water molecule results in synthesis of one NADPH and released six protons (Makita and Hastings 2016; Govindjee et al. 2017). The proton gradient is used to drive ATP production via ATP synthase (Fig. 3.3).

Besides noncyclic electron transport, photosynthetic organism also has cyclic electron transport (CET), which involves only PS I (Fig. 3.2a). Cyclic electron flow is only responsible for synthesis of ATP by generating proton gradient across thylakoid membrane instead of NADPH and thereby improved the ATP/NADPH ratio. Further, shifting of noncyclic to cyclic is an adaptation of organism to protect the photosynthesis against various environmental stresses, such as high light (Wang et al. 2016). Major route of CET in cyanobacteria involves NDH-1 complex participated in both photosynthetic and respiratory electron transports. Respiratory NDH-1 having new NADH-oxidizing module composed of three subunits and oxidizing NADH (Efremov and Sazanov 2012), while photosynthetic NDH-1 having electron acceptor molecule that accepts electron from Fd (Battchikova et al. 2011). Besides this, NdhS is an accessory subunit that can bind with reduced Fd and make link with PS I complex and initiates the NDH-1-dependent cyclic electron transport (He et al. 2015). The cyclic electron transport is considered as the highest quantum yield that comprises transfer of electrons from NADPH to the PQ pool via the NDH-1 complex. Under certain conditions, cyclic electron flow from the reducing side of photosystem I, through the b6f complex and back to P700, is known to occur. This cyclic electron flow is coupled to proton pumping into the lumen. During the recycling of electron, there is production of ATP not NADPH. Thus, it has been proposed that CET pathways are critical for achieving the appropriate balance of ATP and NADPH to power CO_2 fixation (Nogales et al. 2012). However, these electron transport pathways must also power other cellular processes such as nitrogen assimilation, macromolecule synthesis, and the carbon-concentrating mechanism.

3.5 Conclusions

In this chapter, we have briefly described several aspects of the photosynthetic process in cyanobacteria (considered as first oxygen-evolving prokaryotes) with special reference to electron flow. Cyanobacteria have some salient points that differ than other photosynthetic organisms as thylakoid membrane has both photosynthetic and respiratory ET components, high ratio of PSI/PSII (3–5:1) than plants, possess water-soluble PBS function as accessory antenna, and have carotenoids that protect the photosynthetic apparatus against excess light. Thylakoid membranes have PS II, PS I, oxygen-evolving complex (OEC), Cyt b_6f, ATPase, Cyt oxidase, and other mobile electron carriers (plastoquinone and plastocyanin). The difference between linear vs. cyclic electron flow regulated by the redox state of plastoquinon due to mobility of phycobilisome is added by a slower regulation of electron transport pathways through relocation of NDH-1. In this chapter, we have also described structure and mechanism of electron and proton transfer, with special reference to cyclic and noncyclic electron flow and mechanism of Cyt b_6f still the detailing is needed to be illuminated.

References

Barber J (2016) Photosystem II: the water splitting enzyme of photosynthesis and the origin of oxygen in our atmosphere. Q Rev Biophys 49:e14

Battchikova N, Wei L, Du L, Bersanini L, Aro EM, Ma W (2011) Identification of novel Ssl0352 protein (NdhS), essential for efficient operation of cyclic electron transport around photosystem I, in NADPH: plastoquinone oxidoreductase (NDH-1) complexes of *Synechocystis* sp. PCC 6803. J Biol Chem 286:36992–37001

Bricker TM, Roose JL, Fagerlund RD, Frankel LK, Eaton-Rye JJ (2012) The extrinsic proteins of photosystem II. Biochim Biophys Acta 1817:121–142

Chakdar H, Pabbi S (2017) Algal pigments for human health and cosmeceuticals. In: Algal green chemistry. ICAR, New Delhi, pp 171–188

Cooley JW, Vermaas WF (2001) Succinate dehydrogenase and other respiratory pathways in thylakoid membranes of *Synechocystis* sp. strain PCC 6803: capacity comparisons and physiological function. J Bacteriol 183:4251–4258

Cramer WA, Zhang H (2006) Consequences of the structure of the cytochrome b6f complex for its charge transfer pathways. Biochim Biophys Acta 1757:339–345

Croce R, van Grondelle R, van Amerongen H, van Stokkum IHM (2018) Light harvesting in photosynthesis. Foundations of biochemistry and biophysics. Taylor and Francis Group/CRC Press, London, p 611

Dekker JP, Boekema EJ (2005) Supramolecular organization of thylakoid membrane proteins in green plants. Biochim Biophys Acta 1706:12–39

DeRuyter YS, Fromme P (2008) Molecular structure of the photosynthetic apparatus. In: Herrero A, Flores E (eds) The cyanobacteria: molecular biology, genomics, and evolution. Caister Academic Press, Norfolk, UK, pp 217–270

Dietzel L, Brautigam K, Pfannschmidt T (2008) Photosynthetic acclimation: state transitions and adjustment of photosystem stoichiometry. Functional relationships between short-term and long-term light quality acclimation in plants. FEBS J 275:1080–1088

Dismukes GC, Klimov VV, Baranov SV, Kozlov YN, DasGupta J, Tyryshkin A (2001) The origin of atmospheric oxygen on earth: the innovation of oxygenic photosynthesis. Proc Natl Acad Sci U S A 98:2170–2175

Dumay J, Morancais M (2016) Proteins and pigments. In: Fleurence J, Levine I (eds) Seaweed in health and disease prevention. Academic, San Diego, pp 275–318

Efremov RG, Sazanov LA (2012) The coupling mechanism of respiratory complex I: a structural and evolutionary perspective. Biochim Biophys Acta 1817:1785–1795

Ferreira KN, Iverson TM, Maghlaoui K, Barber J, Iwata S (2004) Architecture of the photosynthetic oxygen-evolving center. Science 303:1831–1838

Folea IM, Zhang P, Aro EM, Boekema EJ (2008) Domain organization of Photosystem II in membranes of the cyanobacterium *Synechocystis* sp. *PCC*6803 investigated by electron microscopy. FEBS Lett 582:1749–1754

Gan F, Bryant DA (2015) Adaptive and acclimative responses of cyanobacteria to far-red light. Environ Microbiol 17(10):3450–3465

Glazer AN (1994) Phycobiliproteins - a family of valuable, widely used fluorophores. J Appl Phycol 6:105–112

Govindjee, Shevela D, Björn L (2017) Evolution of the Z-scheme of photosynthesis: a perspective. Photosynth Res 133:5–15

Grossman A, Schaefer MR, Chiang GG, Collier JL (1993) The phycobilisomes, a light-harvesting complex responsive to environmental conditions. Microbiol Rev 57(3):725–749

Guskov A, Kern J, Gabdulkhakov A, Broser M, Zouni A, Saenger W (2009) Cyanobacterial photosystem II at 2.9-Å resolution and the role of quinones, lipids, channels and chloride. Nat Struct Mol Biol 16:334–342

Gwizdala M, Wilson A, Kirilovsky D (2011) In vitro reconstitution of the cyanobacterial photoprotective mechanism mediated by the orange carotenoid protein in Synechocystis PCC 6803. Plant Cell 23:2631–2643

He Z, Zheng F, Wu Y, Li Q, Lv J, Fu P, Mi H (2015) NDH-1L interacts with ferredoxin via the subunit NdhS in *Thermosynechococcus elongatus*. Photosynth Res 126:341–349

Holt TK, Krogmann DW (1981) A carotenoid-protein from cyanobacteria. Biochim Biophys Acta 637:408–414

Huang GJ, Harris MA, Krzyaniak MD, Margulies EA, Dyar SM, Lindquist RJ, Wu Y, Roznyatovskiy VV, Wu YL, Young RM, Wasielewski MR (2016) Photoinduced charge and energy transfer within meta-and para-linked chlorophyll a-Perylene-3,4:9,10-bis (dicarboximide) donor–acceptor dyads. J Phys Chem B 120(4):756–765

Jensen PE, Haldrup A, Rosgaard L, Scheller HV (2003) Molecular dissection of photosystem I in higher plants: topology, structure and function. Physiol Plant 119:313–321

Jensen PE, Haldrup A, Zhang S, Scheller HV (2004) The PSI-O subunit of plant photosystem I is involved in balancing the excitation pressure between the two photosystems. J Biol Chem 279:24212–24217

Jiang L, Wang Y, Yin Q, Liu G, Liu H, Huang Y, Li B (2017) Phycocyanin: a potential drug for cancer treatment. J Cancer 8:3416–3429

Jordan P, Fromme P, Witt HT, Klukas O, Saenger W, Krauss N (2001) Three-dimensional structure of cyanobacterial photosystem I at 2.5 Å resolution. Nature 411:909–917

Kannaujiya VK, Kumar D, Richa, Pathak J, Sonker AS, Singh VR, Sundaram S (2017) Recent advances in production and the biotechnological significance of phycobiliproteins. In: Sinha RP,

Sinha RP, Richa (eds) New approaches in biological research. Nova Science Publisher, New York, pp 1–34

Karapetyan NV, Bolychevtseva YV, Yurina NP, Terekhova IV, Shubin VV, Brecht M (2014) Long-wavelength chlorophylls in photosystem I of cyanobacteria: origin, localization, and functions. Biochemistry (Mosc) 79:213–220

Keren N, Ohkawa H, Welsh EA, Liberton M, Pakrasi HB (2005) Psb29, a conserved 22-kD protein, functions in the biogenesis of photosystem II complexes in *Synechocystis* and *Arabidopsis*. Plant Cell 17:2768–2781

Kerfeld CA (2004) Structure and function of the water-soluble carotenoid-binding proteins of cyanobacteria. Photosynth Res 81:215–225

Kerfeld CA, Sawaya MR, Brahmandam V, Cascio D, Ho KK, Trevithick-Sutton CC, Krogmann DW, Yeates TO (2003) The crystal structure of a cyanobacterial water-soluble carotenoid binding protein. Structure 11:55–65

Kirilovsky D (2010) The photoactive orange carotenoid protein and photoprotection in cyanobacteria. Adv Exp Med Biol 675:139–159

Kok B, Forbush B, McGloin M (1970) Cooperation of charges in photosynthetic O_2 evolution: a linear four step mechanism. Photochem Photobiol 11:457–475

Kühl M, Chen M, Ralph PJ, Schreiber U, Larkum AW (2005) Ecology: a niche for cyanobacteria containing chlorophyll d. Nature 433(7028):820

Kumar J, Parihar P, Singh R, Singh VP, Prasad SM (2015) UV-B induces biomass production and nonenzymatic antioxidant compounds in three cyanobacteria. J Appl Phycol 28:131–140

Lane N (2017) Serial endosymbiosis or singular event at the origin of eukaryotes? J Theor Biol 434:58–67

Leverenz RL, Jallet D, Li MD, Mathies RA, Kirilovsky D, Kerfeld CA (2014) Structural and functional modularity of the orange carotenoid protein: distinct roles for the N- and C-terminal domains in cyanobacterial photoprotection. Plant Cell 26:426–437

Leverenz RL, Sutter M, Wilson A, Gupta S, Thurotte A, Bourcier de Carbon C, Petzold CJ, Ralston C, Perreau F, Kirilovsky D (2015) A 12 Å carotenoid translocation in a photoswitch associated with cyanobacterial photoprotection. Science 348:1463–1466

Liu LN (2016) Distribution and dynamics of electron transport complexes in cyanobacterial thylakoid membranes. Biochim Biophys Acta 1857:256–265

Lundell DJ, Glazer AN (1981) Allophycocyanin B. A common b subunit in *Synechococcus* allophycocyanin B and allophycocyanin. J Biol Chem 256:12600–12606

Makita H, Hastings G (2016) Modeling electron transfer in photosystem I. Biochim Biophys Acta 1857:723–733

Mamedov M, Govindjee, Nadtochenko V, Semenov A (2015) Primary electron transfer processes in photosynthetic reaction centers from oxygenic organisms. Photosynth Res 125:51–63

Manirafasha E, Ndikubwimana T, Zeng X, Lu Y, Jing K (2016) Phycobiliprotein: potential microalgae derived pharmaceutical and biological reagent. Biochem Eng J 109:282–296

Mar T, Govindjee (1972) Kinetic models of oxygen evolution in photosynthesis. J Theor Biol 36:427–446

Masojidek J, Torzillo G, Koblížek M (2013) Photosynthesis in microalgae. In: Handbook of microalgal culture. Wiley, Hoboken, NJ, pp 21–36

Mazor Y, Nataf D, Toporik H, Nelson N (2014) Crystal structures of virus-like photosystem I complexes from the mesophilic cyanobacterium *Synechocystis* PCC 6803. eLife 3:e01496

Melnicki MR, Leverenz RL, Sutter M, López-Igual R, Wilson A, Pawlowski EG, Perreau F, Kirilovsky D, Kerfeld CA (2016) Structure, diversity, and evolution of a new family of soluble carotenoid-binding proteins in cyanobacteria. Mol Plant 9:1379–1394

Moraes CC, Kalil SJ (2009) Strategy for a protein purification design using C-phycocyanin extract. Bioresour Technol 100:5312–5537

Mukherjee S, Stull JA, Yano J, Stamatatos TC, Pringouri K, Stich TA (2012) Synthetic model of the asymmetric $[Mn_3CaO_4]$ cubane core of the oxygen-evolving complex of photosystem II. Proc Natl Acad Sci U S A 109:2257–2262

Nelson N (2011) Photosystems and global effects of oxygenic photosynthesis. Biochim Biophys Acta 1807:856–863

Nelson N, Junge W (2015) Structure and energy transfer in photosystems of oxygenic photosynthesis. Annu Rev Biochem 84:659–683

Nelson N, Yocum CF (2006) Structure and function of photosystems I and II. Annu Rev Plant Biol 57:521–565

Nogales J, Gudmundsson S, Knight EM, Palsson BO, Thiele I (2012) Detailing the optimality of photosynthesis in cyanobacteria through systems biology analysis. Proc Natl Acad Sci U S A 109:2678–2683

Nowicka B, Kruk J (2016) Powered by light: phototrophy and photosynthesis in prokaryotes and its evolution. Microbiol Res 186–187:99–118

Ohkawa H, Pakrasi HB, Ogawa T (2000) Two types of functionally distinct NAD(P)H dehydrogenases in *Synechocystis* sp. strain PCC6803. J Biol Chem 275:31630–31634

Ou Y, Yuan Z, Li K, Yang X (2012) Phycocyanin may suppress d-galactose-induced human lens epithelial cell apoptosis through mitochondrial and unfolded protein response pathways. Toxicol Lett 215:25–30

Page LE, Liberton M, Pakrasi HB (2012) Reduction of photoautotrophic productivity in the cyanobacterium Synechocystis sp. strain PCC 6803 by phycobilisome antenna truncation. J Appl Environ Microbiol 78:6349–6351

Pandey VD, Pandey A, Sharma V (2013) Biotechnological applications of cyanobacterial phycobiliproteins. Int J Curr Microbiol App Sci 2:89–97

Patel SN, Sonani RR, Jakharia K, Bhastana B, Patel HM, Chaubey MG, Singh NK, Madamwar D (2018) Antioxidant activity and associated structural attributes of *Halomicronema* phycoerythrin. Int J Biol Macromol 111:359–369

Punginelli C, Wilson A, Routaboul JM, Kirilovsky D (2009) Influence of zeaxanthin and echinenone binding on the activity of the orange carotenoid protein. Biochim Biophys Acta 1787:280–288

Rakhimberdieva MG, Bolychevtseva YV, Elanskaya IV, Karapetyan NV (2007) Protein-protein interactions in carotenoid triggered quenching of phycobilisome fluorescence in Synechocystis sp. PCC 6803. FEBS Lett 581:2429–2433

Rast A, Heinz S, Nickelsen J (2015) Biogenesis of thylakoid membranes. Biochim Biophys Acta 1847:821–830

Ravi M, Tentu S, Baskar G, Rohan Prasad S, Raghavan S, Jayaprakash P, Jeyakanthan J, Rayala SK, Venkatraman G (2015) Molecular mechanism of anticancer activity of phycocyanin in triple-negative breast cancer cells. BMC Cancer 15:768

Retegan M, Neese F, Panatazis DA, Lubitz W (2014) Electronic structure of the oxygen-evolving complex in photosystem II prior to O-O bond formation. Science 345:804–808

Rexroth S, Mullineaux CW, Ellinger D, Sendtko E, Rögner M, Koenig F (2011) The plasma membrane of the cyanobacterium *Gloeobacter violaceus* contains segregated bioenergetic domains. Plant Cell 23:2379–2390

Schluchter WM, Shen G, Alvey RM, Biswas A, Saunee NA, Williams SR, Mille CA, Bryant DA (2010) Phycobiliprotein biosynthesis in cyanobacteria: structure and function of enzymes involved in post-translational modification. In: Hallenbeck PC (ed) Advances in experimental medicine and biology. Springer, New York, pp 211–228

Shevela D, Eaton-Rye JJ, Shen JR, Govindjee (2012) Photosystem II and the unique role of bicarbonate: a historical perspective. Biochim Biophys Acta 1817:1134–1151

Shi LX, Hall M, Funk C, Schroder WP (2012) Photosystem II, a growing complex: updates on newly discovered components and low molecular mass proteins. Biochim Biophys Acta 1817:13–25

Smith D, Howe CJ (1993) The distribution of photosystem-I and photosystem-II polypeptides between the cytoplasmic and thylakoid membranes of cyanobacteria. FEMS Microbiol Lett 110:341–347

Sonani RR, Singh NK, Anjali A, Prasad B, Kumar J, Madamwar D (2014a) Phycoerythrin extends lifespan and healthspan in *Caenorhabditis elegans*. Age 36:9717

Sonani RR, Singh NK, Kumar J, Thakar D, Madamwar D (2014b) Concurrent purification and antioxidant activity of phycobiliproteins from *Lyngbya* sp. A09DM: an antioxidant and anti-aging potential of phycoerythrin in *Caenorhabditis elegans*. Process Biochem 49:1757–1766

Song C, Velazquez Escobar F, Xu XL, Narikawa R, Ikeuchi M, Siebert F, Gärtner W, Matysik J, Hildebrandt P (2015) A red/green cyanobacteriochrome sustains its color despite a change in the bilin chromophore's protonation state. Biochemistry 54(38):5839–5848

Stock CA, Dunne JP, John JG (2014) Global-scale carbon and energy flows through the marine planktonic food web: an analysis with a coupled physical-biological model. Prog Oceanogr 120:1–28

Summerfield TC, Toepel J, Sherman LA (2008) Low-oxygen induction of normally cryptic psbA genes in cyanobacteria. Biochemistry 47:12939–12941

Tavanandi HA, Mittala R, Chandrasekhar J, Raghavarao KSMS (2018) Simple and efficient method for extraction of C-phycocyanin from dry biomass of *Arthrospira platensis*. Algal Res 31:239–251

Telford WG, Moss MW, Moreseman JP, Allnutt FC (2001) Cyanobacterial stabilized phycobilisomes as fluorochromes for extracellular antigen detection by flow cytometry. J Immunol Methods 254:13–30

Tiwari S, Patel A, Prasad SM (2018) Kinetin alleviates chromium toxicity on growth and PS II photochemistry in Nostoc muscorum by regulating antioxidant system. Ecotoxicol Environ Saf 161:296–304

Tsunoyama Y, Bernát G, Dyczmons NG, Schneider D, Rögner M (2009) Multiple rieske proteins enable short- and long-term light adaptation of *Synechocystis* sp. PCC 6803. J Biol Chem 284:27875–27883

Umena Y, Kawakami K, Shen JR, Kamiya N (2011) Crystal structure of oxygen-evolving photosystem II at a resolution of 1.9 Å. Nature 473:55–60

Vermaas WF (2001) Photosynthesis and respiration in cyanobacteria. Encyclopedia of life sciences

Vogt L, Vinyard DJ, Khan S, Brudvig GW (2015) Oxygen evolving complex of photosystem II: an analysis of second-shell residues and hydrogen-bonding networks. Curr Opin Chem Biol 25C:152–158

Wang X, Gao F, Zhang J, Zhao J, Ogawa T, Ma W (2016) A cytoplasmic protein Ssl3829 is important for NDH-1 hydrophilic arm assembly in *Synechocystis* sp. strain PCC 6803. Plant Physiol 171:864–877

Wilson A, Ajlani G, Verbavatz JM, Vass I, Kerfeld CA, Kirilovsky D (2006) A soluble carotenoid protein involved in phycobilisome-related energy dissipation in cyanobacteria. Plant Cell 18:992–1007

Wilson A, Punginelli C, Gall A, Bonetti C, Alexandre M, Routaboul JM, Kerfeld CA, van Grondelle R, Robert B, Kennis JT, Kirilovsky D (2008) A photoactive carotenoid protein acting as light intensity sensor. Proc Natl Acad Sci U S A 105:12075–12080

Zehr JP (2011) Nitrogen fixation by marine cyanobacteria. Trends Microbiol 19:162–173

Zheng J, Inoguchi T, Sasaki S, Maeda Y, McCarty MF, Fujii M, Ikeda N, Kobayashi K, Sonoda N, Takayanagi R (2013) Phycocyanin and phycocyanobilin from *Spirulina platensis* protect against diabetic nephropathy by inhibiting oxidative stress. AJP Regul Integr Comp Physiol 304:R110–R120

Zhou ZP, Liu LN, Chen XL, Wang JX, Chen M, Zhang YZ, Zhou BC (2005) Factors that affect antioxidant activity of c-phycocyanins from *Spirulina platensis*. J Food Biochem 29:313–322

Zouni A, Witt HT, Kern J, Fromme P, Krauss N, Saenger W, Orth P (2001) Crystal structure of photosystem II from *Synechococcus elongatus* at 3.8 Å resolution. Nature 409:739–743

4 Impacts of Environmental Stress on Physiology and Biochemistry of Cyanobacteria

Aparna Pandey, Garima Singh, Neeraj Pandey, Anuradha Patel, Sanjesh Tiwari, and Sheo Mohan Prasad

Abstract

Cyanobacteria are the first oxygen-evolving organisms on the earth and involved in atmospheric carbon and nitrogen fixation. Besides this, in the current scenario they hold an important position in biotechnology because they are used as biofuels, biopolymers, and secondary metabolites. Rapid industrialization and urbanization including indiscriminate use of fertilizers and pesticides in agricultural fields potentially contaminate the ecosystems (aquatic and terrestrial) and various life forms. Major contributor of pollution is the industrial effluents and alters the growth and development of microbiota associated with agricultural lands. The toxic effects include generation of reactive oxygen species (ROS) that affects the physiological, biochemical, and metabolic processes by causing oxidative stress. The increased oxidative stress significantly alters the composition of lipids of membrane and also reacts with macromolecules (DNA, RNA, and protein). Cyanobacteria have great ability to adapt and adjust against harsh environmental conditions because they are endowed with the well-developed antioxidants, i.e., enzymatic and nonenzymatic that mitigate the ROS-induced negative effects and upregulate the physiological processes. This chapter deals with the variable impact of various environmental stresses on physiology and biochemistry of cyanobacteria.

Keywords

Antioxidant system · Environmental stress · Growth · Oxidative stress · Reactive oxygen species

A. Pandey · G. Singh · N. Pandey · A. Patel · S. Tiwari · S. M. Prasad (✉)
Ranjan Plant Physiology and Biochemistry Laboratory, Department of Botany, University of Allahabad, Allahabad, India

R. P. Rastogi (ed.), *Ecophysiology and Biochemistry of Cyanobacteria*,
https://doi.org/10.1007/978-981-16-4873-1_4

4.1 Introduction

Cyanobacteria are the first photosynthetic prokaryotes occur in moist land, fresh or marine water, ranging from unicellular to filamentous and colonial forms considered as primary producers of aquatic ecosystem and adapted to diverse environmental conditions. Their adaptive role and tolerance ability for extreme situation show its ubiquitous nature (Leaf et al. 2020). Cyanobacteria are the first oxygen-evolving prokaryotes and a major biomass contributor of the aquatic ecosystem. Besides this, they mediate the nitrogen fixation in the agriculture during cultivation of *Oryza sativa* (rice) and *Phaseolus vulgaris* (beans). They also have been enormously important in modeling the course of evolution and ecological modifications throughout the olden times. The cyanobacteria have gained more attention in these days as they are economically very important. Some nitrogen (N_2) fixator species such as *Anabaena* and *Nostoc* provide nitrate (NO_3^-) to plants and serve as an important biofertilizers for growth of plants particularly rice and wheat (Nweze 2009; Patel et al. 2020). Biomass of cyanobacteria is widely used in food, cosmetic, and pharmaceutical industries. Due to the presence of various bioactive compounds, they serve as valuable bioenergy resource for mankind (Chittora et al. 2020), whereas uncontrolled exploitation of environmental heritage due to industrialization contributed to the pollution of environment and thus making it unfit for growth and existence of the cyanobacteria (Tiwari et al. 2019).

Increased industrialization and ultimate discharge of toxic wastes from industries to rivers that further through irrigation practices acquaint various heavy metals in agricultural fields and thus imposing a huge stress to paddy fields cyanobacteria (Patel et al. 2018; Tiwari et al. 2018) (Fig. 4.1). Moreover, enhanced usage of pesticides in agricultural field's full-fill the food demand is causing great damage to cyanobacterial habitat (Tiwari and Prasad 2019). Contaminated environment affects the growth associated with reduction in photosynthetic pigments content, alteration in nutrient uptake, photosynthesis, and their nitrogen fixing ability (Patel et al. 2018, 2020; Tiwari et al. 2020a, b). Moreover, increased level of toxicity eventually leads to the generation of reactive oxygen species (ROS), which interact with the macromolecules (protein, DNA, and lipids), thereby disrupt the membrane structure that leads oxidative stress. To overcome oxidative stress, cyanobacteria endowed antioxidant defense system involving enzymatic antioxidants, viz. superoxide dismutase (SOD), catalase (CAT), guaiacol peroxidase (POD), ascorbate peroxidase (APX), glutathione reductase (GR), dehydroascorbate reductase (DHAR), glutathione-*S*-transferase (GST), and nonenzymatic antioxidants; and cysteine (Cys), proline (Pro), and NP-SH (Kumar et al. 2018). As discussed already about the eminence of cyanobacteria, it becomes necessary to explore the impact of environmental stresses on various physiological and biochemical parameters in them. Thus, this chapter briefly reviews and presents the recent research works concerning about the impacts of environmental stress on cyanobacteria and their adaptation strategies at the physiological, biochemical, and molecular levels.

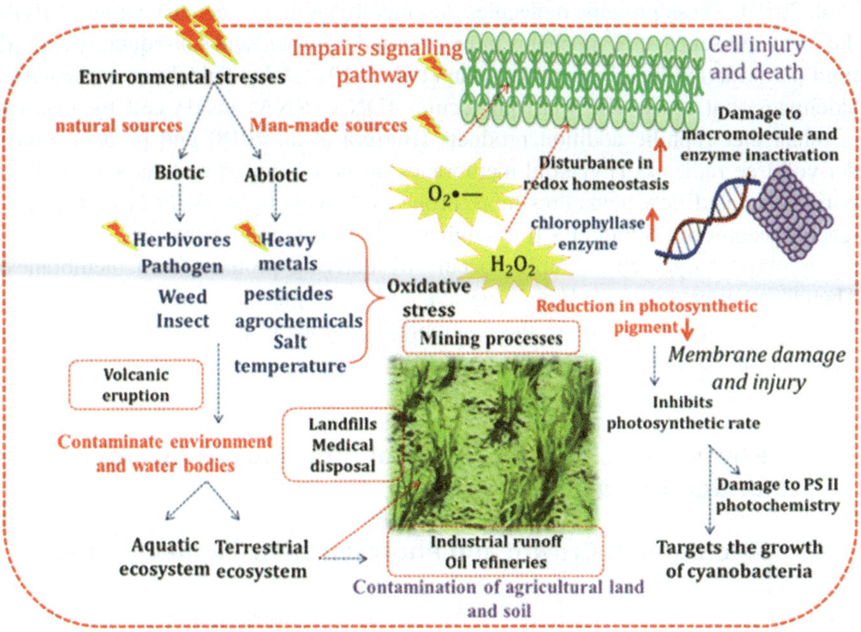

Fig. 4.1 Schematic representation of sources of environmental stress and their impact on cyanobacterial cell

4.2 Impact of Environmental Stress

The different biotic and abiotic factors such as drought, salinity, light, heavy metal, and pesticides make environment challenging for survival of photoautotrophs by inhibiting its growth and development (Tiwari et al. 2018, 2020a, b). Without proper sewage treatment, the industrial sludge that is potent source of heavy metals such as arsenic (As), chromium (Cr), cadmium (Cd), and lead (Pb) directly contaminates the aquatic ecosystem (Patel et al. 2018; Tiwari et al. 2018). Physiological processes such as growth, pigment content (chlorophyll *a*, β-carotene, and phycocyanin; PC), net photosynthetic rate, and nitrogen metabolism are significantly hampered by stress factors. Besides this, oxidative stress is a common end point of abiotic stress factors and several defense mechanisms are executed at different levels in cyanobacterial cell that discussed later.

4.2.1 At Morphological Level (Membrane Structure)

Cyanobacteria serve as best model organisms to study the influence of environmental stress. The cytoplasmic membrane of cyanobacterial mediates the import and export of essential organic substances via channels formed by proteins (Stebegg et al. 2019). These proteins molecules and membrane lipids are easy target of ROS due to the presence of H^+ in chains of unsaturated fatty acid and subsequently caused lipid peroxidation membrane (Rezayian et al. 2019). Malondialdehyde are reactive aldehydes that react with macromolecules (DNA, RNA) inside cell by making covalent electrophilic addition products (Barrera et al. 2018) and produce lipid-derived free radicals. Thylakoid membranes in the cyanobacterial cells are rich in polyunsaturated fatty acids thus more prone to damage by ROS. In cyanobacteria, reaction center of PS I and PS II is synthesized in the plasma membrane; hence, the repair of photosynthetic complex occurs by deriving proteins from membranes. Therefore, expression of *Vipp1* (vesicle inducing plastid protein) is enhanced which plays important role in the assembly of thylakoid membrane during stress (Huang et al. 2006).

4.3 Physiological and Biochemical Alteration in Cyanobacteria Under Stress Condition

4.3.1 Alteration in Growth and Photosynthetic Pigment Contents

Biomass production is negatively affected under various environmental stresses supported by Kumar et al. (2015), Patel et al. (2018), and Tiwari and Prasad (2019) under light, metal, and pesticides stress (Table 4.1). The presence of excessive contaminants results in the competition for uptake of essential nutrients required for enzyme functioning that restrict the cell metabolism and lead to cell death (Patel et al. 2018). Besides growth, the photosynthetic pigment contents, chlorophylls (Chl), carotenoids (Car), and phycobiliproteins (PBPs), involve in photochemistry by capturing light energy from sun and transfer it to the reaction centers (Shah et al. 2017). Cyanobacteria possess different chls such as chl *a*, *b*, *c*, *d*, *e*, and *f*, whereas *a*, *b*, *d*, and *f* are commonly found in most of cyanobacteria (Chittora et al. 2020). Basic structure of Chl is composed of a tetra-pyrrole ring each made of four carbons, one nitrogen atom, and Mg as central atom ion. Chlorophyll *a* is primary pigment as it has dual role and serves as antenna complexes and the reaction centers at PS II and PS I. Growth or biomass production of cyanobacteria totally depends on the pigment contents. The detailed investigation has proved that reduction in pigment content under adverse environmental condition is due to replacement of Mg atom, or inhibition in the enzyme of chl biosynthesis or due to increase in ROS (Chittora et al. 2020). Tiwari et al. (2018) and Patel et al. (2019) also reported that exposure of *Nostoc* to Cr/As declines the Chl *a* content associated with decrease in photosynthetic activity. The phycobiliproteins (PBPs) are arranged in form of rods radiating from a central core; these rods are composed of hexamers, in cyanobacterial system that participating in the absorption of light (Watanabe and Ikeuchi 2013). The PBPs constitute the major part of pigment content, which is water-soluble complexes and found attached to the thylakoid membranes. PBPs are three types: allophycocyanin (APC) core, phycocyanin (PC) rods attached to the core, and phycoerythrin (PE). Abiotic factors severely affect APC and PC contents, which thereby deteriorate the

Table 4.1 Effects of various stress conditions on specific and common physiological and biochemical parameters in cyanobacteria

S. No.	Stress		Specific alterations	Common alterations	Cyanobacteria	References
1.	Heavy metal toxicity	Cd^{2+}	Increase in the thickness of the sheath layer, increase in number and size of polyphosphate granules to incorporate heavy metals into them	Deterioration of thylakoid membranes	*Anabaena flos-aquae*	Surosz and Palinska (2004)
		Pb^{2+}		Decrease in heterocyst frequency	*Anabaena* sp.	Deep et al. (2016)
		Cr^{2+}	Decline in EPS, exo-polysaccharide, the polymer of carbohydrates, and protein contents		*Nostoc muscorum* ATCC 27893 and *Anabaena* sp. PCC 7120	Tiwari et al. (2020a, b)
2.	Metal toxicity	As^{V}	EPS content increased at lower dose, depressions, and grooves on the surface for binding of the metal ions. White crusts over the apertures present more in *Anabaena* sp. than *Nostoc muscorum*	Decline in phycobiliproteins (PBPs)	*Nostoc muscorum, Anabaena* sp.	Patel et al. (2020)
		Al^{3+}		Decrease in EPS secretion, decline in photosynthesis rate	*Anabaena* PCC 7120	Tiwari et al. (2018)
3.	Micronutrient toxicity	Copper toxicity	Excretion of siderophores in huge quantities to alter chemical speciation of surface waters for either sequestration or decrease in copper toxicity source	Nitrogen fixation affected	Cyanobacteria	Rueter and Petersen (1987)
4.	Insecticide	Deltamethrin (2.8% EC)		Induced carbohydrate accumulation, decline in nitrogen fixation	*Calothrix* sp. (GUEco 1002)	Gupta and Baruah (2020)

(continued)

Table 4.1 (continued)

S. No.	Stress		Specific alterations	Common alterations	Cyanobacteria	References
		Cypermethrin	Decline in PS II photochemistry	Enzymatic antioxidant activity increased as defense mechanism	*Nostoc muscorum*	Tiwari et al. (2020a, b)
5.	Salinity	NaCl	Osmotically derived hydrostatic pressure creates tension; thus, cell wall is expanded and stretched at junction of the septum and nascent pole, increased plastoquinone, and a subsequent decrease in chl *a*		*A. cylindrica*	Bhadauriya et al. (2007)
		NaCl		Decrease in heterocyst frequency, increase in carbohydrate contents under low stress as adaptive measure	*A. cylindrica*	Sheikh et al. (2006)
6.	Drought		Accumulation of considerable amount of proline that may act as osmoregulant	Secretion of EPS to protect cell walls from damage during swelling and shrinkage associated with drought stress	*L. boryana*	Lin and Wu (2014)
7.	Radiation stress	UV-B		Decrease in heterocyst frequency	*Anabaena* sp., *Nostoc* sp., *Nostoc carneum*, *Scytonema* sp.	Sinha et al. (1996)
		UV-A + UV-B	Phycobilisomes are disarranged and form amorphous aggregates dispersed in cytoplasm Polyphosphate granules get converted into amorphous structure from round structures	Drastically damaged thylakoids	*Cylindrospermopsis raciborskii* CYRF-01	Noyma et al. (2015)
8.	Microcystin toxicity		PSII is direct target site of microcystin and inhibits it		*Synechococcus elongatus*	Hu et al. (2004)

biomass production. The greater damage to PC content is due to their external localization on the thylakoid membrane (Singh 2014). Not only damaging the pigment content but also stress factor degenerates the surface area of thylakoid membrane, thereby leading to reduction in the photosynthetic activity. Under high light intensity, excitation of RC causes imbalance in PQ redox pool and leads to ROS generation. To overcome this stress situation, cyanobacteria implement a mechanism of state transition to protect the RC of PS II and PS I proteins via movement of their antenna complex when light is preferentially absorbed only by one of the photosystems (Fig. 4.2). Studies have shown that PBPs move on thylakoid membrane when PQ pool gets reduced because of preferential absorption by PS II, and then, antenna complexes detach and attach to PS I called as transition from state I to state II. In plants and green algae, change in fluorescence of PS II and PS I is because of state transitions that have been defined as partial movement of LHC II. In plants and green algae, cytochrome b_6f complex senses the redox states of PQ pool. A high PS II-to-PS I fluorescence ratio is observed at state I (as PSII preferentially absorbs because of antenna complex), whereas a high PS I-to-PS II fluorescence ratio occurs at state II of transition state. The signal cascade involved in the movement of antenna complex based on the redox states of plastoquinone is yet to be known in cyanobacteria and it is might be due to variations in fluorescence (Singh 2014). Besides primary photosynthetic pigments, accessory pigment, i.e., carotenoids perform the function of photoprotection (Shah et al. 2017). Carotenoids present in thylakoid membranes also serve as nonenzymatic antioxidant and scavenge the singlet oxygen; thus, their contents are found to enhance in stress condition. Cars are isoprene derivatives having 40 carbon length chain and are of two kinds, which contain oxygen named xanthophylls (zeaxanthin, astaxanthin, myxoxanthophyll, β-cryptoxanthin, canthaxanthin, lutein, fucoxanthin, and echinenone) and second are carotenes that do not contain oxygen (β-carotene, γ-carotene, ζ-carotene, and lycopene) (Berland et al. 1989). Decrease in contents of xanthophylls leads to enhanced ROS generation in cyanobacteria (Berland et al. 1989).

4.3.2 Alteration in Photosynthetic Activity and Damage to PS II Photochemistry

Cyanobacteria are among the extant lineages, which show oxygen evolution first time and ecologically important as they contribute about 20–30% of the earth's oxygen (Stebegg et al. 2019). Photosynthesis is negatively affected under adverse environmental condition due to alteration in the photosynthetic machinery. Oxygen evolution at PSII site is mediated by oxygen-evolving complex (OEC) and is easy target of heavy metals and decreased the protein synthesis required for the functioning of PS I, PS II, and the light-harvesting components. Further, it also downregulate the genes involved in carbon fixation. The photosynthetic process starts after the absorption of sunlight by antenna pigments and transfer to the RC. The allocation of energy from one PS to another results in formation of ATP and NADPH which then used for fixing the atmospheric carbon. However, the excess absorbed energy is

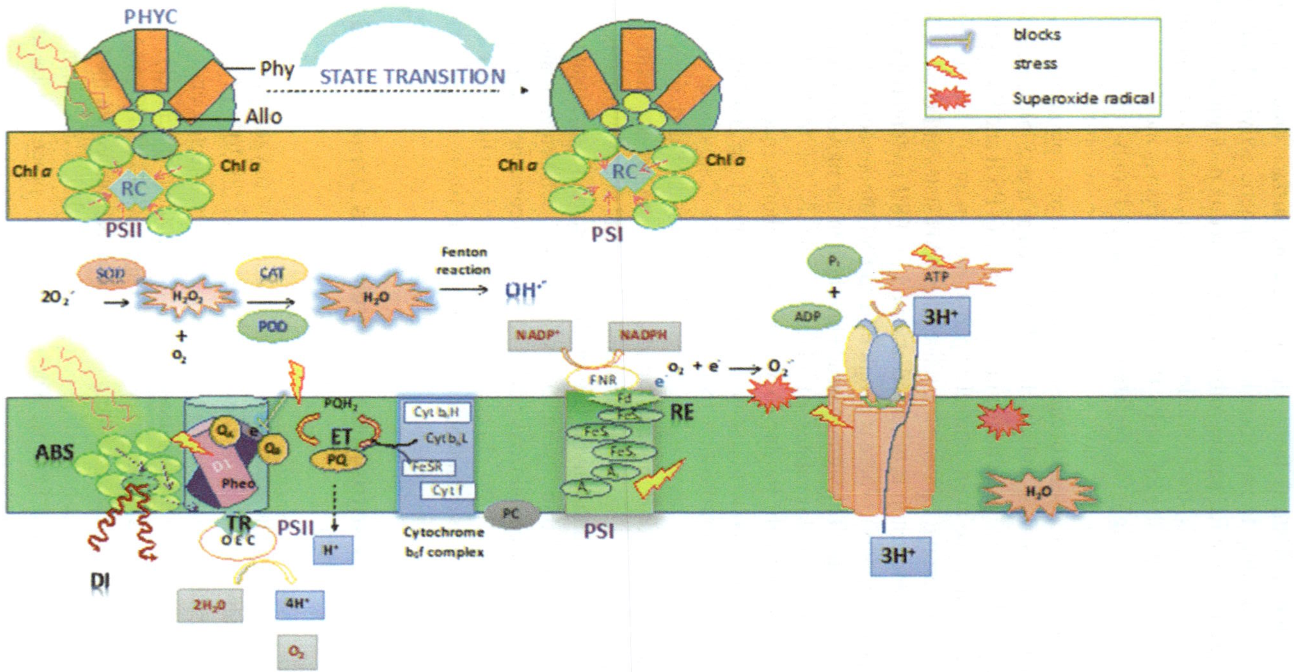

Fig. 4.2 Upper thylakoid membrane represents the partial movement of phycobilisomes (PBPs) from one photosystem to another called state transition, which takes place when preferential excitation of reaction centers causes imbalance in PQ pool. Lower thylakoid membrane represents PSII photochemistry parameters and site, which are blocked due to stress in photosynthetic electron transport chain

dissipated in the form of heat or in form of fluorescence (Strasser et al. 2004). However, environmental pollutants that interrupt the electron flow and consequently energy associated with electrons are dissipated in the form of heat. The D1 protein plays essential role in photochemical quenching, and excessive ROS inhibit their synthesis and other photosystem-related proteins. Sudhir et al. (2005) have reported a loss of proteins such as D1, chlorophyll protein (CP), and other proteins of thylakoid membrane results in decreased photosynthetic rate under salt (NaCl)-stressed *Spirulina platensis*. Under heat stress reaction, centers become inactive and oxygen evolution is hampered. Salt stress inactivated both PS II and PS I in cyanobacteria *Synechococcus*, thus affected photosynthetic rate (Patel et al. 2018). Apart from inactivation of reaction centers, stress factors also affect photosynthetic electron transport. Under stress conditions, when electron transport rate are unable to meet the rate of electron utilization in carbon fixation then it leads to ROS production that eventually decreases oxidation of NADPH and ultimately enhanced reduced ferredoxin (Fd) pool. ROS include singlet oxygen ($^1O_2$), superoxide anion (O_2^-), hydrogen peroxide (H_2O_2), and the hydroxyl radical ($OH^•$; the most toxic). Enhanced ROS leads to decrease in rate of photosynthesis associated with decreased photosynthetic pigments (Tiwari et al. 2020a, b). Cyanobacterial thylakoid membranes are an important site for both oxygenic photosynthesis and respiration; hence, both are affected due to damage of photosynthetic apparatus and thylakoid membrane. Respiratory electron transport chain less known in cyanobacteria comprises of hydrogenases, NAD(P)H dehydrogenases (complex I), succinate dehydrogenase (SDH; complex II), electron transport quinones, cytochrome bc_1 complex (complex III or cytochrome c reductase), cytochrome c_6 plastocyanin, and terminal respiratory oxidase (complex IV) (Lea-Smith et al. 2015). Electrons from respiratory substrate enter through complex I or complex II and reduce PQ pool and then transferred to cytochrome c through cytochrome b_6f complex (Liu 2016). Due to various environmental stress factors, ATP synthesis is blocked, which is essential requirement in various metabolic processes (Allakhverdiev et al. 2005). This leads to excessive production of ROS that damages the photosynthetic apparatus during stress. There is enhancement in ATP requirement; therefore, increased rate of respiration has been found (Patel et al. 2018; Tiwari et al. 2018). The toxicity forced by environmental contaminants was also analyzed via fluorescence transient test, which provides information about the influence on energy fluxes and electron transport.

Photon energy of pigment after got excited have four alternative fates: (1) transfers excitation energy to nearby antenna complex, (2) excited Chl* returns to ground state by release energy in form of heat, (3) transfers its energy to RC and mediate the photochemistry, and (4) emits longer wavelength light called fluorescence. Chlorophyll *a* fluorescence refers to emission of small fraction of energy from Chl *a* by PS II. The energy absorbed by photosynthetic pigment is denoted as absorption flux (ABS), and dissipates energy is denoted as dissipation flux (DI) energy. Energy that utilized to reduce Q_A into Q_B is called trapping flux (TR) and energy associated with electron transport is called electron transport (ET) flux, and energy that reduces final electron acceptors is denoted by RE. In dark-adapted cyanobacterial cells, all Q_A are completely oxidized and RCs are open and fluorescence recorded this time is called zero/minimum fluorescence (F_0). Basic principle that governs chlorophyll *a* fluorescence is that fluorescence of the antenna is high when Q_A in a RC is reduced Q_A (reaction centre is closed) and fluorescence gets quenched when Q_A is oxidized (reaction centre is open). Photochemistry of PS II is studied via JIP-transient parameters such as size and number of active reaction centers (F_v/F_0), quantum yield of PS II (F_v/F_m), yield of electron transport per trapped excitation ($Psi\__0$), performance index (PI_{ABS}) of PS II, efficiency of water splitting complex (F_0/F_v), and quantum yield of electron transport (Phi_E_0). In stressed cells, there is decrease in JIP test parameters that are noticed points toward stress (Jägerbrand and Kudo 2016; Tiwari and Prasad 2019). Besides this, the values of energy fluxes per active RC, absorbance per reaction centre (ABS/RC), energy dissipation flux per reaction centre (DI_0/RC), trapped energy flux per reaction centre (TR_0/RC), and electron transport flux per reaction centre (ET_0/RC) (Fig. 4.2) are studied. TR_0/RC denotes the rate of trapping of exciton by RC that ultimately reduces Q_A to Q_A^-, and it is calculated on the basis of variable fluorescence; therefore, only photochemically active reaction centers that can reduce Q_A is considered and RC refers to be active. In stressed cyanobacterial cells, value of ABS/RC deduces the fraction of the RCs that were active in the healthy state and the fraction that transformed to silent RCs on stress exposure. Thus, values of energy fluxes per active reaction centers, i.e., ABS/RC, DI_0/RC, TR_0/RC, and ET_0/RC, are enhanced under various stresses (Tiwari et al. 2018; Patel et al. 2018).

4.3.3 Inflection on Nitrogen Metabolism (Inorganic Nitrogen Uptake Nitrate and Nitrite Uptake and Ammonia Assimilation)

Nitrogen is an essential component needed for several enzymes, proteins, and nucleic acids that regulate cyanobacterial growth (Forchhammer and Selim 2020). In cyanobacteria, uptake of nitrate and nitrite ions is mediated by ATP-binding cassette (ABC)-type transporter, whereas NrtP permeases carry this function in marine species such as *Trichodesmium* (Herrero et al. 2001). Firstly, nitrate gets reduced to nitrite by consuming two electrons through action of nitrate reductase (NR) and then nitrite is reduced into ammonia through six electrons by action of nitrite reductase (NiR) (Fig. 4.3). Under stress condition, the NR and NiR activity decreased significantly under metal stress in *Nostoc muscorum* and *Anabaena* sp. be due to decreased in nitrate uptake (Tiwari et al. 2020a, b; Patel et al. 2020). For atmospheric N fixation, some cyanobacteria have specialized cells named heterocyst that provides anaerobic condition as nitrogenase is oxygen sensitive, while *Trichodesmium* sp. fix it during dark periods (Capone et al. 1997). Srivastava et al. (2014) reported that low (1 μg/ml) dose of chlorpyrifos improved NO_3^- uptake but decreased NO_2^- in certain paddy fields cyanobacteria, while that of high (2 μg/ml) could not. Ammonium is the most preferred source of nitrogen assimilation in cells. Nitrogen assimilation is highly regulated process in cyanobacteria. The ammonium

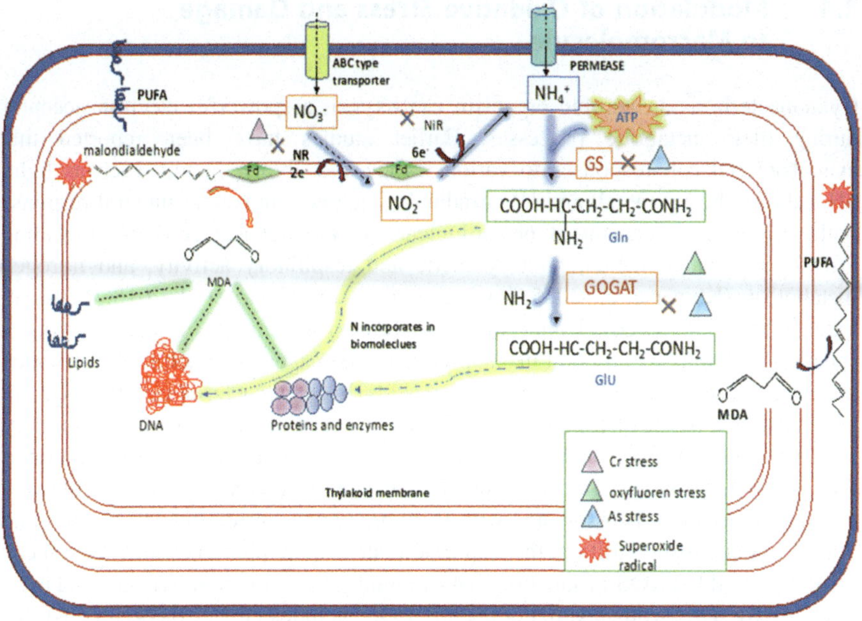

Fig. 4.3 Generation of malondialdehyde (MDA) via peroxidation of polyunsaturated fatty acids in cytoplasmic membrane and thylakoid membrane. MDA thus damages biomolecules. The inhibition of nitrate reductase (NR), nitrite reductase (NiR), glutamine synthetase (GS), and glutamate synthase (GOGAT) activities via heavy metal and pesticide stress

ions are incorporated into carbon skeletons through a cycle called GS-GOGAT pathway named on basis of two enzymes glutamine synthetase (GS) and glutamate synthase (GOGAT), which participate in a preceding manner. The GS catalyzes amination of glutamate and converts into glutamine, which is an ATP-dependent step. However, under stress condition the GS activity is inhibited by direct interaction of two proteins IF7 and IF17. Under stress situation, the ATP requirement is subsequently enhanced because of which available ATP does not meet the demand, thus leading to decreased GS activity as reported under metal stress in *Nostoc muscorum* and *Anabaena* sp. PCC 7120 on exposition to As stress (Patel et al. 2020) (Fig. 4.3). Further, then next two molecules of glutamate are produced by reductive transfer of ammonium group to iso-citrate through action of GOGAT. In the cyanobacteria, two kinds of GOGAT are present on the basis of electron donor: One is Fd-GOGAT and second is NADH-GOGAT (Patel et al. 2020). Sheeba et al. (2020) reported a significant reduction in the activities of ammonia assimilating in oxyfluoren-stressed cyanobacteria.

4.4 Modulation of Oxidative Stress and Damage to Macromolecule

Cyanobacteria as an aerobic organism use oxygen as powerful electron acceptor during their metabolic processes. Earlier studies have been reported that cyanobacterial cell responses to various biotic and abiotic factors such as light, pH, salinity, heavy metal, and UV irradiation include a universal integral response with a transient alteration of physiological and biochemical activities including growth, pigment contents, photosynthetic and respiratory activity, and nitrogen metabolism (Kumar et al. 2015; Rastogi et al. 2015; Tiwari et al. 2018, 2020a, b; Patel et al. 2018, 2020). Under stress conditions, generation of reactive oxygen species is a common end point that subsequently damages the macromolecules such as protein, carbohydrate, DNA, and amino acids. In cyanobacterial cell, the ROS are generated as a result of leakage of electrons during photosynthetic and respiratory electron transport chain or by O_2 reduction or by its energization. Respiration is thought to be a major metabolic process in cyanobacteria (in other aerobically organisms also) that is associated with ROS production inside the cell (Latifi et al. 2009). Under stress condition, the respiration rate is quite high in cyanobacterial cell resulting in added ROS production that eventually leads to oxidative stress (Tiwari et al. 2018). The strong oxidizing agent's ROS includes singlet oxygen ($^1O_2$), the superoxide anion ($O_2^{\bullet-}$), and hydrogen peroxide (H_2O_2). Due to high mobility and reactivity, the H_2O_2 easily crosses the membrane and rapidly reacts with Fe^{2+} or Cu^{2+} and forms more toxic hydroxyl radical ($^{\bullet}OH$) that damages the membrane lipids and causes lipid peroxidation as measured in terms of malondialdehyde equivalent content (MDA) (Patel et al. 2018). The reactive species have different reaction capability and features that target the molecules and show own toxic nature. Among all the reactive species, singlet oxygen is very reactive and have short life period in cells that produced by energy input to oxygen (Gorman and Rodgers 1992) and they immediately react its neighborhood target molecules that is proteins, pigments, and lipids. So, both O_2 and $^{\bullet}OH$ have an unpaired electron, which give high potential to them and interact with surrounding molecules. The superoxide radical is negatively charged and therefore incapable of trans-membrane diffusion. The species oxidizes $[4Fe–4S]^{2+}$ clusters to $[3Fe–4S]^{1+}$ to release divalent iron (Fe^{2+}). The hydroxyl radical is also characterized by extremely high reactivity and a short life. H_2O_2 is less reactive, but can be reduced to the hydroxyl radical in the Fenton reaction and thereby damage the cell (Latifi et al. 2009).

The inhibition of electron transport leads to over-reduction of many components of the electron transfer chain (ETC), first, at the acceptor side of PS II; this might be caused by decreased activity of Rubisco, a key enzyme in the Calvin–Benson cycle, under abiotic stress conditions such as chilling or drought (Asada 2006). The amount of ROS production has to be carefully evaluated since the noninvasive detection techniques like the signal of most ROS markers, for example, quenching of dansyl 2,2,5,5-tetramethyl-2,5-dihydro-1*H*-pyrrole (DanePy), respond in a nonlinear fashion to the ROS level. Therefore, it is difficult to compare the absolute amount of different ROS monitored with different techniques. The major site of ROS production in cyanobacteria is thylakoid membrane where the photosynthetic machinery exists. The reduction in O_2 results in the generation of both H_2O_2 and the hydroxyl radical (Asada 2006), and they are most abundantly produced on acceptor side of PS I as a result of the photosynthetic transport of electrons. They are also generated in illuminated PS II. The $^1O_2$ is generated by the transfer of energy from excited pigments, such as chlorophylls, in the light-harvesting complexes, from excited Fe–S centers in PS I and from photodamaged PS II in which the oxygen-evolving system has been damaged. The generation of various ROS is promoted when the photosynthetic machinery absorbs excess light and also when the availability of CO_2 or of NADP is limited (Asada 2006). The life period of singlet oxygen is about 3.5 μs in an aqueous solution (Egorov et al. 1989) and because of its high reactivity it binds rapidly to all relevant biomolecules resulting in damage to the cyanobacterial cell (Hideg et al. 2007; Mishra et al. 1994). Inside the cell, the life period of singlet oxygen can be decreased down to 200 ns and during this period, depending on physiological conditions they can diffuse over a range of 10 nm or more (Sies and Menck 1992). So, they can easily diffuse through the membranes of cyanobacterial cell. The major target is proteins with the rate constants in the range of 10^8–10^9 M^{-1} S^{-1} (Wilkinson et al. 1995). There are five natural amino acids (Tyr, His, Trp, Met, and Cys) that are primarily attacked by the singlet oxygen and the reaction with Trp results in the formation of peroxides, which is further degraded into different products like *N*-formylkynurenine and becomes stable (Gracanin et al. 2009), and the reaction ability of Trp in proteins has been dependent on local environment of the target (Jensen et al. 2012), while on the other hand, carotenoids play significant role in chlorophyll suppression and quenching (Pogson et al. 2005), under the light stress (Carbonera et al. 2012). The reaction between singlet oxygen and singlet ground state carotenoid leads to the photochemical quenching and also oxidizes the carotenoids by formation of product that can act as signal molecule for stress response (Ramel et al. 2012). Besides this, the oxidation of polyunsaturated fatty acid generates lipid (hydro) peroxides that can acts as initiator of signal pathways and propagation of cellular damage (Triantaphylides and Havaux 2009).

It is supposed that some O_2 radicals are formed via electron dismutation, which is catalyzed by superoxide dismutase (SOD) transformed into H_2O_2 (Asada 2006). In earlier studies, Halliwell (1977) well-versed that O_2 has both oxidizing and reducing properties. Because of the negative charge, it inhibits electrophilic properties in electron-rich molecules, while molecules with less electron are oxidized. $O_2^{\bullet-}$ oxidizes enzymes containing the 4Fe–4S clusters (Imlay 2003), while cytochrome c is reduced (McCord et al. 1977). In amino acids, mainly histidine, methionine, and tryptophan can be oxidized by $O_2^{\bullet-}$ (Dat et al. 2000). In thylakoid membrane, the reduction of H_2O_2 to H_2O is mediated by ascorbate (AsA) under the catalysis of soluble stromal ascorbate peroxidase (APX). The oxidation of AsA to monodehydroascorbate radical (MDHA) is by reduction of MDHA, and this reaction occurs directly either by ferredoxin (Fd) or by NAD(P)H catalyzed by MDHA reductase (MDHAR). The MDHA radical always decays partially into dehydroascorbate (DHA), which is reduced by DHA reductase (DHAR). In that step, reduced glutathione (GSH) is oxidized to glutathione disulfide (GSSG). The reduction of GSSG to GSH occurs by NAD(P)H by glutathione reductase (GR) (Asada 2006; Vranova et al. 2002).

The $^{\bullet}OH$ radical gives rise to the oxidative degradation of proteins and lipids by reacting with them very fast immediately on the site where they are produced (Halliwell 2006). The inhibition of $^{\bullet}OH$ radical production is led by the suppression of H_2O_2 formation by the cell in the presence of Fe^{2+} using metal-binding proteins like ferritins or metallothioneins (Hintze and Theil 2006). On the other hand, $^{\bullet}OH$ radicals are produced in programmed cell death (PCD) as part of defense mechanisms to pathogenic infections.

4.4.1 Influence on Macromolecules

Biomolecules like lipids, proteins, and exopolysaccharides (EPS) after reacting with ROS undergo oxidative damage. As mentioned previously, the oxidative degradation of membrane lipids occurs that results in various degradation products that are formed particularly aldehydes like malanoaldehyde. It is well known that under heavy metal stress the lipids undergo peroxidation and it is also the marker for oxidative stress (Tiwari et al. 2018). Lipids and fatty acids in cyanobacterial cells are said to be the first molecule that significantly plays a role for tolerating various environmental stresses like desiccation, salt-induced damage, low temperature, and high light-induced photo-inhibition (Singh et al. 2002). Structurally, the membrane is rich in polyunsaturated fatty acid (PUFA) that is highly sensitive to oxidative damage, hence resulting in alteration in membrane fluidity, permeability, and cellular metabolic functions (Tiwari et al. 2018). The alteration in plasma membrane affects the nutrient uptake that results in the limiting growth rate in cyanobacterial cell. The decrease in PUFA content is directly associated with increase in MDA levels that is in response of high osmotic stress. Besides this, the nitrogen-assimilating enzymes are also associated with plasma membrane; i.e., the nitrate transporter is integrated in the plasma membrane, and nitrate and nitrite reductase are components of the thylakoid membranes. So, the nitrogen metabolism is also checked (Tiwari et al. 2020a, b). Phenols have antioxidative properties; i.e., they provide protection against oxidative damage to cyanobacterial cells. They accumulate in stress condition and act as metal chelators (Sgherri et al. 2003; Zagoskina et al. 2005; Prasad and Singh 2011). They also scavenge the free radicals generated under different environmental stress. Their chemical structure, type and position, and number of functional groups also affect their bioactive properties (Kanski et al. 2002). Proline is also an antioxidant compound that accumulates under stress conditions. It plays significant role in protection of enzymes and stabilization of the machinery of protein synthesis, and acts as an effective singlet oxygen quencher (Szabados and Savoure 2009). There are two types of exopolysaccharides (EPSs) synthesized by cyanobacterial cells. First category remains attached to the cell wall, whereas second is secreted in the surrounding. The significance of EPS is to form microbial mats under stress condition; this microbial mat is species-specific. So, EPS is the first protective barrier for the cyanobacteria. In mild stress condition, the secretion of EPS is increased to prevent the entry of toxic metals because of their anionic properties and having uronic acids and sulfate groups. However, with the increase in the intensity of stress, increased inhibition of EPS secretion has been reported (Pereira et al. 2009; De Philippis et al. 2011; Tchounwou et al. 2012; Jittawuttipoka et al. 2013; Patel et al. 2020). The synthesis of EPS is directly associated with the carbohydrate contents as it acts as a substrate for EPS biosynthesis and the carbohydrate content is positive proportional to the rate of photosynthesis and degradation of photosynthetic pigments. Under stress condition, the reduction in rate of photosynthesis and degradation of photosynthetic pigments reduce the carbohydrate content in cyanobacterial cell (Tiwari et al. 2020a, b; Patel et al. 2020).

4.5 Tolerance Mechanism in Cyanobacterial System

During the course of evolution, cyanobacteria are being a first oxygen-evolving organism that adapts itself according to the changing environmental conditions (Singh 2014). Thus, they are also called as highly adjustable biological organism, which sustains under severe conditions such as high light intensity, high or low temperature, UV radiation, salinity, and heavy metal pollution (Tiwari et al. 2019). The phenomenon or mechanisms by which cyanobacterial cells reduce the toxicity of various stress factors and maintain the basal growth and developmental processes of the cells are broadly categorized in to two levels (1) at morphological or biochemical level, and (2) at molecular level.

4.5.1 Tolerance Mechanism at Morphological and Biochemical Level

At morphological level, plants protect themselves against harsh environmental conditions and bacterial infection via epidermis/waxy layer (layer) that acts as mechanical barrier (Tiwari et al. 2019). Similar to plants, cyanobacteria form first protective barrier outside the cell that made up of polymeric substance, i.e., polysaccharides and commonly known as exopolysaccharides (EPSs) that performs defensive mechanism against metals stress or antimicrobial infections (Heindl et al. 2014; Patel et al. 2020). The EPS is organic compound, natural polymer having high molecular weight secreted by micro-algae and cyanobacteria and thus maintains the physiological metabolic function of the cell via forming biofilm (Gutnick and Bach 2000). The EPS acts as primary protective barrier against several types of heavy metal contamination such as cadmium (Cd), cobalt (Co), copper (Cu), iron (Fe), nickel (Ni), cerium IV oxide (CeO_2), and titanium dioxide (TiO_2) and thus leads to heavy metal detoxification and it also shows the metal-binding phenomenon due to the presence of hydroxyl ions, carboxyl group, and amide group and thereby leads to metal chelation (Yin et al. 2011a, b; Planchon et al. 2013; Tiwari et al. 2020a, b).

At biochemical level, cyanobacteria protect themselves against oxidative stress via array of antioxidative enzymes (Fig. 4.4; Table 4.1), chiefly classified into two

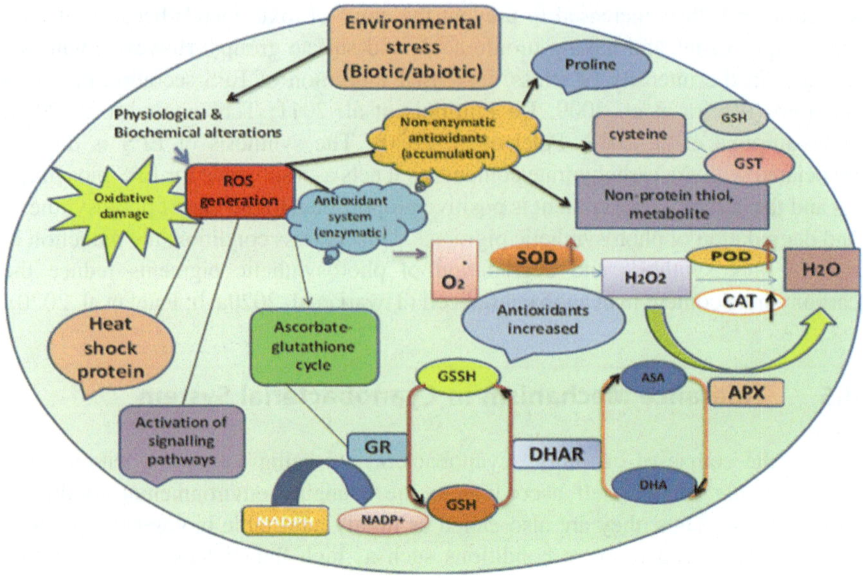

Fig. 4.4 Representation of activation of enzymes and protein under stress condition

groups, i.e., enzymatic antioxidant and nonenzymatic antioxidant (Patel et al. 2018). The enzymatic antioxidants includes SOD (superoxide dismutase), POD (peroxidase), CAT (catalase), GST (glutathione-*S*-transferase), APX (ascorbate peroxidase), GR (glutathione reductase), and DHAR (dehydroascorbate reductase), while nonenzymatic antioxidants are cellular metabolites such as AsA (ascorbate), GSH (reduced glutathione), proline, cysteine, phenolic compounds, alkaloids, nonprotein thiol, α-tocopherols, and carotenoids, which play a vital role in reducing oxidative stress (Latifi et al. 2009; Gill and Tuteja 2010). Under stress conditions, antioxidants convert SOR into hydrogen peroxide (H_2O_2) and water (H_2O) by removing free radical in the presence of various cofactors by multistep reaction mechanism (Latifi et al. 2009). Among various enzymes, SOD is the first line of defense and is a metalloprotein, which catalyzes the dismutation of superoxide radical ($O_2^{\bullet -}$) into H_2O_2 and molecular oxygen (Patel et al. 2018). In earlier studies, it has been also demonstrated that cyanobacterial species such as *Spirulina platensis*, *Chlorella vulgaris*, and *Anabaena* performs defensive phenomenon through upregulation of SOD (Choudhary et al. 2007; Tiwari et al. 2019). Further, CAT is a heme-containing tetrameric enzyme synthesized in peroxisomes of plants and eukaryotic algae and involved in fatty acid oxidation process, purine metabolism, and photorespiration process and converts H_2O_2 into H_2O and O_2 and also upregulation of CAT was noticed in *Chlorella vulgaris* under stress condition. Apart from CAT, another enzyme that participated in H_2O_2 detoxification is POD, which is synthesized in plants and in cyanobacteria under stress condition (Huang et al. 2009; Rezayian et al. 2019). Besides this, H_2O_2 detoxification is mainly done by ascorbate–glutathione (AsA-GSH) cycle (Fig. 4.4), also known as Halliwell–Asada pathway, and this pathway shows the oxidation–reduction of NADPH, AsA, and GSH, thereby performing the detoxification of ROS. In cyanobacteria, APX converts H_2O_2 into water molecule through the ascorbate electron donor and also their isoenzymes present in the cytosol (Oesterhelt et al. 2008). However, GR enzyme is a flavoprotein that participates in AsA-GSH cycle and maintains GSH metabolic mechanism and performs antioxidative process in prokaryotes and in eukaryotes through the oxidation–reduction mechanism. In oxidative stress condition, the ROS accumulate in excess form, which is mainly prevented by the nonenzymatic antioxidant. For example, carotenoids and alpha-tocopherol nonenzymatic antioxidants are found in photoautotrophs and act as ROS scavengers (Rezayian et al. 2019). In cyanobacteria, carotenoids are present in the form of beta-carotene and their derivatives like zeaxanthin. These pigments contain the antioxidative properties and dissipate excess energy from the singlet oxygen ($^1O_2$) via excitation of chlorophyll molecules. Earlier studies have demonstrated that zeaxanthin pigment shows the photo-acclimation properties under UV stress situation in *Synechococcus* PCC 7942, while carotenoids pigment performs the protection mechanism against peroxidation process in *Synechococcus* PCC 7942 during high temperature stress (Latifi et al. 2009). The carotenoid pigment contains glycosidic bond that have high affinity of unsaturation, and these properties of carotenoids make them effective antioxidants in nature. Moreover, tocopherol (vitamin E) is soluble in lipid and synthesized in several photoautotrophs like green algae, plants, and in some cyanobacteria. Under stress condition, tocopherol also reduces the toxicity by the oxidation process of ROS molecules, i.e., singlet oxygen ($^1O_2$) (Rezayian et al. 2019). Ascorbic acid (vitamin C) is water-soluble in nature and performs antioxidative mechanism by the elimination of H_2O_2 in ascorbate–glutathione cycle. In *Spirulina platensis,* increased amount of ascorbic acid eliminates the content of H_2O_2 in cell and also regenerates α-tocopherol from α-chromanoxyl radical during lipid peroxidation process (Rezayian et al. 2019). Reduced glutathione (GSH) is a water-soluble, low molecular weight protein found in prokaryotes and eukaryotes and takes participate in quenching H_2O_2 by acts as antioxidants as reported in some cyanobacteria such as *Spirulina platensis*, *Anabaena* sp., and *Nostoc* muscorum (Tiwari et al. 2020a, b). Apart from enzymatic antioxidants, nonenzymatic antioxidants also participate in ROS detoxification. Among them, cysteine a sulfur containing nonprotein thiol and precursor of GSH is synthesized under oxidative stress (Latifi et al. 2009). Further, proline is a nonenzymatic antioxidant, which shows reduction in oxidative ROS mainly hydroxyl radical ($^•OH$) and singlet oxygen ($^1O_2$) that inhibit programmed cell death (PCD) and also provide stability of protein, regulate acidification phenomenon, protect enzyme, and also act as osmoprotectants in *Spirulina* sp., *Chlorella* sp., and *Anacystis nidulans* (Rezayian et al. 2019; Tiwari et al. 2019).

4.5.2 Tolerance Mechanism at Molecular Level

Under adverse environmental conditions, tolerance mechanism of cyanobacteria increases up to a maximum level by the molecular approaches that give a broader analysis by using OMICS tools that regulate the stress condition (Hagemann 2011). OMICS tools involve synthesis of different RNA transcripts (transcriptome) and protein expressions (proteome) and heat shock proteins (Schirmer et al. 2010).

4.5.2.1 Molecular Chaperone

Molecular chaperone is a specialized protein that assists correct folding process of complex proteins, i.e., conversion of polypeptides into oligomeric form. Molecular chaperons stabilize the protein structure and mediate proper folding (Rajaram et al. 2014). On the basis of biogenesis process, chaperone is synthesized under heat stress and there are mainly two types of HSPs: HSPs70s and HSPs60s. There are other chaperones like HSPs90 and HSPs104 also present in the living organism. In prokaryotes, hsp family includes several chaperones such as GroEL, DnaK, HtpG, and ClpB, which play a vital role in constitutive expression, which is commonly induced by the stress and maintains three-dimensional structure of native protein through ATP hydrolysis (Singh et al. 2006). In cyanobacteria, HSPs are upregulated through the distinct group of genes via transcriptional activation mechanism and their induction magnitude also proportional to the temperature (Chatterjee et al. 2020). HSPs are associated with the membrane and controlled the expression via fidelity mechanism, i.e., physical state changes into the lipid state. Moreover, HSPs broadly classified into various distinct groups, i.e., small heat shock proteins, HSP100 (clpB) family, HSP70/HSP40/HSP25 family, HSP90 family, and HSP60/HSP10 family. Of these, small heat shock protein contains α-crystalline domain structure and no need of ATP during synthesis. The HSPs located between cytoplasm and thylakoid membrane and under stress situation provide protein folding and stabilization of membrane protein (Chaurasia and Apte 2009). For example, in *Synechococcus* sp. during high temperature hspA binds and interacts with the 42 different forms of protein and protects the phycobiliproteins and photosystem II against oxidative damage. Also, *clpB* (caseinolytic peptidase) gene provides thermoregulation in *Synechococcus* sp. (Rajaram and Apte 2008; Richter et al. 2010). In bacterial system, HSP70/HSP40/HSP25 family incorporates with the 70 kDa DnaK and 40 kDa DnaJ and present in *Synechocystis* sp. and *Synechococcus* sp. (Haslbeck et al. 2005). However, HSP 90 family of chaperons contains the HtpG protein, which performs a vital role under oxidative stress and provides photosynthetic stability through the interaction with the phycobiliprotein pigment (Chatterjee et al. 2020).

4.5.2.2 Transcriptional and Post-translational Regulation

Genomics is the study of interaction of genes at the transcriptional level, whereas transcription of RNA from genome called as transcriptomics and protein expression at the translational basis proceeds through the proteomic mechanism and overall provides the tolerance system against the oxidative damage in cyanobacteria (Shrivastava et al. 2015). Adverse environmental stress negatively damages the DNA, protein instability, and lipid. With the help of operon system, cyanobacteria reduce the toxicity of stress; for example, arsBHC operon minimizes arsenic heavy metal toxicity. DNA repair and transcriptional regulation mechanism associate with the different forms of protein, which regulate the metabolic mechanism. In

Anabaena sp., DNA-binding proteins are found, which protects DNA by preventing cleavage of strand and base modification or by transcriptional regulation (Babele et al. 2015; Panda et al. 2015). The activation of DNA repair system and transcriptional machinery system performs upregulation phenomenon against stress-responsive process done with DNA-dependent RNA polymerase α-subunit of enzyme. In earlier study, it has been described that *Spirulina* reduces the chilling and heat stress through the aggregation of chromosomal ATPase activity and thereby plays vital role in replication, modification, and DNA repair (Hongsthong et al. 2008, 2009). Under high-temperature condition, genes like *ParA*, *GvrA*, and *PhrA*, *NusB*, *SigD*, and *SYNPCC7002A2523* of *Synechococcus* function as transcriptional regulator (Xiong et al. 2015b). Moreover, antioxidant enzymes like SOD, POD, and CAT also involved in upregulation gene mechanism through the isoenzyme profiling and thereby they control the gene expression (Patel et al. 2018). While on the other hand, survivals of cyanobacteria in harsh conditions are also dependent on the signal transduction pathways by the mechanism of post-translational modification (PTMs). Covalent modification of protein takes place by the PTMs, which help in splicing and modification of amino acids, i.e., acetylation and phosphorylation, thereby performs the conversion of functional properties of protein as signal and regulates the circadian rhythm, photosynthesis, and nitrogen fixation in cyanobacteria under abiotic stress condition (Xiong et al. 2015a). In recent time, proteomic study of PTMs in cyanobacteria is very useful to identify the complex signaling mechanism, which participates in their evolutionary significance related study (Xiong et al. 2015a). Moreover, serine/threonine kinase activities also regulate the photochemistry and nitrogen metabolism in cyanobacteria. The protein P II (GlnB) is a regulatory protein and also known as phosphoprotein that involves the 32 protein in PTMs, which mainly regulate the nitrogen activity and convert the N/C ratio that reported in *Synechocystis* 6803 under salinity stress (Spät et al. 2015).

4.6 Conclusions

Conclusively, the changing environmental conditions such as high light, temperature, salinity, UV rays, heavy metal, and unselective use of agrochemicals in our surroundings are not only contaminating the environment and ecosystem but also making it unfit for the survival of microorganism (cyanobacteria), plant, animals, and human beings. The environmental contaminants not only contaminating terrestrial and aquatic ecosystem but also adversely affecting the biochemistry (ROS level) and cellular physiological mechanisms of cyanobacteria such as photosynthesis, PS II photochemistry, and nitrogen assimilation are very necessary for their survival, and via this, they also contribute to their economic prospects as biofertilizers and reservoir of bioactive compounds. Cyanobacteria show various defense mechanisms (Fig. 4.5), to sustain or mitigate the environmental stress conditions through antioxidant defense system, molecular chaperones, exopolysaccharides, and transcriptional and post-translational molecular basis.

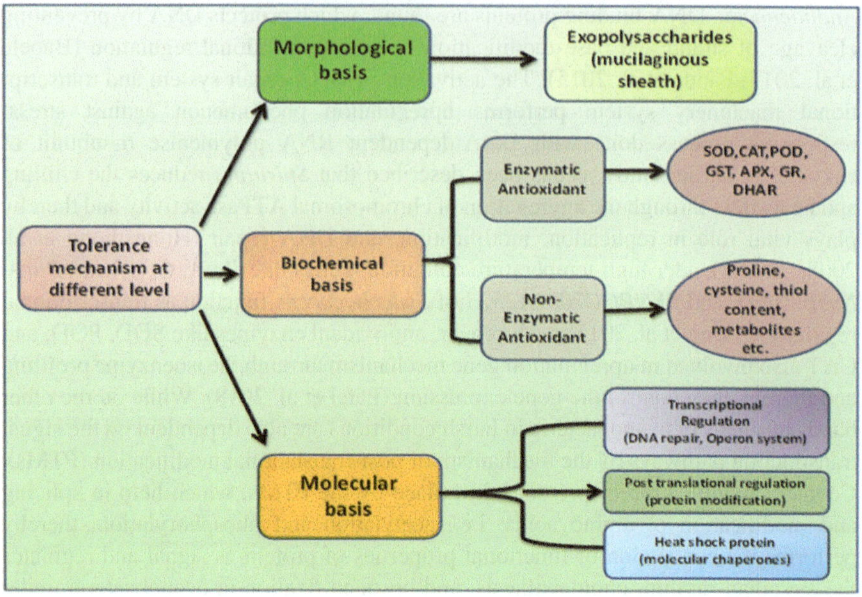

Fig. 4.5 Tolerance behavior at different levels in cyanobacteria

References

Allakhverdiev SI, Nishiyama Y, Takahashi S, Miyairi S, Suzuki I, Murata N (2005) Systematic analysis of the relation of electron transport and ATP synthesis to the photodamage and repair of photosystem II in *Synechocystis*. Plant Physiol 137:263–273

Asada K (2006) Production and scavenging of reactive oxygen species in chloroplasts and their functions. Plant Physiol 141:391–396

Babele PK, Singh G, Kumar A, Tyagi MB (2015) Induction and differential expression of certain novel proteins in *Anabaena* L31 under UV-B radiation stress. Front Microbiol 6:133

Barrera G, Pizzimenti S, Daga M, Dianzani C, Arcaro A, Cetrangolo GP, Giordano G, Cucci MA, Graf M, Gentile F (2018) Lipid peroxidation-derived aldehydes, 4-hydroxynonenal and malondialdehyde in aging-related disorders. Antioxidants 7:102

Berland B, Le-Compion T, Campos-Baeta-Neves MH (1989) Interaction of salinity and temperature on morphology, growth and cell composition of a halotolerant cyanobacterium (*Aphanothece* sp.). Bot Marina 32:317–330

Bhadauriya P, Gupta R, Singh S, Singh BP (2007) Physiological and biochemical alterations in a diazotrophic cyanobacterium *Anabaena cylindrica* under NaCl stress. Curr Microbiol 55:334–338

Capone DG, Zehr JP, Paerl HW, Bergman B, Carpenter EJ (1997) *Trichodesmium*, a globally significant marine cyanobacterium. Science 276:1221–1229

Carbonera D, Gerotto C, Posocco B, Giacometti GM, Morosinotto T (2012) NPQ activation reduces chlorophyll triplet state formation in the moss *Physcomitrella patens*. Biophys Acta 1817:1608–1615

Chatterjee A, Rajarshi K, Ghosh H, Singh MK, Roy OP, Ray S (2020) Molecular chaperones in protein folding and stress management in cyanobacteria. In: Advances in cyanobacterial biology. Academic, San Diego, pp 119–128
Chaurasia AK, Apte SK (2009) Overexpression of the groESL operon enhances the heat and salinity stress tolerance of the nitrogen-fixing cyanobacterium *Anabaena* sp. strain PCC7120. Appl Environ Microbiol 75:6008–6012
Chittora D, Meena M, Barupal T, Swapnil P, Sharma K (2020) Cyanobacteria as a source of biofertilizers for sustainable agriculture. Biochem Biophys Rep 22:100737
Choudhary M, Jetley UK, Khan MA, Zutshi S, Fatma T (2007) Effect of heavy metal stress on proline, malondialdehyde and superoxide dismutase activity in the cyanobacterium *Spirulina platensis*. Ecotoxicol Environ Saf 66:204–209
Dat J, Vandenabeele S, Vranova E, Van Montagu M, Inze D, Van Breusegem F (2000) Dual action of the active oxygen species during plant stress responses. Cell Mol Life Sci 57:779–795
De Philippis R, Colica G, Micheletti E (2011) Exopolysaccharide-producing cyanobacteria in heavy metal removal from water: molecular basis and practical applicability of the biosorption process. Appl Microbiol Biotechnol 92:697–708
Deep PR, Bhattacharyya S, Nayak B (2016) Effect on biochemical parameters of cyanobacterium *Anabaena* sp. under lead stress. Int J Curr Adv Res 4:2114–2129
Egorov SY, Kamalov VF, Koroteev NI, Krasnovsky AA, Toleutaev BN, Zinukov SV (1989) Rise and decay kinetics of photosensitized singlet oxygen luminescence in water. Measurements with nanosecond time-correlated single photon counting technique. Chem Phys Lett 163:421–424
Forchhammer K, Selim KA (2020) Carbon/nitrogen homeostasis control in cyanobacteria. FEMS Microbiol Rev 44:33–53
Gill SS, Tuteja N (2010) Reactive oxygen species and antioxidant machinery in abiotic stress tolerance in crop plants. Plant Physiol Biochem 48:909–930
Gorman AA, Rodgers MA (1992) Current perspectives of singlet oxygen detection in biological environments. J Photochem Photobiol B 14:159–176
Gracanin M, Hawkins CL, Pattison DI, Davies MJ (2009) Singlet-oxygen-mediated amino acid and protein oxidation: formation of tryptophan peroxides and decomposition products. Free Radic Biol Med 47:92–102
Gupta K, Baruah PP (2020) Toxicological impact of deltamethrin on growth and nitrogen content of a rice field cyanobacterium *Calothrix* sp. (GUEco 1002). J Gen Appl Microbiol 66:207–214
Gutnick DL, Bach H (2000) Engineering bacterial biopolymers for the biosorption of heavy metals; new products and novel formulation. Appl Microbiol Biotechnol 54:450–460
Hagemann M (2011) Molecular biology of cyanobacterial salt acclimation. FEMS Microbiol Rev 35:87–123
Halliwell B (1977) Generation of hydrogen peroxide, superoxide and hydroxyl radicals during the oxidation of dihydroxy fumaric acid by peroxidase. Biochem J 163:441–448
Halliwell B (2006) Reactive species and antioxidants. Redox biology is a fundamental theme of aerobic life. Plant Physiol 141:312–322
Haslbeck M, Franzmann T, Weinfurtner D, Buchner J (2005) Some like it hot: the structure and function of small heat-shock proteins. Nat Struct Mol Biol 12:842–846
Heindl JE, Wang Y, Heckel BC, Mohari B, Feirer N, Fuqua C (2014) Mechanisms and regulation of surface interactions and biofilm formation in *Agrobacterium*. Front Plant Sci 5:176
Herrero A, Muro-Pastor AM, Flores E (2001) Nitrogen control in cyanobacteria. J Bacteriol 183 (2):411–425
Hideg E, Kos PB, Vass I (2007) Photosystem II damage induced by chemically generated singlet oxygen in tobacco leaves. Physiol Plant 131:33–40
Hintze KJ, Theil EC (2006) Cellular regulation and molecular interactions of the ferritins cell. Mol Life Sci 63:591–600
Hongsthong A, Sirijuntarut M, Prommeenate P, Lertladaluck K, Porkaew K, Cheevadhanarak S (2008) Proteome analysis at the subcellular level of the cyanobacterium *Spirulina platensis* in response to low-temperature stress conditions. FEMS Microbiol Lett 288:92–101

Hongsthong A, Sirijuntarut M, Yutthanasirikul R, Senachak J, Kurdrid P, Cheevadhanarak S (2009) Subcellular proteomic characterization of the high-temperature stress response of the cyanobacterium *Spirulina platensis*. Proteome Sci 7:1–19

Hu ZQ, Liu YD, Li DH (2004) Physiological and biochemical analyses of microcystin-RR toxicity to the cyanobacterium *Synechococcus elongatus*. Environ Toxicol 19(6):571–577

Huang F, Fulda S, Hagemann M, Norling B (2006) Proteomic screening of salt-stress induced changes in plasma membranes of *Synechocystis* sp. strain PCC 6803. Proteomics 6:910–920

Huang Z, Li L, Huang G, Yan Q, Shi B, Xu X (2009) Growth-inhibitory and metal-binding proteins in *Chlorella vulgaris* exposed to cadmium or zinc. Aquat Toxicol 91:54–61

Imlay JA (2003) Pathways of oxidative damage. Annu Rev Microbiol 57:395–418

Jägerbrand AK, Kudo G (2016) Short-term responses in maximum quantum yield of PSII (Fv/Fm) to ex-situ temperature treatment of populations of bryophytes originating from different sites in Hokkaido, northern Japan. Plants 5:22

Jensen RL, Arnbjerg J, Ogilby PR (2012) Reaction of singlet oxygen with tryptophan in proteins: a pronounced effect of the local environment on the reaction rate. J Am Chem Soc 134:9820–9826

Jittawuttipoka T, Planchon M, Spalla O, Benzerara K, Guyot F, Cassier-Chauvat C (2013) Multidisciplinary evidences that *Synechocystis* PCC6803 exopolysaccharides operate in cell sedimentation and protection against salt and metal stresses. PLoS One 8:e55564

Kanski J, Aksenova M, Stoyanova A, Butterfield DA (2002) Ferulic acid antioxidant protection against hydroxyl and peroxyl radical oxidation in synaptosomal and neuronal cell culture systems in vitro: structure–activity studies. J Nutr Biochem 13:273–281

Kumar J, Singh VP, Prasad SM (2015) NaCl-induced physiological and biochemical changes in two cyanobacteria *Nostoc muscorum* and *Phormidium foveolarum* acclimatized to different photosynthetically active radiation. J Photochem Photobiol B 15:221–232

Kumar J, Singh VP, Prasad SM (2018) An investigation on involvement of the ascorbate-glutathione cycle in modulating NaCl toxicity in two cyanobacteria photoacclimatized to different photosynthetic active radiation. Algal Res 32:7–78

Latifi A, Ruiz M, Zhang CC (2009) Oxidative stress in cyanobacteria. FEMS Microbiol Rev 33:258–278

Leaf MC, Gay Jessica SA, Newbould Matthew J, Hewitt Owen R, Rogers Steven L (2020) Calcareous algae and cyanobacteria. Geol Today 36:2

Lea-Smith DJ, Bombelli P, Vasudevan R, Howe CJ (2015) Photosynthetic, respiratory and extracellular electron transport pathways in cyanobacteria. BBA-Bioenergetics 1857(3):247–255

Lin C, Wu J (2014) Tolerance of soil algae and cyanobacteria to drought stress. J Phycol 50:131–139

Liu LN (2016) Distribution and dynamics of electron transport complexes in cyanobacterial thylakoid membranes. BBA-Bioenergetics 1857:256–265

McCord JM, Crapo JD, Fridovich I, Michelson AM, McCord JM, Fridovich I (1977) Superoxide and superoxide dismutases. Academic, London, pp 11–17

Mishra NP, Francke C, van Gorkom HJ, Ghanotakis DF (1994) Destructive role of singlet oxygen during aerobic illumination of the photosystem II core complex. BBA-Bioenergetics 1186:81–90

Noyma NP, Silva TP, Chiarini-Garcia H, Amado AM, Roland F, Melo RCN (2015) Potential effects of UV radiation on photosynthetic structures of the bloom-forming cyanobacterium *Cylindrospermopsis raciborskii* CYRF-01. Front Microbiol 6:1202

Nweze N (2009) Ecological implications and roles of Cyanobacteria (Cyanophyta) in food security – a review. Plant Prod Res J 13:8–14

Oesterhelt C, Vogelbein S, Shrestha RP, Stanke M, Weber AP (2008) The genome of the thermoacidophilic red microalga *Galdieria sulphuraria* encodes a small family of secreted class III peroxidases that might be involved in cell wall modification. Planta 227:353–362

Panda B, Basu B, Rajaram H, Apte SK (2015) Comparative proteomics of oxidative stress response in three cyanobacterial strains native to Indian paddy fields. J Proteomics 127:152–160

Patel A, Tiwari S, Prasad SM (2018) Toxicity assessment of arsenate and arsenite on growth, chlorophyll a fluorescence and antioxidant machinery in *Nostoc muscorum*. Ecotoxicol Environ Saf 157:369–379

Patel A, Tiwari S, Prasad SM (2019) Auxin (IAA) up regulates growth, photosynthetic pigment and antioxidant machinery in paddy field cyanobacteria *N. muscorum*. Think India J 22 (17):3808–3822

Patel A, Tiwari S, Prasad SM (2020) Effect of time interval on arsenic toxicity to paddy field cyanobacteria as evident by nitrogen metabolism, biochemical constituent, and exopolysaccharide content. Biol Trace Elem Res 20(1):206

Pereira S, Zille A, Micheletti E, Moradas-Ferreira P, De Philippis R, Tamagnini P (2009) Complexity of cyanobacterial exopolysaccharides: composition, structures, inducing factors and putative genes involved in their biosynthesis and assembly. FEMS Microbiol Rev 33:917–941

Planchon M, Jittawuttipoka T, Cassier-Chauvat C, Guyot F, Gelabert A, Benedetti MF, Chauvat F, Spalla O (2013) Exopolysaccharides protect *Synechocystis* against the deleterious effects of titanium dioxide nanoparticles in natural and artificial waters. J Colloid Interface Sci 405:35–43

Pogson BJ, Rissler HM, Frank HA, Wyrdzynski T, Satoh K (2005) The light-driven water–plastoquinone oxidoreductase. Springer, Dordrecht, The Netherlands, pp 515–537

Prasad SM, Singh A (2011) Metabolic responses of *Azollapinnata* to cadmium stress: photosynthesis, antioxidative system and phytoremediation. Chem Ecol 27:543–555

Rajaram H, Apte SK (2008) Nitrogen status and heat-stress-dependent differential expression of the cpn60 chaperonin gene influences thermotolerance in the cyanobacterium *Anabaena*. Microbiology 154:317–325

Rajaram H, Chaurasia AK, Apte SK (2014) Cyanobacterial heat-shock response: role and regulation of molecular chaperones. Microbiology 160:647–658

Ramel F, Birtic S, Ginies C, Soubigou-Taconnat L, Triantaphylides C, Havaux M (2012) Carotenoid oxidation products are stress signals that mediate gene responses to singlet oxygen in plants. Proc Natl Acad Sci U S A 109:5535–5540

Rastogi RP, Sonani RR, Madamwar D (2015) Effects of PAR and UV radiation on the structural and functional integrity of phycocyanin, phycoerythrin and allophycocyanin isolated from the marine cyanobacterium *Lyngbya* sp. A09DM. Photochem Photobiol 91:837–844

Rezayian M, Niknam V, Ebrahimzadeh H (2019) Oxidative damage and antioxidative system in algae. Toxicol Rep 6:1309–1313

Richter K, Haslbeck M, Buchner J (2010) The heat shock response: life on the verge of death. Mol Cell 40:253

Rueter JG, Petersen RR (1987) Micronutrient effects on cyanobacterial growth and physiology. N Z J Mar Freshwater Res 21(3):435–445

Schirmer K, Fischer BB, Madureira DJ, Pillai S (2010) Transcriptomics in ecotoxicology. Anal Bioanal Chem 397:917–923

Sgherri C, Cosi E, Navari-Izzo F (2003) Phenols and antioxidative status of *Raphanus sativus* grown in copper excess. Physiol Plant 118:21–28

Shah SH, Houborg R, McCabe Matthew F (2017) Response of chlorophyll, carotenoid and SPAD-502 measurement to salinity and nutrient stress in wheat (*Triticum aestivum* L.). Agronomy 7 (3):61

Sheeba, Ruhil K, Prasad SM (2020) *Nostoc muscorum* and *Phormidium foveolarum* differentially respond to butachlor and UV-B stress. Physiol Mol Biol Plants 26:841–856

Sheikh TA, Baba ZA, Sofi P (2006) Effect of NaCl on growth and physiological traits of *Anabaena cylindrica* L. Pak J Biol Sci 9(13):2528–2530

Shrivastava AK, Chatterjee A, Yadav S, Singh PK, Singh S, Rai L (2015) UV-B stress induced metabolic rearrangements explored with comparative proteomics in three *Anabaena* species. J Proteomics 127:122–133

Sies H, Menck CF (1992) Singlet oxygen induced DNA damage. Mutat Res 275:367–375

Singh S (2014) A review on possible elicitor molecules of cyanobacteria: their role in improving plant growth and providing tolerance against biotic or abiotic stress. J Appl Microbiol 117:1221–1244

Singh SC, Sinha RP, Hader DP (2002) Role of lipids and fatty acids in stress tolerance in cyanobacteria. Acta Protozool 41:297–308

Singh AK, Summerfield TC, Li H, Sherman LA (2006) The heat shock response in the cyanobacterium *Synechocystis* sp. strain PCC 6803 and regulation of gene expression by *HrcA* and *SigB*. Arch Microbiol 186:273

Sinha RP, Singh N, Kumar A, Kumar HD, Häder M, Häder DP (1996) Effects of UV irradiation on certain physiological and biochemical processes in cyanobacteria. J Photochem Photobiol B 32 (1–2):107–113

Spät P, Maèek B, Forchhammer K (2015) Phosphoproteome of the cyanobacterium *Synechocystis* sp. PCC 6803 and its dynamics during nitrogen starvation. Front Microbiol 6:248

Srivastava PK, Singh VP, Prasad SM (2014) Low and high doses of UV-B differentially modulate chlorpyrifos-induced alterations in nitrogen metabolism of cyanobacteria. Ecotoxicol Environ Saf 107:342–350

Stebegg R, Schmetterer G, Rompel A (2019) Transport of organic substances through the cytoplasmic membrane of cyanobacteria. Phytochemistry 157:206–218

Strasser RJ, Tsimilli-Michael M, Srivastava A (2004) Analysis of the chlorophyll a fluorescence transient. In: Papageorgiou GC, Govindjee (eds) Chlorophyll fluorescence: a signature of photosynthesis, Current climate change reports, vol 19. Springer, Netherlands, pp 321–362

Sudhir PR, Pogoryelov D, Kovacs L, Garab G, Murthy SDS (2005) The effect of salt stress on photosynthetic electron transport and thylakoid membrane proteins in the cyanbacterium *Spirulina platensis*. BMB Rep 38:481–485

Surosz W, Palinska K (2004) Effects of heavy-metal stress on cyanobacterium *Anabaena flos-aquae*. Arch Environ Contam Toxicol 48:40–48

Szabados L, Savoure A (2009) Proline: a multifunctional amino acid. Trends Plant Sci 15:89–97

Tchounwou PB, Yedjou CG, Patlolla AK, Sutton DJ (2012) Heavy metal toxicity and the environment. EXS 101:133–164

Tiwari S, Prasad SM (2019) Regulation of insecticide toxicity by kinetin in two paddy field cyanobacteria: physiological and biochemical assessment. Environ Pollut 259:113806

Tiwari S, Patel A, Prasad SM (2018) Kinetin alleviates chromium toxicity on growth and PS II photochemistry in *Nostoc muscorum* by regulating antioxidant system. Ecotoxicol Environ Saf 161:296–304

Tiwari S, Parihar P, Patel A, Singh R, Prasad SM (2019) Metal in cyanobacteria: physiological and molecular regulation. In: Cyanobacteria. Elsevier, Amsterdam, pp 261–276

Tiwari S, Patel A, Prasad SM (2020a) Phytohormone up-regulates the biochemical constituent, exopolysaccharide and nitrogen metabolism in paddy-field cyanobacteria exposed to chromium stress. BMC Microbiol 20:206

Tiwari S, Verma N, Prasad SM, Singh V (2020b) Cytokinin alleviates cypermethrin toxicity in Nostoc muscorum by involving nitric oxide: regulation of exopolysaccharides secretion, PS II photochemistry and reactive oxygen species homeostasis. Chemosphere 259:127356

Triantaphylides C, Havaux M (2009) Singlet oxygen in plants: production, detoxification and signalling trends. Plant Sci 14:219–228

Vranova E, Inze D, Van Breusegem F (2002) Signal transduction during oxidative stress. J Exp Bot 53:1227–1236

Watanabe M, Ikeuchi M (2013) Phycobilisome: architecture of a light-harvesting supercomplex. Photosynth Res 116:265–276

Wilkinson F, Helman W, Ross AB (1995) Rate constants for the decay and reactions of the lowest electronically excited singlet state of molecular oxygen in solution. An expanded and revised compilation. J Phys Chem Ref Data 24:663–678

Xiong Q, Chen Z, Ge F (2015a) Proteomic analysis of post translational modifications in cyanobacteria. J Proteomics 134:57–64

Xiong Q, Feng J, Li ST, Zhang GY, Qiao ZX, Chen Z (2015b) Integrated transcriptomic and proteomic analysis of the global response of *Synechococcus* sp. PCC 7002 to high light stress. Mol Cell Proteomics 14:1038–1053

Yin XX, Chen J, Qin J, Sun GX, Rosen BP, Zhu YG (2011a) Biotransformation and volatilization of arsenic by three photosynthetic cyanobacteria. Plant Physiol 156:1631–1638

Yin YR, Hu YY, Xiong F (2011b) Sorption of Cu (II) and Cd (II) by extracellular polymeric substances (EPS) from *Aspergillus fumigatus*. Int Biodeterior Biodegrad 65:1012–1018

Zagoskina N, Olenichenko N, Klimov S, Astakhova N, Zhivukhina E, Trunova T (2005) The effects of cold acclimation of winter wheat plants on changes in CO_2-exchange and phenolic compound formation. Russ J Plant Physiol 52:320–325

Photosynthesis Under Abiotic Stress

5

Kinga Kłodawska

Abstract

Cyanobacteria are prokaryotic organisms dependent on performance of oxygenic photosynthesis. They inhabit a wide diversity of environments, in which they are sometimes exposed to extreme conditions. Such abiotic stresses may affect photosynthesis reactions of both light-dependent and light-independent phases. In this chapter, the impact of extreme temperatures, high-intensity illumination, and nutrient starvation on the efficiency of photosynthesis will be discussed. Focus will be put on protective mechanisms of light-dependent reactions.

Keywords

Cold · Heat · Nutrient starvation · High light · Abiotic stress · Protective mechanisms

5.1 Introduction

During millions of years that cyanobacteria have been present on the planet, they must have encounter a vast variety of environmental conditions. They have evolved multiple ways to cope with unfavorable circumstances while maintaining photosynthesis. In this chapter, responses of photosynthetic process to different abiotic stresses were described.

K. Kłodawska (✉)
Faculty of Biochemistry, Biophysics and Biotechnology, Jagiellonian University in Krakow, Krakow, Poland
e-mail: kinga.klodawska@uj.edu.pl

R. P. Rastogi (ed.), *Ecophysiology and Biochemistry of Cyanobacteria*,
https://doi.org/10.1007/978-981-16-4873-1_5

5.2 Light

Photosynthetic organisms have to play very dangerous game with light. Light enables photosynthesis, but when present in excess it can cause severe damage to photosynthetic apparatus. When there is not enough light, carbon fixation rates may not be able to sustain cell's metabolic needs. That is why all phototrophs have evolved complex protective mechanisms that allow them to survive in nonoptimal conditions.

5.2.1 High Light Intensity

Decrease in photosynthetic activity under high irradiation is primary due to singlet oxygen formation in photosystem II (PSII) (Krieger-Liszkay et al. 2008), resulting in damage of the D1 protein of the reaction center. The result is a state in which the rate of photodamage exceeds the rate of PSII repair—photoinhibition (Adir et al. 2003). Severity of this state is proportional to the difference between the rate of primary charge separation reactions in PSII and photosystem I quantum efficiency or the rate of carbon fixation reactions (Tyystjärvi and Aro 1996; Vass et al. 2007). Studies have shown that the irradiation with high light is responsible for a direct damage to PSII in oxygen-evolving complex and the photosystems (Ohnishi et al. 2005; Nishiyama et al. 2006), while other abiotic stress conditions influence primarily the efficiency of repair mechanisms (Gombos et al. 1994; Allakhverdiev et al. 2002; Nishiyama et al. 2004; Aminaka et al. 2006). Such synergic damage to photosynthetic machinery results in a greater negative effect than the irradiation itself (Allakhverdiev and Murata 2004; Guyet et al. 2020).

Sequencing of entire genome of *Synechocystis* sp. PCC6803 (Kaneko and Tabata 1997) enabled further high-throughput analyses of the response to a change in ambient conditions on the level of gene expression (Hihara et al. 2001; Huang et al. 2002; Tu et al. 2004; Singh et al. 2008). Observed variations in certain transcripts levels are in accordance with experimental data on physiological acclimation to high light (Muramatsu and Hihara 2012). The D1 protein turnover is accelerated in high photon flux density (Ohad et al. 1984; Wünschmann and Brand 1992). This is mediated by an FtsH protease (Lindahl et al. 2000; Silva et al. 2003; Nixon et al. 2005). These steps are reflected in an increase in *psbA* and *ftsH* transcript levels (Huang et al. 2002). Cyanobacteria limit transfer of excitation energy from the light-harvesting complexes to the photosynthetic reaction centers by reduction in light-harvesting capability of the cell on the level of the size of cell's phycobilisome (PBS) pool, as well as by energetic decoupling of PBSs from photosystems. This is achieved partly by downregulation of genes encoding some phycobilisome proteins (*cpc* and *apc*), thus alternating PBS protein composition (Nomsawai et al. 1999; Tamary et al. 2012). Excitation transfer from PBSs to photosystems is additionally reduced by nonphotochemical quenching (NPQ) dependent on an orange carotenoid protein (OCP) system (Kirilovsky and Kerfeld 2016; Sluchanko et al. 2018). *Ocp* transcript was found to be upregulated in light-stressed *Synechocystis* sp. cells

(Singh et al. 2008). Change in PSI/PSII stoichiometry is another hallmark of excess irradiation (Sonoike et al. 2001; Dietzel et al. 2008). On gene expression level, it is evidenced by downregulation in PSI proteins encoding genes and genes encoding enzymes involved in pigment biosynthesis (*hem* and *chl*) (Hihara et al. 2001; Singh et al. 2008). It has been shown that the limited availability of chlorophyll is the most probable cause for the decreased level of PSI complexes in high light-treated *Synechocystis* sp. (Muramatsu et al. 2009). Pressure exerted by high photon flux density on photosynthetic electron transport chain is reflected in an increase in the level of transcripts of genes encoding protective proteins, such as flavodiiron proteins (*flv*) and chlorophyll binding high light-inducible proteins (*hli*). More general stress response set of peroxidases (*aphC* and *gpx2*) and heat shock proteins (*groESL*, *clpB*, *dnaK2*, and *htpG*) was also found to be upregulated (Mary et al. 2004).

Parallel actions on all of these levels of protection enable cyanobacteria to cope with exposition to high light in their natural habitats.

5.2.2 Low Light Intensity

Low irradiance does not pose a threat to photosynthetic apparatus. It might, however, limit carbon assimilation to nonsustainable levels in strictly photosynthetic, glucose-intolerant strains. Glucose-tolerant cyanobacterial strains when supplied with glucose are able to grow heterotrophically in continuous very weak light (0.5 μmol photons m^{-2} s^{-1}) or in darkness in a metabolic mode known as light-activated heterotrophic growth (LAHG) (Anderson and McIntosh 1991). LAHG requires at least 5-min daily illumination with blue light of moderate intensity to activate cell division processes. NADH-2 type enzymes NdbA and NdbC were shown to be important regulators of carbon and iron metabolism under LAHG conditions (Huokko et al. 2017, 2019). Another protein required for LAHG was found to be a tetratricopeptide repeat (TRP) family protein encoded by *sll0886* gene in *Synechocystis* sp. (Kong et al. 2003). Detailed characterization of *Synechocystis* sp. glucose-tolerant strain grown under LAHG conditions revealed that of photosynthetically important proteins only PSII-related proteins were significantly downregulated. There was no active PSII complex in thylakoid membranes of these cyanobacteria. On the other hand, PSI and phycobilisomes were present in similar amounts to that found in autotrophically grown cells. Moreover, PSI complexes were photosynthetically active, while PBSs were shown to be largely energetically disconnected (Barthel et al. 2013; Plohnke et al. 2015). Thylakoid structure was also altered in these cells since PSII complexes together with PBSs and lipids of thylakoid membrane are responsible for normal thylakoid structure (Collins et al. 2012; Barthel et al. 2013; Plohnke et al. 2015).

Autotrophic growth under continuous weak light requires occurrence of certain adaptational changes in thylakoid membranes in order to maximize the efficiency of light harvesting. This process is called state transitions (state1–state2 transitions) albeit it was found to have dissimilar functions to the state transitions known from higher plants. This mechanism is based on spatial relocation of PBSs along cytosolic surface of thylakoid membrane (Mullineaux et al. 1997; Sarcina et al. 2001) that is accompanied by the change in excitation transfer from one type of photosystem to another (Van Thor et al. 1998). State 1 occurs when majority of PBS complexes transfer excitation to PSII, whereas State 2 is defined by excitation transfer from majority of PBS complexes to PSI. These changes depend on the quality of light. If PSI complexes enter overexcitation photosynthetic machinery counteracts by transitioning to State 1, to prevent photoinhibition of PSI. When light conditions change and PSII becomes in danger of photoinhibition, photosynthetic machinery transitions to State 2, to mitigate pressure on PSII. *RpaC* gene, designated *sll1926* in *Synechocystis* sp. genome, was identified to be necessary for state transitions. Deletion mutant showed specific phenotype with no alteration in photosynthetic complex assembly and function, and functional photosynthetic and oxidative electron transport but no ability to perform state transitions. It did not show impairment in growth rate at high and normal light intensities, but did divide slower than the wild type when grown under very weak white or yellow light (Emlyn-Jones et al. 1999). Thus, it was concluded that state transition in cyanobacteria is physiologically significant only at low light intensities (Mullineaux and Emlyn-Jones 2005).

Light availability might be crucial growth-limiting factor for cyanobacterial species coexisting in shallow turbid water of natural lakes (Havens et al. 1998).

5.3 Temperature

Photosynthetic activity is temperature-dependent. It increases with the increase in an ambient temperature until it reaches the maximum, and declines rapidly at higher temperatures. Optimal temperatures and survival temperature ranges differ among photosynthetic species. Cyanobacteria colonized and adapted to a wide variety of environments, both cold and hot.

5.3.1 Cold

Temperature has a strong effect on the saturation state of fatty acid components of the membrane lipids, as well as membrane lipid class ratio. It has been shown for multiple bacterial and plant species (Murata 1989; Somerville 1995; Badea and Basu 2009). There are four main lipid species in cyanobacterial membrane: monogalactosyldiacylglycerol (MGDG), digalactosyldiacylglycerol (DGDG), sulfoquinovosyldiacylglycerol (SQDG), and phosphatidylglycerol (PG) (Murata and Nishida 1987). In addition, monoglucosyldiacylglycerol (MGlcDG) is present in cyanobacterial membranes in small amounts as it is an intermediate in MGDG and further DGDG biosynthetic pathway (Sato and Murata 1982). These lipid classes differ in molecule geometry, what is reflected in the physical properties of membranes formed by them (Shipley et al. 1973; Tilcock 1986).

Since photosynthetic complexes are embedded in thylakoid membrane, the membrane properties, such as fluidity and thickness, may affect photosynthetic performance. Interactions between lipid and protein components of biological membranes have been extensively studied, as numerous enzymes are membrane-spanning proteins (Andersen and Koeppe 2007; Phillips et al. 2009). Detailed studies of this phenomenon in cyanobacteria were carried since late 1970s (Fork et al. 1979). Using *Anabaena variabilis,* Sato and Murata (1980) have shown that temperature shift from 38 to 22 °C temporarily slows down lipid synthesis and results in fatty acid desaturation. Opposite temperature change was followed by lipid synthesis intensification and fatty acid saturation. MGDG was assigned to be the class of lipids primarily responsible for the acclimation. Localization of photosynthetic complexes in thylakoid membranes that contain 90% of cyanobacterial cell lipids prompted researchers to examine both lipid-phase transitions and photosynthetic activity simultaneously. Suppression of photosynthetic process as a result of exposure to chilling temperatures was found to be either reversible or irreversible, dependent on the degree of membrane disintegration due to lipid-phase separation (Murata et al. 1984). Identification of genes encoding fatty acid desaturases (Wada and Murata 1989) allowed for systematic studies of knockout mutants in *Synechocystis* sp. PCC6803. Mutant of *Anacystis nidulans* bearing *Synechocystis* sp. PCC6803 desaturase gene (*desA*) was found to cope better with chilling stress than the nontransformed wild-type strain (Wada et al. 1990). The anti-photoinhibitory effect of polyunsaturated fatty acids was proved in the study that involved wild-type *Synechocystis* sp. PCC6803 strain and its derivatives defective in either desaturase acting at Δ12 position of a fatty acid (DesA), or two desaturases, DesA and desaturase acting at Δ6 position (Fad6) (Wada et al. 1992). The lack of polyunsaturated fatty acids in membrane lipids resulted in increased sensitivity to photoinhibition of photosynthesis, even at relatively moderate-light intensities. The effect was particularly pronounced at low temperatures (Gombos et al. 1992) and was attributed to the decreased rate of the D1 protein turnover process (Gombos et al. 1994). Studies on whole cells, thylakoid membranes, and isolated PSII complexes from *Synechocystis* sp. PCC6803 imply that the changes in the desaturation levels in lipid component of thylakoid membrane are responsible for thermosensitivity of photosystem II, not the changes in protein complexes themselves (Aminaka et al. 2006).

Low temperature may also affect photosynthetic pigments accumulation in cyanobacteria. Chlorophyll content of thylakoid membranes decreased when cells of *Cylindrospermopsis raciborskii* were cultivated at low temperature, and at a given temperature, it was lower at higher light intensities (Várkonyi et al. 2002). In *Synechocystis* sp. PCC6803, chlorophyll-to-protein ratio was lower in cells cultivated at suboptimal temperature (15 °C), while total carotenoids were found to be accumulated in such conditions (Kłodawska et al. 2015). Conversely, *Anabaena* sp. PCC7120 cells exhibit decrease in both chlorophyll and carotenoid accumulation at 15 °C, as compared to 23 and 30 °C (Kłodawska et al. 2019).

5.3.2 Heat

Cyanobacteria and other oxygenic phototrophs inhabit environments of temperatures not exceeding 73 °C. Since some nonphotosynthetic organisms can live at higher ambient temperatures, it was concluded that the condition limiting expansion of phototrophs into hotter sites is the stability of photosynthetic apparatus (Brock 1967). Studies have shown that some thermophilic strains can live in environments where they are constantly under high-temperature stress, as long as it allows them to avoid competitor species (Miller et al. 1998). Such extreme environments may be populated by communities of phototrophic and heterotrophic species dependent on each other in a phenomenon referred to as "community metabolism" (Anderson et al. 1987; Nold and Ward 1996; Steunou et al. 2006).

Detailed study of thermophilic *Synechococcus* sp. from a hot spring (Miller et al. 1998) has shown that this species, although living at temperature about 70 °C, have its photosynthetic activity peaking at 63–67 °C. It was also shown that in cells incubated at suboptimal (55 °C) and supraoptimal (70 °C) temperatures photosynthesis saturated at lower light intensities, than at optimal temperature. Moreover, cells incubated at both nonoptimal temperatures were more prone to inhibition of photosynthesis by UV radiation than cells incubated at optimal temperature. Relation between stress-inducing temperatures and high irradiation was also reported for freshwater *Anabaena flos-aquae* (Ibelings 1996) and other bloom-forming cyanobacteria (Robarts and Zohary 1987). Marine cyanobacterium *Arthrospira* sp. (*Spirulina* sp.) subjected to heat stress exhibited over-reduction of photosystem II acceptor side, damage of its donor side, and decrease in energetic connectivity between photosystem II protein subunits. At the same time, photosystem I activity and oxygen evolution were observed to be enhanced (Zhang and Liu 2016). Additionally, it was shown that increasing unsaturation of fatty acids in thylakoid lipids, together with the accumulation of xanthophylls, can stabilize PSI trimers in *Synechocystis* sp. PCC6803 (Zakar et al. 2017).

Pigment composition of cyanobacterial cells may also undergo changes due to high-temperature treatment. Both *Synechocystis* sp. PCC6803 and *Anabaena* sp. PCC7120 cells cultivated at 37 °C exhibited chlorophyll-to-protein ratio lower than at optimal temperature (30 °C), but higher than at suboptimal temperature (15 °C). Total carotenoid pigments were on roughly the same level as at optimal temperature, but significantly lower than in cold-stressed cells in *Synechocystis* sp. (Kłodawska et al. 2015). In *Anabaena* sp. high-temperature cultures, carotenoids were accumulated at a level twofold lower than at 15 °C and 66-fold lower than at 23 °C (Kłodawska et al. 2019).

Dryland cyanobacteria often encounter combined high-temperature and high irradiation stress conditions, and they might have evolved different and unexpected approach to such environment. *Microcoleus vaginatus,* a dominant species of biological soil crusts, was reported to be protected against high-temperature irreversible photosynthesis inhibition by the process of desiccation. While hydrated bacteria exposed to stress suffered from permanent decline of photosynthesis, desiccated cells were able to restore photosynthesis upon rehydration (Lan et al.

2014). Cyanobacteria in crusts appear to have lower sensitivity to photoinhibition than aquatic species (Harel et al. 2004).

5.4 Nutrient Starvation

Carbon, nitrogen, sulfur, and phosphorus are necessary macronutrients for all microorganisms. Additionally, for cyanobacteria iron is an important element, due to its involvement in photosynthetic electron transport. In case of unavailability of a certain element organisms response in general and specific ways. General response is similar for different stresses in variety of microorganisms (Hecker and Völker 1998), while specific response leads to compensation for a particular limitation. This subject was extensively reviewed by Schwarz and Forchhammer (2005).

5.4.1 Carbon and Nitrogen

Carbon and nitrogen metabolism are co-regulated, as both of these elements are needed for protein synthesis. Their ratio is crucial for cell homeostasis (Zhang et al. 2018; Forchhammer and Selim 2020; Veaudor et al. 2020).

5.4.1.1 Inorganic Carbon

For all photosynthetic cyanobacteria, inorganic carbon (C_i) availability is a growth-limiting factor when they operate in a photoautotrophic mode (Karlsen et al. 2018). CO_2 is the primary carbon source for cyanobacteria. To adapt to possible low C_i concentrations, cyanobacteria possess carbon concentrating mechanism (CCM) that involves specialized cell compartments, carboxysomes (Price et al. 1998; Liran et al. 2018). This mechanism is constitutively active on a certain level but becomes highly upregulated in C_i limiting conditions. In cyanobacteria, C_i uptake is a light-dependent process involving cyclic electron transport around PSI (Ogawa and Inoue 1983; Ogawa et al. 1985).

In *Synechocystis* sp., cultures arrest in anabolic reactions, 80% drop in photosynthetic activity (measured as oxygen evolution), and inhibition of cell division was observed within 6 h after initiation of CO_2 depletion. After 24 h, protein synthesis was decreased to 20% of control level, while total abundance of ribosomes did not change significantly (Karlsen et al. 2018).

5.4.1.2 Nitrogen

Some cyanobacterial strains are able to fix gaseous dinitrogen in specialized cells called heterocysts (Thiel 2004; Haselkorn 2007). This feature saves them from the danger of nitrogen unavailability. Most cyanobacteria, however, acquire nitrogen as ammonium, nitrate, and nitrite. Additionally, some strains are able to use urea, cyanate, and amino acids (Valladares et al. 2002; García-Fernández et al. 2004; Flores and Herrero 2005).

Since nitrogen is crucial for such basic metabolic processes as synthesis of proteins and nucleic acids, its depletion imposes rapid and severe effects on cyanobacterial cells. *Synechocystis* sp. cultures exhibited growth rate decrease after 12 h and growth cessation after 48 h of nitrogen depletion (Krasikov et al. 2012). Cyanobacteria are able to recycle intracellular nitrogen in the process that involves degradation of PBS complexes (Yamanaka and Glazer 1980; Duke et al. 1989). PBSs can constitute up to 50% of cyanobacterial cytosolic proteins; hence, they can be viewed as a large nitrogen storage (Görl et al. 1998; Von Wobeser et al. 2011). Additional advantage is the resulting decrease in cell's light-harvesting capability that protects against over-excitation and leads to limitation of the rate of photosynthesis (Sauer et al. 2001). Carbon assimilated under nitrogen depletion may be stored as poly-β-hydroxybutyrate (PHB) inclusions until stress is relieved (Allen 1984; Hai et al. 2001). *Synechococcus* PCC 7942 was shown to be able to regenerate pigmentation and reinitiate growth upon nitrogen repletion, even after prolonged starvation (Görl et al. 1998). This rapid recovery is possible partly because though during starvation cells inhibit most metabolic activities, PSI is kept active and can quickly commence energy generation (Krasikov et al. 2012).

5.4.2 Iron

Iron availability limits photosynthesis in significant part of both marine and freshwater habitats (reviewed by Gledhill and Buck 2012). Iron depletion strongly affects cyanobacterial phototrophic metabolism because iron atoms (in hems or iron-sulfur clusters) are crucial for redox reactions of electron transport chain. Some Fe-containing redox proteins can be substituted by non-Fe-dependent ones. For example, flavin-containing flavodoxin was shown to replace ferredoxin, and copper-containing plastocyanin was shown to replace cytochrome c_6. In several studies, the pool of redox proteins decreased during iron starvation, what was reflected in the decrease in electron transfer rate and oxygen evolution (Sandmann and Malkin 1983; Sandmann 1985). Prolonged iron limitation led to cell division retardation and chlorophyll degradation resulting in culture bleaching (Sandström et al. 2002; Fraser et al. 2013). However, despite bleaching, after 9 days in iron-depleted medium cells remained viable and have recovered when transferred to an iron-repleted conditions (Sandström et al. 2002). Chlorophyll *a*-binding IsiA (iron stress-inducible) protein was shown to be intensively expressed in iron-starved cyanobacteria (Burnap et al. 1993). This 36 kDa protein exhibits sequence homology to CP43 (PsbC) subunit of PSII. It is accumulated in form of rings surrounding PSI complexes (Bibby et al. 2001; Boekema et al. 2001). IsiA is supposed to have a broad range of functions including light harvesting and excitation transfer to PSI (Andrizhiyevskaya et al. 2002; Melkozernov et al. 2003), excitation quenching when energetically disconnected from PSI (Wilson et al. 2007; Chen et al. 2017), chlorophyll *a* storage (Sarcina and Mullineaux 2004), and protection against photo-oxidative stress (Havaux et al. 2005). Global gene expression in iron-deficient conditions was extensively studied in *Synechocystis* sp. PCC6803. It was shown that iron availability regulates expression of genes encoding proteins involved in iron uptake and storage (such as bacterioferritins), genes encoding subunits of the photosystems and photosynthetic electron transport chain, and other genes, for example, those encoding proteins involved in nitrogen metabolism (Singh et al. 2003; Shcolnick et al. 2009; Hernández-Prieto et al. 2012).

5.4.3 Phosphorus

Phosphorus limitation seems to have less severe effects on cyanobacteria than nitrogen limitation. Bloom-forming cyanobacterium *Microcystis aeruginosa* was observed to retain nearly control growth rate levels for 7 days of cultivation in phosphorus-deficient medium (Yue et al. 2015). Chlorophyll *a* and phycocyanin contents per cell were also similar to that in control cultures. Maximum quantum yield of PSII (F_v/F_m) was slightly lower throughout the experiment course in treated cells, relative to the control. These quite mild effects were attributed to the capability of cyanobacteria to store large amounts of excess phosphorus in form of polyphosphate granules that can be utilized when needed (Kulaev and Vagabov 1983; Gómez-García et al. 2003). Proteomic profiling of phosphorus-starved *M. aeruginosa* showed downregulation of protein synthesis-related and carbon fixation-related proteins, as well as ferredoxin–thioredoxin reductase and ferredoxin–NADP reductase, suggesting general reduction in metabolic rate (Yue et al. 2015).

Phosphorus starvation induced differential gene expression in two *Prochlorococcus* ecotypes: high-light-adapted MED4 and low-light-adapted MIT9313 (Martiny et al. 2006). Only MIT9313 ecotype exhibited general reduction in the metabolic rate. Both ecotypes upregulated some of their genes involved in phosphorus uptake and storage (*pstS*, *phoB*, *pstABCS*, *phoE*). Based on analysis of the occurrence of different sets of phosphorus acquisition genes in 11 *Prochlorococcus* ecotypes, it was hypothesized that genome variability is a consequence of phosphorus availability in areas from which the strains were isolated (Martiny et al. 2006).

Proteomic study of *Prochlorococcus* MED4 strain confirmed accumulation of uptake-related proteins in P-limited conditions (Fuszard et al. 2010). Additionally, regulation of photosynthesis-related proteins was addressed. Proteins involved in maintaining structural integrity of both PSI and PSII were accumulated (PsaD and Mn-stabilizing protein), while important components of electron transport chain were downregulated (PsaA, PsaF, CP43), suggesting simultaneous reduction in photosynthesis rate and stabilization of photosystems structure. Metabolic rate of MED4 cells was overall reduced, as evidenced by downregulation of enzymes involved in glycolysis, carbon fixation, amino acid, and protein biosynthesis (Fuszard et al. 2010). Similar but not identical results were obtained by full-genome microarray for *Synechococcus* sp. WH8102 (Tetu et al. 2009).

5.5 Conclusions

Photosynthesis is crucial for survival of cyanobacteria. That is why they must try to maintain it in all environmental conditions. Several abiotic stresses, such as extreme temperatures, high irradiation, and nutrient unavailability, impact photosynthetic process. Cyanobacteria respond to them addressing either particular stressor itself (e.g., by increasing myxoxanthophyll levels in high light) or by engaging secondary stress protective mechanism that fight against oxidative pressure resulting from uncoupling of otherwise tightly regulated photosynthetic electron transport chain. All of these signals are supposed to be perceived by two main routes: (1) cyanobacterial two-component signal transduction pathways and (2) a mechanism mediated through the redox poise of Q_B/PQH_2 (Ritter et al. 2020).

References

Adir N, Zer H, Shochat S, Ohad I (2003) Photoinhibition – a historical perspective. Photosynth Res 76(1–3):343

Allakhverdiev SI, Murata N (2004) Environmental stress inhibits the synthesis de novo of proteins involved in the photodamage–repair cycle of photosystem II in *Synechocystis* sp. PCC 6803. Biochim Biophys Acta Bioenergetics 1657(1):23–32

Allakhverdiev SI, Nishiyama Y, Miyairi S, Yamamoto H, Inagaki N, Kanesaki Y, Murata N (2002) Salt stress inhibits the repair of photodamaged photosystem II by suppressing the transcription and translation of *psbA* genes in *Synechocystis*. Plant Physiol 130(3):1443–1453

Allen MM (1984) Cyanobacterial cell inclusions. Annu Rev Microbiol 38(1):1–25

Aminaka R, Taira Y, Kashino Y, Koike H, Satoh K (2006) Acclimation to the growth temperature and thermosensitivity of photosystem II in a mesophilic cyanobacterium, *Synechocystis* sp. PCC6803. Plant Cell Physiol 47(12):1612–1621

Andersen OS, Koeppe RE (2007) Bilayer thickness and membrane protein function: an energetic perspective. Annu Rev Biophys Biomol Struct 36:107–130

Anderson SL, McIntosh L (1991) Light-activated heterotrophic growth of the cyanobacterium *Synechocystis* sp. strain PCC 6803: a blue-light-requiring process. J Bacteriol 173(9):2761–2767

Anderson KL, Tayne TA, Ward DM (1987) Formation and fate of fermentation products in hot spring cyanobacterial mats. Appl Environ Microbiol 53(10):2343–2352

Andrizhiyevskaya EG, Schwabe TM, Germano M, D'Haene S, Kruip J, van Grondelle R, Dekker JP (2002) Spectroscopic properties of PSI–IsiA supercomplexes from the cyanobacterium *Synechococcus* PCC 7942. Biochim Biophys Acta Bioenergetics 1556(2–3):265–272

Badea C, Basu SK (2009) The effect of low temperature on metabolism of membrane lipids in plants and associated gene expression. Plant Omics 2(2):78

Barthel S, Bernát G, Seidel T, Rupprecht E, Kahmann U, Schneider D (2013) Thylakoid membrane maturation and PSII activation are linked in greening *Synechocystis* sp. PCC 6803 cells. Plant Physiol 163(2):1037–1046

Bibby TS, Nield J, Barber J (2001) Iron deficiency induces the formation of an antenna ring around trimeric photosystem I in cyanobacteria. Nature 412(6848):743–745

Boekema EJ, Hifney A, Yakushevska AE, Piotrowski M, Keegstra W, Berry S, Kruip J (2001) A giant chlorophyll–protein complex induced by iron deficiency in cyanobacteria. Nature 412(6848):745–748

Brock TD (1967) Micro-organisms adapted to high temperatures. Nature 214(5091):882–885

Burnap RL, Troyan T, Sherman LA (1993) The highly abundant chlorophyll-protein complex of iron-deficient *Synechococcus* sp. PCC7942 (CP43′) is encoded by the *isiA* gene. Plant Physiol 103(3):893–902

Chen HYS, Liberton M, Pakrasi HB, Niedzwiedzki DM (2017) Reevaluating the mechanism of excitation energy regulation in iron-starved cyanobacteria. Biochim Biophys Acta Bioenergetics 1858(3):249–258

Collins AM, Liberton M, Jones HD, Garcia OF, Pakrasi HB, Timlin JA (2012) Photosynthetic pigment localization and thylakoid membrane morphology are altered in *Synechocystis* 6803 phycobilisome mutants. Plant Physiol 158(4):1600–1609

Dietzel L, Bräutigam K, Pfannschmidt T (2008) Photosynthetic acclimation: state transitions and adjustment of photosystem stoichiometry–functional relationships between short-term and long-term light quality acclimation in plants. FEBS J 275(6):1080–1088

Duke CS, Cezeaux A, Allen MM (1989) Changes in polypeptide composition of *Synechocystis* sp. strain 6308 phycobilisomes induced by nitrogen starvation. J Bacteriol 171(4):1960–1966

Emlyn-Jones D, Ashby MK, Mullineaux CW (1999) A gene required for the regulation of photosynthetic light-harvesting in the cyanobacterium *Synechocystis* 6803. Mol Microbiol 33: 1050–1058

Flores E, Herrero A (2005) Nitrogen assimilation and nitrogen control in cyanobacteria. Biochem Soc Trans 33(1):164–167

Forchhammer K, Selim KA (2020) Carbon/nitrogen homeostasis control in cyanobacteria. FEMS Microbiol Rev 44(1):33–53

Fork DC, Murata N, Sato N (1979) Effect of growth temperature on the lipid and fatty acid composition, and the dependence on temperature of light-induced redox reactions of cytochrome f and of light energy redistribution in the thermophilic blue-green alga *Synechococcus lividus*. Plant Physiol 63(3):524–530

Fraser JM, Tulk SE, Jeans JA, Campbell DA, Bibby TS, Cockshutt AM (2013) Photophysiological and photosynthetic complex changes during iron starvation in *Synechocystis* sp. PCC 6803 and *Synechococcus elongatus* PCC 7942. PLoS One 8(3):e59861

Fuszard MA, Wright PC, Biggs CA (2010) Cellular acclimation strategies of a minimal picocyanobacterium to phosphate stress. FEMS Microbiol Lett 306(2):127–134

García-Fernández JM, de Marsac NT, Diez J (2004) Streamlined regulation and gene loss as adaptive mechanisms in *Prochlorococcus* for optimized nitrogen utilization in oligotrophic environments. Microbiol Mol Biol Rev 68(4):630–638

Gledhill M, Buck KN (2012) The organic complexation of iron in the marine environment: a review. Front Microbiol 3:69

Gombos Z, Wada H, Murata N (1992) Unsaturation of fatty acids in membrane lipids enhances tolerance of the cyanobacterium *Synechocystis* PCC6803 to low-temperature photoinhibition. Proc Natl Acad Sci U S A 89(20):9959–9963

Gombos Z, Wada H, Murata N (1994) The recovery of photosynthesis from low-temperature photoinhibition is accelerated by the unsaturation of membrane lipids: a mechanism of chilling tolerance. Proc Natl Acad Sci U S A 91(19):8787–8791

Gómez-García MR, Losada M, Serrano A (2003) Concurrent transcriptional activation of ppa and ppx genes by phosphate deprivation in the cyanobacterium *Synechocystis* sp. strain PCC 6803. Biochem Biophys Res Commun 302(3):601–609

Görl M, Sauer J, Baier T, Forchhammer K (1998) Nitrogen-starvation-induced chlorosis in *Synechococcus* PCC 7942: adaptation to long-term survival. Microbiology 144(9):2449–2458

Guyet U, Nguyen NA, Doré H, Haguait J, Pittera J, Conan M, Hoebeke M (2020) Synergic effects of temperature and irradiance on the physiology of the marine *Synechococcus* strain WH7803. Front Microbiol 11:1707

Hai T, Hein S, Steinbüchel A (2001) Multiple evidence for widespread and general occurrence of type-III PHA synthases in cyanobacteria and molecular characterization of the PHA synthases from two thermophilic cyanobacteria: *Chlorogloeopsis fritschii* PCC 6912 and *Synechococcus* sp. strain MA19. Microbiology 147(11):3047–3060

Harel Y, Ohad I, Kaplan A (2004) Activation of photosynthesis and resistance to photoinhibition in cyanobacteria within biological desert crust. Plant Physiol 136(2):3070–3079

Haselkorn R (2007) Heterocyst differentiation and nitrogen fixation in cyanobacteria. In: Associative and endophytic nitrogen-fixing bacteria and cyanobacterial associations. Springer, Dordrecht, pp 233–255

Havaux M, Guedeney G, Hagemann M, Yeremenko N, Matthijs HC, Jeanjean R (2005) The chlorophyll-binding protein IsiA is inducible by high light and protects the cyanobacterium *Synechocystis* PCC6803 from photooxidative stress. FEBS Lett 579(11):2289–2293

Havens KE, Phlips EJ, Cichra MF, Li BL (1998) Light availability as a possible regulator of cyanobacteria species composition in a shallow subtropical lake. Freshw Biol 39(3):547–556

Hecker M, Völker U (1998) Non-specific, general and multiple stress resistance of growth-restricted *Bacillus subtilis* cells by the expression of the σB regulon. Mol Microbiol 29(5):1129–1136

Hernández-Prieto MA, Schön V, Georg J, Barreira L, Varela J, Hess WR, Futschik ME (2012) Iron deprivation in *Synechocystis*: inference of pathways, non-coding RNAs, and regulatory elements from comprehensive expression profiling. G3 2(12):1475–1495

Hihara Y, Kamei A, Kanehisa M, Kaplan A, Ikeuchi M (2001) DNA microarray analysis of cyanobacterial gene expression during acclimation to high light. Plant Cell 13(4):793–806

Huang L, McCluskey MP, Ni H, LaRossa RA (2002) Global gene expression profiles of the cyanobacterium *Synechocystis* sp. strain PCC 6803 in response to irradiation with UV-B and white light. J Bacteriol 184(24):6845–6858

Huokko T, Muth-Pawlak D, Battchikova N, Allahverdiyeva Y, Aro EM (2017) Role of type 2 NAD (P)H dehydrogenase NdbC in redox regulation of carbon allocation in Synechocystis. Plant Physiol 174(3):1863–1880

Huokko T, Muth-Pawlak D, Aro EM (2019) Thylakoid localized type 2 NAD(P)H dehydrogenase NdbA optimizes light-activated heterotrophic growth of *Synechocystis* sp. PCC 6803. Plant Cell Physiol 60(6):1386–1399

Ibelings BW (1996) Changes in photosynthesis in response to combined irradiance and temperature stress in cyanobacterial surface waterblooms. J Phycol 32(4):549–557

Kaneko T, Tabata S (1997) Complete genome structure of the unicellular cyanobacterium *Synechocystis* sp. PCC6803. Plant Cell Physiol 38(11):1171–1176

Karlsen J, Asplund-Samuelsson J, Thomas Q, Jahn M, Hudson EP (2018) Ribosome profiling of *Synechocystis* reveals altered ribosome allocation at carbon starvation. mSystems 3(5):1–12

Kirilovsky D, Kerfeld CA (2016) Cyanobacterial photoprotection by the orange carotenoid protein. Nat Plants 2(12):1–7

Kłodawska K, Kovács L, Várkonyi Z, Kis M, Sozer Ö, Laczkó-Dobos H, Malec P (2015) Elevated growth temperature can enhance photosystem I trimer formation and affects xanthophyll biosynthesis in cyanobacterium *Synechocystis* sp. PCC6803 cells. Plant Cell Physiol 56(3):558–571

Kłodawska K, Bujas A, Turos-Cabal M, Żbik P, Fu P, Malec P (2019) Effect of growth temperature on biosynthesis and accumulation of carotenoids in cyanobacterium *Anabaena* sp. PCC 7120 under diazotrophic conditions. Microbiol Res 226:34–40

Kong R, Xu X, Hu Z (2003) A TPR-family membrane protein gene is required for light-activated heterotrophic growth of the cyanobacterium *Synechocystis* sp. PCC 6803. FEMS Microbiol Lett 219(1):75–79

Krasikov V, Aguirre von Wobeser E, Dekker HL, Huisman J, Matthijs HC (2012) Time-series resolution of gradual nitrogen starvation and its impact on photosynthesis in the cyanobacterium *Synechocystis* PCC 6803. Physiol Plant 145(3):426–439

Krieger-Liszkay A, Fufezan C, Trebst A (2008) Singlet oxygen production in photosystem II and related protection mechanism. Photosynth Res 98(1–3):551–564

Kulaev IS, Vagabov VM (1983) Polyphosphate metabolism in micro-organisms. In: Advances in microbial physiology, vol 24. Academic, San Diego, pp 83–171

Lan S, Wu L, Zhang D, Hu C (2014) Desiccation provides photosynthetic protection for crust cyanobacteria *Microcoleus vaginatus* from high temperature. Physiol Plant 152(2):345–354

Lindahl M, Spetea C, Hundal T, Oppenheim AB, Adam Z, Andersson B (2000) The thylakoid FtsH protease plays a role in the light-induced turnover of the photosystem II D1 protein. Plant Cell 12(3):419–431

Liran O, Shemesh E, Tchernov D (2018) Investigation into the CO_2 concentrating step rates within the carbon concentrating mechanism of *Synechocystis* sp. PCC6803 at various pH and light intensities reveal novel mechanistic properties. Algal Res 33:419–429

Martiny AC, Coleman ML, Chisholm SW (2006) Phosphate acquisition genes in *Prochlorococcus* ecotypes: evidence for genome-wide adaptation. Proc Natl Acad Sci U S A 103(33):12552–12557

Mary I, Tu CJ, Grossman A, Vaulot D (2004) Effects of high light on transcripts of stress-associated genes for the cyanobacteria *Synechocystis* sp. PCC 6803 and *Prochlorococcus* MED4 and MIT9313. Microbiology 150(5):1271–1281

Melkozernov AN, Bibby TS, Lin S, Barber J, Blankenship RE (2003) Time-resolved absorption and emission show that the CP43′ antenna ring of iron-stressed *Synechocystis* sp. PCC6803 is efficiently coupled to the photosystem I reaction center core. Biochemistry 42(13):3893–3903

Miller SR, Wingard CE, Castenholz RW (1998) Effects of visible light and UV radiation on photosynthesis in a population of a hot spring cyanobacterium, a *Synechococcus* sp., subjected to high-temperature stress. Appl Environ Microbiol 64(10):3893–3899

Mullineaux CW, Emlyn-Jones D (2005) State transitions: an example of acclimation to low-light stress. J Exp Bot 56(411):389–393

Mullineaux CW, Tobin MJ, Jones GR (1997) Mobility of photosynthetic complexes in thylakoid membranes. Nature 390:421–424

Muramatsu M, Hihara Y (2012) Acclimation to high-light conditions in cyanobacteria: from gene expression to physiological responses. J Plant Res 125(1):11–39

Muramatsu M, Sonoike K, Hihara Y (2009) Mechanism of downregulation of photosystem I content under high-light conditions in the cyanobacterium *Synechocystis* sp. PCC 6803. Microbiology 155(3):989–996

Murata N (1989) Low-temperature effects on cyanobacterial membranes. J Bioenerg Biomembr 21(1):61–75

Murata N, Nishida I (1987) Lipids of blue-green algae (cyanobacteria). In: Lipids: structure and function. Academic, San Diego, pp 315–347

Murata N, Wada H, Hirasawa R (1984) Reversible and irreversible inactivation of photosynthesis in relation to the lipid phases of membranes in the blue-green algae (cyanobacteria) *Anacystis nidulans* and *Anabaena variabilis*. Plant Cell Physiol 25(6):1027–1032

Nishiyama Y, Allakhverdiev SI, Yamamoto H, Hayashi H, Murata N (2004) Singlet oxygen inhibits the repair of photosystem II by suppressing the translation elongation of the D1 protein in *Synechocystis* sp. PCC 6803. Biochemistry 43(35):11321–11330

Nishiyama Y, Allakhverdiev SI, Murata N (2006) A new paradigm for the action of reactive oxygen species in the photoinhibition of photosystem II. Biochim Biophys Acta Bioenergetics 1757(7):742–749

Nixon PJ, Barker M, Boehm M, de Vries R, Komenda J (2005) FtsH-mediated repair of the photosystem II complex in response to light stress. J Exp Bot 56(411):357–363

Nold SC, Ward DM (1996) Photosynthate partitioning and fermentation in hot spring microbial mat communities. Appl Environ Microbiol 62(12):4598–4607

Nomsawai P, Tanseau de Marsac N, Thomas JC, Tanticharoen M, Cheevadhanark S (1999) Light regulation of phycocyanin and gene expression in *Spirulina platensis* C1 (*Arthrospira* sp. PCC9438). Plant Cell Physiol 40:1194–1202

Ogawa T, Inoue Y (1983) Photosystem I-initiated postillumination CO_2 burst in a cyanobacterium, *Anabaena variabilis*. Biochim Biophys Acta Bioenergetics 724(3):490–493

Ogawa T, Miyano A, Inoue Y (1985) Photosystem-I-driven inorganic carbon transport in the cyanobacterium, *Anacystis nidulans*. Biochim Biophys Acta Bioenergetics 808(1):77–84

Ohad I, Kyle DJ, Arntzen CJ (1984) Membrane protein damage and repair: removal and replacement of inactivated 32-kilodalton polypeptides in chloroplast membranes. J Cell Biol 99(2):481–485

Ohnishi N, Allakhverdiev SI, Takahashi S, Higashi S, Watanabe M, Nishiyama Y, Murata N (2005) Two-step mechanism of photodamage to photosystem II: step 1 occurs at the oxygen-evolving complex and step 2 occurs at the photochemical reaction center. Biochemistry 44(23):8494–8499

Phillips R, Ursell T, Wiggins P, Sens P (2009) Emerging roles for lipids in shaping membrane-protein function. Nature 459(7245):379–385

Plohnke N, Seidel T, Kahmann U, Rögner M, Schneider D, Rexroth S (2015) The proteome and lipidome of *Synechocystis* sp. PCC 6803 cells grown under light-activated heterotrophic conditions. Mol Cell Proteomics 14(3):572–584

Price GD, Sültemeyer D, Klughammer B, Ludwig M, Badger MR (1998) The functioning of the CO_2 concentrating mechanism in several cyanobacterial strains: a review of general physiological characteristics, genes, proteins, and recent advances. Can J Bot 76(6):973–1002

Ritter SP, Lewis AC, Vincent SL, Lo LL, Cunha APA, Chamot D, Owttrim GW (2020) Evidence for convergent sensing of multiple abiotic stresses in cyanobacteria. Biochim Biophys Acta Gen Subjects 1864(1):129462

Robarts RD, Zohary T (1987) Temperature effects on photosynthetic capacity, respiration, and growth rates of bloom-forming cyanobacteria. N Z J Mar Freshw Res 21(3):391–399

Sandmann G (1985) Consequences of iron deficiency on photosynthetic and respiratory electron transport in blue-green algae. Photosynth Res 6(3):261–271

Sandmann G, Malkin R (1983) Iron-sulfur centers and activities of the photosynthetic electron transport chain in iron-deficient cultures of the blue-green alga *Aphanocapsa*. Plant Physiol 73(3):724–728

Sandström S, Ivanov AG, Park YI, Öquist G, Gustafsson P (2002) Iron stress responses in the cyanobacterium *Synechococcus* sp. PCC7942. Physiol Plant 116(2):255–263

Sarcina M, Mullineaux CW (2004) Mobility of the IsiA chlorophyll-binding protein in cyanobacterial thylakoid membranes. J Biol Chem 279(35):36514–36518

Sarcina M, Tobin MJ, Mullineaux CW (2001) Diffusion of phycobilisomes on the thylakoid membranes of the cyanobacterium *Synechococcus* 7942: effects of phycobilisome size, temperature and membrane lipid composition. J Biol Chem 276:46830–46834

Sato N, Murata N (1980) Temperature shift-induced responses in lipids in the blue-green alga, *Anabaena variabilis*: the central role of diacylmonogalactosylglycerol in thermo-adaptation. Biochim Biophys Acta Lipids Lipid Metab 619(2):353–366

Sato N, Murata N (1982) Lipid biosynthesis in the blue-green alga, *Anabaena variabilis*: I. Lipid classes. Biochim Biophys Acta Lipids Lipid Metab 710(3):271–278

Sauer J, Schreiber U, Schmid R, Völker U, Forchhammer K (2001) Nitrogen starvation-induced chlorosis in *Synechococcus* PCC 7942: low-level photosynthesis as a mechanism of long-term survival. Plant Physiol 126(1):233–243

Schwarz R, Forchhammer K (2005) Acclimation of unicellular cyanobacteria to macronutrient deficiency: emergence of a complex network of cellular responses. Microbiology 151(8):2503–2514

Shcolnick S, Summerfield TC, Reytman L, Sherman LA, Keren N (2009) The mechanism of iron homeostasis in the unicellular cyanobacterium *Synechocystis* sp. PCC 6803 and its relationship to oxidative stress. Plant Physiol 150(4):2045–2056

Shipley GG, Green JP, Nichols BW (1973) The phase behavior of monogalactosyl, digalactosyl, and sulphoquinovosyl diglycerides. Biochim Biophys Acta Biomembranes 311(4):531–544

Silva P, Thompson E, Bailey S, Kruse O, Mullineaux CW, Robinson C, Nixon PJ (2003) FtsH is involved in the early stages of repair of photosystem II in *Synechocystis* sp PCC 6803. Plant Cell 15(9):2152–2164

Singh AK, McIntyre LM, Sherman LA (2003) Microarray analysis of the genome-wide response to iron deficiency and iron reconstitution in the cyanobacterium *Synechocystis* sp. PCC 6803. Plant Physiol 132(4):1825–1839

Singh AK, Elvitigala T, Bhattacharyya-Pakrasi M, Aurora R, Ghosh B, Pakrasi HB (2008) Integration of carbon and nitrogen metabolism with energy production is crucial to light acclimation in the cyanobacterium *Synechocystis*. Plant Physiol 148(1):467–478

Sluchanko NN, Slonimskiy YB, Shirshin EA, Moldenhauer M, Friedrich T, Maksimov EG (2018) OCP–FRP protein complex topologies suggest a mechanism for controlling high light tolerance in cyanobacteria. Nat Commun 9(1):1–15

Somerville C (1995) Direct tests of the role of membrane lipid composition in low-temperature-induced photoinhibition and chilling sensitivity in plants and cyanobacteria. Proc Natl Acad Sci U S A 92(14):6215

Sonoike K, Hihara Y, Ikeuchi M (2001) Physiological significance of the regulation of photosystem stoichiometry upon high light acclimation of *Synechocystis* sp. PCC 6803. Plant Cell Physiol 42(4):379–384

Steunou AS, Bhaya D, Bateson MM, Melendrez MC, Ward DM, Brecht E, Grossman AR (2006) In situ analysis of nitrogen fixation and metabolic switching in unicellular thermophilic cyanobacteria inhabiting hot spring microbial mats. Proc Natl Acad Sci U S A 103(7):2398–2403

Tamary E, Kiss V, Nevo R, Adam Z, Bernát G, Rexroth S, Reich Z (2012) Structural and functional alterations of cyanobacterial phycobilisomes induced by high-light stress. Biochim Biophys Acta Bioenergetics 1817(2):319–327

Tetu SG, Brahamsha B, Johnson DA, Tai V, Phillippy K, Palenik B, Paulsen IT (2009) Microarray analysis of phosphate regulation in the marine cyanobacterium *Synechococcus* sp. WH8102. ISME J 3(7):835–849

Thiel T (2004) Nitrogen fixation in heterocyst-forming cyanobacteria. In: Genetics and regulation of nitrogen fixation in free-living bacteria. Springer, Dordrecht, pp 73–110

Tilcock CP (1986) Lipid polymorphism. Chem Phys Lipids 40(2–4):109–125

Tu CJ, Shrager J, Burnap RL, Postier BL, Grossman AR (2004) Consequences of a deletion in dspA on transcript accumulation in *Synechocystis* sp. strain PCC6803. J Bacteriol 186(12):3889–3902

Tyystjärvi E, Aro EM (1996) The rate constant of photoinhibition, measured in lincomycin-treated leaves, is directly proportional to light intensity. Proc Natl Acad Sci U S A 93(5):2213–2218

Valladares A, Montesinos ML, Herrero A, Flores E (2002) An ABC-type, high-affinity urea permease identified in cyanobacteria. Mol Microbiol 43(3):703–715

Van Thor JJ, Mullineaux CW, Matthijs HCP, Hellingwerf KJ (1998) Light harvesting and state transitions in cyanobacteria. Bot Acta 111(6):430–443

Várkonyi Z, Masamoto K, Debreczeny M, Zsiros O, Ughy B, Gombos Z, Szalontai B (2002) Low-temperature-induced accumulation of xanthophylls and its structural consequences in the photosynthetic membranes of the cyanobacterium *Cylindrospermopsis raciborskii*: an FTIR spectroscopic study. Proc Natl Acad Sci U S A 99(4):2410–2415

Vass I, Cser K, Cheregi O (2007) Molecular mechanisms of light stress of photosynthesis. Ann N Y Acad Sci 1113:114

Veaudor T, Blanc-Garin V, Chenebault C, Diaz-Santos E, Sassi JF, Cassier-Chauvat C, Chauvat F (2020) Recent advances in the photoautotrophic metabolism of cyanobacteria: biotechnological implications. Life 10(5):71

Von Wobeser EA, Ibelings BW, Bok J, Krasikov V, Huisman J, Matthijs HC (2011) Concerted changes in gene expression and cell physiology of the cyanobacterium *Synechocystis* sp. strain PCC 6803 during transitions between nitrogen and light-limited growth. Plant Physiol 155(3):1445–1457

Wada H, Murata N (1989) *Synechocystis* PCC6803 mutants defective in desaturation of fatty acids. Plant Cell Physiol 30(7):971–978

Wada H, Combos Z, Murata N (1990) Enhancement of chilling tolerance of a cyanobacterium by genetic manipulation of fatty acid desaturation. Nature 347(6289):200–203

Wada H, Gombos Z, Sakamoto T, Murata N (1992) Genetic manipulation of the extent of desaturation of fatty acids in membrane lipids in the cyanobacterium *Synechocystis* PCC6803. Plant Cell Physiol 33(5):535–540

Wilson A, Boulay C, Wilde A, Kerfeld CA, Kirilovsky D (2007) Light-induced energy dissipation in iron-starved cyanobacteria: roles of OCP and IsiA proteins. Plant Cell 19(2):656–672

Wünschmann G, Brand JJ (1992) Rapid turnover of a component required for photosynthesis explains temperature dependence and kinetics of photoinhibition in a cyanobacterium, *Synechococcus* 6301. Planta 186(3):426–433

Yamanaka G, Glazer AN (1980) Dynamic aspects of phycobilisome structure. Arch Microbiol 124(1):39–47

Yue D, Peng Y, Yin Q, Xiao L (2015) Proteomic analysis of *Microcystis aeruginosa* in response to nitrogen and phosphorus starvation. J Appl Phycol 27(3):1195–1204

Zakar T, Herman E, Vajravel S, Kovacs L, Knoppová J, Komenda J, Laczko-Dobos H (2017) Lipid and carotenoid cooperation-driven adaptation to light and temperature stress in *Synechocystis* sp. PCC6803. Biochim Biophys Acta Bioenergetics 1858(5):337–350

Zhang L, Liu J (2016) Effects of heat stress on photosynthetic electron transport in a marine cyanobacterium *Arthrospira* sp. J Appl Phycol 28(2):757–763

Zhang CC, Zhou CZ, Burnap RL, Peng L (2018) Carbon/nitrogen metabolic balance: lessons from cyanobacteria. Trends Plant Sci 23(12):1116–1130

UV Stress Responses in Cyanobacteria

6

Donat P. Häder and Rajesh P. Rastogi

Abstract

Cyanobacteria are the oldest group of prokaryotes with oxygen-evolving photosynthesis. They are supposed to have evolved in an atmosphere with little or no oxygen and therefore no protecting stratospheric ozone layer. Since cyanobacteria have to utilize sunlight for photosynthesis, they are simultaneously exposed to deleterious solar UV radiation. In order to survive, they had to develop countermeasures. One strategy is fast reproduction in order to make up for losses due to radiation damage. Another mechanism is mat and crust formation, which protects the organisms in lower levels while sacrificing the ones in the top layer. Vertical migration in the water column using changing buoyancy helps to bring the organisms out of the danger zone. Likewise, gliding cyanobacteria have been found to move to a position deeper in the water to avoid excessive UV exposure. Efficient repair mechanisms have been developed to replace damaged proteins in the photosynthetic apparatus and to repair damage in the cellular DNA. Many cyanobacteria synthesize UV-absorbing pigments such as mycosporine-like amino acids and scytonemin, deposited in the outer cell layers or extracellularly, which absorb UV photons before they can damage vital biomolecules within the cell.

Keywords

Cyanobacteria · UV stress · Repair mechanisms · UV-absorbing pigments · Migration · Buoyancy

D. P. Häder (✉)
Department of Biology, Emeritus of Friedrich-Alexander University, Erlangen, Germany
e-mail: donat@dphaeder.de

R. P. Rastogi
Ministry of Environment, Forest & Climate Change, New Delhi, India

R. P. Rastogi (ed.), *Ecophysiology and Biochemistry of Cyanobacteria*,
https://doi.org/10.1007/978-981-16-4873-1_6

6.1 Introduction

Bacteria were the first organisms on earth using solar light to harvest energy. Most modern photosynthetic bacteria use a single photosystem (of two possible ones), which operates under anoxygenic conditions since oxygen is toxic to many prokaryotes. The active pigment is one of several bacteriochlorophylls. In contrast, cyanobacteria were the first organisms to develop oxygenic photosynthesis based on two photosystems, which operate in tandem. Oxygen is produced by splitting water. This process is thought to have started about 2.3 billion years ago and since then oxygen accumulates in the atmosphere (Soo et al. 2017). Cyanobacterial photosynthesis uses chlorophyll *a* in both reaction centers which eukaryotic plants also utilize; therefore, the latter are thought to have evolved from the prokaryotic ancestors, which have been taken up in the form of endosymbiosis (Cihlář et al. 2019).

Cyanobacteria can be unicellular, floating in fresh or marine water, or growing on terrestrial or underwater surfaces. While most have diameters of a few micrometers, some are so minute that they have long been overlooked in marine plankton communities because of the too large pore size of the common plankton nets. However, during the past few decades it was found that some of them, especially the picoplanktonic genera *Prochlorococcus* and *Synechococcus*, which are in the 0.1–1 μm diameter size class, form major components of the marine ecosystems (Casey et al. 2019). Assessments of the contribution of picoplankton to the total biomass in the top 150 m of the water column indicated that they may account for up to 50% or more with *Prochlorococcus* being the most abundant and responsible for 70% of the picoplankton population (Linacre et al. 2019). Figure 6.1 represents the morphological structure of some unicellular and filamentous cyanobacteria.

Unicellular cyanobacteria may form colonies, which are held together by the extracellular polysaccharide slime which the cells produce and excrete (Sato et al. 2017). *Microcystis* cells collected from Lake Mead, Nevada, USA, were found to produce an outer sheath up to 30 μm thick (He and Wert 2016). Other genera form unbranched, pseudo-branched, or truly branched uniseriate filaments, which are covered by a cylindrical slime tube (Kabirnataj et al. 2018; Singh 2017).

Cyanobacteria are found in almost all habitats on earth. Aquatic forms live both in fresh and marine waters. Terrestrial cyanobacteria are found from the tropics to polar regions; they can cover rocks, salt marshes and even the barks of trees. They are adapted to low temperatures and desiccation (Jimel 2020), while others can survive in hot thermal springs (Cheng et al. 2020). Several species of cyanobacteria have been reported from a hypersaline desert (Patel et al. 2019). A wide variety of cyanobacteria lives in symbioses with all kinds of plants and animals. Lichens are a symbiosis of algae and/or cyanobacteria with fungi (Pankratov et al. 2017), and the aquatic fern *Azolla* harbors cyanobacteria in thallus cavities (Sánchez-Baracaldo and Cardona 2020). Cyanobacteria have been found to form symbioses with diatoms, bryophytes, gymnosperms and angiosperms, and they are even found in symbiosis with animals such as marine sponges and worms (Rai 2018).

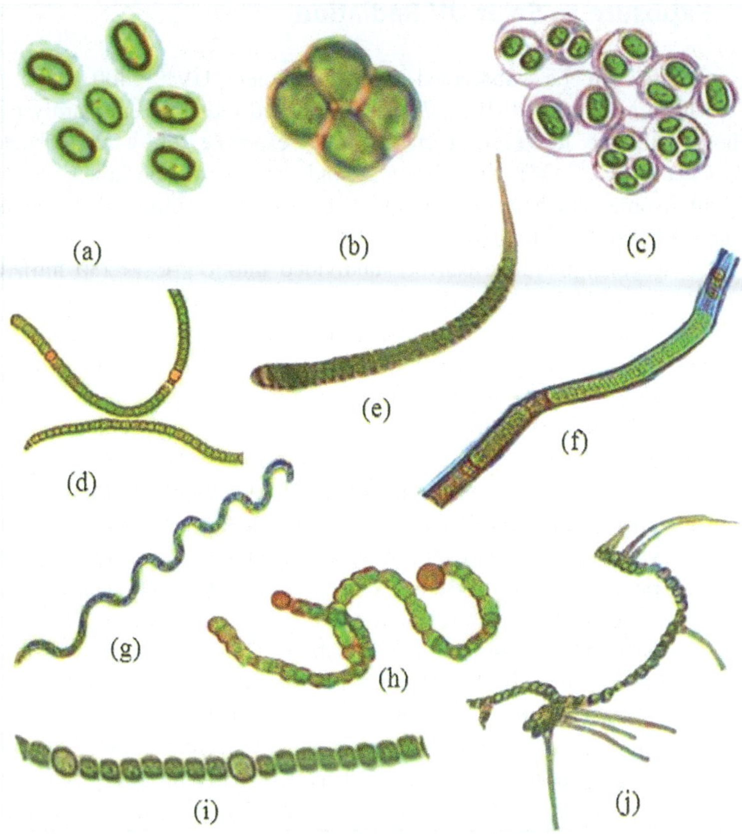

Fig. 6.1 Photographs showing the morphology of some cyanobacteria (**a**, *Gloeocapsa* sp.; **b**, *Chroococcus* sp.; **c**, *Cyanothece* sp.; **d**, *Scytonema* sp.; **e**, *Calothrix* sp.; **f**, *Lyngbya* sp.; **g**, *Arthrospira* sp.; **h**, *Nostoc* sp.; **i**, *Anabaena* sp.; **j**, *Fischerella* sp.). (Images by RP Rastogi)

In contrast to some earlier reports, cyanobacteria cannot swim in water (Menon et al. 2020). They do not have cilia, flagella or other moving organelles such as bacteria, flagellates and other eukaryotes (Miyata et al. 2020). But many are motile using a slow gliding movement. Some uniseriate filaments such as *Anabaena* or *Phormidium* glide within their sheath, which they may shed at the rear. They may also reverse their direction of movement triggered by external light or chemical stimuli. Motility has been studied in the model cyanobacterium *Synechocystis* sp., PCC 6803. In this organism, motility has been identified to be based on the presence of thick TFP pili, which can be extended, retracted and adhered to the substratum (Chen et al. 2020). Even though not capable of active swimming, planktonic cyanobacteria can undergo vertical migrations in the water column by changing their buoyancy (Kai and Lan 2020). This can be achieved by the production and collapse of gas vesicles (Dyer and Needoba 2020).

6.2 Exposure to Solar UV Radiation

Solar radiation can be subdivided into ultraviolet (UV, <400 nm), visible (400–700 nm), and infrared (IR, >700 nm). Infrared radiation can hardly be used for photosynthesis; however, there is one example of a cyanobacterium (*Synechococcus* PCC7335) which has a second core-membrane linker (ApcE2) of the phycobilisome which is noncovalently bound which allows the organism to utilize near IR (Miao et al. 2016).

The UV wavelength range can be subdivided into UV-C (<280 nm), UV-B (280–315 nm), and UV-A (315–400 nm) (Aphalo 2017). Photosynthesis is mainly supported by visible radiation, but under certain conditions (low radiation under cloud cover) UV-A can be utilized by some macroalgae (Xu and Gao 2016). Generally speaking, UV radiation is detrimental for organisms, especially at excessive intensities. Today, UV-C is quantitatively absorbed by oxygen and ozone in the atmosphere. UV-B is also significantly filtered out mainly by stratospheric ozone. But before the atmospheric oxygenation, organisms were exposed to and had to cope with much higher surface UV-B and in addition UV-C than today. This was the situation for cyanobacteria during their Achaean evolution even though the presence of Fe(III)-Si precipitates absorbed up to 70% of the incoming UV-C radiation. However, it is assumed that the remaining UV-C caused high mortality rates and limited cyanobacterial expansion in marine habitats (Mloszewska et al. 2018).

Solar irradiances strongly depend on a number of physical conditions on our planet. The solar zenith angle (SZA) determines the irradiance in all wavelength bands, which are highest in the tropics and gradually decrease toward the poles as monitored by 17 stations of the Eldonet network (Table 6.1). Much higher irradiances have been measured during a recent 1-year campaign in the high Andes near Laguna Lejia (Chile, latitude 23° 26′ 23.30″ S, longitude 67° 38′ 14.29″ W) at an elevation of 4715 m (Häder and Cabrol 2020). Figure 6.2 shows the mean monthly irradiances of PAR, UV-A, UV-B, and short-wavelength UV-B (295–310 nm).

Since today solar UV-C radiation does not hit the surface of the earth, UV-B is the most detrimental wavelength band for organisms exposed to solar radiation. In addition to latitude, UV-B radiation at ground level is controlled by the atmospheric water content (especially clouds), albedo and aerosols, and total column ozone (Häder and Cabrol 2020). At the same latitude, irradiances in the Southern Hemisphere are higher than in the Northern Hemisphere, because of the different earth–sun distances (Cordero et al. 2014). The stratospheric ozone concentration is lower in the tropics than at mid- and higher latitudes, resulting in higher solar UV-B irradiances. UV radiation increases with elevation (Blumthaler et al. 1997). In Northern Chile, IR increases by 27%, PAR by 6%, and UV by 20% from sea level to 5100 m altitude (Cordero et al. 2016). Clouds can reduce or enhance solar irradiation by absorption or scattering, quantified by the cloud modification factor (CMF) (Feister et al. 2015).

Stratospheric ozone depletion by chlorofluorocarbons (CFCs) and other anthropogenic trace gases such as organobromides and chlorocarbons has increased

Table 6.1 Location, latitude, longitude, elevation above sea level, and mean summer irradiances in PAR, UV-A, and UV-B measured by 17 ELDONET stations (after Häder et al. 2007)

				Summer irradiances [W m^{-2}]		
Location	Latitude	Longitude	Elevation m a.s.l.	PAR ± S.D.	UV-A ± S.D.	UV-B ± S.D.
Abisko	68° 50′ N	19° 00′ E	385	284.69 ± 34.13	33.99 ± 9.05	0.77 ± 0.32
Lund	55° 07′ N	13° 04′ E	50	380.78 ± 32.52	59.92 ± 7.26	1.55 ± 0.47
Helgoland	54° 10′ N	07° 51′ E	61	353.11 ± 54.24	44.08 ± 6.32	0.77 ± 0.34
Erlangen	49° 35′ N	11° 00′ E	280	393.30 ± 35.40	52.70 ± 9.38	1.33 ± 0.30
Karlsruhe	49° 03′ N	08° 23′ E	200	385.14 ± 12.21	49.55 ± 7.14	1.07 ± 0.49
Ljubljana	46° 04′ N	14° 33′ E	300	412.84 ± 27.13	59.97 ± 2.89	1.52 ± 0.15
Bonassola	44° 10′ N	09° 30′ E	10	411.60 ± 37.38	61.23 ± 8.89	1.60 ± 0.29
Pisa	43° 43′ N	10° 23′ E	100	390.28 ± 0.08	55.46 ± 0.61	1.05 ± 0.03
Logrono	42° 28′ N	02° 27′ W	380	387.44 ± 26.45	57.48 ± 5.23	1.53 ± 0.21
Lisbon	38° 42′ N	09° 10′ W	105	398.67 ± 31.68	62.08 ± 8.55	1.60 ± 0.41
Athens	37° 58′ N	23° 46′ E	110	393.82 ± 49.42	55.91 ± 8.03	1.67 ± 0.85
Sierra Nevada	37° 04′ N	03° 20′ W	2850	430.87 ± 25.89	61.52 ± 3.76	1.88 ± 0.32
Malaga	36° 43′ N	04° 23′ W	18	414.21 ± 13.32	61.88 ± 2.96	1.90 ± 0.25
Gran Canaria	27° 55′ N	15° 35′ W	8	419.84 ± 20.31	64.26 ± 5.32	2.05 ± 0.24
Joinville	26° 15′ S	48° 55′ W	120	413.81 ± 0.19	55.31 ± 4.77	1.41 ± 0.36
Playa Union	43° 15′ S	65° 00′ W	20	424.26 ± 46.71	62.33 ± 3.68	1.89 ± 0.15
Lauder	45° 01′ S	169° 41′ E	370	429.08 ± 27.43	61.31 ± 5.23	1.70 ± 0.30

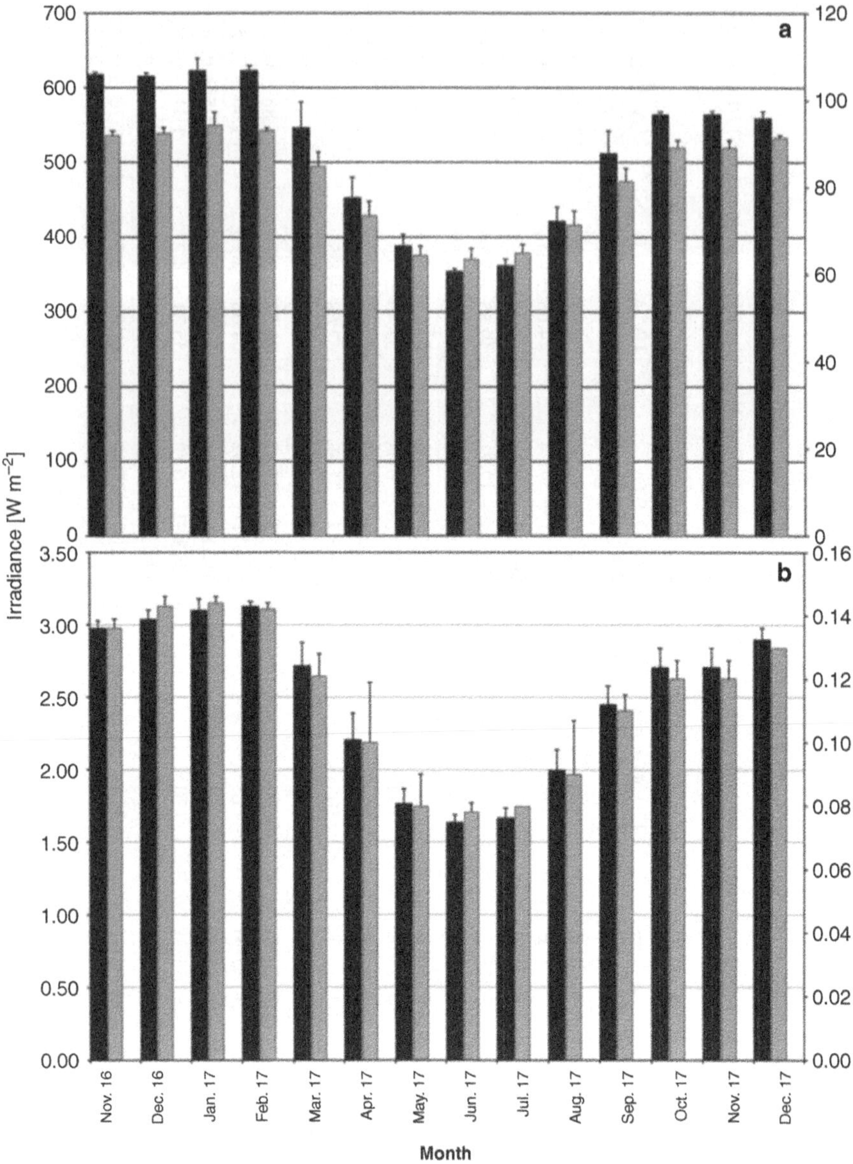

Fig. 6.2 Mean monthly irradiances of PAR, UV-A, UV-B, and short-wavelength UV-B (295–310 nm) monitored over a year in the high Andes (Laguna Lejia, Chile, latitude 23° 26′ 23.30″ S, longitude 67° 38′ 14.29″ W at an elevation of 4715 m) (Häder and Cabrol 2020)

terrestrial UV-B radiation, but due to the Montreal Protocol and its amendments this effect is stopped and slowly reverses (Bais et al. 2018). But a recovery to pre-1980 levels is predicted only for or after mid-century due to the long lifetimes of CFCs in the stratosphere, which can be decades (Hoffmann et al. 2014). Global climate change alters total column ozone and therefore UV irradiances (Williamson et al. 2014; Schnell et al. 2016; Meul et al. 2016).

6.3 UV Effects on Cyanobacteria

There are a several nonphotosynthetic cyanobacteria whose diversity, distribution, and ecology are currently hardly known (Monchamp et al. 2019). Some are found in dark, deep terrestrial habitats such as rocks using a hydrogen-based lithoautotrophic metabolism (Puente-Sánchez et al. 2018). In contrast, all photosynthetic cyanobacteria require solar radiation for their energy harvesting. Therefore, they are inevitably exposed to solar UV radiation. Solar UV radiation affects several key cellular biomolecules and machinery (e.g., DNA and proteins), cellular morphology, photosynthesis, growth, survival, pigmentation, and nitrogen metabolism enzymes in cyanobacteria (Sinha et al. 1995a, 1998; Kumar et al. 1996; Rastogi et al. 2014a, b).

6.3.1 Damage and Repair of DNA

The most deleterious UV-B radiation is absorbed by important biomolecules including proteins, nucleic acids, and lipids, resulting in considerable damage of exposed organisms and affecting physiological, biochemical, and ecological functions, such as morphology, differentiation, growth, development, pigmentation, and motility and orientation (Häder 1993a, b; Pathak et al. 2018). Absorption of UV-B photons by the cellular DNA results in the formation of cyclobutane pyrimidine dimers (CPDs), which are the most notable lesions (about 75–80%) induced by solar UV radiations (Pathak et al. 2019b; Rastogi et al. 2010a) (Fig. 6.3). Besides CPDs, 6-4 photoproducts (6-4PPs), are the second most frequently occurring DNA lesions (about 20–25%), which are formed mainly under UV-C and readily converted into their Dewar valence isomers upon exposure to UV radiation (Rastogi et al. 2010a) (Fig. 6.4).

These dimers are induced between two adjacent pyrimidine bases (thymine, cytosine, and uracil). This defect seems like a minor change in the structure of the DNA, but may have far-reaching consequences for the biochemical processes in the cell since the DNA reproduction and transcription into RNA are stopped there.

CPDs are repaired by the cells in a process called photoreactivation, which involves the enzyme DNA photolyase (Rastogi et al. 2011). This enzyme possesses two noncovalently linked cofactors such as $FADH_2$ and absorbs blue or UV-A photons and uses their energy to split the dimer (Pathak et al. 2019b; Rastogi et al. 2020). If the lesion is not repaired, it results in s signature mutation (Brash and Seidman 2020). Photolyases are very old enzymes found in bacteria all the way to vertebrates (Sinha and Häder 2002; Zhang et al. 2013). In cyanobacteria, this process has been studied, e.g., in *Anacystis nidulans*, and the enzyme has been purified (Eker

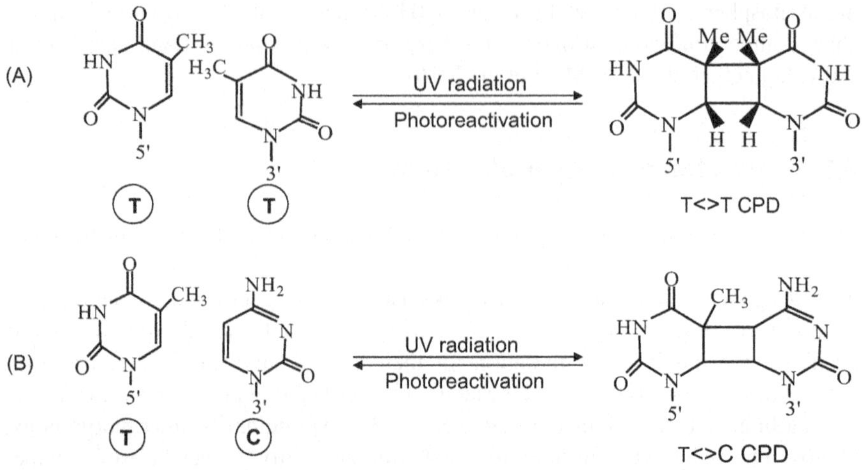

Fig. 6.3 Formation of cyclobutane–pyrimidine dimers (CPDs) induced mainly by UV-B on DNA having adjacent thymine/cytosine bases. (**a**) Thymine–thymine cyclobutane–pyrimidine dimer (T<>T CPD) and (**b**) thymine–cytosine cyclobutane–pyrimidine (T<>C CPD) dimer. Both T<>T and T<>C CPDs split to form two canonical thymine/cytosine bases by means of photoreactivation in the presence of the photolyase enzyme (Rastogi et al. 2010a)

et al. 1990). However, it is interesting to note that placental mammals including humans lack this repair mechanism and must rely on other DNA repair mechanisms such as nucleotide excision repair (see below) (Jans et al. 2005). UV-C mainly induces the formation of thymine–pyrimidone (6-4) dimers. These lesions are also repaired by a photolyase (Kavakli et al. 2019).

Other UV-induced DNA damages include DNA–protein crosslinks (Rastogi 2010; Richa et al. 2015; Rajneesh et al. 2018) and 8-oxo-7,8-dihydroguanyl, 8-oxo-Ade, 2,6-diamino-4-hydroxy-5-formamidoguanine and oxazolone, which result from oxidations products of purine bases of the DNA (Doetsch et al. 1995; Hall et al. 1996).

DNA lesions, which are not repaired by a photolyase during photoreactivation, can be mended by excision repair (see review by Pathak et al. 2019b). This mechanism is independent of light and uses several enzymes (Bergi and Trivedi 2020). It is based on the removal of a small number of bases, e.g., after a single-strand break, which are subsequently resynthesized and inserted using the complementary strand. One form of excision repair is the base excision repair pathway in which one or two bases are removed and substituted after, e.g., desiccation or radiation stress (Singh 2018). The alternative is nucleotide excision repair, which removes DNA lesions including CPDs or 6,4 photoproducts (6,4 PPs), DNA intrastrand crosslinks, chemical adducts, or by oxidative damage by reactive oxygen species (Sinha 2017).

Recombinational repair is a powerful mechanism to restore the correct DNA sequence after single- or double-strand DNA breaks. This pathway is fairly complex involving more than 20 gene products in *E. coli*. Initially, an exonuclease enlarges

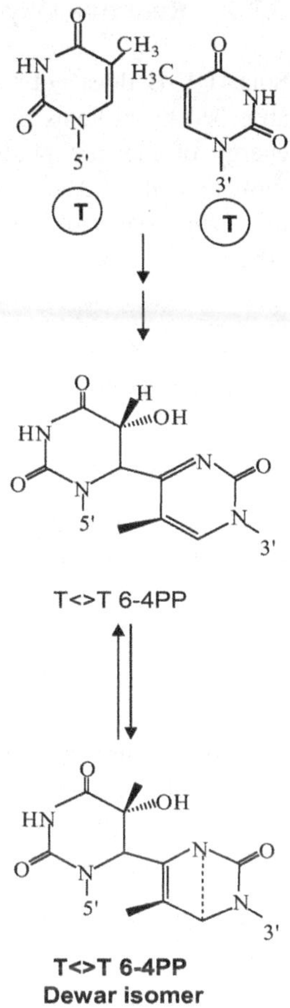

Fig. 6.4 Formation of DNA lesion 6-4 photoproducts (6-4PPs) and their Dewar valence isomers (Rastogi et al. 2010a)

the DNA break and the gap is identified by RecFOR proteins. Subsequently, RecBCD and RecFOR perform the repair by homologous recombination (Rastogi et al. 2015). If all fails, cells retreat to the last resort, called SOS repair. This is initiated by different and substantial DNA damages or when the DNA replication is inhibited as studied in the cyanobacterium *Anabaena* sp. (Kumar et al. 2018). This pathway relies on the interaction of several repressor proteins including RecA and LexA, which block the 40 or so SOS response genes. Once the blockage is released, the SOS genes, each consisting of a 20-nucleotide-long SOS box, start their work. One of them codes for the SulA protein, which delays the cell division until all damages are repaired. However, many differences in the components of the SOS repair mechanism exist between bacteria and cyanobacteria and between species (Kumar et al. 2018).

6.3.2 Reactive Oxygen Species

Solar UV-B does not have to exert direct effects on cellular targets. It may be absorbed by proteins or other biomolecules in the cell upon which the excitation energy of the UV photon is transferred to, e.g., oxygen, which results in the formation of reactive oxygen species (ROS). The reduction of molecular oxygen results in superoxide, which may lead to the production of most other ROS (Turrens 2003).

$$O_2 + e^- \rightarrow {}^{\bullet}O_2^-$$

Dismutation of superoxide results in hydrogen peroxide

$$2H^+ + {}^{\bullet}O_2^- + {}^{\bullet}O_2^- \rightarrow H_2O_2 + O_2$$

which may be partially reduced to a hydroxide ion and a hydroxyl radical or may be fully reduced to water

$$H_2O_2 + e^- \rightarrow HO^- + {}^{\bullet}OH$$

$$2H^+ + 2e^- + H_2O_2 \rightarrow 2H_2O$$

Another pathway transfers the excited energy of an absorbing molecule, such as chlorophyll, to a nearby oxygen (the ground state of which is a triplet state, ${}^3O_2$) which is converted to singlet oxygen (${}^1O_2$) which is highly reactive and destructs nearby biomolecules and structures, even though its lifetime is rather short (on the order of 10–40 ns) (Moan and Berg 1991). In this case, the chlorophyll acts as a photosensitizer (Ph). This response is a major damaging mechanism in photosynthesis (Krieger-Liszkay 2005) but also occurs in mitochondria (Thomas et al. 1992). As an aside: This photodynamic reaction induced by introduced photosensitizers such as hematoporphyrin is used in medical treatment of superficial cancers in humans (Lv et al. 2016).

$$Ph + h\nu \rightarrow {}^3Ph$$

$${}^3Ph + {}^3O_2 \rightarrow Ph + {}^1O_2$$

Oxygen is toxic at higher concentrations. After the development of an oxygenic atmosphere on our planet, many early life forms such as bacteria had to find protection from the increasing oxygen concentration. Today, many of these bacteria are confined to anoxic environments such as sediments (Valentine 2002). All other organisms were forced by the environmental pressure to develop mechanisms to protect themselves from ROS. This is mainly achieved by two different mechanisms. One is the employment of passive antioxidants such as ascorbic acid, α-tocopherol, glutathione, lycopene, lutein, and isoflavones (Sindhi et al. 2013). Such ROS scavengers are also found in many cyanobacteria (Radyukina et al. 2019; He and

Häder 2002b). The alternative strategy to counter the stress of ROS is the involvement of antioxidant enzymes such as superoxide dismutase (SOD), catalase, glutathione peroxidase, ascorbate peroxidase, monodehydroascorbate reductase, dehydroascorbate reductase, or glutathione reductase found, e.g., in *Anabaena* sp. (He and Häder 2002a) and other cyanobacteria (Aráoz and Häder 1999a). The production of ROS induced by UV-B radiation can be shown and monitored by employing the ROS-sensitive, oxygen-sensing probe 2′,7-′-dichlorodihydrofluorescein diacetate (DCFH-DA) (He and Häder 2002a; Rastogi et al. 2010b).

6.4 UV Damage of the Photosynthetic Apparatus and Repair

The photosynthetic apparatus is a key target of damaging solar UV radiation. In addition to unspecific effects on proteins, lipids, membranes, and other biologically important molecules and structures, UV-B affects the photosynthetic electron transport, quantum yield, and oxygen production (Xue et al. 2005). Exposure to UV radiation bleaches the photosynthetic pigments in the cyanobacterium *Phormidium uncinatum* (Donkor and Häder 1997; Sinha et al. 2005). Phycobilisomes containing the phycobiliprotein accessory pigments are broken down into smaller components by increasing exposure (Sinha et al. 1995c, b). In the initial phase of exposure, the phycobiliprotein fluorescence increases indicating that they can no longer transfer the excitation energy to the photosynthetic reaction centers (Donkor and Häder 1996). After prolonged exposure to UV, the fluorescence of the accessory phycobiliproteins decreases (Rastogi et al. 2015). Photodegradation of phycobilisomes by UV radiation was also confirmed in *Nostoc* sp. and *Aulosira fertilissima* (Aráoz García 1998; Banerjee et al. 1998). Exposure to UV radiation also impairs the translation activity in the cyanobacterium *Nostoc* sp. (Araoz et al. 1998). In contrast, low-level UV-B irradiances induce phycoerythrin synthesis in *Nostoc* (Aráoz and Häder 1999b).

As indicated above, ROS generated by photodynamic reactions are a potential mechanism to damage biomolecules and structures within the photosynthetic apparatus. Solar energy is absorbed by accessory pigments such as phycobilins in the phycobilisomes of cyanobacteria (as well as by chlorophyll *b*, chlorophyll *c*, or chlorophyll *d* in algae and higher plants) and transferred to the photosynthetic reaction centers of photosystems I and II (Jaiswal et al. 2018). An excited electron from the special chlorophyll *a* dimer P680 in PS II is transferred to a primary acceptor (pheophytin) (Khaing et al. 2019) from where it is handed along a chain of redox components via P700, the reaction center PS I (where the electron is again excited to a higher energetic level) until it is finally utilized to reduce NADP (Tikhonov and Subczynski 2019). The missing electron in P680 is subsequently replaced by an electron generated by the photolytic splitting of water on the inside of the thylakoids by an enzymatic Mn complex (Böhmer et al. 2017).

$$H_2O \rightarrow \frac{1}{2}O_2 + 2H^+ + 2e^-$$

The oxygen is released as a waste product and the protons are used to drive an ATPase, which generates ATP, which is used together with the reduction equivalents (reduced NADP) to reduce CO_2 to sugar in the Calvin cycle (Michelet et al. 2013; Janasch et al. 2019). Under high light conditions, the electron transport chain is fully reduced and cannot accept any more electrons from P680 (Lea-Smith et al. 2016). Therefore, the excitation energy can be transferred to a nearby oxygen resulting in singlet oxygen (s. above) (Lee and Min 2010). In order to avoid this, potential damaging situation-specific carotenoids are arranged in close vicinity to the chlorophyll so that the excitation energy can be transferred to the carotenoids, which relax the energy in the form of heat (Pospíšil and Prasad 2014; Schäfer et al. 2005).

Another target of solar UV is the D1 protein located in PS II encoded by the *psbAI* gene (Rexroth et al. 2017). This protein is responsible to transfer the excited electrons from P680 to pheophytin (Khaing et al. 2019). This protein is easily kinked by excessive visible or UV radiation, which stops the electron transport. This lesion is rapidly repaired by proteolytic removal of the damaged protein and subsequent replacement by a newly synthesized protein (Campbell et al. 1998; Ehling-Schulz and Scherer 1999).

6.5 Motility and Orientation

As indicated above, many unicellular and multicellular cyanobacteria are motile. Some show a gliding motility, others use changes in buoyancy to realize vertical migration. These mechanisms are used to avoid areas of deleterious solar radiation. Picoplanktons, such as *Synechococcus* and *Prochlorococcus*, which are major biomass producers in the oceans, use fast reproduction to overcome the population losses due to excessive UV (Häder and Gao 2018). In addition, there is a pronounced seasonal variability and changes in vertical distribution (Al-Otaibi et al. 2020). In the Red Sea, *Synechococcus* was found close to the surface, while *Prochlorococcus* was responsible for a chlorophyll maximum between 40 and 76 m. *Prochlorococcus* populations had a maximum in summer and a minimum in winter, while *Synechococcus* showed the opposite temporal distribution. In addition, there are low light- and high light adapted genetically different populations dwelling at different depths (Linacre et al. 2019).

Cyanobacteria with gliding motility respond to visible and UV radiation to select a suitable habitat. The coccoid *Synechocystis* sp. secretes a mixture of complex polysaccharides to drive their motility (Varuni et al. 2017) and shows a pronounced phototaxis, which can be positive (toward the light source) at low irradiances or negative (away from the light source) at high irradiances, while the rod-shaped *Synechococcus elongatus* PCC 7942 has no phototactic motility (Yang et al. 2018). In *Synechocystis* sp., PCC6803 a blue light-dependent signal cascade controls positive and negative phototaxis (Sugimoto et al. 2017). The cyanobacterial phytochrome 2 regulates the expression of motility-related genes via the second messenger cyclic-GMP (Wallner et al. 2020).

Also, some filamentous cyanobacteria show a gliding motility when in contact with a surface (Qiu et al. 2019). While the exact mechanism has not yet been revealed, several distinct structural features such as specifically arranged protein fibrils and organelle-like structures have been identified, which are thought to be involved in the secretion of mucilage (Hoiczyk 2000). The unbranched heterotrichous *Anabaena variabilis* has been found to move either as straight filaments or in a U-shaped form (Nultsch and Wenderoth 1983; Nultsch et al. 1979). When irradiated, the trichomes bend toward the light at low fluence rates (1.35 W m^{-2}) and away from the light source at higher (27 W m^{-2}). These authors assumed that the switch from positive to negative phototaxis is controlled by the intracellular level of singlet oxygen since gassing the moving filaments with N_2 or Ar shifts the transition from positive to negative phototaxis to higher irradiances (Nultsch and Schuchart 1985). The photoreceptor for photoorientation is assumed to consist of a superfamily of tetrapyrrole-binding molecules, cyanobacteriochromes (Ikeuchi and Ishizuka 2008). Exposure to UV-B radiation delays differentiation of vegetative cells into heterocysts and akinetes (Blakefield and Harris 1994), induces bleaching of the phycobilins (Agel et al. 1987), and affects productivity and nitrogen fixation in *Anabaena* (Lesser 2008). It also causes a significant decrease in the quantum yield of PSII. The effects on photosynthesis are thought to be due to the production of ROS, since exposure to UVR results in an increase in the level of superoxide dismutase.

The nonheterocystous filamentous *Phormidium uncinatum* does not show phototaxis, but orients with respect to light using photophobic responses. When a trichome enters a bright light field from a dark area, e.g., when it leaves the shade under a leaf, it reverses the direction of movement and glides back; this response is called a step-up photophobic reaction. In contrast, when leaving a low irradiance area moving into a dark area, it may also reverse the direction of movement (step-down photophobic response) (Nultsch and Häder 1970). The organisms respond even to small differences in the irradiances of two adjacent light fields as low as 4%. This can be demonstrated by projecting a photographic negative onto a population of *Phormidium* trichomes, which accumulate in areas of appropriate irradiance, forming a photographic positive (Häder 1984) (Fig. 6.5). The photosynthetic pigments are responsible for the photophobic responses in this cyanobacterium (Nultsch and Häder 1974; Häder 1974). The direction of movement is controlled by an electric potential gradient along the length of the trichome. During a photophobic reversal of movement, this gradient inverts (Häder 1978; Häder and Burkart 1978).

Motility of *Phormidium* is strongly impaired by solar and monochromatic UV irradiation (Häder et al. 1986). The action spectrum shows a strong response in the UV-B. In contrast, the photophobic response was not impaired by solar or artificial UV radiation (Häder and Häder 1990). Bleaching kinetics indicate that the accessory phycobilins, D-phycoerythrin, is easily bleached by UV radiation, followed by the

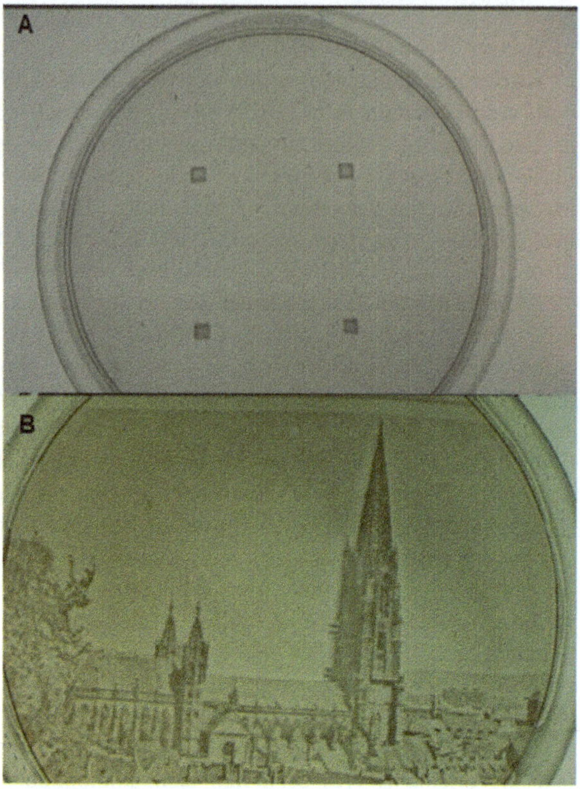

Fig. 6.5 *Phormidium* trichomes accumulate in low irradiance light fields projected onto a Petri dish (**A**). A photographic negative of the Münster in Freiburg has been projected onto a suspension of *Phormidium* trichomes, which accumulate in areas of appropriate irradiances forming a photographic positive (**B**) (after (Häder 1984))

carotenoids, while chlorophyll *a* was found to be the most resistant pigment to bleaching.

Several filamentous cyanobacteria (*Phormidium uncinatum*, two strains, *Anabaena variabilis and Oscillatoria tenuis*) protect themselves by vertical migration. The filaments were suspended in an agar layer inside a slanting groove made from dark PVC, which was placed in a pond reaching from 10 to 100 cm. After 4-h exposure to solar radiation, the organisms had moved to a position at about 50–60 cm below the water surface (Donkor and Häder 1995). Mat-forming *Oscillatoria* on Antarctica's McMurdo Ice Shelf have also been found to show vertical migration controlled by solar visible and UV radiation: At low irradiances (<8 W m^{-2}, no UV), the filaments migrated completely to the surface, while higher irradiances (>60 W m^{-2}, including UV-A and UV-B) induced downward migration (Nadeau et al. 1999). Similar vertical migrations were also found in *Microcoleus* and *Halomicronema* in microbial mats in the French Camargue (Fourçans et al. 2006) and in coastal microbial mats (Lichtenberg et al. 2020).

6.6 UV-Screening Pigments

In response to the pressure of solar UV radiation, many cyanobacteria (but also eukaryotic phytoplankton and macroalgae) have developed UV-absorbing pigments such as mycosporine-like amino acids (MAAs) and scytonemin (Scy) to screen out deleterious radiation before it can hit essential biomolecules and cellular structures (Sinha et al. 2007). Picoplanktons, such as the marine *Synechococcus* and *Prochlorococcus*, are too small (<1 μm) to use UV-screening pigments since the concentration would have to be too high to be effective over very small transmission distances (Garcia-Pichel 1994); therefore, these organisms rely on vertical migration, repair mechanisms, and fast replication to counter the challenge of solar UV radiation.

Mycosporine-like amino acids (MAAs) are small, hydrophilic, and colorless molecules with a cyclohexenimine or cyclohexenone chromophore attached to the nitrogen substituent of an amino acid or its imino alcohol (Pathak et al. 2019a). More than 20 different MAAs are known today, which are characterized by their high molar extinction coefficients (28,100–50,000 M^{-1} cm^{-1}) and a strong absorption in the UV between 310 and 362 nm (Pathak et al. 2017a; Rastogi et al. 2020) (Fig. 6.6). The absorbed UV photon energy is dissipated as heat and does not result in ROS generation (Conde et al. 2007). Some MAAs have even be found to possess free radical scavenging capacity (Rastogi et al. 2016).

Some cyanobacteria—and only this group of organisms—have developed another group of UV-absorbing pigments, i.e., scytonemins (Rastogi et al. 2012, 2014c). These molecules are heterocyclic phenolic dimers, which are excreted into the extracellular polysaccharide sheath (Pathak et al. 2017a, b) (Fig. 6.7). In addition to a major peak at 386 nm, the oxidized form shows peaks at 252 and 300 nm. The UV-C peak may be a reminder of the life history of cyanobacteria in an anoxygenic atmosphere with no stratospheric ozone layer. In *Nostoc punctiforme,* the molecule is coded by a gene cluster of 18 ORFs (Soule et al. 2007), and a possible biosynthesis pathway has been suggested by Balskus and Walsh (2009).

6.7 Conclusions

Photosynthetic organisms require solar energy for their metabolism. Simultaneously, they are exposed to deleterious UV-A and UV-B radiation, since cyanobacteria have started their development on our planet when the atmosphere contained only traces of oxygen, and consequently, no stratospheric ozone layer existed to protect them from the even more energetic UV-C radiation. At moderate irradiances, UV-A can be utilized to drive photosynthesis in some phytoplankton, but shorter wavelengths are always detrimental for living organisms. UV photons are absorbed by lipids, proteins, and other biologically important molecules and are consequently prone to modify these components and destroy cellular structures. In addition, absorption of solar UV radiation can induce reactive oxygen species (ROS).

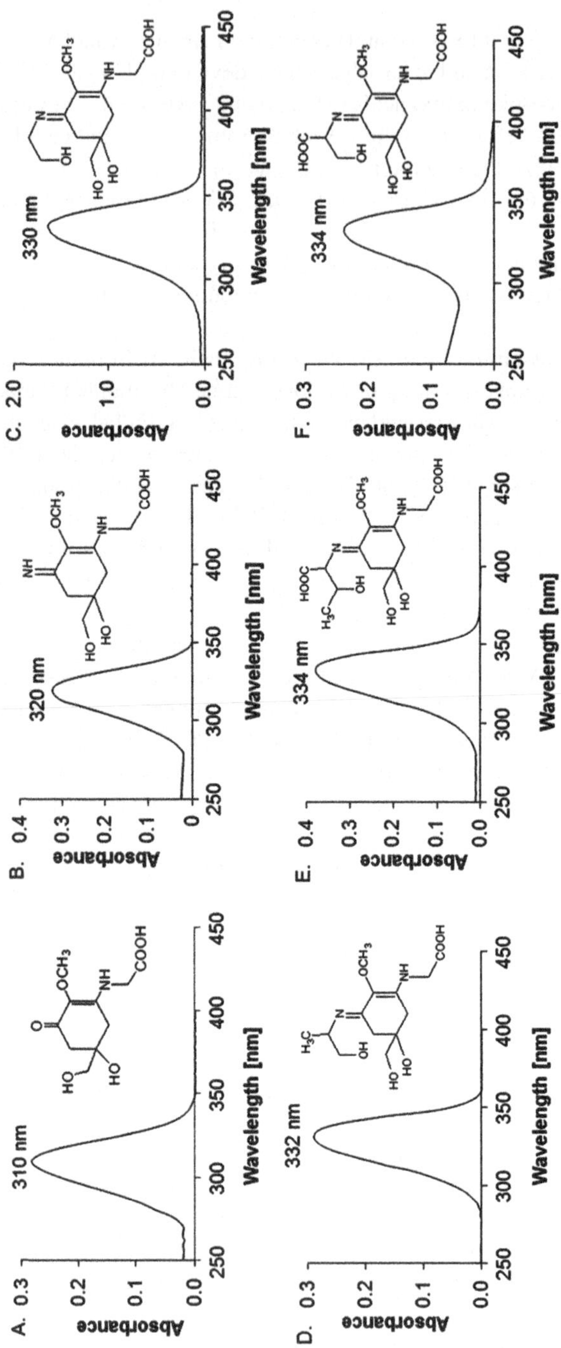

Fig. 6.6 Chemical structure of some common MAAs and their absorption spectra reported in different cyanobacteria. (**a**) M-Glycine; (**b**) palythine; (**c**) asterina; (**d**) palythinol; (**e**) porphyra-334; and (**f**) shinorine

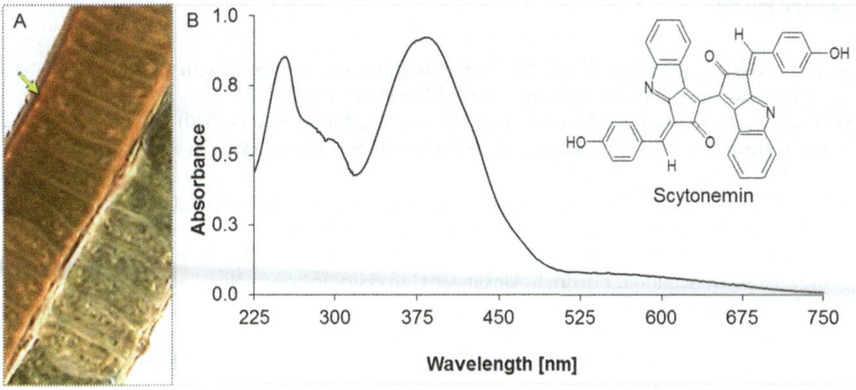

Fig. 6.7 Occurrence of scytonemin in the extracellular sheath (shown by arrow) of *Lyngby* sp. (**a**), their chemical structure and UV absorption maxima (**b**)

In order to protect the cells from deleterious solar UV radiation, organisms have developed a plethora of mechanisms and strategies against induced damage. The DNA is a key target of solar UV-B radiation, which is of vital importance since its integrity warrants the correct transmission of information to the next generation. Therefore, a large number of concepts have been developed to repair any UV-induced damage and modification including the involvement of photolyases, which remove dimers in the nucleotide strand. Other mechanisms include excision, recombination, and SOS repair pathways. A likewise important target of solar UV radiation is the photosynthetic apparatus. Short-wavelength photons bleach accessory pigments and chlorophyll *a* and induce ROS such as singlet oxygen, which in turn destroys biologically important structures. Cells have developed enzymatic and nonenzymatic strategies to quench ROS production. Damage of the redox elements of the photosynthetic electron transport chain is repaired by removal of mutilated proteins and replacement by newly synthesized molecules.

Other strategies to avoid excessive exposure to solar radiation include mat formation and vertical migration to bring organisms out of the danger zone. This can be achieved by using phototaxis or photophobic responses. One important mechanism is the production and incorporation of UV-absorbing pigments such as mycosporine-like amino acids and scytonemins, which prevent the transmission of damaging photons to vital biomolecules and structures in the center of the cell. In addition, some of these substances have antioxidant properties. These molecules have a potential to serve humans as UV protectants in suntan lotions as a replacement of artificial organic molecules (Guillerme et al. 2017; Richa and Sinha 2013).

References

Agel G, Nultsch W, Rhiel E (1987) Photoinhibition and its wavelength dependence in the cyanobacterium *Anabaena variabilis*. Arch Microbiol 147:370–374

Al-Otaibi N, Huete-Stauffer TM, Calleja ML, Irigoien X, Morán XAG (2020) Seasonal variability and vertical distribution of autotrophic and heterotrophic picoplankton in the Central Red Sea. PeerJ 8:e8612

Aphalo PJ (2017) Quantification of UV radiation. In: Jordan BR (ed) UV-B radiation and plant life: molecular biology to ecology. CAB International, Boston, MA

Aráoz García LR (1998) Tolerance mechanisms against ultraviolet-B radiation in phytoplankton organisms. Dissertation, Friedrich-Alexander University Erlangen-Nürnberg, Germany

Aráoz R, Häder D-P (1999a) Enzymatic antioxidant activity in two cyanobacteria species exposed to solar radiation. Recent Res Dev Photochem Photobiol 3:123–132

Aráoz R, Häder D-P (1999b) Phycoerythrin synthesis is induced by solar UV-B in the cyanobacterium *Nostoc*. Plant Physiol Biochem 37:223–229

Araoz R, Lebert M, Häder D-P (1998) Translation activity under ultraviolet radiation and temperature stress in the cyanobacterium *Nostoc* sp. J Photochem Photobiol B Biol 47:115–120

Bais AF, Lucas RM, Bornman JF, Williamson CE, Sulzberger B, Austin AT, Wilson SR, Andrady AL, Bernhard G, McKenzie RL et al (2018) Environmental effects of ozone depletion, UV radiation and interactions with climate change: UNEP Environmental Effects Assessment Panel, update 2017. Photochem Photobiol Sci 17:127–179

Balskus EP, Walsh CT (2009) An enzymatic cyclopentyl [b] indole formation involved in scytonemin biosynthesis. J Am Chem Soc 131:14648–14649

Banerjee M, Sinha RP, Häder D-P (1998) Biochemical and spectroscopic changes in phycobiliproteins of the cyanobacterium, *Aulosira fertilissima*, induced by UV-B radiation. Acta Protozool 37:145–148

Bergi J, Trivedi R (2020) Bioremediation of saline soil by cyanobacteria. Microbial bioremediation & biodegradation. Springer, New York

Blakefield MK, Harris DO (1994) Delay of cell differentiation in *Anabaena aequalis* caused by UV-B radiation and the role of photoreactivation and excision repair. Photochem Photobiol 59: 204–208

Blumthaler M, Ambach W, Ellinger R (1997) Increase of solar UV radiation with altitude. J Photochem Photobiol B Biol 39:130–134

Böhmer S, Köninger K, Gómez-Baraibar Á, Bojarra S, Mügge C, Schmidt S, Nowaczyk MM, Kourist R (2017) Enzymatic oxyfunctionalization driven by photosynthetic water-splitting in the cyanobacterium *Synechocystis* sp. PCC 6803. Catalysts 7:240

Brash DE, Seidman MM (2020) Defective post-replication repair of UV photoproducts in melanoma: a mutator phenotype. Mol Oncol 14:5–7

Campbell D, Eriksson MJ, Öquist G, Gustafsson P, Clarke AK (1998) The cyanobacterium *Synechococcus* resists UV-B by exchanging photosystem II reaction-center D1 proteins. Proc Natl Acad Sci U S A 95:364–369

Casey JR, Björkman KM, Ferrón S, Karl DM (2019) Size dependence of metabolism within marine picoplankton populations. Limnol Oceanogr 64:1819–1827

Chen Z, Li X, Tan X, Zhang Y, Wang B (2020) Recent advances in biological functions of thick pili in the cyanobacterium *Synechocystis* sp. PCC 6803. Front Plant Sci 11:241

Cheng Y-I, Chou L, Chiu Y-F, Hsueh H-T, Kuo C-H, Chu H-A (2020) Comparative genomic analysis of a novel strain of Taiwan hot-spring cyanobacterium *Thermosynechococcus* sp. CL-1. Front Microbiol 11:82

Cihlář J, Füssy Z, Oborník M (2019) Evolution of tetrapyrrole pathway in eukaryotic phototrophs, Advances in botanical research. Elsevier, Amsterdam

Conde FR, Churio MS, Previtali CM (2007) Experimental study of the excited-state properties and photostability of the mycosporine-like amino acid palythine in aqueous solution. Photochem Photobiol Sci 6:669–674

Cordero RR, Seckmeyer G, Damiani A, Riechelmann S, Rayas J, Labbe F, Laroze D (2014) The world's highest levels of surface UV. Photochem Photobiol Sci 13:70–81

Cordero R, Damiani A, Seckmeyer G, Jorquera J, Caballero M, Rowe P, Ferrer J, Mubarak R, Carrasco J, Rondanelli R (2016) The solar spectrum in the Atacama Desert. Sci Rep 6:22457

Doetsch PW, Zastawny TH, Martin AM, Dizdaroglu M (1995) Monomeric base damage products from adenine, guanine, and thymine induced by exposure of DNA to ultraviolet radiation. Biochemistry (USA) 34:737–742

Donkor VA, Häder D-P (1995) Protective strategies of several cyanobacteria against solar radiation. J Plant Physiol 145:750–755

Donkor VA, Häder D-P (1996) Effects of ultraviolet irradiation on photosynthetic pigments in some filamentous cyanobacteria. Aquat Microbial Ecol 11:143–149

Donkor VA, Häder D-P (1997) Ultraviolet radiation effects on pigmentation in the cyanobacterium *Phormidium uncinatum*. Acta Protozool 36:49–55

Dyer SW, Needoba JA (2020) Use of high-resolution pressure nephelometry to measure gas vesicle collapse as a means of determining growth and turgor changes in planktonic cyanobacteria. Appl Environ Microbiol 86(2):e01790–e01719

Ehling-Schulz M, Scherer S (1999) UV protection in cyanobacteria. Eur J Phycol 34:329–338

Eker A, Kooiman P, Hessels J, Yasui A (1990) DNA photoreactivating enzyme from the cyanobacterium *Anacystis nidulans*. J Biol Chem 265:8009–8015

Feister U, Cabrol N, Häder D-P (2015) UV irradiance enhancements by scattering of solar radiation from clouds. Atmosphere 5:1211–1228

Fourçans A, Solé A, Diestra E, Ranchou-Peyruse A, Esteve I, Caumette P, Duran R (2006) Vertical migration of phototrophic bacterial populations in a hypersaline microbial mat from Salins-de-Giraud (Camargue, France). FEMS Microbiol Ecol 57:367–377

Garcia-Pichel F (1994) A model for internal self-shading in planktonic organisms and its implications for the usefulness of ultraviolet sunscreens. Limnol Oceanogr 39:1704–1717

Guillerme J-B, Couteau C, Coiffard L (2017) Applications for marine resources in cosmetics. Cosmetics 4:35

Häder D-P (1974) Participation of two photosystems in the photo-phobotaxis of *Phormidium uncinatum*. Arch Microbiol 96:255–266

Häder D-P (1978) Evidence of electrical potential changes in photophobically reacting blue-green algae. Arch Microbiol 118:115–119

Häder D-P (1984) Wie orienteren sich Cyanobakterien im Licht. Biologie in unserer Zeit 14:78–83

Häder D-P (1993a) Effects of enhanced solar ultraviolet radiation on aquatic ecosystems. In: Tevini M (ed) UV-B radiation and ozone depletion. Effects on humans, animals, plants, microorganisms, and materials. Lewis Publ., Boca Raton, FL

Häder D-P (1993b) Risks of enhanced solar ultraviolet radiation for aquatic ecosystems. In: Round FE, Chapman DJ (eds) Progress in phycological research. Biopress Ltd., Bristol

Häder D-P, Burkart U (1978) Mathematical model for photophobic accumulations of blue-green algae in light traps. J Math Biol 5:293–304

Häder DP, Cabrol NA (2020) Monitoring of solar irradiance in the high Andes. Photochem Photobiol 96(5):1133–1139

Häder D-P, Gao K (2018) Phytoplankton responses to ocean climate change drivers. In: Aquatic ecosystems in a changing climate, vol 62. CRC Press, Boca Raton, FL

Häder D-P, Häder M (1990) Effects of solar radiation on motility, photomovements and pigmentation in two strains of the cyanobacterium, *Phormidium uncinatum*. Acta Protozool 29:291–303

Häder D-P, Watanabe M, Furuya M (1986) Inhibition of motility in the cyanobacterium, *Phormidium uncinatum*, by solar and monochromatic UV irradiation. Plant Cell Physiol 27: 887–894

Häder D-P, Lebert M, Schuster M, del Ciampo L, Helbling EW, McKenzie R (2007) ELDONET - a decade of monitoring solar radiation on five continents. Photochem Photobiol 83:1348–1357

Hall DB, Holmlin RE, Barton JK (1996) Oxidative DNA damage through long-range electron transfer. Nature 382:731–735

He YY, Häder D-P (2002a) Involvement of reactive oxygen species in the UV-B damage to the cyanobacterium *Anabaena* sp. J Photochem Photobiol B Biol 66:73–80
He YY, Häder D-P (2002b) Reactive oxygen species and UV-B: effect on cyanobacteria. Photochem Photobiol Sci 1:729–736
He X, Wert EC (2016) Colonial cell disaggregation and intracellular microcystin release following chlorination of naturally occurring *Microcystis*. Water Res 101:10–16
Hoffmann L, Hoppe C, Müller R, Dutton G, Gille J, Griessbach S, Jones A, Meyer C, Spang R, Volk C (2014) Stratospheric lifetime ratio of CFC-11 and CFC-12 from satellite and model climatologies. Atmos Chem Phys 14:12479–12497
Hoiczyk E (2000) Gliding motility in cyanobacteria: observations and possible explanations. Arch Microbiol 174:11–17
Ikeuchi M, Ishizuka T (2008) Cyanobacteriochromes: a new superfamily of tetrapyrrole-binding photoreceptors in cyanobacteria. Photochem Photobiol Sci 7:1159–1167
Jaiswal A, Koli DK, Kumar A, Kumar S, Sagar S (2018) Pigments analysis of cyanobacterial strains. Int J Chem Stud 6:1248–1251
Janasch M, Asplund-Samuelsson J, Steuer R, Hudson EP (2019) Kinetic modeling of the Calvin cycle identifies flux control and stable metabolomes in *Synechocystis* carbon fixation. J Exp Bot 70:973–983
Jans J, Schul W, Sert Y-G, Rijksen Y, Rebel H, Eker AP, Nakajima S, van Steeg H, de Gruijl FR, Yasui A (2005) Powerful skin cancer protection by a CPD-photolyase transgene. Curr Biol 15: 105–115
Jimel M (2020) Algal and cyanobacterial adaptations to low temperature and desiccation. Bachelor, Karlova
Kabirnataj S, Nematzadeh GA, Talebi AF, Tabatabaei M, Singh P (2018) *Neowestiellopsis* gen. nov, a new genus of true branched cyanobacteria with the description of *Neowestiellopsis persica* sp. nov. and *Neowestiellopsis bilateralis* sp. nov., isolated from Iran. Plant Syst Evol 304:501–510
Kai W, Lan SS (2020) Vertical migration by bulk phytoplankton sustains biodiversity and nutrient input to the surface ocean. Sci Rep 10:1142
Kavakli IH, Ozturk N, Gul S (2019) DNA repair by photolyases, Advances in protein chemistry and structural biology. Elsevier, Amsterdam
Khaing EP, Zhong V, Eaton-Rye JJ (2019) Impairment of photosystem II assembly and acceptor side electron transfer following mutation of Thr243 and Lys264 of D2 in cyanobacteria. Photosynth Hydrogen Energy Res Sustain 46
Krieger-Liszkay A (2005) Singlet oxygen production in photosynthesis. J Exp Bot 56:337–346
Kumar A, Sinha RP, Häder D-P (1996) Effect of UV-B on enzymes of nitrogen metabolism in the cyanobacterium *Nostoc calcicola*. J Plant Physiol 148:86–91
Kumar A, Kirti A, Rajaram H (2018) Regulation of multiple abiotic stress tolerance by LexA in the cyanobacterium *Anabaena* sp. strain PCC7120. Biochim Biophys Acta Gene Regul Mech 1861: 864–877
Lea-Smith DJ, Bombelli P, Vasudevan R, Howe CJ (2016) Photosynthetic, respiratory and extra-cellular electron transport pathways in cyanobacteria. Biochim Biophys Acta Bioenergetics 1857:247–255
Lee J, Min DB (2010) Analysis of volatile compounds from chlorophyll photosensitized linoleic acid by headspace solid-phase microextraction (HS-SPME). Food Sci Biotechnol 19:611–616
Lesser MP (2008) Effects of ultraviolet radiation on productivity and nitrogen fixation in the cyanobacterium, *Anabaena* sp. (Newton's strain). Hydrobiologia 598:1–9
Lichtenberg M, Cartaxana P, Kühl M (2020) Vertical migration optimizes photosynthetic efficiency of motile cyanobacteria in a coastal microbial mat. Front Mar Sci 7:359
Linacre L, Durazo R, Camacho-Ibar V, Selph K, Lara-Lara J, Mirabal-Gómez U, Bazán-Guzmán C, Lago-Lestón A, Fernández-Martín E, Sidón-Ceseña K (2019) Picoplankton carbon biomass assessments and distribution of *Prochlorococcus* ecotypes linked to Loop Current Eddies during summer in the southern Gulf of Mexico. J Geophys Res Oceans 124(11):8342–8359

Lv W, Zhang Z, Zhang KY, Yang H, Liu S, Xu A, Guo S, Zhao Q, Huang W (2016) A mitochondria-targeted photosensitizer showing improved photodynamic therapy effects under hypoxia. Angew Chem Int Ed 55:9947–9951

Menon SN, Varuni P, Menon GI (2020) Information integration and collective motility in phototactic cyanobacteria. PLoS Comput Biol 16:e1007807

Meul S, Dameris M, Langematz U, Abalichin J, Kerschbaumer A, Kubin A, Oberländer-Hayn S (2016) Impact of rising greenhouse gas concentrations on future tropical ozone and UV exposure. Geophys Res Lett 43:2919–2927

Miao D, Ding W-L, Zhao B-Q, Lu L, Xu Q-Z, Scheer H, Zhao K-H (2016) Adapting photosynthesis to the near-infrared: non-covalent binding of phycocyanobilin provides an extreme spectral red-shift to phycobilisome core-membrane linker from *Synechococcus* sp. PCC7335. Biochim Biophys Acta Bioenergetics 1857:688–694

Michelet L, Zaffagnini M, Morisse S, Sparla F, Pérez-Pérez ME, Francia F, Danon A, Marchand C, Fermani S, Trost P (2013) Redox regulation of the Calvin–Benson cycle: something old, something new. Front Plant Sci 4:470

Miyata M, Robinson RC, Uyeda TQ, Fukumori Y, Si F, Haruta S, Homma M, Inaba K, Ito M, Kaito C (2020) Tree of motility–a proposed history of motility systems in the tree of life. Genes Cells 25:6–21

Mloszewska AM, Cole DB, Planavsky NJ, Kappler A, Whitford DS, Owttrim GW, Konhauser KO (2018) UV radiation limited the expansion of cyanobacteria in early marine photic environments. Nat Commun 9:1–8

Moan J, Berg K (1991) The photodegradation of porphyrins in cells can be used to estimate the lifetime of singlet oxygen. Photochem Photobiol 53:549–553

Monchamp M-E, Spaak P, Pomati F (2019) Long term diversity and distribution of non-photosynthetic cyanobacteria in peri-alpine lakes. Front Microbiol 9:3344

Nadeau T-L, Howard-Williams C, Castenholz RW (1999) Effects of solar UV and visible irradiance on photosynthesis and vertical migration of *Oscillatoria* sp.(Cyanobacteria) in an Antarctic microbial mat. Aquat Microb Ecol 20:231–243

Nultsch W, Häder D-P (1970) Bestimmungen der photo-phobotaktischen Unterschiedsschwelle bei *Phormidium uncinatum*. Berichte der Deutschen Botanischen Gesellschaft 83:185–192

Nultsch W, Häder D-P (1974) Über die Rolle der beiden Photosysteme in der Photo-phobotaxis von *Phormidium uncinatum*. Berichte der Deutschen Botanischen Gesellschaft 87:83–92

Nultsch W, Schuchart H (1985) A model of the phototactic reaction chain of the cyanobacterium *Anabaena variabilis*. Arch Microbiol 142:180–184

Nultsch W, Wenderoth K (1983) Partial irradiation experiments with *Anabaena variabilis* (Kütz). Z Pflanzenphysiol 111:1–7

Nultsch W, Schuchart H, Höhl M (1979) Investigations on the phototactic orientation of *Anabaena variabilis*. Arch Microbiol 122:85–91

Pankratov T, Kachalkin A, Korchikov E, Dobrovol'skaya T (2017) Microbial communities of lichens. Microbiology 86:293–309

Patel HM, Rastogi RP, Trivedi U, Madamwar D (2019) Cyanobacterial diversity in mat sample obtained from hypersaline desert, Rann of Kachchh. 3 Biotech 9:304

Pathak J, Ahmed H, Sinha RP (2017a) Metabolomic profiling of cyanobacterial UV-protective compounds. Curr Metabolomics 5:138–163

Pathak J, Sonker AS, Richa, Rajneesh R, Kannaujiya VK, Singh V, Ahmed H, Sinha RP (2017b) Screening and partial purification of photoprotective pigment scytonemin from cyanobacterial crusts dwelling on the historical monuments in and around Varanasi, India. Microbiol Res 8(1):4–12

Pathak J, Häder D-P, Sinha RP (2018) Impacts of ultraviolet radiation on certain physiological and biochemical processes in cyanobacteria inhabiting diverse habitats. Environ Exp Bot 161:375–387

Pathak J, Ahmed H, Singh PR, Singh SP, Häder D-P, Sinha RP (2019a) Mechanisms of photoprotection in cyanobacteria, Cyanobacteria. Elsevier, Amsterdam

Pathak J, Singh PR, Häder DP, Sinha RP (2019b) UV-induced DNA damage and repair: a cyanobacterial perspective. Plant Gene 19:100194

Pospíšil P, Prasad A (2014) Formation of singlet oxygen and protection against its oxidative damage in photosystem II under abiotic stress. J Photochem Photobiol B Biol 137:39–48

Puente-Sánchez F, Arce-Rodríguez A, Oggerin M, García-Villadangos M, Moreno-Paz M, Blanco Y, Rodríguez N, Bird L, Lincoln SA, Tornos F (2018) Viable cyanobacteria in the deep continental subsurface. Proc Natl Acad Sci U S A 115:10702–10707

Qiu Y, Tian S, Gu L, Hildreth M, Zhou R (2019) Identification of surface polysaccharides in akinetes, heterocysts and vegetative cells of *Anabaena cylindrica* using fluorescein-labeled lectins. Arch Microbiol 201:17–25

Radyukina N, Mikheeva L, Karbysheva E (2019) Low molecular weight antioxidants in cyanobacteria and plant cells. Biol Bull Rev 9:520–531

Rai AN (2018) CRC handbook of symbiotic cyanobacteria. CRC Press, Boca Raton, FL

Rajneesh, Chatterjee A, Pathak J, Ahmed H, Singh V, Singh DK, Pandey A, Singh SP, Richa, Häder D-P, Sinha RP (2018) Ultraviolet radiation-induced DNA damage and mechanisms of repair in cyanobacteria: an overview. In: Sinha RP, Shrivastava UP (eds) Biotechnology in agriculture, industry and medicine. Trends in life science research. Nova Biomedical, New York

Rastogi RP (2010) UV-B-induced DNA damage and repair in cyanobacteria. PhD, Banaras Hindu University

Rastogi RP, Kumar A, Tyagi MB, Sinha RP (2010a) Molecular mechanisms of ultraviolet radiation-induced DNA damage and repair. J Nucleic Acids 2010:592980

Rastogi RP, Singh SP, Häder D-P, Sinha RP (2010b) Detection of reactive oxygen species (ROS) by the oxidant-sensing probe 2′,7′-dichlorodihydrofluorescein diacetate in the cyanobacterium *Anabaena variabilis* PCC 7937. Biochem Biophys Res Commun 397:603–607

Rastogi RP, Singh SP, Häder D-P, Sinha RP (2011) Ultraviolet-B-induced DNA damage and photorepair in the cyanobacterium *Anabaena variabilis* PCC 7937. Environ Exp Bot 74:280–288

Rastogi RP, Kumari S, Richa, Han T, Sinha RP (2012) Molecular characterization of hot spring cyanobacteria and evaluation of their photoprotective compounds. Can J Microbiol 58:719–727

Rastogi RP, Incharoensakdi A, Madamwar D (2014a) Responses of a rice-field cyanobacterium *Anabaena siamensis* TISTR-8012 upon exposure to PAR and UV radiation. J Plant Physiol 171:1545–1553

Rastogi RP, Sinha RP, Moh SH, Lee TK, Kottuparambil S, Kim Y-J, Rhee J-S, Choi E-M, Brown MT, Häder D-P (2014b) Ultraviolet radiation and cyanobacteria. J Photochem Photobiol B Biol 141:154–169

Rastogi RP, Sonani RR, Madamwar D (2014c) The high-energy radiation protectant extracellular sheath pigment scytonemin and its reduced counterpart in the cyanobacterium *Scytonema* sp. R77DM. Bioresour Technol 171:396–400

Rastogi RP, Sonani RR, Madamwar D (2015) Effects of PAR and UV radiation on the structural and functional integrity of phycocyanin, phycoerythrin and allophycocyanin isolated from the marine cyanobacterium *Lyngbya* sp. A09DM. Photochem Photobiol 91(4):837–844

Rastogi RP, Sonani RR, Madamwar D, Incharoensakdi A (2016) Characterization and antioxidant functions of mycosporine-like amino acids in the cyanobacterium *Nostoc* sp. R76DM. Algal Res 16:110–118

Rastogi RP, Madamwar D, Nakamoto H, Incharoensakdi A (2020) Resilience and self-regulation processes of microalgae under UV radiation stress. J Photochem Photobiol C Photochem Rev 43:100322

Rexroth S, Nowaczyk MM, Rögner M (2017) Cyanobacterial photosynthesis: the light reactions, Modern topics in the phototrophic prokaryotes. Springer, New York

Richa, Sinha RP (2013) Biomedical applications of mycosporine-like amino acids. In: Kim SK (ed) Marine microbiology, bioactive compounds and biotechnological applications. Wiley-VCH Publishers, Germany

Richa, Sinha RP, Häder D-P (2015) Physiological aspects of UV-excitation of DNA. In: Barbatto M, Borin AC, Ullrich S (eds) Topics in current chemistry: photoinduced phenomena in nucleic acids II: DNA fragments and phenomenological aspects. Springer, Berlin

Sánchez-Baracaldo P, Cardona T (2020) On the origin of oxygenic photosynthesis and cyanobacteria. New Phytol 225:1440–1446

Sato M, Omori K, Datta T, Amano Y, Machida M (2017) Influence of extracellular polysaccharides and calcium ion on colony formation of unicellular *Microcystis aeruginosa*. Environ Eng Sci 34: 149–157

Schäfer L, Vioque A, Sandmann G (2005) Functional in situ evaluation of photosynthesis-protecting carotenoids in mutants of the cyanobacterium *Synechocystis* PCC6803. J Photochem Photobiol B Biol 78:195–201

Schnell JL, Prather MJ, Josse B, Naik V, Horowitz LW, Zeng G, Shindell DT, Faluvegi G (2016) Effect of climate change on surface ozone over North America, Europe, and East Asia. Geophys Res Lett 43:3509–3518

Sindhi V, Gupta V, Sharma K, Bhatnagar S, Kumari R, Dhaka N (2013) Potential applications of antioxidants–a review. J Pharm Res 7:828–835

Singh K (2017) Taxonomy and morphology of cyanobacteria the genus *Hapalosiphon* (Stigonematales). J Nat Resour Dev 12:5–10

Singh H (2018) Desiccation and radiation stress tolerance in cyanobacteria. J Basic Microbiol 58: 813–826

Sinha RP (2017) Role of photosynthetically active radiation and dark conditions on the repair of ultraviolet-B radiation-induced damages in the cyanobacterium *Nostoc* sp. strain HKAR-2. Focus Med Sci J 3

Sinha RP, Häder D-P (2002) UV-induced DNA damage and repair: a review. Photochem Photobiol Sci 1:225–236

Sinha RP, Kumar HD, Kumar A, Häder D-P (1995a) Effects of UV-B irradiation on growth, survival, pigmentation and nitrogen metabolism enzymes in cyanobacteria. Acta Protozool 34: 187–192

Sinha RP, Lebert M, Kumar A, Kumar HD, Häder D-P (1995b) Disintegration of phycobilisomes in a rice field cyanobacterium *Nostoc* sp. following UV irradiation. Biochem Mol Biol Int 37:697–706

Sinha RP, Lebert M, Kumar A, Kumar HD, Häder D-P (1995c) Spectroscopic and biochemical analyses of UV effects on phycobiliproteins of *Anabaena* sp. and *Nostoc carmium*. Bot Acta 108:87–92

Sinha RP, Krywult M, Häder D-P (1998) Effects of ultraviolet, monochromatic and PAR waveband on nitrate reductase activity and pigmentation in a rice field cyanobacterium, *Anabaena* sp. Acta Hydrobiol 40:105–112

Sinha RP, Kumar A, Tyagi MB, Häder D-P (2005) Ultraviolet-B-induced destruction of phycobiliproteins in cyanobacteria. Physiol Mol Biol Plants 11:313–319

Sinha RP, Singh SP, Häder D-P (2007) Database on mycosporines and mycosporine-like amino acids (MAAs) in fungi, cyanobacteria, macroalgae, phytoplankton and animals. J Photochem Photobiol B Biol 89:29–35

Soo RM, Hemp J, Parks DH, Fischer WW, Hugenholtz P (2017) On the origins of oxygenic photosynthesis and aerobic respiration in Cyanobacteria. Science 355:1436–1440

Soule T, Stout V, Swingley WD, Meeks JC, Garcia-Pichel F (2007) Molecular genetics and genomic analysis of scytonemin biosynthesis in *Nostoc punctiforme* ATCC 29133. J Bacteriol 189:4465–4472

Sugimoto Y, Nakamura H, Ren S, Hori K, Masuda S (2017) Genetics of the blue light-dependent signal cascade that controls phototaxis in the cyanobacterium *Synechocystis* sp. PCC6803. Plant Cell Physiol 58:458–465

Thomas C, MacGill RS, Miller GC, Pardini RS (1992) Photoactivation of hypericin generates singlet oxygen in mitochondria and inhibits succinoxidase. Photochem Photobiol 55:47–53

Tikhonov AN, Subczynski WK (2019) Oxygenic photosynthesis: EPR study of photosynthetic electron transport and oxygen-exchange, an overview. Cell Biochem Biophys 77:47–59

Turrens JF (2003) Mitochondrial formation of reactive oxygen species. J Physiol 552:335–344

Valentine DL (2002) Biogeochemistry and microbial ecology of methane oxidation in anoxic environments: a review. Antonie Van Leeuwenhoek 81:271–282

Varuni P, Menon SN, Menon GI (2017) Phototaxis as a collective phenomenon in cyanobacterial colonies. Sci Rep 7:1–10

Wallner T, Pedroza L, Voigt K, Kaever V, Wilde A (2020) The cyanobacterial phytochrome 2 regulates the expression of motility-related genes through the second messenger cyclic di-GMP. Photochem Photobiol Sci 19(5):631–643

Williamson CE, Zepp RG, Lucas RM, Madronich S, Austin AT, Ballaré CL, Norval M, Sulzberger B, Bais A, McKenzie RL, Robinson SA, Häder D-P, Paul ND, Bornman JF (2014) Solar ultraviolet radiation in a changing climate. Nat Clim Chang 4:434–441

Xu J, Gao K (2016) Photosynthetic contribution of UV-A to carbon fixation by macroalgae. Phycologia 55:318–322

Xue L, Zhang Y, Zhang T, An L, Wang X (2005) Effects of enhanced ultraviolet-B radiation on algae and cyanobacteria. Crit Rev Microbiol 31:79–89

Yang Y, Lam V, Adomako M, Simkovsky R, Jakob A, Rockwell NC, Cohen SE, Taton A, Wang J, Lagarias JC (2018) Phototaxis in a wild isolate of the cyanobacterium *Synechococcus elongatus*. Proc Natl Acad Sci U S A 115:E12378–E12387

Zhang F, Scheerer P, Oberpichler I, Lamparter T, Krauß N (2013) Crystal structure of a prokaryotic (6-4) photolyase with an Fe-S cluster and a 6,7-dimethyl-8-ribityllumazine antenna chromophore. Proc Natl Acad Sci U S A 110:7217–7222

Molecular Mechanisms of Stress Tolerance in Cyanobacteria

7

Nedeljka Rosic

Abstract

Organisms exposed to diverse environmental conditions have developed mechanisms that enable them to adjust, adapt and survive. Understanding the adaptation strategies employed by different species will allow us to better comprehend and predict future biodiversity alterations under changing climate conditions.

Cyanobacteria are prokaryotes found in various aquatic habitats, including terrestrial, as well as freshwater and saltwater systems. The aim of this study is to portray the diversity of habitats occupied by cyanobacteria and the molecular mechanisms employed by cyanobacterial species in response to exposure to, often extreme, abiotic stressors such as elevated water temperatures and light stress. Specifically, the focus of this review is to summarise the underlining mechanisms utilised by cyanobacteria allowing for their survival over billions of years. The climate changes that have been occurring over the past few decades in the environment have resulted in global warming, increased solar radiation and hypersalinity, and are predicted to continue rising in the future. Increased environmental pressure on marine cyanobacteria and other organisms is especially happening due to the prolonged and irregular periods of warm sea temperature, so-called marine heatwaves (MHWs). These periods of MHW are becoming more frequent and are putting additional pressure on marine organisms and diverse ecosystems. Many organisms adjust by changing the profile of genes expressed and performing complex cellular changes in response to stress. Understanding the consequences of elevated temperatures, hypersalinity and the impact

N. Rosic (✉)
Faculty of Health, Southern Cross University, Gold Coast, QLD, Australia

Marine Ecology Research Centre, Southern Cross University, Lismore, NSW, Australia
e-mail: nedeljka.rosic@scu.edu.au

R. P. Rastogi (ed.), *Ecophysiology and Biochemistry of Cyanobacteria*,
https://doi.org/10.1007/978-981-16-4873-1_7

of damaging ultraviolet radiation is not completely clear in terms of the ability of cyanobacteria to adapt/acclimatise. Under heat stress, changes in gene regulations in cyanobacteria affect the expression of different genes such as the one encoding molecular chaperones, photosynthetic and oxidative stress-related genes. An overview of various molecular mechanisms is discussed here, as well as the ability of cyanobacteria to endure thermal and light stress conditions.

Keywords

Cyanobacteria · Adaptation · Abiotic stress · Heat stress · Light stress · Genetics · Epigenetics

7.1 Introduction

Cyanobacteria (also known as blue-green algae) are photosynthetic prokaryotes that belong to the domain of bacteria and the phylum Cyanophyta. They evolved approximately 2.5–3.5 billion years ago when the development of oxygenic photosynthesis allowed the transition from anaerobic to aerobic life on the earth. These largest groups of gram-negative photosynthetic bacteria have substantial similarities to plants regarding their lipid content and chloroplast membranes (Los and Murata 1999). Like plants and eukaryotic algae, photosynthetic cyanobacteria are able to produce sugar via carbon fixation using CO_2 and solar energy, while releasing O_2 into the atmosphere (Stanier and Cohen-Bazire 1977).

Cyanobacteria are found all around the globe in various terrestrial and aquatic ecosystems. They are capable of surviving in extreme environmental conditions, including very low temperatures like the ones found in the arctic region or extremely hot environments such as the ones seen in hot springs and deserts (Rai et al. 2013). For example, the diverse linage of the Nostocales was found in both hot and hold habitats, such as specialist *Dapistostemon* living in tropical regions and *Nostoc* living in polar regions, while the members of the family *Gomphospaericeae* are found mainly in warm waters (Dvorak et al. 2017; Vincent 2007). Cyanobacteria have a key role in the global environment via the nutrient cycle as they are able to fix nitrogen from the atmosphere, regulate the total nitrogen budget and efficiently function at high-salt concentrations (Capone et al. 1997; Karl 2002; Řeháková et al. 2009; Schneider et al. 2013). They are the primary producer in oceans (Hagemann 2011) and are capable of surviving under high salinity conditions like those found in hypersaline lakes and lagoons (Oren 2015). In hypersaline waters with salt concentrations above 10%, where many species such as eukaryotic micro- and macro-algae would struggle to survive, cyanobacteria are capable of enduring challenging environmental conditions (Schneider et al. 2013; Oren 2015). The example of species living in hypersaline conditions includes different morphological types such as unicellular halotolerant alkaliphilic cyanobacterium *Aphanothece halophytica* (Laloknam et al. 2006) and nonheterocystous filamentous species such as *Halospirulina tapeticola* (Oren 2015; Nübel et al. 2000).

7.2 Biotechnology and the Importance of Cyanobacteria

In the past, natural products were used for treating different diseases. For example, ancient Egyptians used the bark of willow trees to treat pain (Pitt 2015). Many natural cures were discovered following the recipes from ancient remedies, like the newly developed antimalaria medicine (Liu and Liu 2016). The list of biological properties of natural bioproducts is increasing at constant rate including the importance of these compounds for numerous biotechnological applications. The number of novel marine natural products (MNPs) has been increasing at a yearly rate, with 643 new compounds in 2016 (Blunt et al. 2018); 723 new compounds in 2017 (Carroll et al. 2019); and 1884 new compounds in 2018 (Carroll et al. 2020). Such an increase in novel bioactive molecules from the marine environments has especially been coming from marine microorganisms and phytoplankton, including marine bacteria, cyanobacteria, marine fungi and dinoflagellates. Invertebrates are at the top of this list, with the second number of new compounds being isolated from the sponge, cnidarians, molluscs, tunicates and others (Carroll et al. 2019). In invertebrates, most of the MNPs are microbe-driven and are a result of the symbiotic relationship between an invertebrate host and the microbial endosymbionts (Jiménez 2018). For example, cyanobacteria build many symbiotic relationships providing important bioactive compounds to the marine invertebrate hosts (Namikoshi and Rinehart 1996).

Cyanobacteria have significant potential for biotechnological applications ranging from use as biofuels for renewable energy (e.g. hydrogen or alcohols) to bioactive compounds with antibacterial, antiviral, antifungal, photoprotective and anticancer activities (Abed et al. 2009; Chen et al. 2019). These photosynthetic prokaryotes have been used as a food source rich in vitamins and proteins (Singh et al. 2005). Among many nutritious compounds identified in various cyanobacterial species, there are also pigments (e.g., carotenoids and chlorophyll) (Encarnação et al. 2015; Singh et al. 2005), minerals (Kamennaya et al. 2012) and, in some cases, toxins including hepatotoxins (i.e., microcystins and nodularins) and neurotoxins (i.e., anatoxins and saxitoxins) (Namikoshi and Rinehart 1996). Especially rich in bioactive compounds is marine species *Lyngbya majuscule producing* polyketides, lipopeptides and toxin such as apratoxin A that shows potent anticancer activity (Luesch et al. 2001; Singh et al. 2005).

The biotechnological applications of cyanobacteria are increasing over time, including the use of their bioactive compounds in the food industry for the production of protein-based food ingredients (Grossmann et al. 2020). In agriculture, cyanobacteria are used as biofertilisers (i.e. *Anabaena* sp.; *Nostoc* sp.) to improve soil fertility and as a food source such as in the case of *Spirulina*, which is characterised by anti-inflammatory properties (Karkos et al. 2011; Garlapati et al. 2019). In medicine, cyanobacteria bioproducts are used as natural drugs (Bin-Meferij and Hamida 2019; Amador-Castro et al. 2020), and in cosmetics, their bioactive compounds have shown promising photoprotective and antioxidant potential (Morone et al. 2019). Among the other very essential bioproducts, cyanobacteria contain UV-absorbing mycosporine-like amino acids (MAAs) and scytonemin (Rastogi and Incharoensakdi 2014a, b), which are promising molecules for the creation of organics sunscreens (Rosic 2019). The bioremediation or detoxification of hazardous chemicals from the environment is successfully done by certain cyanobacterial species (e.g., diazotrophic cyanobacteria) capable of removing or degrading heavy metals from the contaminated soils (Priyanka et al. 2020).

7.3 The Diversity and Abundance of Cyanobacteria

Cyanobacteria are capable of living in a diverse range of environmental conditions, from the coldest to the hottest conditions and from a sole independent state to a symbiotic partnership. There are over 2500 cyanobacterial species reported in the CyanoDB database, although an accurate estimation of true biodiversity is still to be determined for some taxonomic groups (Nabout et al. 2013). There are other recent databases that provide information about the diversity of cyanobacterial species including the NCBI Taxonomy database are available via https://www.ncbi.nlm.nih.gov/taxonomy (Schoch et al. 2020); *JGI Genome Portal* offers information about sequencing projects and related data that are available via https://genome.jgi.doe.gov/portal/cyanobacteria/cyanobacteria.info.html (Nordberg et al. 2014); and *Cyanobacterial KnowledgeBase* (CKB) is the database that includes genomic and proteomic data about 74 completely sequenced cyanobacterial species including the tools for sequence analyses http://nfmc.bdu.ac.in/ckb/ (Peter et al. 2015). There are also *Cyanobacteria Gene Annotation Database* (http://cyano.genome.jp/), community database and *CyanoType database* that includes taxonomic, phylogenetic or genomic data for some cyanobacterial strains (http://lege.ciimar.up.pt/cyanotype) (Ramos et al. 2017).

7.3.1 Growth of Cyanobacteria

In various ecosystems, the ability of cyanobacteria to grow is limited by the presence of nutrients and specifically the levels of nitrogen (N) and/or phosphorus (P), as well as light and temperature conditions. The P availability is known to be a major limiting factor in freshwater ecosystems (Schindler 1977), impacting cyanobacteria abundance in lakes (Downing et al. 2001). Culturing conditions can be modified to improve cyanobacterial growth and the yield of isolated bioactive compounds including the changes in the external conditions (Bode et al. 2002). In the case of MNPs, this would include the manipulation of temperature, salinity and light, as major abiotic factors that may significantly impact the synthesis of bioproducts (Ingebrigtsen et al. 2016).

7.3.2 Hot vs. Cold Environments

Cyanobacteria are found in very cold environments, such as the Antarctic region. The ice-based habitats occupied by cyanobacteria include the meltwater ponds during summer that are entirely frozen during winter. Consequently, for species of cyanobacteria to be able to survive these extreme conditions of freezing, desiccation and high salinity stress need to be highly flexible, such as the representatives from the orders Nostocales and Oscillatoriales (Howard-Williams et al. 1989; Jungblut et al. 2005). Cyanobacteria are also found in the soils and rocks of Antarctica in areas of melted water (Zakhia et al. 2008) and as biofilms within these rocks (Vincent 1989; Hughes et al. 2004). The presence of cryptoendolithic cyanobacteria was confirmed using molecular analyses in these rock and soil samples while biofilms and benthic mats were discovered in rivers and in different Antarctic lakes (Vincent 1989; Vopel and Hawes 2006).

Similar to Antarctica, cyanobacteria are found in the Arctic region in various lakes, streams and ponds, with the most common species from the order Oscillatoriales and Nostocales forming the benthic microbial mats (Bonilla et al. 2005). Terrestrial cyanobacteria found in Arctic soils and rocks are represented by diverse communities of cryptoendolithic and hypolithic cyanobacteria (Cockell and Stokes 2004; Omelon et al. 2006). Some cyanobacteria are found in the Arctic marine environment, influenced by freshwater inflows and are still less abundant compared to warmer sea waters (Waleron et al. 2007). Cyanobacteria are also found in Alpine habitats, including freshwater lakes, rocks and soil crusts (Zakhia et al. 2008).

In high-temperature environments, such as hot springs and geysers, cyanobacteria are able to survive despite the extremely high temperatures. Temperatures above 45 °C in hot springs result in mat formation, as the top layers are made from these photosynthetic microorganisms (Boomer et al. 2009). These mats or biofilms are usually 1–2 mm thick, resulting in a severe reduction in visible light in deeper water layers (Castenholz 2009). In the acidic hot spring pools in the Waiotapu geothermal area (Rotorua, New Zealand) with water temperature between 50 and 80 °C and pH of 1–3 (Jones and Renaut 2006), cyanobacterial mats were formed above water at the side of the spring pools and channels (Fig. 7.1). The sampling done in Fairy Geyser, Yellowstone National Park, indicated the seasonal shift in the population of varied cyanobacterial species (Boomer et al. 2009). The cyanobacterium of the genus *Synechococcus* found in alkaline hot springs in Yellowstone National Park and many other hot springs is a commonly studied group to better understand the evolution of thermotolerance (Miller and Castenholz 2000). The species coming from this genus are able to form biofilms at temperatures reaching ~73 °C (Boomer et al. 2009), although the majority of cyanobacterial species prefer warm temperatures ranging between 30 and 40 °C (Dodds and Whiles 2010). For that reason, cyanobacterium *Synechococcus* is the most commonly found at higher temperatures (54–74 °C), while the motile filamentous species of the genus *Oscillatoria* are found to dominate in hot springs with a slightly lower temperature range of 47–54 °C (Dodds and Whiles 2010).

Fig. 7.1 Waiotapu geothermal area of Rotorua (New Zealand) with acidic hot springs characterised by the formation of tick green cyanobacterial mats alongside Champagne pool (**a**) and channels (**b**)

7.3.3 Freshwater vs. Saltwater Environments

Cyanobacteria occur in freshwater environments, and under certain conditions, the algal overgrowth results in their overabundance, known as algal blooms. These harmful algal blooms (HABs) usually happen due to fluctuations in nutrient content, typically the elevation of nitrogen and phosphorous. The additional factors contributing to HAB are extreme pH conditions—acidic or alkaline—and low oxygen levels (Dvorak et al. 2017). According to the World Health Organization, there is a substantial worldwide occurrence of toxin-producing cyanobacteria (World Health Organization 2003). Massive algal blooms result in toxin production that could impact drinking water (Codd et al. 2005; He et al. 2016; Zamyadi et al. 2012), as well as other human activities, including fisheries, and also result in ecological shifts (Dvorak et al. 2017). Consequently, harmful cyanobacteria producing toxin substances are now recognised as a global problem requiring improved and novel strategies for HAB prediction, monitoring and suppression (Schmidt et al. 2014; Paerl and Otten 2016). Cyanobacterial toxins are grouped based on their toxicology to hepatotoxins, neurotoxins, cytotoxin irritants and dermatotoxins (Wiegand and Pflugmacher 2005). Examples of highly toxic cyanotoxins include microcystins, which are cyclic heptapeptides that have acute and chronic toxicities and are found in many genera, including *Microcystis*, *Nostoc* and *Anabaena* (Bláha et al. 2009). Another example is anatoxin-a, which is a neurotoxin that has a negative impact on nerve transmission, acting as a postsynaptic neuromuscular blocker and is also found in multiple genera (Bláha et al. 2009).

Cyanobacteria are found in saltwater and are able to tolerate different levels of salinity based on the external concentration of NaCl. In nutrient-poor oceans, from tropic to subtropical regions, cyanobacteria are the dominating phytoplankton (in terms of species numbers and biomass produced), making cyanobacteria an important primary producer in oceans (Sigman and Hain 2012). The high levels of salinity, as well as the variability in salt concentrations, are often limiting factors for microbial growth (Hagemann 2011). The ability to acclimatise to the high-salt environments was studied in the model organism *Synechocystis* sp. 6803 (Hagemann 2011). The subsequent phases were identified to be important for the process of salt acclimation, involving hundreds of differential expressed genes, toxic inorganic ions extrusion via active processes and the accumulation of compatible organic solutes.

7.3.4 Solitary vs. Symbiotic Life

Cyanobacteria are found in solitary and symbiotic forms. Symbiosis is a partnership between two organisms that can be beneficial, neutral, but also negative for some partners in symbiosis (Demoulin et al. 2019). The symbiosis between cyanobacteria and other organisms includes plants, fungi and also algae (Whitton and Potts 2002). The majority of these symbioses are based on the ability of cyanobacteria to fix nitrogen and transfer it to the host (Whitton and Potts 2002). The mutualistic symbiotic relationship between fungi and cyanobacteria (and/or eukaryotic

Fig. 7.2 Orange (**a**) and green (**b**) lichens growing on the rocks in Cania Gorge dry rainforest in Queensland (Australia)

unicellular algae) has been successfully built in lichens (Hyvärinen et al. 2002). Lichens are capable of surviving in various environmental conditions but usually occupy the surfaces of tree bark and rocks (Fig. 7.2). These lichens can adapt to prolonged periods of dehydration and hydration, like those reported in high mountains (Aubert et al. 2007).

Certain cyanobacterial species from the genus *Synechococcus* that tolerate high salinity concentrations are the most common cyanobacteria found in mutualistic symbiosis with marine sponges (Erwin and Thacker 2008). The species from the order Nostocales are found in different environments widely distributed in terrestrial and aquatic habitats and also form symbiotic relationships (Dvorak et al. 2017). Some species from the genus *Chroococcidiopsis* are very diverse and found in hot and cold environments, as a bluish-green layer under the rock debris (Boison et al. 2004), as well as in hot springs with different pH and temperature conditions (Sompong et al. 2005) and in symbiosis with various algae (Büdel et al. 2009). The cyanobacterium *Richelia intracellularis* forms a cyanobacterial–diatom symbiosis with several diatom species providing critical nitrogen supply to its partner in symbiosis (Foster and Zehr 2006).

7.4 Stressful Environments Occupied by Cyanobacteria

Identifying 'planetary boundaries' important for controlling human activities is needed to sustain the natural balance and protect the earth's limited resources (Rockstrom et al. 2009). Over the past 10,000 years, in the period known as the Holocene, the environmental conditions on our planet have been very stable (Petit et al. 1999). However, after the industrial revolution, anthropogenic activities challenged the global environment by having a detrimental impact on various ecosystems. So far, cyanobacteria have successfully survived various environmental conditions over the past 3 billion years, starting with the extremely harsh period characterised by very high UV levels and anaerobic conditions (Cockell and Knowland 1999).

The ability of an organism to tolerate stress varies and is influenced by multiple factors, including the organisms' past and present environmental conditions (Middlebrook et al. 2008; Rosic and Dove 2011; Rai et al. 2013). Consequently, the same external conditions could be potentially beneficial for one organism and detrimental for another, highlighting the importance of understanding the definition of stress as a condition leading to a decrease in an organism's functionality and fitness (Bijlsma and Loeschcke 2005). Cyanobacteria, as ancient organisms, have been able to survive and adapt to various highly hostile environments (Rastogi et al. 2020). Current climate changes are resulting in the deterioration of many different ecosystems around the world (Gardner et al. 2003). Due to increasing seawater temperatures and the current trajectory of greenhouse gas emission scenarios (IPCC 2014), there is an alarming decline in marine species biodiversity (De'ath et al. 2009). Marine ecosystems such as coral reefs are predicted to deteriorate further and to be severely impaired by 2040–2050 (Hoegh-Guldberg et al. 2017). World global initiatives are trying to enforce industrial limits in carbon emission with an aim to control the progress of global warming and further reduce temperature increases (Seneviratne et al. 2018). This book chapter will provide an overview of the diversity of molecular mechanisms applied by cyanobacteria to adjust, survive and tolerate shifts in environmental conditions.

7.4.1 Cyanobacterial Coping Mechanisms During Stress

Stress can be defined as a condition where disruptive changes in an organism's functionality lead to reduced fitness (Bijlsma and Loeschcke 2005). There are two types of environmental stresses, abiotic and biotic stressors. Abiotic stressors include many external factors, such as high levels of solar radiation, high and low temperatures, high salinity and drought, while biotic stressors encompass the negative impact of living organisms such as various pathogens (e.g., fungi and bacteria). Both stress factors contribute to functional challenges organisms may face during their lifetime. If stressful conditions are overreaching the functional threshold that the organism can tolerate over a prolonged time, adaptability and acclimatisation may not be possible and stress exposure may potentially lead to detrimental consequences. Here, we will discuss the impacts of major abiotic factors impacting the functionality of cyanobacteria such as temperature stress, UV stress and hypersalinity (Table 7.1).

7.4.2 Temperature Variations and Adaptive Mechanisms

Cyanobacteria adapted to cold temperatures are capable of maintaining the osmotic balance and membrane fluidity needed to survive freezing conditions (Zakhia et al. 2008). This membrane flexibility is possibly due to the accumulation of shorter chain polyunsaturated fatty acids within the membrane (Vincent 2007). The osmotic balance that allows for a reduction in the freezing temperature of intracellular fluids and promotes survival at low temperatures is possibly due to the accumulation of compatible solutes such as trehalose. This osmotic equilibrium also reduces the impact of cell desiccation that cells are exposed to during periods of very low temperatures due to the production of extracellular polymeric substances (Zakhia et al. 2008). These extracellular compounds, such are extracellular polysaccharides (ECP), protect cold-adapted cyanobacterial species from ice formation around the cells (Vincent 2007) and improve cold and desiccation tolerance (Tamaru et al. 2005).

Due to climate change, prolonged periods of marine heatwaves (MHWs) have been reported and characterised by increased seawater temperatures (Oliver et al. 2018; Hayashida et al. 2020). As a result, increased thermal stress on cyanobacteria and other organisms, due to more frequent MHW, is putting additional pressure on the survival of many species and ocean biodiversity. As a result, many organisms will be able to adjust by modifying their gene expression profiles. Under high-temperature conditions, changes are happening in molecular mechanisms involving the production of certain molecular chaperones such as heat shock proteins (HSPs) important for protein folding, unfolding, aggregation, degradation and transport. HSPs prevent damage done to cells under different environmental stress conditions (Rosic et al. 2011). Major abiotic factors influencing HSP induction include temperature stress, UVR and salinity, as well as exposure to pollutants such as pesticides and heavy metals. It is critical to understand how cyanobacteria can adapt and

Table 7.1 Molecular mechanisms employed by cyanobacteria for adaptation to major abiotic stress factors including high and low temperatures, UV radiation and salinity stress

Abiotic stressor	Mechanisms	Molecule	Features	References
Temperature	Osmotic balance; membrane fluidity	Shorter chain polyunsaturated fatty acids	Accumulation of shorter chain polyunsaturated fatty acids within the membrane	Vincent (2007)
	Osmotic balance	Trehalose	Accumulations of compatible solutes allow tolerance to cold by decreasing the intracellular fluid freezing temperature and cell desiccation	
	Adhesion, cellular protection	Extracellular polysaccharides (EPS)	*Nostoc commune* macroscopic colonies have cells embedded within the EPS improving cold and desiccation tolerance	Tamaru et al. (2005), Rossi and De Philippis (2015)
	Differential gene expression/ protein synthesis	Heat shock proteins (HSPs)	HSPs affect protein folding, unfolding, aggregation, degradation and transport	
UV radiation	Photoprotection; scavenging ROS	Scytonemin	Absorbing radiation in the UV-A range (315–400 nm) with max at 370 nm; antiproliferative, anti-inflammatory and antioxidant activities	Vincent (2007), Sinha and Häder (2008), Fuentes-Tristan et al. (2019)
		Mycosporine-like amino acids (MAAs)	Absorbing radiation in the UVA and UVB (280–315 nm) ranges with max from 310 to 362 nm; antiproliferative, anti-inflammatory; antiageing features; *antioxidant* activities	Shick and Dunlap (2002), Singh et al. (2010), Rosic (2019)
		Carotenoid pigments	Absorbing radiation in the UVR and visible range 300–600 nm; light-harvesting molecules; antioxidant, anti-inflammatory and antiproliferative activities	Klassen (2010), Rastogi et al. (2010), Lopes et al. (2020)
	Scavenging ROS	Polyphenols	Antioxidant properties	Singh et al. (2017)
Hypersalinity	Osmotic balance (via accumulation of compatible solutes)	Disaccharides (sucrose and trehalose)	In freshwater cyanobacteria (low-salt tolerance strains), protection against desiccation	Klähn and Hagemann (2011), Oren (2015)

(continued)

Table 7.1 (continued)

Abiotic stressor	Mechanisms	Molecule	Features	References
		Glucosylglycerol (GG)	In cyanobacteria from marine environments (moderate-salt tolerance); filamentous (e.g. Coleofasciculus) and unicellular (*Synechocystis*) halophilic cyanobacteria	Borowitzka et al. (1980), Klähn and Hagemann (2011)
		Glycine betaine (GB)	In hypersaline environments, GB has accumulated plus GG, disaccharides at lower concentrations	Klähn and Hagemann (2011), Oren (2015)
		MAAs	MAA secondary metabolites in the cytoplasm play a role in osmotic regulation and drought protection	Oren and Gunde-Cimerman (2007)

potentially acclimatise to elevated temperatures associated with MHW, as they are anticipated to be more prominent in the future. Other studies suggested that biomarkers based on differential transcriptomics profiles may be used to detect and monitor early changes in gene regulations and the onset of thermal stress (Rosic et al. 2010, 2013). In cyanobacteria, the ability of HSP to maintain protein homeostasis via protein folding, unfolding, aggregation and degradation processes has been proposed to be critically important for their adaptability over billions of years (Rajaram et al. 2014). High temperatures were found to impact photosynthetic processes and PSII activity decreased as a result (Zhang and Liu 2016). As a result of elevated temperatures, changes were detected at all levels, including transcriptomics, proteomics and lipidomics. The effect of heat stress on the expression of molecular chaperones, photosynthetic and oxidative stress-related genes was reported in the model organism cyanobacterium *Synechocystis* sp. PCC 6803 (Murata et al. 2007; Červený et al. 2015). The enzyme histidine kinase Hik34 was found to play an important role in heat tolerance via the regulation of expression of heat shock genes (Suzuki et al. 2005). During short-term stress exposure, the acclimation of cyanobacterial photosystem II (Murata et al. 2007) and other cellular structures such as cytoplasmic and thylakoid membranes was also reported (Inoue et al. 2001; Balogi et al. 2005). A high level of HSP expression was found to be critical for the adaptation of *Synechocystis* sp. model cyanobacterium to prolonged heat stress (over 24 h), including the presence of functional histidine kinase Hik34 (Červený et al. 2015).

7.4.3 UV Stress and the Mechanisms of Protection

Over a long period of time, organisms on the planet Earth developed various and multiple strategies to reduce the negative impacts of damaging UV radiation. Various strategies were implemented to fix DNA damage like nucleotide excision repair, base excision and the mismatch nucleotide repair mechanism (Korbee et al. 2010; Rosic 2012). The accumulation of antioxidants capable of mopping reactive oxygen species (ROS) and UV-absorbing molecules capable of absorbing solar radiation are additional mechanisms of UV protection (Korbee et al. 2010; Ikehata and Ono 2011). Cyanobacteria have been exposed to high irradiance and, specifically, high levels of damaging ultraviolet radiation (UVR) over a long period of history (Shick and Dunlap 2002). The photoprotective mechanisms in cyanobacteria include the presence of UV-absorbing molecule scytonemin, mycosporine-like amino acids (MAAs) and also carotenoid pigments (Korbee et al. 2010; Rastogi et al. 2010; Singh et al. 2010). Cyanobacteria are also excellent sources of antioxidants important for scavenging ROS (Ismaiel et al. 2014), such as polyphenolic compounds (Singh et al. 2017). Cyanobacterial species that accumulate these antioxidant polyphenols are capable of a higher level of flexibility and adaptation when exposed to abiotic stressors like UVR and also adaptability to various habitats (Singh et al. 2017).

Scytonemin is a pigment of yellow-brown colour, unique to cyanobacteria and capable of absorbing within the UVA range with the maximum absorption at 370 nm (Sinha and Häder 2008; Rastogi et al. 2014). This UV-absorbing compound was first time discovered in some terrestrial cyanobacteria as reported in 1849 by Nageli and colleagues (see review by Sinha and Häder 2008).

The molecule of scytonemin (molecular mass 544 Da) is a dimer made of indolic and phenolic units connected to produce a unique ring so-called 'the scytonemin skeleton' (Proteau et al. 1993). This pigment was identified in 13 from 20 analysed strains of cyanobacteria including *Synechococcus* sp., *Gloeocapsa* sp. and *Lyngbya aestuarii*, but not in *Nostoc microscopicum*, *Spirulina subsalsa* and other species (Proteau et al. 1993). Beyond UV-absorbing properties, this pigment also has an antioxidant capacity in scavenging ROS, as well as antiproliferative and anti-inflammatory activities (Fuentes-Tristan et al. 2019). High levels of scytonemin have been found to result in dark (even black) cyanobacterial mat coloration (Vincent 2007).

Carotenoid pigments are accessory pigments in the photosynthetic pathway found in cyanobacteria and are also found in other organisms capable of photosynthesis (Klassen 2010). They play a role in photoprotection by absorbing solar radiation within the range of 300–600 nm and are necessary as light-harvesting molecules in photosynthesis (Klassen 2010; Rastogi et al. 2010). In cyanobacteria, carotenoid pigments play a role as antioxidants scavenging free radicals, such as the superoxide anion radical and with the highest concentration reported in terrestrial and freshwater strains (Lopes et al. 2020). As powerful antioxidants, carotenoids also showed the potential for the treatment of psoriasis (Lin and Huang 2016).

In cyanobacteria and many other species, including aquatic and terrestrial animals, phytoplankton, fungi, and macro- and micro-algae accumulation of MAAs has multiple protective roles (Shick and Dunlap 2002; Sinha et al. 2007). The UV-absorbing 'S-320' compounds were first discovered in various coral species and a blue-green alga from the Great Barrier Reef in 1969 (Shibata 1969). MAAs were identified to be small molecules (<400 Da), colourless and hydrophilic compounds that in core contain a ring (cyclohexenone or cyclohexenimine) conjugated to an amino acid residue or imino alcohol (Nakamura et al. 1982; Dunlap and Chalker 1986; Carreto and Carignan 2011). MAAs are a diverse group of small amino acid-based molecules playing the role of antioxidants and photoprotective molecules (Singh et al. 2010) and are the most commonly found secondary metabolites among aquatic species (Rastogi et al. 2010). These UV-absorbing molecules are aquatic sunscreen molecules shared among many marine and freshwater species that can absorb UVR in both UVA and UVB ranges, with maximum absorbance between 310 and 362 nm (Shick and Dunlap 2002). MAAs were for the first time identified in 13 from 20 analysed cyanobacterial species, including *Synechococcus* sp., *Gloeocapsa* sp. and *Spirulina subsalsa* (Garcia-Pichel and Castenholz 1993).

The characteristic of MAA and scytonemin pigments are the ability not only to absorb UVR and in that way prevent ROS production, but also to act as antioxidants scavenging already produced ROS (Wada et al. 2013; Rastogi et al. 2016). The double UV-absorbing set of molecules, including MAAs and scytonemin that is found in cyanobacteria, has been thought to be an important factor for protection against extremely high UV radiation that was characteristic of the earths' atmosphere when still deprived of oxygen (Cockell and Knowland 1999). In that way, the multifunctional roles of UV-absorbing compounds are achieved in the protection of cyanobacterial well-being during extreme exposure to UVR.

7.4.4 Adapting to Hypersalinity Stress and Other Coping Mechanisms

Under hypersaline conditions, microbial organisms need to keep the osmotic balance between internal and external environments (Table 7.1). The mechanisms employed in balancing water permeability are happening due to the massive accumulation of ions such K^+ and Cl^- in some other microorganisms (Oren 2006). Cyanobacteria accumulate organic osmotic solutes to balance out high salinity conditions (Hagemann 2011). Many diverse cyanobacterial species are reported to live at high salinity conditions, even reaching NaCl saturation levels (Oren 2015), and many are capable of adapting to changing salinity conditions (Golubic 1980).

In the case of cyanobacteria, which are less tolerant of high salinity conditions like many freshwater cyanobacteria, under salt stress, they accumulate additional quantities of disaccharides, such as sucrose and trehalose, as compatible solutes to maintain osmotic balance (Oren 2015). The accumulation of sucrose in cyanobacteria has been applied in other photosynthetic organisms, such as algae and land plants and was an essential mechanism for adaptation to salt stress (Kolman et al. 2015). The accumulation of disaccharides allows improved osmotic balance under hypersalinity conditions and an increase in adaptability to other abiotic stresses such as desiccation and high temperatures (Klähn and Hagemann 2011). However, these disaccharide molecules provide only initial protection under low-salt stress, while with further increases in salinity (e.g. Coleofasciculus), osmotic protection is achieved by the accumulation of organic solutes such as glucosylglycerol (Karsten 1996). Glucosylglycerol is an osmotic solute found in filamentous (e.g. Coleofasciculus) and unicellular (from the genus *Synechocystis*) halophilic cyanobacteria (Borowitzka et al. 1980).

In marine environments, many cyanobacteria have demonstrated moderate salt tolerance due to the accumulation of glucosylglycerol, which provides higher salt tolerance than sucrose and/or trehalose on its own (Klähn and Hagemann 2011; Oren 2006, 2015). Under extreme salinity conditions, cyanobacteria can survive the high osmotic pressure, thanks to additional adaptation mechanisms. These salt-tolerant cyanobacteria accumulate organic solutes such as glycine betaine plus glucosylglycerol (Klähn and Hagemann 2011). The solute glycine betaine is most commonly found in halophilic cyanobacteria living in high salinity conditions and is often accompanied by other compatible solutes, for example glucosylglycerol and disaccharides at lower concentrations (Klähn and Hagemann 2011).

Additional adaptation mechanisms to the hypersaline conditions include the accumulation of UV-absorbing MAA secondary metabolites in the cytoplasm, which, as low molecular weight neutral compounds, play a role in osmotic regulation and drought (Oren and Gunde-Cimerman 2007). In some cases, these high salinity waters are deprived of oxygen resulting in the development of an additional adaptation mechanism that involves the use of sulphide as an electron donor for the anoxygenic photosynthesis (Klatt et al. 2015). Instead of oxygenic photosynthesis, where water is used as an electron donor, in anoxygenic photosynthesis, the electron supply is from sulphide, which allows for CO_2 fixation and the generation of photosynthetic products (Padan 1979).

7.4.5 Redox Controlling Mechanisms

All cyanobacteria are capable of using redox controlling mechanisms to mitigate the negative impacts of various biotic and abiotic stress factors, including hypersalinity. The redox system in cyanobacteria comprised of redox enzymes glutaredoxins and thioredoxins is critical for reducing the negative impact of oxidative stress and for cellular protection from free radicals (Grant 2001). For example, glutaredoxins, which are divided into six classes based on the active site and the region of glutathione binding (Couturier et al. 2009), are heat-stable oxidoreductases important for cellular redox homoeostasis in cyanobacteria (Begas et al. 2017).

7.5 Perspective and Conclusion

Surviving species in today's environment are the ones that managed to adapt and to adjust to changing environments through the history of the planet Earth. Prokaryotic cyanobacteria are an example of ancient organisms that, after over 2 billion years of history, manage to adjust to changes happening in their surroundings and to employ various mechanisms to survive. Exposed to extreme levels of UVR, these early organisms accumulate different types of sunscreens and get a sufficient level of protection from damaging UV levels. Cyanobacteria have some shared UV-absorbing compounds that are found in many other species such as MAAs and carotenoids, and a unique cyanobacterial scytonemin, making them potentially the most adapted organisms on the planet to extreme levels of UV radiation. The variability of external conditions in terms of high and low temperatures did not prevent cyanobacteria's astonishing adaptability to accumulate molecules such as extracellular polysaccharides, to protect them from freezing polar temperatures by decreasing intracellular freezing point and increasing desiccation tolerance. Various molecular mechanisms are employed under high temperatures, including molecular chaperones to allow cyanobacteria survival even in hot springs where temperatures are reaching over 60 °C.

Cyanobacteria occupy habitats where the environmental conditions are impossible for survival for almost any other species, such as extremely hypersalinity conditions. By the accumulation of different osmotically compatible solutes, cyanobacteria are able to accommodate and preserve intracellular content from dehydration, even in hostile saline environments. Finally, these versatile cyanobacterial species have a robust redox system and many bioactive components that can be used in biotechnology, utilising their UV-absorbing, antioxidant, antiproliferative and anti-inflammatory properties. Therefore, the winners that are here to stay despite the impacts of climate change in existing environments and can adjust to almost any stressors are certainly miscellaneous cyanobacterial species.

Acknowledgement The author would like to sincerely thank the reviewers, as well as Ms Isidora Skrlin, for their critical reviews of this book chapter.

References

Abed RMM, Dobretsov S, Sudesh K (2009) Applications of cyanobacteria in biotechnology. J Appl Microbiol 106:1–12

Amador-Castro F, Rodriguez-Martinez V, Carrillo-Nieves D (2020) Robust natural ultraviolet filters from marine ecosystems for the formulation of environmental friendlier bio-sunscreens. Sci Total Environ 749:141576

Aubert S, Juge C, Boisson A-M, Gout E, Bligny R (2007) Metabolic processes sustaining the reviviscence of lichen *Xanthoria elegans* (Link) in high mountain environments. Planta 226: 1287–1297

Balogi Z, Török Z, Balogh G, Jósvay K, Shigapova N, Vierling E, Vígh L, Horváth I (2005) "Heat shock lipid" in cyanobacteria during heat/light-acclimation. Arch Biochem Biophys 436:346–354

Begas P, Liedgens L, Moseler A, Meyer AJ, Deponte M (2017) Glutaredoxin catalysis requires two distinct glutathione interaction sites. Nat Commun 8:14835

Bijlsma R, Loeschcke V (2005) Environmental stress, adaptation and evolution: an overview. J Evol Biol 18:744–749

Bin-Meferij MM, Hamida RS (2019) Biofabrication and antitumor activity of silver nanoparticles utilizing novel Nostoc sp. Bahar M. Int J Nanomedicine 14:9019–9029

Bláha L, Babica P, Maršálek B (2009) Toxins produced in cyanobacterial water blooms - toxicity and risks. Interdiscip Toxicol 2:36–41

Blunt JW, Carroll AR, Copp BR, Davis RA, Keyzers RA, Prinsep MR (2018) Marine natural products. Nat Prod Rep 35:8–53

Bode HB, Bethe B, Höfs R, Zeeck A (2002) Big effects from small changes: possible ways to explore nature's chemical diversity. ChemBioChem 3:619–627

Boison G, Mergel A, Jolkver H, Bothe H (2004) Bacterial life and dinitrogen fixation at a gypsum rock. Appl Environ Microbiol 70:7070

Bonilla S, Villeneuve V, Vincent W (2005) Benthic and planktonic algal communities in a High Arctic Lake: pigment structure and contrasting responses to nutrient enrichment. J Phycol 41: 1120–1130

Boomer SM, Noll KL, Geesey GG, Dutton BE (2009) Formation of multilayered photosynthetic biofilms in an alkaline thermal spring in Yellowstone National Park, Wyoming. Appl Environ Microbiol 75:2464

Borowitzka LJ, Demmerle S, Mackay MA, Norton RS (1980) Carbon-13 nuclear magnetic resonance study of osmoregulation in a blue-green alga. Science 210:650–651

Büdel B, Darienko T, Deutschewitz K, Dojani S, Friedl T, Mohr KI, Salisch M, Reisser W, Weber B (2009) Southern African biological soil crusts are ubiquitous and highly diverse in drylands, being restricted by rainfall frequency. Microbial Ecol 57:229–247

Capone DG, Zehr JP, Paerl HW, Bergman B, Carpenter EJ (1997) Trichodesmium, a globally significant marine cyanobacterium. Science 276:1221

Carreto JI, Carignan MO (2011) Mycosporine-like amino acids: relevant secondary metabolites. Chemical and ecological aspects. Mar Drugs 9:387–446

Carroll AR, Copp BR, Davis RA, Keyzers RA, Prinsep MR (2019) Marine natural products. Nat Prod Rep 36:122–173

Carroll AR, Copp BR, Davis RA, Keyzers RA, Prinsep MR (2020) Marine natural products. Nat Prod Rep 37:175–223

Castenholz RW (2009) Mats, microbial. In: Schaechter M (ed) Encyclopedia of microbiology, 3rd edn. Academic, Oxford, pp 278–292

Červený J, Sinetova MA, Zavřel T, Los DA (2015) Mechanisms of high temperature resistance of *Synechocystis* sp. PCC 6803: an impact of histidine kinase 34. Life (Basel, Switzerland) 5:676–699

Chen H, Li T, Wang Q (2019) Ten years of algal biofuel and bioproducts: gains and pains. Planta 249:195–219

Cockell CS, Knowland J (1999) Ultraviolet radiation screening compounds. Biol Rev Camb Philos Soc 74:311–345

Cockell C, Stokes M (2004) Widespread colonization by polar hypoliths. Nature 431:414

Codd GA, Morrison LF, Metcalf JS (2005) Cyanobacterial toxins: risk management for health protection. Toxicol Appl Pharmacol 203:264–272

Couturier J, Jacquot J-P, Rouhier N (2009) Evolution and diversity of glutaredoxins in photosynthetic organisms. Cell Mol Life Sci 66:2539–2557

De'ath G, Lough JM, Fabricius KE (2009) Declining coral calcification on the great barrier reef. Science 323:116–119

Demoulin CF, Lara YJ, Cornet L, François C, Baurain D, Wilmotte A, Javaux EJ (2019) Cyanobacteria evolution: insight from the fossil record. Free Radic Biol Med 140:206–223

Dodds WK, Whiles MR (2010) Unusual or extreme habitats. In: Dodds WK, Whiles MR (eds) Freshwater ecology, 2nd edn. Academic, London, pp 375–398

Downing JA, Watson SB, McCauley E (2001) Predicting Cyanobacteria dominance in lakes. Can J Fish Aquat Sci 58:1905–1908

Dunlap WC, Chalker BE (1986) Identification and quantitation of near-UV absorbing compounds (S-320) in a hermatypic scleractinian. Coral Reefs 5:155–159

Dvorak P, Casamatta D, Hasler P, Jahodářová E, Norwich A, Poulíčková A (2017) Diversity of the Cyanobacteria. In: Hallenbeck P (ed) Modern topics in the phototrophic prokaryotes: environmental and applied aspects. Springer, Cham, pp 3–46

Encarnação T, Pais AACC, Campos MG, Burrows HD (2015) Cyanobacteria and microalgae: a renewable source of bioactive compounds and other chemicals. Sci Prog 98:145–168

Erwin PM, Thacker RW (2008) Cryptic diversity of the symbiotic cyanobacterium *Synechococcus* spongiarum among sponge hosts. Mol Ecol 17:2937–2947

Foster R, Zehr J (2006) Characterization of diatom-cyanobacteria symbioses on the basis of nifH, hetR and 16S rRNA sequences. Environ Microbiol 8:1913–1925

Fuentes-Tristan S, Parra-Saldivar R, Iqbal HMN, Carrillo-Nieves D (2019) Bioinspired biomolecules: mycosporine-like amino acids and scytonemin from *Lyngbya* sp. with UV-protection potentialities. J Photochem Photobiol B Biol 201:111684

Garcia-Pichel F, Castenholz RW (1993) Occurrence of UV-absorbing, mycosporine-like compounds among cyanobacterial isolates and an estimate of their screening capacity. Appl Environ Microbiol 59:163–169

Gardner TA, Cote IM, Gill JA, Grant A, Watkinson AR (2003) Long-term region-wide declines in Caribbean corals. Science 301:958–960

Garlapati D, Chandrasekaran M, Devanesan A, Mathimani T, Pugazhendhi A (2019) Role of cyanobacteria in agricultural and industrial sectors: an outlook on economically important byproducts. Appl Microbiol Biotechnol 103:4709–4721

Golubic S (1980) Halophily and halotolerance in cyanophytes. Orig Life 10:169–183

Grant CM (2001) Role of the glutathione/glutaredoxin and thioredoxin systems in yeast growth and response to stress conditions. Mol Microbiol 39:533–541

Grossmann L, Hinrichs J, Weiss J (2020) Cultivation and downstream processing of microalgae and cyanobacteria to generate protein-based technofunctional food ingredients. Crit Rev Food Sci Nutr 60:2961–2989

Hagemann M (2011) Molecular biology of cyanobacterial salt acclimation. FEMS Microbiol Rev 35:87–123

Hayashida H, Matear RJ, Strutton PG, Zhang X (2020) Insights into projected changes in marine heatwaves from a high-resolution ocean circulation model. Nat Commun 11:4352

He X, Liu YL, Conklin A, Westrick J, Weavers LK, Dionysiou DD, Lenhart JJ, Mouser PJ, Szlag D, Walker HW (2016) Toxic cyanobacteria and drinking water: impacts, detection, and treatment. Harmful Algae 54:174–193

Hoegh-Guldberg O, Poloczanska ES, Skirving W, Dove S (2017) Coral reef ecosystems under climate change and ocean acidification. Front Mar Sci. https://doi.org/10.3389/fmars.2017.00158

Howard-Williams C, Pridmore R, Downes MT, Vincent WF (1989) Microbial biomass, photosynthesis and chlorophyll a related pigments in the ponds of the McMurdo Ice Shelf, Antarctica. Antarct Sci 1:125–131

Hughes K, McCartney HA, Lachlan-Cope T, Pearce D (2004) A preliminary study of airborne microbial biodiversity over Peninsular Antarctica. Cell Mol Biol (Noisy-le-Grand, France) 50: 537–542

Hyvärinen M, Härdling R, Tuomi J (2002) Cyanobacterial lichen symbiosis: the fungal partner as an optimal harvester. Oikos 98:498–504

Ikehata H, Ono T (2011) The mechanisms of UV mutagenesis. J Radiat Res 52:115–125

Ingebrigtsen RA, Hansen E, Andersen JH, Eilertsen HC (2016) Light and temperature effects on bioactivity in diatoms. J Appl Phycol 28:939–950

Inoue N, Taira Y, Emi T, Yamane Y, Kashino Y, Koike H, Satoh K (2001) Acclimation to the growth temperature and the high-temperature effects on photosystem II and plasma membranes in a mesophilic cyanobacterium, *Synechocystis* sp. PCC6803. Plant Cell Physiol 42:1140–1148

IPCC (2014) Climate change 2014: impacts, adaptation, and vulnerability, Part B: Regional aspects. Contribution of Working Group II to the fifth assessment report of the Intergovernmental Panel on Climate Change. IPCC, Cambridge

Ismaiel MMS, El-Ayouty YM, Piercey Normore MD (2014) Antioxidants characterization in selected cyanobacteria. Ann Microbiol 64:1223–1230

Jiménez C (2018) Marine natural products in medicinal chemistry. ACS Med Chem Lett 9:959–961

Jones B, Renaut RW (2006) Growth of siliceous spicules in acidic Hot Springs, Waiotapu geothermal area, North Island, New Zealand. PALAIOS 21:406–423

Jungblut AD, Hawes I, Mountfort D, Hitzfeld B, Dietrich DR, Burns BP, Neilan BA (2005) Diversity within cyanobacterial mat communities in variable salinity meltwater ponds of McMurdo Ice Shelf, Antarctica. Environ Microbiol 7:519–529

Kamennaya NA, Ajo-Franklin CM, Northen T, Jansson C (2012) Cyanobacteria as biocatalysts for carbonate mineralization. Minerals 2:338–364

Karkos PD, Leong SC, Karkos CD, Sivaji N, Assimakopoulos DA (2011) *Spirulina* in clinical practice: evidence-based human applications. Evid Based Complement Altern Med 2011: 531053

Karl DM (2002) Nutrient dynamics in the deep blue sea. Trends Microbiol 10:410–418

Karsten U (1996) Growth and organic osmolytes of geographically different isolates of *Microcoleus chthonoplastes* (cyanobacteria) from benthic microbial mats: response to salinity change. J Phycol 32:501–506

Klähn S, Hagemann M (2011) Compatible solute biosynthesis in cyanobacteria. Environ Microbiol 13:551–562

Klassen JL (2010) Phylogenetic and evolutionary patterns in microbial carotenoid biosynthesis are revealed by comparative genomics. PLoS One 5:e11257

Klatt JM, Al-Najjar MAA, Yilmaz P, Lavik G, de Beer D, Polerecky L (2015) Anoxygenic photosynthesis controls oxygenic photosynthesis in a cyanobacterium from a sulfidic spring. Appl Environ Microbiol 81:2025

Kolman MA, Nishi CN, Perez-Cenci M, Salerno GL (2015) Sucrose in cyanobacteria: from a salt-response molecule to play a key role in nitrogen fixation. Life (Basel, Switzerland) 5:102–126

Korbee N, Teresa Mata M, Figueroa FL (2010) Photoprotection mechanisms against ultraviolet radiation in *Heterocapsa* sp. (Dinophyceae) are influenced by nitrogen availability: mycosporine-like amino acids vs. xanthophyll cycle. Limnol Oceanogr 55:899–908

Laloknam S, Tanaka K, Buaboocha T, Waditee R, Incharoensakdi A, Hibino T, Tanaka Y, Takabe T (2006) Halotolerant cyanobacterium *Aphanothece halophytica* contains a betaine transporter active at alkaline pH and high salinity. Appl Environ Microbiol 72:6018–6026

Lin X, Huang T (2016) Oxidative stress in psoriasis and potential therapeutic use of antioxidants. Free Radic Res 50:585–595

Liu W, Liu Y (2016) Youyou Tu: significance of winning the 2015 Nobel Prize in Physiology or Medicine. Cardiovasc Diagn Therapy 6:1–2

Lopes G, Clarinha D, Vasconcelos V (2020) Carotenoids from cyanobacteria: a biotechnological approach for the topical treatment of psoriasis. Microorganisms 8:302

Los DA, Murata N (1999) Responses to cold shock in cyanobacteria. J Mol Microbiol Biotechnol 1: 221–230

Luesch H, Yoshida WY, Moore RE, Paul VJ, Corbett TH (2001) Total structure determination of apratoxin A, a potent novel cytotoxin from the marine cyanobacterium *Lyngbya majuscula*. J Am Chem Soc 123:5418–5423

Middlebrook R, Hoegh-Guldberg O, Leggat W (2008) The effect of thermal history on the susceptibility of reef-building corals to thermal stress. J Exp Biol 211:1050–1056

Miller SR, Castenholz RW (2000) The evolution of thermotolerance in hot spring cyanobacteria of the genus *Synechococcus*. J Phycol 36:48

Morone J, Alfeus A, Vasconcelos V, Martins R (2019) Revealing the potential of cyanobacteria in cosmetics and cosmeceuticals—a new bioactive approach. Algal Res 41:101541

Murata N, Takahashi S, Nishiyama Y, Allakhverdiev SI (2007) Photoinhibition of photosystem II under environmental stress. Biochim Biophys Acta 1767:414–421

Nabout JC, da Silva Rocha B, Carneiro FM, Sant'Anna CL (2013) How many species of Cyanobacteria are there? Using a discovery curve to predict the species number. Biodivers Conserv 22:2907–2918

Nakamura H, Kobayashi J, Hirata Y (1982) Separation of mycosporine-like amino acids in marine organisms using reversed-phase high-performance liquid chromatography. J Chromatogr A 250:113–118

Namikoshi M, Rinehart KL (1996) Bioactive compounds produced by cyanobacteria. J Ind Microbiol 17:373–384

Nordberg H, Cantor M, Dusheyko S, Hua S, Poliakov A, Shabalov I, Smirnova T, Grigoriev IV, Dubchak I (2014) The genome portal of the Department of Energy Joint Genome Institute: 2014 updates. Nucleic Acids Res 42:D26–D31

Nübel U, Garcia-Pichel F, Muyzer G (2000) The halotolerance and phylogeny of cyanobacteria with tightly coiled trichomes (*Spirulina turpin*) and the description of Halospirulina tapeticola gen. nov., sp. nov. Int J Syst Evol Microbiol 50(Pt 3):1265–1277

Oliver ECJ, Donat MG, Burrows MT, Moore PJ, Smale DA, Alexander LV, Benthuysen JA, Feng M, Sen Gupta A, Hobday AJ, Holbrook NJ, Perkins-Kirkpatrick SE, Scannell HA, Straub SC, Wernberg T (2018) Longer and more frequent marine heatwaves over the past century. Nat Commun 9:1324

Omelon CR, Pollard WH, Ferris FG (2006) Environmental controls on microbial colonization of high Arctic cryptoendolithic habitats. Polar Biol 30:19–29

Oren A (2006) Life at high salt concentrations. In: Dworkin M, Falkow S, Rosenberg E, Schleifer KH, Stackebrandt E (eds) The prokaryotes. Springer, New York, pp 263–282

Oren A (2015) Cyanobacteria in hypersaline environments: biodiversity and physiological properties. Biodivers Conserv 24:781–798

Oren A, Gunde-Cimerman N (2007) Mycosporines and mycosporine-like amino acids: UV protectants or multipurpose secondary metabolites? FEMS Microbiol Lett 269:1–10

Padan E (1979) Facultative anoxygenic photosynthesis in cyanobacteria. Annu Rev Plant Physiol 30:27–40

Paerl HW, Otten TG (2016) Duelling 'CyanoHABs': unravelling the environmental drivers controlling dominance and succession among diazotrophic and non-N_2-fixing harmful cyanobacteria. Environ Microbiol 18:316–324

Peter AP, Lakshmanan K, Mohandass S, Varadharaj S, Thilagar S, Abdul Kareem KA, Dharmar P, Gopalakrishnan S, Lakshmanan U (2015) Cyanobacterial knowledge base (CKB), a compendium of cyanobacterial genomes and proteomes. PLoS One 10:e0136262

Petit JR, Jouzel J, Raynaud D, Barkov NI, Barnola JM, Basile I, Bender M, Chappellaz J, Davis M, Delaygue G, Delmotte M, Kotlyakov VM, Legrand M, Lipenkov VY, Lorius C, Pépin L, Ritz C, Saltzman E, Stievenard M (1999) Climate and atmospheric history of the past 420,000 years from the Vostok ice core, Antarctica. Nature 399:429

Pitt C (2015) Five old remedies that are still healing us today. BBC News

Priyanka, Kumar C, Chatterjee A, Wenjing W, Yadav D, Singh PK (2020) Cyanobacteria: potential and role for environmental remediation. In: Singh P, Kumar A, Borthakur A (eds) Abatement of environmental pollutants. Elsevier, Amsterdam, pp 193–202

Proteau PJ, Gerwick WH, Garcia-Pichel F, Castenholz R (1993) The structure of scytonemin, an ultraviolet sunscreen pigment from the sheaths of cyanobacteria. Experientia 49:825–829

Rai S, Pandey S, Shrivastava AK, Singh PK, Agrawal C, Rai LC (2013) Understanding the mechanisms of abiotic stress management in cyanobacteria with special reference to proteomics. In: Srivastava AK, Rai AN, Neilan BA (eds) Stress biology of cyanobacteria: molecular mechanisms to cellular responses. CRC Press Taylor & Francis Group, Boca Raton, FL, pp 93–112

Rajaram H, Chaurasia A, Apte S (2014) Cyanobacterial heat shock response: role and regulation of molecular chaperones. Microbiology (Reading, England) 160(Pt 4):647–658

Ramos V, Morais J, Vasconcelos VM (2017) A curated database of cyanobacterial strains relevant for modern taxonomy and phylogenetic studies. Sci Data 4:170054

Rastogi RP, Incharoensakdi A (2014a) Characterization of UV-screening compounds, mycosporine-like amino acids, and scytonemin in the cyanobacterium *Lyngbya* sp. CU2555. FEMS Microbiol Ecol 87:244–256

Rastogi RP, Incharoensakdi A (2014b) UV radiation-induced biosynthesis, stability and antioxidant activity of mycosporine-like amino acids (MAAs) in a unicellular cyanobacterium *Gloeocapsa* sp. CU2556. J Photochem Photobiol B Biol 130:287–292

Rastogi RP, Richa, Sinha RP, Singh SP, Häder D-P (2010) Photoprotective compounds from marine organisms. J Ind Microbiol Biotechnol 37:537–558

Rastogi RP, Sonani RR, Madamwar D (2014) The high-energy radiation protectant extracellular sheath pigment scytonemin and its reduced counterpart in the cyanobacterium *Scytonema* sp. R77DM. Bioresour Technol 171:396–400

Rastogi RP, Sonani RR, Madamwar D, Incharoensakdi A (2016) Characterization and antioxidant functions of mycosporine-like amino acids in the cyanobacterium *Nostoc* sp. R76DM. Algal Res 16:110–118

Rastogi RP, Madamwar D, Nakamoto H, Incharoensakdi A (2020) Resilience and self-regulation processes of microalgae under UV radiation stress. J Photochem Photobiol C: Photochem Rev 43:100322

Řeháková K, Zapomělová E, Prášil O, Veselá J, Medová H, Oren A (2009) Composition changes of phototrophic microbial communities along the salinity gradient in the solar saltern evaporation ponds of Eilat, Israel. Hydrobiologia 636:77–88

Rockstrom J, Steffen W, Noone K, Persson A, Chapin FS III, Lambin EF, Lenton TM, Scheffer M, Folke C, Schellnhuber HJ, Nykvist B, de Wit CA, Hughes T, van der Leeuw S, Rodhe H,

Sorlin S, Snyder PK, Costanza R, Svedin U, Falkenmark M, Karlberg L, Corell RW, Fabry VJ, Hansen J, Walker B, Liverman D, Richardson K, Crutzen P, Foley JA (2009) A safe operating space for humanity. Nature 461:472–475

Rosic NN (2012) Phylogenetic analysis of genes involved in mycosporine-like amino acid biosynthesis in symbiotic dinoflagellates. Appl Microbiol Biotechnol 94:29–37

Rosic NN (2019) Mycosporine-like amino acids: making the foundation for organic personalised sunscreens. Mar Drugs 17(11):638

Rosic NN, Dove S (2011) Mycosporine-like amino acids from coral dinoflagellates. Appl Environ Microbiol 77:8478–8486

Rosic NN, Pernice M, Dunn S, Dove S, Hoegh-Guldberg O (2010) Differential regulation by heat stress of novel cytochrome P450 genes from the dinoflagellate symbionts of reef-building corals. Appl Environ Microbiol 76:2823–2829

Rosic NN, Pernice M, Dove S, Dunn S, Hoegh-Guldberg O (2011) Gene expression profiles of cytosolic heat shock proteins Hsp70 and Hsp90 from symbiotic dinoflagellates in response to thermal stress: possible implications for coral bleaching. Cell Stress Chaperones 16:69–80

Rosic NN, Leggat W, Kaniewska P, Dove S, Hoegh-Guldberg O (2013) New-old hemoglobin-like proteins of symbiotic dinoflagellates. Ecol Evol 3:822–834

Rossi F, De Philippis R (2015) Role of cyanobacterial exopolysaccharides in phototrophic biofilms and in complex microbial mats. Life (Basel, Switzerland) 5:1218–1238

Schindler DW (1977) Evolution of phosphorus limitation in lakes. Science 195:260–262

Schmidt JR, Wilhelm SW, Boyer GL (2014) The fate of microcystins in the environment and challenges for monitoring. Toxins (Basel) 6:3354–3387

Schneider D, Arp G, Reimer A, Reitner J, Daniel R (2013) Phylogenetic analysis of a microbialite-forming microbial mat from a hypersaline lake of the Kiritimati atoll, Central Pacific. PLoS One 8:e66662

Schoch CL, Ciufo S, Domrachev M, Hotton CL, Kannan S, Khovanskaya R, Leipe D, McVeigh R, O'Neill K, Robbertse B, Sharma S, Soussov V, Sullivan JP, Sun L, Turner S, Karsch-Mizrachi I (2020) NCBI Taxonomy: a comprehensive update on curation, resources and tools. Database (Oxford) 2020:baaa062

Seneviratne SI, Rogelj J, Séférian R, Wartenburger R, Allen MR, Cain M, Millar RJ, Ebi KL, Ellis N, Hoegh-Guldberg O, Payne AJ, Schleussner CF, Tschakert P, Warren RF (2018) The many possible climates from the Paris Agreement's aim of 1.5°C warming. Nature 558:41–49

Shibata K (1969) Pigments and a UV-absorbing substance in corals and a blue-green alga living in the Great Barrier Reef1. Plant Cell Physiol 10:325–335

Shick JM, Dunlap WC (2002) Mycosporine-like amino acids and related gadusols: biosynthesis, accumulation, and UV-protective functions in aquatic organisms. Annu Rev Physiol 64:223–262

Sigman D, Hain M (2012) The biological productivity of the ocean. Nat Educ 3:1–16

Singh S, Kate BN, Banerjee UC (2005) Bioactive compounds from cyanobacteria and microalgae: an overview. Crit Rev Biotechnol 25:73–95

Singh SP, Klisch M, Sinha RP, Häder D-P (2010) Genome mining of mycosporine-like amino acid (MAA) synthesizing and non-synthesizing cyanobacteria: a bioinformatics study. Genomics 95: 120–128

Singh DP, Prabha R, Verma S, Meena KK, Yandigeri M (2017) Antioxidant properties and polyphenolic content in terrestrial cyanobacteria. 3 Biotech 7:134

Sinha RP, Häder D-P (2008) UV-protectants in cyanobacteria. Plant Sci 174:278–289

Sinha RP, Singh SP, Häder D-P (2007) Database on mycosporines and mycosporine-like amino acids (MAAs) in fungi, cyanobacteria, macroalgae, phytoplankton and animals. J Photochem Photobiol B Biol 89:29–35

Sompong U, Hawkins PR, Besley C, Peerapornpisal Y (2005) The distribution of cyanobacteria across physical and chemical gradients in hot springs in northern Thailand. FEMS Microbiol Ecol 52:365–376

Stanier RY, Cohen-Bazire G (1977) Phototrophic prokaryotes: the cyanobacteria. Annu Rev Microbiol 31:225–274

Suzuki I, Kanesaki Y, Hayashi H, Hall JJ, Simon WJ, Slabas AR, Murata N (2005) The histidine kinase Hik34 is involved in thermotolerance by regulating the expression of heat shock genes in synechocystis. Plant Physiol 138:1409–1421

Tamaru Y, Takani Y, Yoshida T, Sakamoto T (2005) Crucial role of extracellular polysaccharides in desiccation and freezing tolerance in the terrestrial cyanobacterium *Nostoc commune*. Appl Environ Microbiol 71:7327–7333

Vincent WF (1989) Microbial ecosystems of Antarctica. Antarct Sci 1:179–180

Vincent WF (2007) Cold tolerance in cyanobacteria and life in the cryosphere. In: Seckbach J (ed) Algae and cyanobacteria in extreme environments. Springer Netherlands, Dordrecht, pp 287–301

Vopel K, Hawes I (2006) Photosynthetic performance of benthic microbial mats in Lake Hoare, Antarctica. Limnol Oceanogr 51:1801–1812

Wada N, Sakamoto T, Matsugo S (2013) Multiple roles of photosynthetic and sunscreen pigments in cyanobacteria focusing on the oxidative stress. Metabolites 3:463–483

Waleron M, Waleron K, Vincent W, Wilmotte A (2007) Allochthonous inputs of riverine picocyanobacteria to coastal waters in the Arctic Ocean. FEMS Microbiol Ecol 59:356–365

Whitton B, Potts M (2002) The ecology of cyanobacteria: their diversity in time and space. Springer Netherlands, Dordrecht

Wiegand C, Pflugmacher S (2005) Ecotoxicological effects of selected cyanobacterial secondary metabolites: a short review. Toxicol Appl Pharmacol 203:201–218

World Health Organization (2003) Guidelines for safe recreational water environments. Volume 1: Coastal and fresh waters. http://www.who.int/water_sanitation_health/bathing/srwe2full.pdf

Zakhia F, Jungblut A, Taton A, Vincent W, Wilmotte A (2008) Cyanobacteria in cold ecosystems. In: Margesin R, Schinner F, Marx JC, Gerday C (eds) Psychrophiles: from biodiversity to biotechnology. Springer, Berlin, pp 121–135

Zamyadi A, Ho L, Newcombe G, Bustamante H, Prévost M (2012) Fate of toxic cyanobacterial cells and disinfection by-products formation after chlorination. Water Res 46:1524–1535

Zhang L, Liu J (2016) Effects of heat stress on photosynthetic electron transport in a marine cyanobacterium Arthrospira sp. J Appl Phycol 28:757–763

Stress Proteins and Signal Transduction in Cyanobacteria

8

Ruchi Rai, Krishna Kumar Rai, Shilpi Singh, Alka Raj, and L. C. Rai

Abstract

Stress-induced proteomics unveils various adaptive mechanisms evolved by cyanobacteria to combat different environmental stresses. Acclimation strategies to these abiotic stresses, i.e., heat, salinity, desiccation, UV-B, and metals, involve expression of stress-specific proteins or metabolites via transcriptional activation of stress-responsive genes. Among the various stress-responsive upregulated proteins, two-component system proteins (TCSs) are most crucial and predominantly involved in signaling pathways, which are comprised of sensory kinase (histidine kinases) and a response regulator (Rre). Serine/threonine kinase (STK) genes and phosphatases play essential role in regulating cellular activities in cyanobacteria. Reactive oxygen species generated as a result of stresses also act as ubiquitous signal molecules and are a central component associated with signaling transduction pathway. This chapter summarizes how cyanobacteria sense and respond to ever-changing environment by employing two-component signal transduction system in conjunction with other signaling components such as kinases, phosphatases, RNA polymerase sigma factors, and transcription factors as integral network in the regulation of the responses of cyanobacterial cells to various types of stress.

Keywords

Stress proteins · Signal transduction · Two-component system · Histidine kinases · Response regulators · STKs

R. Rai · K. K. Rai · S. Singh · A. Raj · L. C. Rai (✉)
Molecular Biology Section, Centre of Advanced Study in Botany, Institute of Science, Banaras Hindu University, Varanasi, India
e-mail: lcrai@bhu.ac.in

R. P. Rastogi (ed.), *Ecophysiology and Biochemistry of Cyanobacteria*,
https://doi.org/10.1007/978-981-16-4873-1_8

8.1 Introduction

Cyanobacteria are oxygenic photoautotrophs originated approximately 3.5 billion years ago and are endowed with indispensable metabolic features of evolutionary significance. These microbes have led to the oxygenation of the earth's atmosphere (Demoulin et al. 2019; Soo et al. 2017) and aid in origin of plastids in algae and higher plants through endosymbiosis (de Vries and Archibald 2017). They are significant contributors to global photosynthetic productivity (Garcia-Pichel et al. 2003) and potential biofertilizer (Vaishampayan et al. 2001; Chatterjee et al. 2017) that helps to improve crop productivity and soil fertility in economical and eco-friendly manner, hence beneficial in the development of sustainable agriculture (Singh et al. 2016, 2017). Cyanobacteria can be considered as one of the most primeval and ecologically successful group inhabiting several extreme conditions (Abed et al. 2009) including hot and cold deserts, hot springs, hypersaline environments, terrestrial environments (soil, deserts, and glaciers), symbioses (Ashby and Houmard 2006), and even in the absence of combined nitrogen (Castenholz 2001). In view of the similarity between plasma and thylakoid membranes of cyanobacterial cells with the chloroplast of higher plants in terms of lipid composition and assembly of membranes (Rodriguez-Ezpeleta et al. 2005; Los et al. 2010), they are considered as powerful model systems for studying the molecular mechanisms of the responses and acclimation to stress (Los and Murata 2004; Jensen and Leister 2014).

The diversity of cyanobacteria is also reflected in the complexity of their stress-responsive regulatory proteins (Ashby and Houmard 2006), which enable them to adapt and survive under extreme conditions (Tandeau de Marsac and Houmard 1993). The knowledge on molecular mechanism of stress signals perception and expression of target genes for producing specific metabolites/proteins leading to adaptive responses has still some lacunae. However, stress-induced proteomics helps in dissection and characterization of cluster of "adaptation proteins" associated with the abiotic stress mitigation in cyanobacteria along with their role in metabolic pathways (Babele et al. 2019).

In this overview, an attempt has been made to briefly summarize major stress-responsive proteins in cyanobacteria under various abiotic stresses and also to recapitulate the role of two-component systems (TCSs) proteins and serine–threonine kinases (STKs) in integrated network of complex cyanobacterial signal transduction pathways.

8.2 Stress-responsive Proteins Under Various Abiotic Stresses

A large number of whole-genome sequencing projects for cyanobacterial species are still going on with around 400 cyanobacterial genomes available in public databases (Alvarenga et al. 2017). Although proteomics of large number of cyanobacteria has been investigated till date, e.g., *Synechococcus, Cyanothece, Trichodesmium, Prochlorococcus, Microcystis, and Nostoc, Synechocystis* PCC 6803 and *Anabaena*

PCC7120 have emerged as model organisms for studying various physiological, biochemical, and molecular stress responses due to availability of fully sequenced genome (Kaneko et al. 1996, 2001). After genomic investigation, transcriptomic and proteomic analyses including post-translational modification (PTM) systems needed to be performed to reveal mechanism involved in triggering stress protein expression and the signaling networks linked with modified physiology of cyanobacteria. Gel-based proteomics has been very well adapted to analyze the proteomic alterations (Jungblut et al. 1996) in response to various abiotic and biotic stresses (Castielli et al. 2009). The most extensively used gel-based technique for protein identification is two-dimensional gel electrophoresis (2-DE), in which a large number of cellular proteins appear in the form of spots on the gel matrix and can be clearly visualized after proper protein staining (Mikkat et al. 2014). Although several methods have been developed to enhance the quantity and quality of protein spots in a 2-D gel, these improvements are still not adequate to describe the complete proteomic profile. Recently, gel-free isobaric tags for relative and absolute quantitation (iTRAQ) technique have emerged as new trend in this field of research and used frequently due to its better reliability as compared to 2-DE (Aggarwal et al. 2006). Together with the advancements in traditional biology, molecular biology aided with computational biology and system biology tools/techniques has also been operated for the development of high-throughput techniques (HTT) to explore the transcriptomic and metabolomic profiling (Ow and Wright 2009) for better understanding of stress adaptive mechanism in cyanobacteria. High-throughput proteomic approaches loaded with a huge range of proteomic tools such as high-resolution liquid chromatography (LC) and more advanced tandem mass spectrometers (MS) have significant role to understand the cellular functions because it can correlate the actual function of genes and its translational products, which reflect the cellular protein profile under defined stress conditions (Li et al. 2011). Advent of LC-MS based tagging approaches such as isobaric tags for relative and absolute quantitation (iTRAQ) (Stensjö et al. 2007), stable isotope labeling by amino acids in cell culture (SILAC) (Dephoure et al. 2013), and isotope-coded affinity tags (ICAT) (Sandh et al. 2014) has unraveled several new sets of proteins involved in stress management with absolute quantification of complete protein profile.

Genomics and transcriptomics integrated with proteomics advocate that the modification in actual cellular physiology and metabolic state of the cells in long-term cellular adaptation is governed by the synthesis of "stress-responsive proteins." These abiotic stress-induced proteins can be categorized as DNA repair/protection and transcription regulators, heat shock proteins (HSPs), and other stress-related proteins, cellular antioxidative enzymes, proteins of lipid, and other cellular metabolisms, two-component system proteins (TCSs), and hypothetical proteins, which are briefly summarized in Table 8.1. However, enzymes of amino acid, fatty acid, and cofactor biosynthesis, and proteins of energy metabolism (photosynthesis, respiration, and carbon/nitrogen fixation) were observed under category of downregulated stress-responsive proteins (Babele et al. 2019).

Stress-repressed protein-like enzymes primarily involved in amino acids, lipid, and cofactor biosynthetic pathways and energy metabolisms such as photosynthesis,

Table 8.1 List of some common and specific stress-responsive proteins on the basis of their roles under different mechanisms

Types of Stress-induced proteins	Organism	Name of Proteins	Roles	References
1. DNA repair/ protection and transcription regulator				
	Anabaena sp.	DNA-binding proteins (Dps)	Against oxidative stress and homeostasis. Prevent DNA base modification and strand cleavage but not interfering with normal DNA metabolism. Capable to provide comprehensive defense due to three integral features of the protein–DNA binding, sequestration of iron, and tolerance against ROS	Shcolnick et al. (2007), Babele et al. (2015), Panda et al. (2015), Shrivastava et al. (2015)
	Spirulina platensis	SbcC exonuclease	Showed upregulation during cold stress and high temperature and has significant role in DNA repair and genome stability	Hongsthong et al. (2008, 2009)
	Synechococcus	ParA, GvrA, and PhrA, NusB, SigD, and SYNPCC7002_A2523	Upregulated enzymes and transcriptional regulator under high light stress	Xiong et al. (2015)
	Synechococcus strain WH8102, strain BL107, and *Synechocystis* PCC6803	AbrB	Common transcriptional regulator in these strains upregulated under low-temperature stress and might have a significant role in the uptake of carbon and nitrogen	Kaniya et al. (2013), Varkey et al. (2016)

2. Heat shock proteins				
	Synechococcus sp.	DnaK chaperone	Role in the enhancement of membrane fluidity by modulating membrane lipids under heat stress and also helps in protein translocation and translational machinery	Katano et al. (2006)
	Synechocystis PCC6803	GroEL, GroEs, 60KD chaperonin 1, DnaK protein 2	Molecular chaperon induced under salt, acid, and UV stresses	Fulda et al. (2006), Kurian et al. (2006), Gao et al. (2009)
	Synechocystis PCC6803	Peptidyl-prolyl isomerase	Act as trigger factor also known as cold shock protein. Acquaintance with the ribosome and assists in proper folding of newly synthesized polypeptides by increasing the affinity of GroEL with unfolded polypeptides	Prakash et al. (2010)
	Anabaena sp.	GrpE, chaperonin GroEL, and DnaK type molecular chaperone	Upregulated chaperons and stress proteins in response to UV-B stress	Shrivastava et al. (2015)
3. Antioxidative and cellular defense reaction proteins				
	Anabaena sp. PCC7120	AhpC (An+ahpC)	Overexpression of the protein augmented photosynthesis, nitrogen fixation, and modulated regulatory network of antioxidative proteins	Shrivastava et al. (2016)

(continued)

Table 8.1 (continued)

Types of Stress-induced proteins	Organism	Name of Proteins	Roles	References
	Anabaena sp.	AhpC, catalase, peroxiredoxin	Showed upregulation during oxidative stress when significant accumulation of H_2O_2 seen into the cell	Babele et al. (2015)
	Synechocystis PCC6803	Thioredoxin (Trx)	An antioxidant enzyme having role in suppressing cell death provides reducing equivalent to antioxidant system. Also known to effectively decrease the intramolecular disulfide bridges in different target proteins	Pandey et al. (2012)
	Anabaena sp.	Catalases, oxidoreductase	These enzymes and their homologs implicit resistance to the cell against oxidative stress and also in case of desiccation stress	Katoh et al. (2004)
	Synechocystis PCC6803	Glutaredoxins (Grx-s)	A group of small ubiquitous proteins maintain the cytoplasmic thiol-redox state. Their oxidation is carried out by substrate and nonenzymatic reduction by glutathione. *Synechocystis* mutants with glutaredoxin knockout of genes showed enhanced sensitivity against peroxides	Holmgren (1989), Latifi et al. (2009)

	Anabaena sp. PCC7120	Alr 2321, All0580	GlyI and GlyII (glyoxalases) of cyanobacteria involved in methylglyoxal detoxification and multiple abiotic stress tolerance in *E. coli*	(Rai et al. 2019a, b)
4. Proteins of lipid metabolism and transporters				
	Spirulina platensis	(3R)-hydroxymyristoyl-[acyl-carrier-protein] -dehydratase or FabZ	First reported dehydratase upregulated under low-temperature stress which helps in biosynthesis of unsaturated fatty acids	Jeamton et al. (2008)
	Spirulina platensis	3-oxoacyl-[acyl-carrier protein] reductase	A protein of lipid metabolism might help to promote photosynthesis by accelerating the activity of Na^+/H^+ transporter	Wang et al. (2013)
	Anabaena sp.	UDP-sulfoquinovose synthase (SQD1)	Associated with the biosynthesis of sulfolipids during cadmium toxicity	Singh et al. (2015)
	Anabaena sp. PCC7120	Aldo-keto reductase (AKR17A1)	Multiple abiotic stress tolerance and helps in degradation of herbicide butachlor	Agrawal et al. (2015)
	Synechocystis PCC6803	Slr1962, Slr0681	Upregulated under acid stress as ion transporter	Kurian et al. (2006)

(continued)

Table 8.1 (continued)

Types of Stress-induced proteins	Organism	Name of Proteins	Roles	References
	Synechocystis PCC6803	FutA1(Slr1295), FutA2 (Slr0513)	FutA1 is a part of ABC transporter concerned with the $Fe3^{+}$ uptake by the cells and a major iron-binding protein found upregulated after exposing high light or heat shock. FutA2 has essential role in the protection of photosystem II following iron depleting condition	Miranda et al. (2013)
	Anabaena sp.	PstB2	An ATP-binding protein found to be induced upon cadmium exposure	Singh et al. (2015)
	Synechocystis PCC6803	ABC transporter subunit ycf24, ABC transporter permease protein, nitrate/nitrite-binding protein (NrtA), phosphate-binding protein (PstS1), putative SbtB, phosphate transport ATP-binding protein (PstB1), and iron-binding protein (FutA1)	Found to be localized in periplasmic region and significantly upregulated under high pH stress. It is supposed that high pH causes nutrient deficiency, thus inducing the expression of phosphate and the ATP-binding proteins	Zhang et al. (2009)
5. Hypothetical proteins				
	Synechocystis PCC6803	Sll1863, Sll1762, Sll0596, Slr1485, Sll1549, Slr2144, Slr1535, Sll0595, and Slr0711	Accumulated under salt stress inside the cell. slr0711 has role in folate biosynthesis, sll0595 has a role in signaling, while few have unknown functions	Fulda et al. (2006)

	Anabaena doliolum	Alr 0803	Role in Cu homeostasis	Bhargava et al. (2008)
	Anabaena sp. PCC7120	Alr0893	Upregulated under UV-B stress and might be involved in nullifying the damaging effect of UV radiation with its combined ATP-independent repair activity of PfpI protease and ferritin domain	Shrivastava et al. (2015)
		Alr0882, All5218, All3014, Alr3199, Alr4050, Alr3904, and Alr3090	Accumulation in response under high salt stress Alr0882 functions as universal stress protein. Crystal structure of All3014 revealed it as fosfomycin resistance protein	Rai et al. (2013)
	Anabaena sp. PCC7120	All1122 and Alr0750 Alr0765, All4894	Universal stress proteins (UspA) Provide multiple abiotic stress tolerance when heterologously expressed in *E. coli*	Sen et al. (2019), Chatterjee et al. (2020), Rai et al. (2020)

respiration, and nitrogen fixation are significantly downregulated under variety of abiotic stresses. Few important examples are enzymes required for amino acid metabolism cysteine synthase (*all2521*) (3-phosphoshikimate 1-carboxyvinyl transferase), fatty acid and phospholipid metabolism (CTP synthetase), and cell envelope proteins (dTDP-glucose 4–6-dehydratase, glucose-1-phosphate thymidyl transferase), which are found to be downregulated in *Synechocystis* sp. PCC6803 under strong UV-B irradiation (Gao et al. 2009). Also, prolonged UV-B exposure to cells of *Synechocystis* sp. downregulated many enzymes involved in energy metabolic pathways, i.e., amino methyltransferase, GDP-mannose pyrophosphorylase, phosphoglycerate mutase, and large subunit of carbamoyl-phosphate synthase (Gao et al. 2009). UV-B stress can also severely damage photosynthetic proteins like allophycocyanin beta subunit, phycocyanin subunit, and phycoerythrocyanin alpha chain and downregulate the respiratory enzymatic activity such as F_0F_1 ATP synthase (beta subunit) and D-3-phosphoglycerate dehydrogenase catalyzing the conversion of 3-phospho-D-glycerate into 3 phosphohydroxypyruvate in *Anabaena* L31 cells (Babele et al. 2015). Glucokinase and glyceraldehyde 3-phosphate dehydrogenase-2 showed reduced levels in methyl viologen-treated *Anabaena* PCC7120 cells (Panda et al. 2014). Reduced levels of most common PSII (ApcA, CpcA, CpcG4, CpcB, and PecC) proteins and ATP synthase alpha subunit were observed in three *Anabaena* species (*A. doliolum*, *A.* PCC 7120, and *A.* L31) under the influence of UV-B radiation exposure (Shrivastava et al. 2015) and cadmium toxicity (Singh et al. 2015). In this connection, enzymes of Calvin cycle [ribulose 1,5-bisphosphate carboxylase/oxygenase (RuBisCO) large subunit, and phosphoribulokinase], pentose phosphate pathway (6-phosphogluconate dehydrogenase, transketolase, and fructose bisphosphate aldolase), and glycolysis (glucose-6-phosphate isomerase, glyceraldehyde 3-phosphate dehydrogenase, fructose-1,6-bisphosphate aldolase, and phosphoglycerate kinase) were found depressed in the *Anabaena* sp. exposed to cadmium (Singh et al. 2015).

8.3 Two-component Signal Transduction Pathways: Histidine Kinases (Hiks) and Response Regulators (Rre)

Two-component regulatory systems are ubiquitously found in almost all domains of life and serve as coupling mechanism in organisms to sense and respond to changes in environmental conditions (Stock et al. 2000). Histidine protein kinases (Hiks) together with their partner response regulators form two-component signaling systems that undoubtedly constitute the most extensively used one of all signal-transduction enzymes in cyanobacteria (Marin et al. 2003). The typical Hiks are a transmembrane receptor with an amino-terminal extracellular sensing domain and a carboxy-terminal cytosolic signaling domain. When the sensor (Hiks) is activated via specific environmental stimuli, it sends a signal to the kinase domain of sensor. Hiks catalyze the transfer of phosphate from ATP to a unique histidine residue, which then transfers phosphate to the conserved aspartate of 'receiver domain' in response regulators (Rre). Upon phosphorylation, the protein enhances its binding to

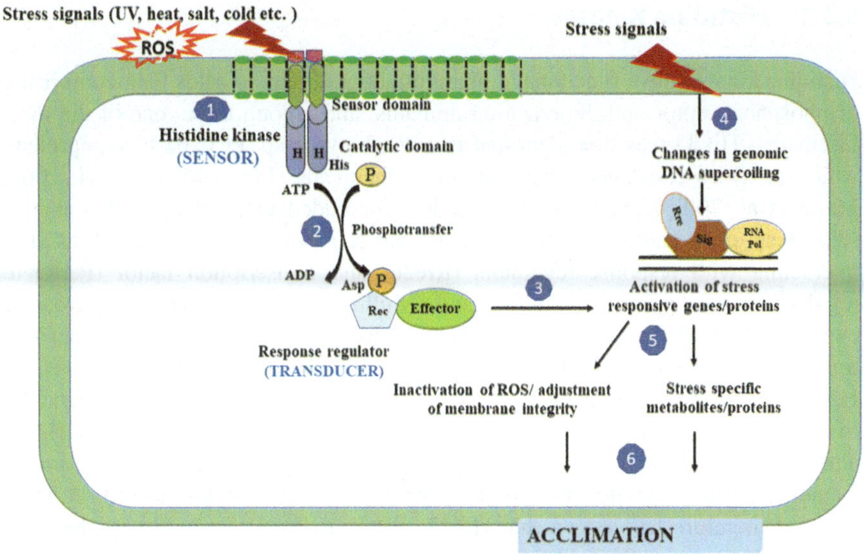

Fig. 8.1 A general scheme showing signaling pathway of a cyanobacterial cell executing two-component system and subsequently leading to adaptive responses to environmental stress. (1) Histidine kinase is activated via specific environmental stimuli and sends signal to the kinase domain of sensor. (2) It catalyzes the transfer of phosphate from ATP to a unique histidine residue and then to the conserved aspartate of "receiver domain" in response regulators (Rre). (3) Phosphorylation modifies the transcription of stress-responsive genes. (4) Environmental stresses lead to genomic DNA supercoiling that causes alterations in transcription of many genes. (5) It results in generation of stress-specific protein and metabolites. (6) Leading to adaptive responses (acclimation)

DNA and modifies the transcription of genes that control the adaptive responses. These sensors also perceive the changes in membrane fluidity caused by temperature fluctuations resulting in gene expression related to temperature acclimation (e.g., lipid unsaturation) (Los et al. 2013). In addition to various environmental stimuli, reactive oxygen species (ROS) and more specifically H_2O_2 act as a universal signal enabling cyanobacteria to regulate the expression of a number of genes including genes required for ROS inactivation resulting in protection from various environmental stresses (Schmitt et al. 2014; Mironov et al. 2019). Environmental stresses may also directly affect the structure of DNA (chromosome packing) and cause alterations in transcription of many genes (Fig. 8.1). DNA microarray and RNA-seq technologies suggest the induction of similar sets of genes in cyanobacteria in response to different types of stress (Los 2004; Mironov et al. 2019) or activation of "multisensory proteins" that can perceive abiotic stresses regardless of their nature (Sinetova and Los 2016).

8.3.1 Histidine Kinases

Histidine kinases have functional domain of about 60 amino acids long comprising of phosphoacceptor and dimerization domains, and among these, one of the most fascinating Hik33 was first identified in *Synechocystis* sp. PCC 6803 as a protein, which provides resistance against harmful chemicals (Lopez-Maury et al. 2002; Marin et al. 2003). Later on, more studies concluded that Hik33 is also able to sense heat, cold, salt, and drought signals and activates specific set of genes after interacting with response regulators (Rre26) and transcription factor (PerR) to regulate stress response (Ashby and Mullineaux 1999). DNA microarray experiments have shown that Hik33, Hik34, Hik16, Hik41, and peroxide-sensitive transcriptional regulator PerR are involved in the perception of elevated levels of H2O2 and in the subsequent regulation of H_2O_2-induced expression of genes in *Synechocystis* (Marin et al. 2003). Apart from the perception of stress signals, Hik33 alters cell responses to cold, salt, hyperosmotic, and oxidative stresses in the cyanobacterium *Synechocystis* sp. PCC 6803 (Murata and Los 2006). A large body of literature has shown that Hik33 protein exerted responsive mechanisms after interacting with SipA-like protein (Ssl3451) as confirmed by yeast two-hybrid system (Imamura et al. 2003; Sato et al. 2007; Giner-Lamia et al. 2014). Furthermore, another group of researchers have indicated that Hik33 coupled with SipA-like protein (Ssl3451) enhances its phosphorylation under cold, heat, and salt stress (Sakayori et al. 2009). Another histidine kinases that perform multifarious role as sensor and activator are Hik31 that induces response against various abiotic stresses, thus protecting photosynthetic electron transport (Imamura et al. 2003; Giner-Lamia et al. 2014). In addition, researchers have also identified an orthologous protein Slr1285 in *Synechocystis* sp. PCC 6803, which is a complete histidine kinase (Hik34, Chk34) having HisKA and HATOase domains exclusively involved in the perception of salt and drought stress after interacting with the response regulator Slr1783 (Rre1, Crr1) (Shoumskaya et al. 2005). Likewise, another histidine kinase Hik10 (Chk10) has also been found to regulate several abiotic stress tolerance responses after interacting with different response regulators (Rre3) (Paithoonrangsarid et al. 2004). Genome-wide transcriptome analysis has revealed that these histidine kinases such as Hik33 and Hik34 are also able to regulate DNA supercoiling under various abiotic stress conditions either by modifying promoter region or by activating specific DNA-binding transcription factors; however, the exact mechanism has yet to be deciphered (Los 2004; Dorman 2006).

8.3.2 Response Regulators

Response regulators (Rre) are the basic players that regulate the activity of histidine kinases via phosphorylation. In general, some response regulators contain an approximately 110 amino acid long sequence known as "receiver domain" rich in Asp phosphorylation site, whereas some possess two conjoint domains where receiver domain is fused with a type of output domain having effector activity

(Yoshihara et al. 2002). Several studies have indicated that in most cases this output domain functions as DNA-binding domain thus enabling response regulator to functions as transcription factors and along with Asp phosphorylation domain bind to target gene or interact with two-component system to initiate stress response (Yeh et al. 1997). Similar kinds of proteins have also been reported in cyanobacteria with known functions, i.e., PilH (Rre7) of *Synechocystis* sp. PCC 6803 that stimulates cell motility (Yoshihara et al. 2002). Orthologs of this protein have been found in *Anabaena* PCC 7120 where researchers have identified a DevR (Alr0442) protein, which, upon interacting with HepK (All4496) a member of two-component system, regulates the biosynthesis of polysaccharide under extreme climatic conditions (Zhou and Wolk 2003). Concomitantly, orthologs have also been found in *Synechococcus elongatus* 7942 and *Synechocystis* sp. PCC 6803 where Crr42 and DivK response regulators stimulate cell division and are also able to encode DNA gyrase/topoisomerase IV, thus regulating chromatin structure under different abiotic stresses (Zhou and Wolk 2003). Another type of response regulator that has been identified in cyanobacteria belongs to the group of OmpR type, of which two are identified in *Synechocystis* sp. PCC 6803, i.e., RpaA (Rre31) and RpaB (Rre26) both of which have been predicted to play regulatory role in the biosynthesis and distribution of phycobilisomes (Paithoonrangsarid et al. 2004; Shoumskaya et al. 2005). Furthermore, it is now known that RpaA could also partner up with HiK33 and Rre26 to stimulate response in *Synechocystis* sp. PCC 6803 under hyperosmotic stress (Paithoonrangsarid et al. 2004). Interestingly, a histidine kinase Chk10 has been identified that functions downstream of Rrr3 in all the cyanobacterial strain except *Synechocystis* sp. PCC 6803, which is upregulated under N_2-deprivation condition (Shoumskaya et al. 2005). Researchers have also identified a response regulator in *Anabaena* PCC 7120 named RRII-NarL OrrA (Alr3768) that is shown to play exceptional role in regulation under osmotic stress (Schwartz et al. 1998). In some cyanobacteria, response regulators with three functional domains have also been reported; i.e., together with Treg and Hpt (histidine phosphotransferase) they also contain an additional domain GGDEF, which has been thought to stimulate heterocyst formation and nitrogen fixation (Chiang et al. 1992). In another study, a knockout of Slr2100 and Sll1624 in *Synechocystis* sp. PCC 6803 differentially regulated the cGMP level in response to UV-B radiation and exhibited similar phenotype as Crr20 mutant (Cadoret et al. 2005). There are numerous response regulators, which exert their function after interacting with other domains such as GAF, PAS, and IF2. However, their structural and functional characterization is needed for proper understanding of their role in abiotic stress tolerance.

8.3.3 Hybrid Kinases

Hybrid kinases were among the most abundant and uncharacterized signaling domains present in cyanobacteria as revealed by whole-genome sequencing. One of the most interesting features of these multidomain proteins is their complete absence from open ocean non-N_2-fixing species (Marin et al. 2003). Hybrid kinases

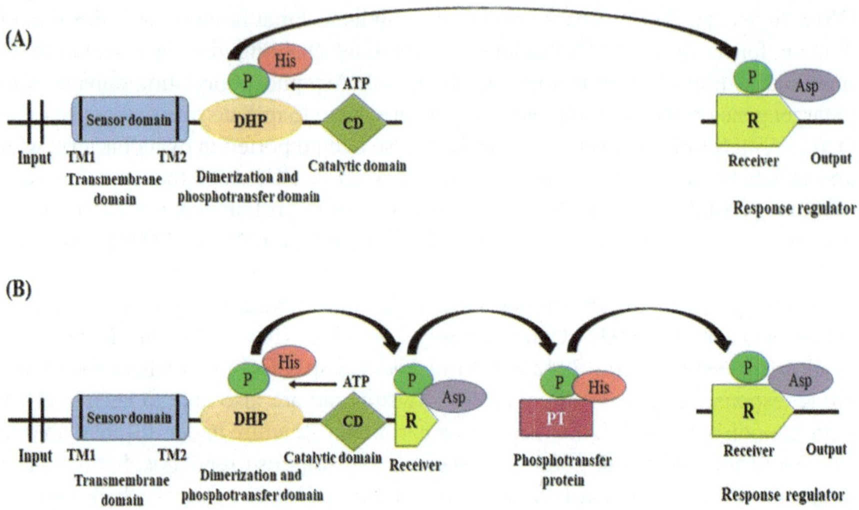

Fig. 8.2 Overview of functional differences in histidine kinases and hybrid kinases in cyanobacterial two-component systems (TCSs). (**a**) Stress signal perception promotes binding of ATP to the catalytic domain (CA) of histidine kinases, which then phosphorylate conserved histidine residue into the dimerization and phosphotransfer domain (DHP). The phosphoryl group (P) is then imparted onto a conserved aspartic acid residue present in the receiver domain of the response regulator (RR). (**b**) Hybrid kinases involve multistep phosphorelay systems where an additional receiver domain is fused with the catalytic domain along with an additional phosphotransfer protein

autophosphorylate and then transfer the phosphoryl group to their own internal receiver domain, rather than to a separate response regulator (RR) protein. The phosphoryl group is then shuttled to histidine phosphotransferase (HPT) and subsequently to a terminal RR, which can evoke the desired response via phosphorelay system (Fig. 8.2). These hybrid kinases possess a RR domain along with HisKA-HATPase domains and divided into seven classes ranging from HYI to HYVII. The HYI class has open reading frames (ORFs) constituting single RR domain analogous to N terminal to the HisK, whereas HYII classes are those which have single RR domain analogous to C terminal to the HisK and HYIII classes are those which possess ORFs with either two or three RR C terminal linked to single HisKA-HATPase domain (Paithoonrangsarid et al. 2004; Shoumskaya et al. 2005). The HYIV type contains a single histidine kinase with at least one RR domain on each side, HYVs possess one RR domain in between two histidine kinase, HYVI contains one HATP, CheW, and RR domain along with Hpt/Hkd domain, and HYVII is cataloged as incomplete hybrids having HATPase domains (Cann 2004). Among cyanobacteria, *Synechocystis* sp. PCC 6803 has been documented to contain five HYI-type hybrid kinases, which are hyperaccumulated in response to osmotic stress (Shoumskaya et al. 2005). Similarly, *Anabaena* PCC 7120 is known to contain genes (AphC and CyaC) that encode HYI-type hybrid kinases possessing Per-Arnt-Sim (PAS) and Per-Arnt-Sim C-terminal (PAC) domains (Paithoonrangsarid et al. 2004).

In addition, S. elongatus is reported to contain a HYII-type GAF domain that has the ability to revamp the circadian clock under various abiotic stress conditions. Furthermore, detailed characterization of these hybrid kinases has revealed some striking functions such as ability to bind phycobilisomes even in the absence of ligand-binding residues, ability to auto-phosphorylate HK domain, and ability to modulate auto-kinase activity (Mika and Hengge 2005). *Anabaena* PCC 7120 also contains one subclass of HYIII-type hybrid kinases encoded by Alr2279 (Chy133) with an additional N-terminal HNOBA domain capable of sensing heme-dependent gaseous molecules (Iyer et al. 2003). Similarly, HYIV-type hybrid kinases (HK19) have also been found in *Synechocystis* sp. PCC 6803, which are activated in response to temperature fluctuations. The two-component systems, i.e., histidine kinases and hybrid kinases, are also known to trigger the transcription of stress-responsive genes/proteins for effective scavenging of ROSs such as hydrogen peroxide, methyl-viologen, 3-(3,4-dichlorophenyl-1,1)-dimethylurea (DCMU), and 2,5-dibromo-3-methyl-6-isopropyl-p-benzoquinone (DBMIB), thus ameliorating the toxic effect of abiotic stress-induced oxidative damages (Shoumskaya et al. 2005). List of some potential histidine and hybrid kinases involved in the activation of stress-responsive genes under various abiotic stress conditions is shown in Table 8.2.

8.4 Serine/Threonine Kinases (STKs): Phosphorylation on Ser, Thr, and Tyr Residues in Cyanobacteria

Although cyanobacterial signaling relies to a large extent on two-component systems, serine/threonine protein kinases (STKs) supplement the well-known role of two-component systems. These protein kinases, together with their cognate phosphatases, play a central role in signal transduction by catalyzing reversible protein phosphorylation. Phosphorylation in serine, threonine, and tyrosine residues has been mainly associated with cellular signaling in eukaryotes until it was discovered in cyanobacteria by radioactive labeling of proteins (Mann 1994). After complete genome sequencing of few cyanobacterial sp., the putative STK proteins were identified by comparison of deduced amino acid sequences encoded by open reading frames with known amino acid sequences of eukaryotic protein kinases (Mizuno et al. 1996; Zhang et al. 1998). The number of STKs in an organism varies with genome size, ecophysiology, and physiological properties, and their evolution is the result of gene gain or loss and shuffling and insertion of domains (Zhang et al. 2007). STKs show homology in their catalytic domains to eukaryotic Ser/Thr kinases and are known to be present unevenly in cyanobacteria with their complete absence in marine unicellular *Prochlorococcus* and *Synechococcus* WH8102. The presence of FHA, WD40, PAS, and GAF domains beside conserved catalytic domains attributed additional complicated functions to these STKs (Wang et al. 2002). It has been proposed that some Ser/Thr kinases or phosphatases could be coupled to two-component systems in the signal transduction pathways (Zhang 1996; Zhang et al. 1998; Wang et al. 2002).

Table 8.2 List of some potential two-component systems involved in the regulation of abiotic stress tolerance in cyanobacteria

Two-component system	Gene name	Organisms and locus ID	Functional domain	Stress	Regulatory role	References
Histidine kinases						
HisKA	Hik34, Chk34	*Synechocystis* sp. strain PCC 6803 (Slr1285, Slr1783)	HKA	Hyperosmotic and salt stress	Regulation of expression of stress-responsive genes	Paithoonrangsarid et al. (2004)
HATPase	Chk112	*N. punctiforme* (Crr87), *Anabaena* sp. strain PCC 7120 (All7129), *Synechocystis* sp. PCC 6803 (Slr2098)	HATP	Hyperosmotic and salt stress	Activation of stress signaling cascades	Shoumskaya et al. (2005)
HKI	Chk2, Chk7, Chk9	*Synechocystis* sp. strain PCC 6803 (Slr1147)	HKA + HATP	Hyperosmotic	Activation of SigB genes	Shoumskaya et al. (2005)
HKII	Chk24, Chk26, Chk35	*S. elongatus* 7942, *Anabaena* sp. PCC 7120 DevR (Alr0442, Crr42)	GAF + HKA +HATP	Light stress	Biosynthesis of phycobilisomes and polysaccharide	Fischer and Lagarias (2004)
HKIII	Chk33, Chk95	*Synechocystis* sp. PCC 6803 (Hik10), *Anabaena* sp. PCC 7120 and *A. variabilis*	HMP + PAS+ HKA + HATP	Hyperosmotic, cold stress	Regulation of cold-responsive genes	Anantharaman and Aravind (2000)
Response regulators						
RRI CheY	Crr7–13	*Synechocystis* sp. strain PCC 6803, *Anabaena* sp. PCC 7120 DevR, and *S. elongatus* 7942	RR	Hyperosmotic	Regulation of DNA gyrase/ topoisomerase IV activity	Zhou and Wolk (2003)
RRII OmpR	Crr26, 31, 37	*Synechocystis* sp. PCC 6803, *Anabaena* sp. PCC 7120 (*all4312*)	RR + TREG	Hyperosmotic and salt stress	Cellular responses to nitrogen deprivation	Ashby and Mullineaux (1999)
RRIV+HD	Crr18, Crr8	*Synechocystis* sp. PCC 6803 (*sll1624*, Slr1305), *Anabaena* sp. strain PCC 7120 (Alr3768, Alr2280)	RR + HD	Osmotic and salt stress	Regulation of purine nucleotide cyclase activity	Cadoret et al. (2005)

Hybrid kinases						
HYI	Chy23	*Synechocystis* sp. strain PCC 6803 (Chk16), *Anabaena* sp. strain PCC 7120 (AphC), *N. punctiforme* (NpF1799)	RR + HKA +HATP	Hyperosmotic	Regulation of physiological functions	Paithoonrangsarid et al. (2004)
HYII	Chy21	*Synechocystis* sp. strain PCC 6803	RR + HKA +HATP	UV-B stress	Regulation of cGMP biosynthesis	Ochoa de Alda and Houmard (2000), Galperin et al. (2001)
HYIII	Chy44–46	*Synechocystis* sp. strain PCC 6803 (Slr2098, Slr1759), *Anabaena* sp. strain PCC 7120 (Alr2279), *N. punctiforme*	RR + HKA +Hpt	Hyperosmotic and salt stress	Regulation of expression of stress-responsive genes	Ochoa de Alda and Houmard (2000)
HYIV	Chy89	*Synechocystis* sp. strain PCC 6803 (Slr1905)	RR + HKA +HATP+Hpt	Hyperosmotic and salt stress	Regulation of expression of stress-responsive genes	Kasahara et al. (1997)

The presence of Ser/Thr kinases in cyanobacteria has been recognized by whole-genome sequencing and confirmed 52 genes in *Anabaena* sp. PCC 7120 (Wang et al. 2002) and 12 putative Ser/Thr kinases in *Synechocystis* PCC 6803 (Zorina et al. 2011). Seven of these Ser/Thr kinases belong to PKN2 subfamily denoted by SpkA, SpkB, SpkC, SpkD, SpkE, SpkF, and SpkG, and five of them belong to ABC1 subfamily, which is denoted by SpkH, SpkI, SpkJ, SpkK, and SpkL (Zorina et al. 2011). The systematic analysis of all these Ser/Thr kinases has been shown to encode proteins that are involved in the regulation of cell motility, photosynthesis, nitrogen metabolism, phosphorylation of GroES chaperone, and as possible regulators of transcription factors under various abiotic stress conditions (Zorina et al. 2011). Genomic data of *Anabaena* PCC 7120 indicated the presence of 12 genes encoding HstK with both Ser/Thr kinase domain and His kinase domain ensuring the interaction of two-component systems and STKs (Phalip et al. 2001).

The role of STKs in cold response was studied in *Synechocystis* sp. PCC6803 via screening of collection of STK mutants that identified four enzymes SpkB, SpkD, SpkE, and SpkG as possible transcriptional regulators at lower temperatures (Zorina et al. 2014). SpkG helps in sensing the high-salt signal directly rather than mediating signals (Liang et al. 2011), while SpkH is involved in signaling pathway of hyperosmotic stress response (Paithoonrangsarid et al. 2004) and regulated by two-component system Hik34-Rre1. Recently, it has been investigated that STKs of cyanobacteria are also involved in oxidative stress tolerance similar to animals and plants. Ser/Thr kinase SpkB, which is known to be involved in the control of cell motility (Kamei et al. 2003), is also required for survival of the cyanobacterium under increased concentrations of reactive oxygen species. SpkB was found to be inhibited by oxidation and reactivated by thioredoxin-catalyzed reduction and may be subject to redox regulation through modulation of the redox state of its cysteines suggesting the possible roles of the *Synechocystis* STKs in ROS tolerance (Mata-Cabana et al. 2012). In *Anabaena* sp. PCC 7120, several genes encoding STKs are induced upon iron limitation or oxidative stress, e.g., pkn22 (*alr2502*) (Xu et al. 2003) and gene adjacent to pkn22 (*alr2503*), and encode a protein similar to peroxiredoxin predicting the possible function of pkn22 in response to iron and oxidative stress.

Protein phosphatases constitute important role in signal transduction pathway by executing reversible phosphorylation of proteins. They are classified into three categories in eukaryotes on the basis of domain architectures and three-dimensional structures (Yigong 2009): aspartate-based phosphatases (Fcp/Scp), metal-dependent phosphatases (PPMs), and phosphoprotein phosphatases (PPPs). In *Synechocystis*, Sll1033 (SynPPM3) and Sll1387 (SynPPP1) are homologs to the PPM and PPP families of phosphatases. Signal transduction protein P(II) is dephosphorylated by protein phosphatase PphA (*sll1771*), while other putative Ser/Thr phosphatases include *sll1365, slr0114, slr0328, slr1860, slr1983, and slr2031* (Li et al. 2004). Fourteen genes encoding phospho-Ser, Thr, and Tyr phosphatases were identified in *Anabaena* PCC 7120 of which three genes encode phospho-tyrosine phosphatases (PTPs) and eight proteins are Ser/Thr phosphatases of the PP2C family (Wang et al. 2002).

8.5 Other Potential Sensors and Transducers in Cyanobacteria

Several regulators, e.g., sigma factors and transcription factors, receive the information about the changing environment from sensors and couple it to signal transduction chains of cyanobacteria. Promoter binding, initiation of transcription, and efficiency of transcription by the RNA polymerase can be negatively and positively affected by these regulators.

8.5.1 RNA Polymerase Sigma Factors and Transcription Factors

Transcription is performed by the RNA polymerase (RNAP) holoenzyme, comprising a core enzyme and sigma (σ) factor, which are assigned to groups 1, 2, and 3 based on phylogenetic analyses (Imamura and Asayama 2009). In response to stress, sigma factors modulated promoter selectivity via general switching ("σ switching") of multiple RNAP holoenzymes with different σ factors. Group 1 is composed of primary σ factors (PSF) comprising sigA genes that are essential for cell viability, while group 2 σ factors (sigB, sigC, sigD, and sigE genes) are nonessential for cell viability. Group 3 or alternative σ factors (sigF, sigG, sigH, and sigI genes) are involved in the transcription under stress conditions. Attempts to characterize the functional role of each σ factor have been recently made (Imamura and Asayama 2009), which revealed that sigA encodes the PSF, sigB is multifunctional for osmotic, salt, or oxidative stress and also involved in the transcription of the heat shock genes, sigC under nitrogen starvation, and heat acclimation, while sigE participates in positive regulation of sugar catabolic pathways. sigD recognizes the promoter of psbA that encodes the major protein D1 of the PSII alternative σ factors that belong to group 3, and sigF represents a terminal element of a signal-transducing pathway-sensing salt. sigH is responsible for the regulation of heat shock genes, and sigG (*slr1545*) was found to be essential for growth. Coordination among group 1 and group 2 σ factors seems to contribute to the sensing of environmental changes; however, role of cyanobacterial σ factors in response to oxidative stress is yet to be explored.

RNA polymerase specificity is also controlled by sequence-specific DNA-binding proteins called transcription factors (TFs), which help in regulation of transcription of genes involved in stress signaling pathway. Fifty-seven genes for TFs were found in the *Synechocystis* PCC 6803 genome, which includes the DNA-binding domains of seven families. Majority of transcription factors are regulated not only at the level of transcription, but also at a post-translational level. A small LuxR-type transcription factor in *Synechocystis* PCC 6803, PedR (photosynthetic electron transport-dependent regulator; locus ssl0564), was shown to be involved in transcriptional regulation and establishes an important link between perception of changes in photosynthetic activity. HrcA, which is known to interact with sigma factors SigB, SigE, and/or sensory histidine kinase Hik34, regulates the expression of few heat shock genes including groESL and groEL2 (Nakamoto et al. 2003). PerR (Slr1738) regulates a set of genes, which are induced in response to hydrogen peroxide (Li et al. 2004) and also involved in reprogramming of cellular metabolism in response to excess cadmium concentrations (Houot et al. 2007). Another TF, which is involved in regulation of response to oxidative stress, is the autorepressor PrqR (Sll0886), which negatively regulates the prqR-prqA operon and the response to methyl viologen (Kirik et al. 2003).

In the heterocystous cyanobacterium *Anabaena* PCC7120, a transcriptional regulator of the ArsR family, Alr1867, was reported as a redox-active transcriptional repressor of the trxA2 gene (*all1866*) and was designated RexT (redox-sensing transcriptional regulator of thioredoxin A2) (Ehira and Ohmori 2012). Identification of Trx-interacting transcription factors, PedR and RexT, suggests that Trx may be a key component of transcriptional regulation. ChlR is a transcriptional regulator of the MarR family (multiple antibiotic resistance regulator), and at least 11 cyanobacteria possess putative orthologs.

8.5.2 DNA Supercoiling: Role in Perception of Stress Signals and The Regulation of Gene Expression

Changes in DNA supercoiling caused by environmental stress regulate expression of downstream genes confirming successful acclimatization of cells (Prakash et al. 2009). Studies of changes in the supercoiling of DNA were restricted to plasmid DNAs at the start in *E. coli*, *B. subtilis,* and *Salmonella typhimurium*. However, DNA microarray and Northern blotting analyses in cyanobacterium *Synechocystis* sp. PCC 6803 indicated that the function of Hik33 and Hik34 depends on the degree of supercoiling of the genomic DNA (Prakash et al. 2009). Most of the cold-induced genes controlled by Hik33 require negative supercoiling of DNA, whereas most of the heat-induced genes require high-temperature induced relaxation of genomic DNA followed by action of the DNA gyrase, to maintain the extent of degree of supercoiling of DNA for efficient transcription. Cold stress caused an increase in the negative supercoiling of the promoter region of the *desB* gene for a fatty acid desaturase and directly controlled its expression at low temperature (Los 2004; Prakash et al. 2009) pointing that temperature-induced changes in supercoiling of DNA might contribute to the expression of genes for ROS inactivation, ribosomal proteins, RNA chaperons, cell wall, and lipid metabolism (Los and Murata 2004) in *Synechocystis* sp. PCC6803. In addition, salt and hyperosmotic stress also known to affect the negative supercoiling of DNA leading to transcription of stress-responsive genes.

8.6 Conclusions

A large integrated network of signal transduction pathways assists cyanobacteria to sense and respond to continuously changing environmental conditions. Although the mechanism is not yet explored, cyanobacteria possess "multi-stress sensors" that perceive and transduce more than one kind of environmental signal regardless of its nature. Several regulators such as sigma factors and transcription factors ensure the integration of different signals into extensive signaling pathway. ROS-induced oxidative stress initiates array of signaling pathways, which are still to be discovered advocating the basis for future research on the entire pathway of H_2O_2 signal transduction in cyanobacteria. Also, there is need of cross-talk among two-component systems (TCSs) and STK systems to explore the mechanism of multiple phosphorylation schemes combined in a single pathway of cyanobacterial signal transduction system.

Acknowledgements L.C. Rai thanks National Academy of Sciences India for NASI Senior Scientist Platinum Jubilee Fellowship, Indian Council of Agricultural Research-National Bureau of Agriculturally Important Microorganisms (ICAR-NBAIM), and SERB Govt. of India for financial support. Dr. Ruchi Rai is thankful to DST, New Delhi, for Women Scientist Scheme A (WOSA) award, Dr. Shilpi Singh thanks UGC for D.S. Kothari fellowship, and Dr. Krishna Kumar Rai is thankful to NASI for RAship. Alka Raj thanks UGC for Junior Research Fellowship (JRF). We thank Head and Coordinator, CAS in Botany for facilities through FIST and PURSE supports.

References

Abed RMM, Sergey D, Kumar S (2009) Applications of cyanobacteria in biotechnology. J Appl Microbiol 106(1):1–12

Aggarwal K, Choe LH, Lee KH (2006) Shotgun proteomics using the iTRAQ isobaric tags. Brief Funct Genomics Proteomics 5:112–120

Agrawal C, Sen S, Yadav S, Rai S, Rai LC (2015) A novel aldo-keto reductase (AKR17A1) of *Anabaena* sp. PCC 7120 degrades the rice field herbicide butachlor and confers tolerance to abiotic stresses in *E. coli*. PLoS One 10(9):e0137744

Alvarenga DO, Fiore MF, Varani AM (2017) A metagenomic approach to cyanobacterial genomics. Front Microbiol 8:809

Anantharaman V, Aravind L (2000) Cache-a signaling domain common to animal Ca (2+)-channel subunits and a class of prokaryotic chemotaxis receptors. Trends Biochem Sci 25:535–537

Ashby MK, Houmard J (2006) Cyanobacterial two-component proteins: structure, diversity, distribution and evolution. Microbiol Mol Biol Rev 70(2):472–509

Ashby MK, Mullineaux CW (1999) Cyanobacterial ycf27 genes regulate the coupling of phycobilisomes to photosystems I and II. FEMS Microbiol Lett 181:253–260

Babele PK, Singh G, Kumar A, Tyagi MB (2015) Induction and differential expression of certain novel proteins in *Anabaena* L31 under UV-B radiation stress. Front Microbiol 6:133

Babele PK, Kumar J, Chaturvedi V (2019) Proteomic de-regulation in cyanobacteria in response to abiotic stresses. Front Microbiol 10:1315

Bhargava P, Mishra Y, Srivastava AK, Narayan OP, Rai LC (2008) Excess copper induces anoxygenic photosynthesis in *Anabaena doliolum*: a homology based proteomic assessment of its survival strategy. Photosynth Res 96:61–74

Cadoret JC, Rousseau B, Perewoska I, Sicora C, Cheregi O, Vass I, Houmard J (2005) Cyclic nucleotides, the photosynthetic apparatus and response to a UV-B stress in the cyanobacterium *Synechocystis* sp. PCC 6803. J Biol Chem 280:33935–33944

Cann MJ (2004) Signalling through cyclic nucleotide monophosphates in cyanobacteria. New Phytol 159:289–293

Castenholz RW (2001) Phylum BX. Cyanobacteria. In: Boone DR, Castenholz RW (eds) Bergey's manual of systematic bacteriology, 2nd edn. Springer, New York, NY, pp 473–599

Castielli O, De la Cerda B, Navarro JA, Hervás M, De la Rosa MA (2009) Proteomic analyses of the response of cyanobacteria to different stress conditions. FEBS Lett 583:1753–1758

Chatterjee A, Singh S, Agrawal C, Yadav S, Rai R, Rai LC (2017) Role of algae as biofertilizer. In: Rastogi RP, Madamwar D, Pandey A (eds) Algal green chemistry: recent progress in biotechnology. Elsevier, Amsterdam, pp 189–201

Chatterjee A, Singh S, Rai R, Rai S, Rai LC (2020) Functional characterization of Alr0765, a hypothetical protein from *Anabaena* PCC 7120 involved in cellular energy status sensing, iron acquisition and abiotic stress management in *E. coli* using molecular, biochemical and computational approaches. Curr Genomics 21:295–310

Chiang GG, Schaefer MR, Grossman AR (1992) Complementation of a red-light-in different cyanobacterial mutant. Proc Natl Acad Sci U S A 89:9415–9419

de Vries J, Archibald JM (2017) Endosymbiosis: did plastids evolve from a freshwater cyanobacterium? Curr Biol 27:103–105

Demoulin CF, Lara YJ, Cornet L, François C, Baurain D, Wilmotte A, Emmanuelle JJ (2019) Cyanobacteria evolution: insight from the fossil record. Free Radic Biol Med 140:206–223

Dephoure N, Gould KL, Gygi SP, Kellogg DR (2013) Mapping and analysis of phosphorylation sites: a quick guide for cell biologists. Mol Biol Cell 24:535–542

Dorman CJ (2006) DNA supercoiling and bacterial gene expression. Sci Prog 89:151–166

Ehira S, Ohmori M (2012) The redox-sensing transcriptional regulator RexT controls expression of thioredoxin A2 in the cyanobacterium *Anabaena* Sp. strain PCC 7120. J Biol Chem 287(48)

Fischer AJ, Lagarias JC (2004) Harnessing phytochrome's glowing potential. Proc Natl Acad Sci U S A 101:17334–17339

Fulda S, Mikkat S, Huang F, Huckauf J, Marin K, Norling B, Hagemann M (2006) Proteome analysis of salt stress response in the cyanobacterium *Synechocystis* sp. strain PCC 6803. Proteomics 6:2733–2745

Galperin MY, Gaidenko TA, Mulkidjanian AY, Nakano M, Price CW (2001) MHYT, a new integral membrane sensor domain. FEMS Microbiol Lett 205:17–23

Gao Y, Xiong W, Li XB, Gao CF, Zhang YL, Li H, Wu Q (2009) Identification of the proteomic changes in *Synechocystis* sp. PCC 6803 following prolonged UV-B irradiation. J Exp Bot 60: 1141–1154

Garcia-Pichel F, Benlap J, Neuer S, Schanz F (2003) Estimates of global cyanobacterial biomass and its distribution. Algol Stud 109:213–227

Giner-Lamia J, López-Maury L, Florencio FJ (2014) Global transcriptional profiles of the copper response in the cyanobacterium *Synechocystis* sp. PCC 6803. PLoS One 9:e108912

Holmgren A (1989) Thioredoxin and glutaredoxin systems. J Biol Chem 264:13963–13966

Hongsthong A, Sirijuntarut M, Prommeenate P, Lertladaluck K, Porkaew K, Cheevadhanarak S, Tanticharoen M (2008) Proteome analysis at the subcellular level of the cyanobacterium Spirulina platensis in response to low-temperature stress conditions. FEMS Microbiol Lett 288(1):92–101

Hongsthong A, Sirijuntarut M, Yutthanasirikul R, Senachak J, Kurdrid P, Cheevadhanarak S, Tanticharoen M (2009) Subcellular proteomic characterization of the hightemperature stress response of the cyanobacterium Spirulina platensis. Proteome Sci 7(1):1–19

Houot L, Floutier M, Marteyn B, Michaut M, Picciocchi A, Legrain P, Aude JC, Cassier-Chauvat C, Chauvat F (2007) Cadmium triggers an integrated reprogramming of the metabolism of *Synechocystis* PCC6803, under the control of the Slr1738 regulator. BMC Genomics 8:350

Imamura S, Asayama M (2009) Sigma factors for cyanobacterial transcription. Gene Regul Syst Bio 3:65–87

Imamura S, Yoshihara S, Nakano S, Shiozaki N, Yamada A, Tanaka K, Takahashi H, Asayama M, Shirai M (2003) Purification, characterization, and gene expression of all sigma factors of RNA polymerase in a cyanobacterium. J Mol Biol 325:857–872

Iyer LM, Anantharaman V, Aravind L (2003) Ancient conserved domains shared by animal soluble guanylyl cyclases and bacterial signaling proteins. BMC Genomics 4:5

Jeamton W, Mungpakdee S, Sirijuntarut M, Prommeenate P, Cheevadhanarak S, Tanticharoen M et al (2008) A combined stress response analysis of *Spirulina platensis* in terms of global differentially expressed proteins, and mRNA levels and stability of fatty acid biosynthesis genes. FEMS Microbiol Lett 281:121–131

Jensen PE, Leister D (2014) Cyanobacteria as an experimental platform for modifying bacterial and plant photosynthesis. Front Bioeng Biotechnol 2:7

Jungblut P, Thiede B, Zimny-Arndt U, Muller EC, Scheler C, Wittmann-Liebold B et al (1996) Resolution power of two-dimensional electrophoresis and identification of proteins from gels. Electrophoresis 17:839–847

Kamei A, Yoshihara S, Yuasa T, Geng X, Ikeuchi M (2003) Biochemical and functional characterization of a eukaryotic-type protein kinase, SpkB, in the cyanobacterium, *Synechocystis* sp. PCC 6803. Curr Microbiol 46(4):296–301

Kaneko T, Sato S, Kotani H, Tanaka A, Asamizu E, Nakamura Y, Miyajima N, Hirosawa M, Sugiura M, Sasamoto S et al (1996) Sequence analysis of the genome of the unicellular cyanobacterium *Synechocystis* sp. strain PCC 6803. II. Sequence determination of the entire genome and assignment of potential protein-coding regions. DNA Res 3:109–136

Kaneko T, Nakamura Y, Wolk CP, Kuritz T, Sasamoto S, Watanabe A, Iriguchi M, Ishikawa A, Kawashima K, Kimura T, Kishida Y (2001) Complete genomic sequence of the filamentous nitrogen-fixing cyanobacterium *Anabaena* sp. strain PCC 7120. DNA Res 8(5):205–213

Kaniya Y, Kizawa A, Miyagi A, Kawai-Yamada M, Uchimiya H, Kaneko Y, Nishiyama Y, Hihara Y (2013) Deletion of the transcriptional regulator cyAbrB2 deregulates primary carbon metabolism in *Synechocystis* sp. PCC 6803. Plant Physiol 162:1153–1163

Kasahara M, Yashiro K, Sakamoto T, Ohmori M (1997) The *Spirulina platensis* adenylate cyclase gene, cyaC, encodes a novel signal transduction protein. Plant Cell Physiol 38:828–836

Katano Y, Nimura-Matsune K, Yoshikawa H (2006) Involvement of DnaK3, one of the three DnaK proteins of cyanobacterium *Synechococcus* sp. PCC7942, in translational process on the surface of the thylakoid membrane. Biosci Biotechnol Biochem 70:1592–1598

Katoh H, Asthana R, Ohmori M (2004) Gene expression in the cyanobacterium *Anabaena* sp. PCC7120 under desiccation. Microb Ecol 47:164–174

Kirik IA, Zinchenko VV, Shestakov SV, Babykin MM (2003) Trans- and cis-acting autorepressors of the prqR gene in *Synechocystis* cyanobacteria sp. PCC6803. Mol Biol (Mosk) 37:1035–1044

Kurian D, Phadwal K, Mäenpää P (2006) Proteomic characterization of acid stress response in *Synechocystis* sp. PCC 6803. Proteomics 6:3614–3624

Latifi A, Ruiz M, Zhang CC (2009) Oxidative stress in cyanobacteria. FEMS Microbiol Rev 33: 258–278

Li H, Singh AK, McIntyre LM, Sherman LA (2004) Differential gene expression in response to hydrogen peroxide and the putative PerR regulon of *Synechocystis* sp. strain PCC 6803. J Bacteriol 186:3331–3345

Li T, Yang HM, Cui SX, Suzuki I, Zhang LF, Li L, Bo TT, Wang J, Murata N, Huang F (2011) Proteomic study of the impact of Hik33 mutation in *Synechocystis* sp. PCC 6803 under normal and salt stress conditions. J Proteome Res 11:502–514

Liang C, Zhang X, Chi X, Guan X, Li Y, Qin S, Shao HB (2011) Serine/Threonine protein kinase SpkG is a candidate for high salt resistance in the unicellular cyanobacterium *Synechocystis* sp. PCC 6803. PLoS One 6(5):e18718

Lopez-Maury L, Garcia-Dominguez M, Florencio FJ, Reyes JC (2002) A two-component signal transduction system involved in nickel sensing in the cyanobacterium *Synechocystis* sp. PCC 6803. Mol Microbiol 43:247–256

Los DA (2004) The effect of low-temperature-induced DNA supercoiling on the expression of the desaturase genes in *Synechocystis*. Cell Mol Biol 50:605–612

Los DA, Murata N (2004) Membrane fluidity and its roles in the perception of environmental signals. Biochim Biophys Acta 1666:142–157

Los DA, Zorina A, Sinetova M, Kryazhov S, Mironov K, Zinchenko VV (2010) Stress sensors and signal transducers in cyanobacteria. Sensors 10(3):2386–2415

Los DA, Mironov KS, Allakhverdiev SI (2013) Regulatory role of membrane fluidity in gene expression and physiological functions. Photosynth Res 116(2):489–509

Mann NH (1994) Protein phosphorylation in cyanobacteria. Microbiology 140:3207–3215

Marin K, Suzuki I, Yamaguchi K, Ribbeck K, Yamamoto H, Kanesaki Y, Hagemann M, Murata N (2003) Identification of histidine kinases that act as sensors in the perception of salt stress in *Synechocystis* sp. PCC 6803. Proc Natl Acad Sci U S A 100:9061–9066

Mata-Cabana A, Garcia-Dominguez M, Florencio FJ, Lindahl M (2012) Thiol-based redox modulation of a cyanobacterial eukaryotic-type serine/threonine kinase required for oxidative stress tolerance. Antioxid Redox Signal 17(4):521–533

Mika F, Hengge R (2005) A two-component phosphotransfer network involving ArcB, ArcA, and RssB coordinates synthesis and proteolysis of (RpoS) in *E. coli*. Genes Dev 19:2770–2781

Mikkat S, Fulda S, Hagemann M (2014) A 2-D gel electrophoresis based snapshot of the phosphoproteome in the cyanobacterium *Synechocystis* sp. strain PCC 6803. Microbiology 160(Pt 2):296–306

Miranda H, Cheregi O, Netotea S, Hvidsten TR, Moritz T, Funk C (2013) Co-expression analysis, proteomic and metabolomic study on the impact of a Deg/HtrA protease triple mutant in *Synechocystis* sp. PCC 6803 exposed to temperature and high light stress. J Proteome 78: 294–311

Mironov KS, Sinetova MA, Shumskaya M, Los DA (2019) Universal molecular triggers of stress responses in cyanobacterium *Synechocystis*. Life 9(3):67

Mizuno T, Kaneko T, Tabata S (1996) Compilation of all genes encoding bacterial two-component signal transducers in the genome of the cyanobacterium, *Synechocystis* sp. strain PCC 6803. DNA Res 3:407–414

Murata N, Los DA (2006) Histidine kinase Hik33 is an important participant in cold signal transduction in cyanobacteria. Physiol Plant 126:17–27

Nakamoto H, Suzuki M, Kojima K (2003) Targeted inactivation of the hrcA repressor gene in cyanobacteria. FEBS Lett 549:57–62

Ochoa de Alda JAG, Houmard J (2000) Genomic survey of cAMP and cGMP signalling components in the cyanobacterium *Synechocystis* PCC 6803. Microbiology 146:3183–3194

Ow SY, Wright PC (2009) Current trends in high throughput proteomics in cyanobacteria. FEBS Lett 583:1744–1752

Paithoonrangsarid K, Shoumskaya MA, Kanesaki Y, Satoh S, Tabata S, Los DA, Zinchenko VV, Hayashi H, Tanticharoen M, Suzuki I, Murata N (2004) Five histidine kinases perceive osmotic stress and regulate distinct sets of genes in *Synechocystis*. J Biol Chem 279:53078–53086

Panda B, Basu B, Rajaram H, Apte SK (2014) Methyl viologen responsive proteome dynamics of *Anabaena* sp. strain PCC7120. Proteomics 14:1895–1904

Panda B, Basu B, Rajaram H, Apte SK (2015) Comparative proteomics of oxidative stress response in three cyanobacterial strains native to Indian paddy fields. J Proteomics 127(Pt A):152–160

Pandey S, Rai R, Rai LC (2012) Proteomics combines morphological, physiological and biochemical attributes to unravel the survival strategy of *Anabaena* sp. PCC7120 under arsenic stress. J Proteome 75:921–937

Phalip V, Li JH, Zhang CC (2001) HstK, a cyanobacterial protein with both a serine/threonine kinase domain and a histidine kinase domain: implication for the mechanism of signal transduction. Biochem J 360:639–644

Prakash JSS, Sinetova M, Kupriyanovab E, Zorina A, Suzuki I, Murata N, Los DA (2009) DNA supercoiling regulates the stress-inducible expression of genes in the cyanobacterium. Mol BioSyst 5(12):1904–1912

Prakash JS, Krishna PS, Sirisha K, Kanesaki Y, Suzuki I, Shivaji S et al (2010) An RNA helicase, CrhR, regulates the low-temperature-inducible expression of heat-shock genes groES, groEL1 and groEL2 in *Synechocystis* sp. PCC 6803. Microbiology 156:442–451

Rai S, Singh S, Shrivastava AK, Rai LC (2013) Salt and UV-B induced changes in *Anabaena* PCC 7120: physiological, proteomic and bioinformatics perspectives. Photosynth Res 118:105–114

Rai S, Yadav S, Rai R, Chatterjee A, Singh S, Rai LC (2019a) Molecular and biochemical characterization of All0580 as a methylglyoxal detoxifying glyoxalase II of *Anabaena* sp. PCC7120 that confers abiotic stress tolerance in *E. coli*. Int J Biol Macromol 124:981–993

Rai S, Rai R, Singh PK, Rai LC (2019b) Alr2321, a multiple stress inducible glyoxalase I of *Anabaena* sp. PCC7120 detoxifies methylglyoxal and reactive species oxygen. Aquat Toxicol 214:105238

Rai R, Singh S, Chatterjee A, Rai KK, Rai S, Rai LC (2020) All4894 encoding a novel fasciclin (FAS-1 domain) protein of *Anabaena* sp. PCC7120 revealed the presence of a thermostable β-glucosidase. Algal Res 51:102036

Rodriguez-Ezpeleta N, Brinkmann H, Burey SC, Roure B, Burger G, Loffelhardt W et al (2005) Monophyly of primary photosynthetic eukaryotes: green plants, red algae, and glaucophytes. Curr Biol 15:1325–1330

Sakayori T, Shiraiwa Y, Suzuki I (2009) A *Synechocystis* homolog of SipA protein, Ssl3451, enhances the activity of the histidine kinase Hik33. Plant Cell Physiol 50:1439–1448

Sandh G, Ramström M, Stensjö K (2014) Analysis of the early heterocyst Cys-proteome in the multicellular cyanobacterium *Nostoc punctiforme* reveals novel insights into the division of labor within diazotrophic filaments. BMC Genomics 15:1064

Sato S, Shimoda Y, Muraki A, Kohara M, Nakamura Y, Tabata S (2007) A large-scale protein-protein interaction analysis in *Synechocystis* sp. PCC6803. DNA Res 14:207–216

Schmitt FJ, Renger G, Friedrich T, Kreslavski VD, Zharmukhamedov SK, Los DA, Kuznetsov VV, Allakhverdiev SI (2014) Reactive oxygen species: re-evaluation of generation, monitoring and role in stress-signaling in phototrophic organisms. Biochim Biophys Acta 1837:835–848

Schwartz SH, Black TA, Jager K, Panoff JM, Wolk CP (1998) Regulation of an osmoticum-responsive gene in *Anabaena* sp. strain PCC 7120. J Bacteriol 180:6332–6337

Sen S, Rai R, Chatterjee A, Rai S, Yadav S, Agrawal C, Rai LC (2019) Molecular characterization of two novel proteins All1122 and Alr0750 of *Anabaena* PCC 7120 conferring tolerance to multiple abiotic stresses in *Escherichia coli*. Gene 685:230–241

Shcolnick S, Shaked Y, Keren N (2007) A role for mrgA, a DPS family protein, in the internal transport of Fe in the cyanobacterium *Synechocystis* sp. PCC6803. Biochim Biophys Acta 1767: 814–819

Shoumskaya MA, Paithoonrangsarid K, Kanesaki Y, Los DA, Zinchenko VV, Tanticharoen M, Suzuki I, Murata N (2005) Identical Hik-Rre systems are involved in perception and transduction of salt signals and hyperosmotic signals but regulate the expression of individual genes to different extents in *Synechocystis*. J Biol Chem 280:21531–21538

Shrivastava AK, Chatterjee A, Yadava S, Singh PK, Singh S, Rai LC (2015) UV-B stress induced metabolic rearrangements explored with comparative proteomics in three *Anabaena* spp. J Proteome 127:122–133

Shrivastava AK, Pandey S, Yadav S, Mishra Y, Singh PK, Rai R, Singh S, Rai S, Rai LC (2016) Comparative proteomics of wild type, an+ ahpC and AnΔahpC strains of *Anabaena* sp. PCC7120 demonstrates AhpC mediated augmentation of photosynthesis, N_2-fixation and modulation of regulatory network of antioxidative proteins. J Proteome 140:81–99

Sinetova MA, Los DA (2016) Systemic analysis of transcriptomics of *Synechocystis*: common stress genes and their universal triggers. Mol BioSyst 12:3254–3258

Singh PK, Shrivastava AK, Chatterjee A, Pandey S, Rai S, Singh S, Rai LC (2015) Cadmium toxicity in diazotrophic *Anabaena* spp. adjudged by hasty up-accumulation of transporter and signaling and severe down-accumulation of nitrogen metabolism protein. J Proteome 127:134–146

Singh JS, Kumar A, Rai AN, Singh DP (2016) Cyanobacteria: a precious bio-resource in agriculture, ecosystem, and environmental sustainability. Front Microbiol 7:529

Singh R, Parihar P, Singh M, Bajguz A, Kumar J, Singh S, Singh VP, Prasad SM (2017) Uncovering potential applications of cyanobacteria and algal metabolites in biology, agriculture and medicine: current status and future prospects. Front Microbiol 8:515

Soo RM, Hemp J, Parks DH, Fischer WW, Hugenholtz P (2017) On the origins of oxygenic photosynthesis and aerobic respiration in cyanobacteria. Science 355(6332):1436–1440

Stensjö K, Ow SY, Barrios-Llerena ME, Lindblad P, Wright PC (2007) An iTRAQ-based quantitative analysis to elaborate the proteomic response of *Nostoc* sp. PCC 7120 under N_2 fixing conditions. J Proteome Res 6:621–635

Stock AM, Victoria LR, Paul NG (2000) Two-component signal transduction. Annu Rev Biochem 69:183–215

Tandeau de Marsac N, Houmard J (1993) Adaptation of cyanobacteria to environmental stimuli: new steps towards molecular mechanisms. FEMS Microbiol Lett 104(1–2):119–189

Vaishampayan A, Sinha RP, Hader DP, Dey T, Gupta AK, Bhan U, Rao AL (2001) Cyanobacterial biofertilizers in rice agriculture. Bot Rev 67:453–516

Varkey D, Mazard S, Ostrowski M, Tetu SG, Haynes P, Paulsen IT (2016) Effects of low temperature on tropical and temperate isolates of marine *Synechococcus*. ISME J 10:1252–1263

Wang L, Sun YP, Chen WL, Li JH, Zhang CC (2002) Genomic analysis of protein kinases, protein phosphatases and two-component regulatory systems of the cyanobacterium *Anabaena* sp. strain PCC 7120. FEMS Microbiol Lett 217:155–165

Wang H, Yang Y, Chen W, Li D, Li P, Zhao X, Wang X, Li A, Bao Q (2013) Identification of differentially expressed proteins of *Arthrospira (Spirulina) plantensis*-YZ under salt-stress conditions by proteomics and qRT-PCR analysis. Proteome Sci 11:6

Xiong Q, Feng J, Li ST, Zhang GY, Qiao ZX, Chen Z et al (2015) Integrated transcriptomic and proteomic analysis of the global response of *Synechococcus* sp. PCC 7002 to high light stress. Mol Cell Proteomics 14:1038–1053

Xu WL, Jeanjean R, Liu YD, Zhang CC (2003) pkn22 (*alr2502*) encoding a putative ser/thr kinase in the cyanobacterium *Anabaena* sp. PCC 7120 is induced by both iron starvation and oxidative stress and regulates the expression of isiA. FEBS Lett 553(1–2):179–182

Yeh KC, Wu SH, Murphy JT, Lagarias JC (1997) A cyanobacterial phytochrome two-component light sensory system. Science 277:1505–1508

Yigong S (2009) Serine/threonine phosphatases: mechanism through structure. Cell 139:468–484

Yoshihara S, Geng X, Ikeuchi M (2002) PilG gene cluster and split pilL genes involved in pilus biogenesis, motility and genetic transformation in the cyanobacterium *Synechocystis* sp. PCC 6803. Plant Cell Physiol 43:513–521

Zhang CC (1996) Bacterial signalling involving eukaryotic-type protein kinases. Mol Microbiol 20: 9–15

Zhang CC, Gonzalez L, Phalip V (1998) Survey, analysis and genetic organization of genes encoding eukaryotic-like signaling proteins on a cyanobacterial genome. Nucl Acids Res 26: 3619–3625

Zhang X, Zhao F, Guan X, Yang Y, Liang C, Qin S (2007) Genome-wide survey of putative serine/threonine protein kinases in cyanobacteria. BMC Genomics 8:395

Zhang LF, Yang HM, Cui SX, Hu J, Wang J, Kuang TY, Norling B, Huang F (2009) Proteomic analysis of plasma membranes of cyanobacterium *Synechocystis* sp. strain PCC 6803 in response to high pH stress. J Proteome Res 8:2892–2902

Zhou R, Wolk CP (2003) A two-component system mediates developmental regulation of biosynthesis of a heterocyst polysaccharide. J Biol Chem 278:19939–19946

Zorina A, Stepanchenko N, Novikova GV, Sinetova M, Panichkin VB, Moshkov IE, Zinchenko VV, Shestakov SV, Suzuki I, Murata N, Los DA (2011) Eukaryotic-like ser/thr protein kinases SpkC/F/K are involved in phosphorylation of GroES in the cyanobacterium *Synechocystis*. DNA Res 18:137–151

Zorina A, Bedbenov VS, Novikova GV, Panichkin VB, Los DA (2014) Involvement of serine/threonine protein kinases in the cold stress response in the cyanobacterium *Synechocystis* sp. PCC 6803: functional characterization of SpkE protein kinase. Mol Biol 48(3):390–398

Evolution and Diversification of the GroEL/Chaperonin Paralogs in Cyanobacteria

9

Hitoshi Nakamoto

Abstract

Molecular chaperones are involved in maintaining cellular protein homeostasis under normal and stress conditions. They interact with unfolded, misfolded, and aggregated proteins and assembled protein complexes. Upon stress, their cellular levels increase greatly to maintain a functional proteome. Molecular chaperones have been extensively studied in *E. coli,* and the *E. coli* paradigm has greatly contributed to the development of chaperone research. However, there are exceptions to the paradigm, which are observed in *groEL* paralogs in phototrophs. GroEL is a bacterial member of the GroEL/chaperonin/Hsp60 family, which is evolutionarily conserved. In contrast to *E. coli,* which has only a single *groESL* operon, almost all cyanobacterial genomes encode one each of the *groESL1* operon and a monocistronic *groEL2* gene. Accumulating evidence has shown that regulation of gene expression, structure, and function of cyanobacterial GroELs are mutually distinct and different from *E. coli* GroEL. In cyanobacteria, transcription of the *groESL1* operon and *groEL2* is induced not only by heat but also by light. Two highly conserved regulatory elements, CIRCE and K-box, are involved in *groESL1* transcription, whereas the regulatory mechanisms of *groEL2* transcription appear to be more diversified. Studies in *E. coli* and cyanobacterial cells have indicated that GroEL1 is equivalent to *E. coli* GroEL, which is essential. On the other hand, GroEL2 is nonessential, but plays a role under stress. Biochemical studies have shown that GroEL1 and GroEL2 are clearly different. The current status of research thus strongly suggests

H. Nakamoto (✉)
Department of Biochemistry and Molecular Biology, Graduate School of Science and Engineering, Saitama University, Saitama, Japan
e-mail: nakamoto@mail.saitama-u.ac.jp

R. P. Rastogi (ed.), *Ecophysiology and Biochemistry of Cyanobacteria*, https://doi.org/10.1007/978-981-16-4873-1_9

that the *groEL2* gene is the outcome of neofunctionalization after *groESL* gene duplication.

Keywords

Heat shock protein · Molecular chaperone · Heat shock response · Gene expression · Cyanobacteria · Stress · Proteostasis

9.1 Introduction

The three-dimensional structure a protein forms is defined by its amino acid sequence. This is known as Anfinsen's dogma (Anfinsen 1973). The enzyme ribonuclease, investigated in vitro by Christian Anfinsen, is a small and stable protein that re-acquires its native three-dimensional structure spontaneously even after total denaturation. However, some proteins, like ribulose 1,5-bisphosphate carboxylase/oxygenase (Rubisco), the most abundant protein on earth, which fixes atmospheric CO_2 for photosynthesis, do not spontaneously refold once it is denatured (Goloubinoff et al. 1989). R John Ellis (Barraclough and Ellis 1980) discovered that a ~60 kDa protein, which he named chaperonin, binds the Rubisco large subunit (synthesized in chloroplasts) transiently during the assembly of Rubisco large and small subunits (synthesized in cytosol) in higher plant chloroplasts. This was the first report that a protein binds a newly synthesized other protein. Subsequently, Pierre Goloubinoff with George H. Lorimer (Goloubinoff et al. 1989) showed that denatured prokaryotic Rubisco (large subunit) refolds and assembles to its native oligomer only in the presence of GroEL (a prokaryotic homolog of the chloroplast chaperonin), GroES (co-chaperonin), and ATP. Its folding/assembly did not occur at all in the absence of any one of them (see Fig. 9.14). They demonstrated that assembly of some proteins, like Rubisco, requires the assistance of a set of proteins, thus aptly named molecular chaperones. GroEL/chaperonin is one of those molecular chaperones that are highly conserved and ubiquitously distributed among organisms (Table 9.1). These molecular chaperones are involved in a wide variety of cellular processes including regulation of gene expression, de novo folding of proteins that are newly synthesized in ribosomes, protein transport across organelle membranes, and protein degradation. In these processes, molecular chaperones regulate assembly/de-assembly of protein oligomers/complexes, assist/facilitate folding of non-native proteins, maintain a transporting protein in an unfolded or loosely folded conformation, generate a pulling force during protein translocation, cooperate with the ubiquitin/proteasome system, and mediate autophagy. Under normal and especially under stress conditions such as heat shock, molecular chaperones interact with unfolded, misfolded, and aggregated proteins (Fig. 9.1). They assist in the folding of unfolded and/or misfolded polypeptides. They also prevent protein aggregation. Chaperones like Hsp70/DnaK and Hsp104/ClpB can reactivate/solubilize (dis-aggregate) aggregated proteins. In this chapter, I focus on

Table 9.1 Molecular chaperones that are highly conserved, ubiquitously distributed

Chaperone families	Biochemical functions	Homologs in bacteria
Small Hsp (sHsp)	Prevent aggregation	IbpA, IbpB HspA
Hsp60/Cpn60/GroEL	Prevent aggregation Assist protein folding	GroEL, Cpn60
Hsp70/DnaK	Prevent aggregation Assist protein folding Solubilize aggregates	DnaK
Hsp90/HtpG	Prevent aggregation Stabilize/regulate protein kinases, etc. Assist protein folding	HtpG
Hsp104/ClpB	Solubilize aggregates	ClpB

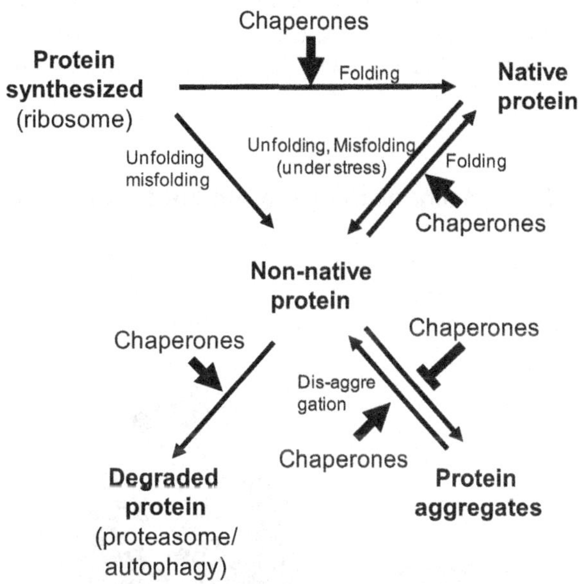

Fig. 9.1 Involvement of molecular chaperones in quality control of cellular proteomes

GroELs in cyanobacteria. Detailed information and discussion concerning other cyanobacterial molecular chaperones are found in Nakamoto (2013).

9.2 The Hsp60/Chaperonin 60/GroEL Family

GroEL is present in the bacterial cytosol and belongs to the Hsp60/chaperonin 60 (Cpn60)/GroEL family (Table 9.1). GroEL is classified as one of type I chaperonins and has homologs in chloroplasts and mitochondria, whereas the cytosol of eukaryotes and archaea contains type II chaperonins (Horwich et al.

2007). Studies in *Escherichia coli* and other bacteria have shown that GroEL is essential (Fayet et al. 1989).

9.2.1 Structure of GroEL and the GroEL-GroES Complex

The structure and action mechanism of *E. coli* GroEL have been studied in great detail (Horwich et al. 2007; Hayer-Hartl et al. 2016). *E. coli* GroEL is a tetradecamer of 57 kDa subunits (Fig. 9.2c) made up of two sevenfold (almost perfectly) rotationally symmetrical rings, which are stacked back to back. Inside each ring, there is a cavity (Fig. 9.2a). The opening of the central cavity exposes hydrophobic residues for the binding of a single non-native polypeptide and GroES. Each GroEL subunit consists of three domains: an equatorial domain, an intermediate hinge domain, and an apical domain. GroES, the 10-kDa co-chaperonin for GroEL, forms a heptameric dome (Fig. 9.2b). It closes the cavity of the GroEL cylinder to generate a folding chamber. The chamber can accommodate a protein of up to ~60-kDa in size.

9.2.2 Mechanism of Protein Folding Assisted by *E. coli* GroEL

The chaperone reaction cycle of *E. coli* GroEL is illustrated in Fig. 9.3 (Horwich et al. 2007; Hayer-Hartl et al. 2016). This cycle is thought to be evolutionarily conserved. A non-native protein captured by GroEL is encapsulated by GroES in an

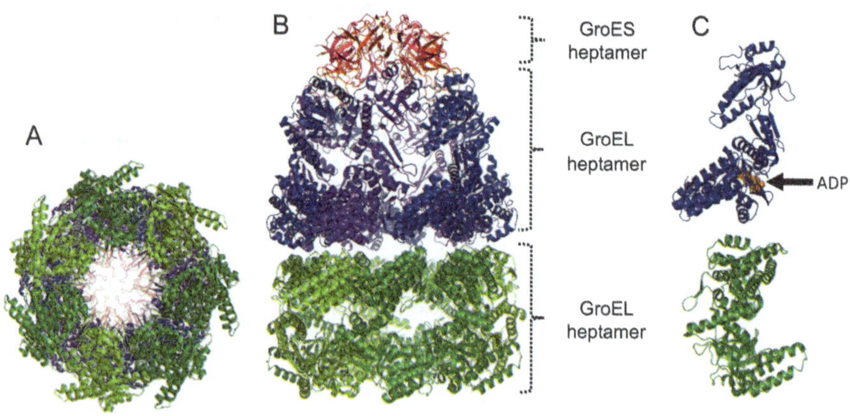

Fig. 9.2 Structure of *E. coli* GroEL heptamer ring, GroEL-GroES-ADP complex, and GroEL subunit. (**a**) GroEL heptamer ring viewed from above down the axis of sevenfold symmetry. The central cavity measures ~45 Å in diameter. (**b**). Double rings of GroEL heptamer capped with GroES heptamer viewed from side of the GroES-GroEL-ADP complex. ADP binds to the upper ring, resulting in conformational changes in the ring. (**c**) GroEL subunits taken from the upper ring (*cis* ring) and the lower ring (*trans* ring) of the GroES-GroEL complex (**b**). Constructed from PDBID: 1pcq (Chaudhry et al. 2003, see also Xu et al. 1997)

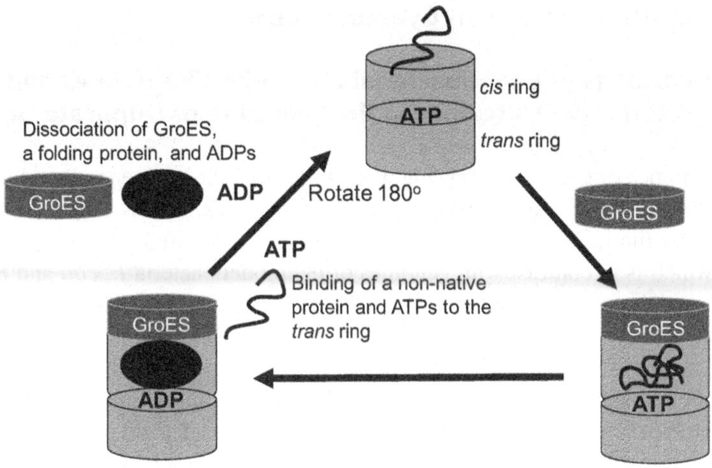

Fig. 9.3 Chaperone cycle of the *E. coli* GroEL. A GroEL-GroES-ADP asymmetric complex binds a non-native substrate polypeptide via hydrophobic interactions with its apical domains of the GroEL ring (*trans* ring, shown at the lower left), which results in discharging GroES, substrate polypeptide, and ADP from the other ring. ATP binding (either before polypeptide or thereafter) is followed by GroES. Concomitant with the GroES binding, the polypeptide is ejected into the central cavity (shown at the lower right), where it begins to fold in this chamber. Folding continues during ATP hydrolysis in the *cis* ring (~10 s). The ATP hydrolysis weakens the *cis* complex and also permits ATP (and non-native polypeptide) binding to the *trans* ring. This discharges ADP, substrate polypeptide, and GroES from the *cis* ring, regardless of the folding state of the substrate polypeptide. The released polypeptide may fold to its native state. When it is unable to reach the native state, it can bind to GroEL again for another folding attempt. The previous *trans* ring forms a new *cis* ring to begin the chaperone cycle again

ATP-dependent manner. The captured non-native protein folds in the cavity during the time when GroEL hydrolyzes its bound ATP to ADP (~10 s). After hydrolysis of ATP in the GroES-bound ring (*cis* ring), ATP binds to the opposite ring (*trans* ring), which triggers discharge of the protein, ADP, and GroES from the *cis* ring. If the released substrate is not correctly folded, it can rebind for further cycles of chaperoning.

The isolation of the non-native protein in the GroEL chamber whose inner wall is hydrophilic is important for the protein to avoid intermolecular interaction with non-native proteins in the cell. When non-native protein molecules are isolated in the GroEL cavity one by one, they can escape from "aggregating environments." The total protein and RNA inside *E. coli* occupy a substantial fraction (300–400 g/L) of the total volume of a cell. This macromolecular crowding results in favorable conditions for aggregation of non-native proteins (Ellis 2001). Proteins are inherently prone to form aggregates, and macromolecular crowding in a cell increases the tendency. GroEL lowers the risk of protein aggregation.

9.3 Multiple GroELs in Cyanobacteria

9.3.1 Paralogs of Cyanobacterial Molecular Chaperones and Three Alternative Outcomes in the Evolution of Duplicate Genes

In the genomes of cyanobacteria (Fig. 9.4), there are multiple genes encoding small Hsp, GroEL, DnaK, and ClpB homologs (Table 9.2). I have been particularly intrigued by the fact that cyanobacterial genomes contain multiple *groEL* and *clpB* genes, which is in contrast with genomes of the model bacteria *E. coli* and *Bacillus subtilis,* which contain only one *groEL* and *clpB* gene (Table 9.2). These multiple genes are likely to have their origin in gene duplication of the ancestor gene. Duplicated genes will have different fates, which may include nonfunctionalization, subfunctionalization, and neofunctionalization (Lynch and Conery 2000; Rensing 2014, see Fig. 9.5). In nonfunctionalization, one copy is silenced or lost by degenerative mutations. In subfunctionalization, two copies lose functions complementarily. In neofunctionalization, one copy acquires a novel, beneficial function, while the other copy retains the original function. The preservation of multiple, functional *groEL* genes in cyanobacteria during evolution suggests that they result from either subfunctionalization or neofunctionalization. In this chapter, I will describe the *groEL* paralogs in cyanobacteria and claim that neofunctionalization has taken place in cyanobacterial *groEL2* genes.

9.3.2 Gene Organization of *groEL1* and *groEL2* in Cyanobacterial Genomes

The *groEL* gene in *E. coli* and *B. subtilis* forms an operon with the *groES* gene, the *groESL* operon. Thus, the expression of the two genes is controlled coordinately in response to heat and other stresses. By contrast, László Vígh's group in Hungary

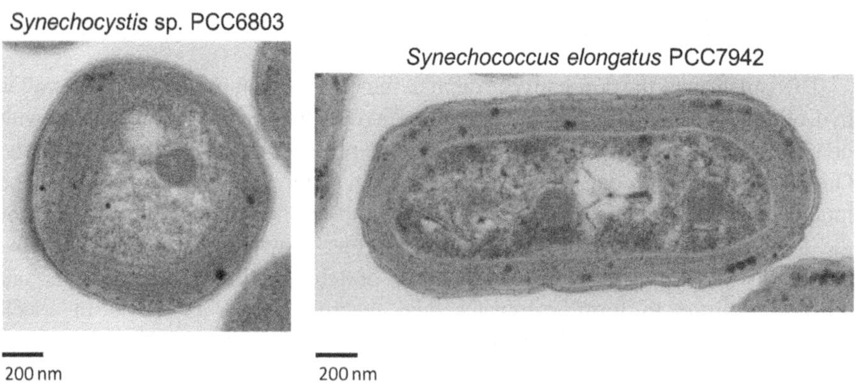

Fig. 9.4 Freshwater unicellular cyanobacteria *Synechocystis* sp. PCC6803 and *Synechococcus elongatus* PCC7942. Courtesy of Professor Yasuko Kaneko

Table 9.2 The number of chaperone genes in cyanobacteria and *E. coli*

Chaperone family	# of homologs in cyanobacteria	# of homologs in *E. coli*
Small Hsp (sHsp)	1 or 2	2
Hsp60/Cpn60/GroEL	2 or 3	1
Hsp70/DnaK	3	3
Hsp90/HtpG	1	1
Hsp104/ClpB	2	1

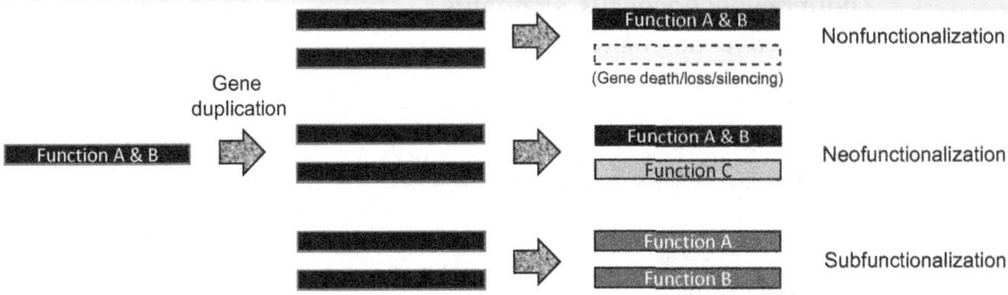

Fig. 9.5 Three alternative outcomes in the evolution of duplicate genes. For details, see text and also Rensing (2014)

(Lehel et al. 1993) and our group (Tanaka et al. 1997) found that there are two *groEL* genes in mesophilic and thermophilic cyanobacteria, respectively (Fig. 9.6). Interestingly, one of the two genes, the *groEL2* gene, is not associated with a *groES* gene. Our transcriptional analysis indicates that the *groEL2* gene is not co-transcribed with any other gene and thus monocistronic (Furuki et al. 1996). The gene organization suggests that GroEL2 is not functionally related to GroES, thus deviating from the *E. coli* paradigm (Fig. 9.3). Among 141 cyanobacterial genomes, 115 genomes encode a single *groESL* operon and a single monocistronic *groEL2* (Weissenbach et al. 2017, see Fig. 9.6b). Most other genomes encode two *groESL* operons and a single *groEL2* gene although a few encode two *groESL* operons and two *groEL2* genes. For example, *Chlorogloeopsis fritschii* PCC6912 and some other filamentous-type cyanobacteria have two *groESL* operons and a single *groEL2* (Fig. 9.7). The phylogenetic tree constructed from all cyanobacterial homologous *groEL* DNA sequences reveals two main clades that correspond to the *groESL1* operon and *groEL2*, indicating that the duplication of *groESL1* and *groEL2* is ancient and occurred at the base of the cyanobacterial phylogenetic tree (Weissenbach et al. 2017, see Fig. 9.7a).

9.3.3 Regulation of Transcription of *groESL1* and *groEL2* in Cyanobacteria

Major molecular chaperones are highly expressed as heat shock proteins upon a sudden high-temperature shift (heat shock). The current view is that enhanced

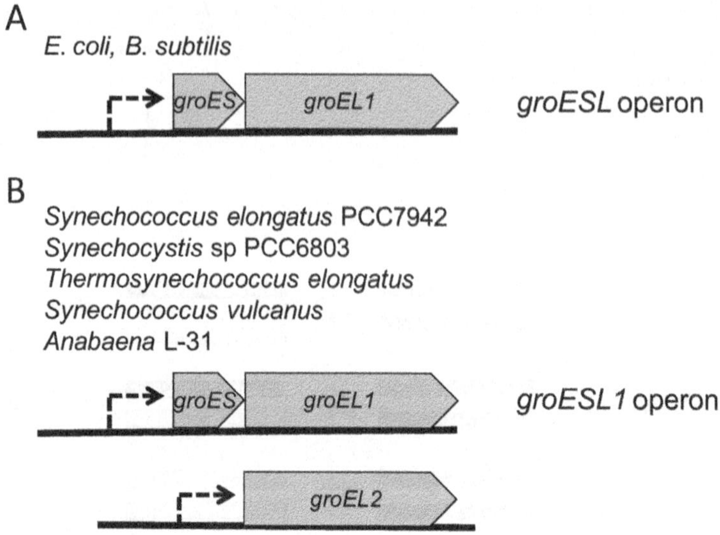

Fig. 9.6 Gene organization of *groEL*s in *E. coli*, *B. subtilis*, and various cyanobacteria. There is only one *groEL* that forms an operon with *groES* in the genomes of *E. coli* and *B. subtilis*. In contrast, there are one *groESL1* operon and one monocistronic *groEL2* in the genomes of many cyanobacterial species

expression of molecular chaperones is physiologically important to sustain cellular protein homeostasis under stress.

In contrast to the heat shock response in *E. coli* and *B. subtilis*, heat shock response in cyanobacteria is modified not only by heat but also by light. A further complication is that the copies of duplicated *groE* genes appear to be regulated differently, indicating that *groEL1* and *groEL2* have acquired mutually different regulatory mechanisms during their evolution. The heat shock response in *E. coli* has been studied in great detail, but investigations in *B. subtilis* have challenged the *E. coli* paradigm, and a totally different regulatory mechanism was demonstrated in *B. subtilis*. Our work has shown that the mechanism in cyanobacteria is similar to that in *B. subtilis*, but more complicated. In the following, I will describe the transcriptional regulation of the *groESL* operon in *E. coli* and *B. subtilis* before turning to regulation in cyanobacteria.

9.3.3.1 Positive Regulation of *groESL* Transcription by the Alternative Sigma Factor σ^{32} in *E. coli*. For Details, See Guisbert et al. (2008)

The heat shock response in *E. coli* is mediated by the transcription factor σ^{32}, a sigma factor that is a subunit of RNA polymerase and facilitates promoter recognition by the polymerase. RNA polymerase holoenzyme containing σ^{32} initiates transcription at heat shock promoters of genes encoding molecular chaperones and other heat shock genes. The heat shock promoter is different from that recognized by the major (or housekeeping) sigma factor σ^{70}.

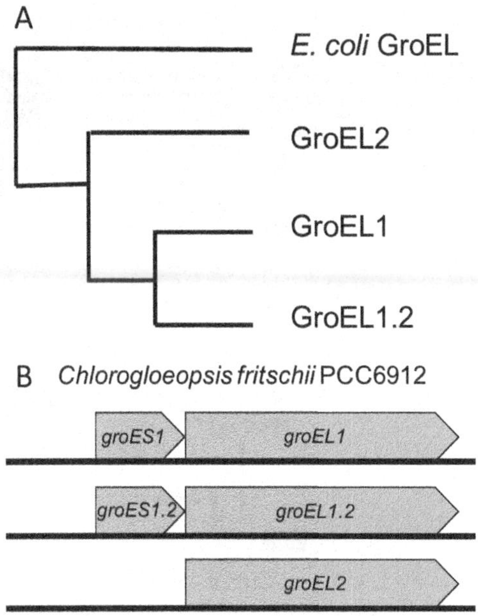

Fig. 9.7 Phylogeny of multiple *groESL* operons and *groEL2* (**a**) and their gene organization in *Chlorogloeopsis fritschii* PCC6912 (**b**). See Weissenbach et al. (2017)

Upon heat shock, synthesis and stabilization of σ^{32} are enhanced, leading to a rapid increase in the σ^{32} level. Since σ^{32} is an extremely unstable protein (half-life, ~1 min), its level is kept very low under nonstressed conditions. The σ^{32} increase results in an increase in the synthesis of GroES/GroEL and DnaK/DnaJ/GrpE. These chaperones interact with unfolded/misfolded proteins. When chaperone levels become abundant relative to unfolded/misfolded proteins in the proteostasis control system, which consists mainly of chaperones and heat shock-induced proteases, GroES/GroEL and DnaK/DnaJ/GrpE are freed from non-native proteins and bind to σ^{32}, leading to its destabilization and inactivation. Since σ^{32} is one of the substrates of the chaperones, these molecular chaperones exert a negative feedback on the heat shock response (Fig. 9.8a).

9.3.3.2 Negative Regulation of *groESL* Transcription by the CIRCE/HrcA System in *B. subtilis*. For Details, See Schumann (2016)

In *B. subtilis*, no stress-responsive alternative sigma (like σ^{32}) is involved in the heat induction of the *groESL* operon. Rather, the operon is preceded by the SigA promoter. SigA is the major sigma factor playing a housekeeping role in *B. subtilis*. The transcription start site is followed by a perfect inverted repeat of 9 bp separated by a 9 bp spacer (TTAGCACTC-N9-GAGTGCTAA), which is designated CIRCE (controlling inverted repeat of chaperone expression). CIRCE is an operator to which the repressor HrcA binds. Under normal conditions, the transcription is repressed by HrcA. In the event of a heat shock, HrcA changes its conformation so that it dissociates from the CIRCE element. GroEL, required to keep HrcA active, is thought to exert a negative feedback on the heat shock response

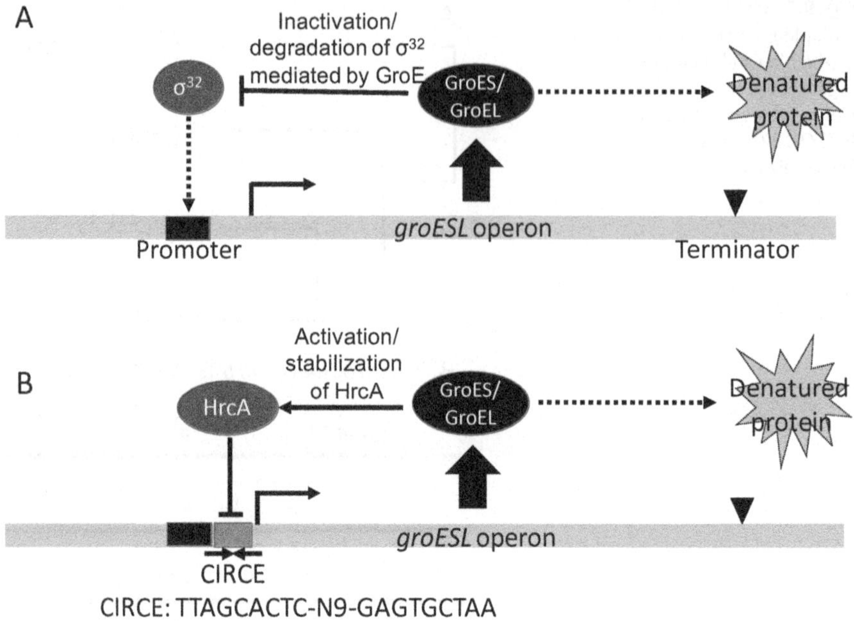

Fig. 9.8 Positive (**a**) and negative (**b**) regulations of the *groESL* transcription in *E. coli* and *B. subtilis*, respectively. (**a**) In *E. coli*, the sigma factor σ^{32} plays a major role in the regulation. Under normal conditions, σ^{32} is directly bound either by GroEL/GroES or DnaK/DnaJ/GrpE (not shown in this figure) chaperone system, resulting in inactivation of σ^{32}. These chaperone systems also participate in degradation of σ^{32} by the FtsH protease. After a heat shock, the chaperones dissociate from σ^{32} and bind to the denatured proteins. Denatured proteins titrate the chaperones from these regulatory roles, allowing active σ^{32} to increase the synthesis of chaperones and proteases under stress. (**b**) In *B. subtilis*, the CIRCE/HrcA system plays a major role in the regulation. Under normal conditions, the HrcA repressor that is stabilized by (free) GroES/GroEL chaperone system binds to the perfect inverted repeats (called as CIRCE). After a heat shock, HrcA will change its conformation, dissociate from CIRCE, and stay in an inactive conformation as long as GroES/GroEL is interacting with the denatured proteins during stress

by binding to HrcA (Fig. 9.8b). Immediately after the heat shock, molecular chaperones including GroEL/GroES and proteases interact with denatured proteins, resulting in their refolding and degradation. The recovery of proteostasis increases the cellular level of free GroEL/GroES, which is thought to reactivate HrcA, and transcription shut-off ensues.

9.3.3.3 Regulation of Transcription of *groESL1* and *groEL2* in Cyanobacteria

A Negative Regulation by the CIRCE/HrcA System

There is no heat shock promoter located upstream from the *groESL1* operon of *Synechocystis* sp. PCC6803 or *Thermosynechococcus elongatus* although the promoter sequence is homologous to the consensus sequence recognized by *E. coli* σ^{70}

(Nakamoto et al. 2003; Sato et al. 2008). The *groESL1* transcription of *Synechocystis* sp. PCC6803 and *T. elongatus* is initiated from the same transcriptional start site under both normal and stress conditions (Nakamoto et al. 2003, Sato et al. 2008). This strongly suggests that the same promoter is used under normal and stress conditions, eliminating involvement of a stress-responsive alternative sigma factor with a special heat shock promoter for *groESL1* heat induction. In cyanobacterial genomes from various species, there are only few genomes, if any, which contain an ORF coding for the σ^{32} homolog.

Thus, I conclude that *E. coli*-type positive regulation does not operate in cyanobacteria. However, immediately after the −10 promoter sequence of *groESL1* the CIRCE operator is present. As far as we could determine by searching the cyanobacterial genomes (Nakamoto and Kojima 2017), CIRCE is conserved in all cyanobacterial *groESL1*. By contrast, CIRCE is not well conserved among *groEL2*s, although there are cyanobacterial species that have it, including *Synechocystis* sp. PCC6803 and *Anabaena* sp. PCC7120 (Nakamoto and Kojima 2017). The gene encoding the HrcA repressor protein that binds to CIRCE is also conserved in cyanobacterial genomes (Saito et al. 2020).

In order to test whether the CIRCE/HrcA system is involved in a negative regulation of cyanobacterial *groEL* gene expression, we disrupted the *hrcA* gene in *Synechocystis* sp. PCC6803 (Nakamoto et al. 2003). The transcriptions of the *groESL1* operon and the *groEL2* gene that contain CIRCE in their upstream regions were de-repressed (or greatly enhanced) in the mutant under normal growth conditions at 30 °C. Similarly, an *hrcA* null mutant of *Anabaena* sp. PCC7120 expressed both GroEL1 and GroEL2 proteins at elevated levels under normal growth conditions at 27 °C as compared with the wild type (Rajaram and Apte 2010). It has been shown by electrophoretic mobility shift assay (EMSA) that *Anabaena* HrcA repressor protein specifically binds to a DNA fragment containing CIRCE. Furthermore, the negative regulation of *Anabaena groESL1* and *groEL2* transcriptions by the CIRCE/HrcA system was confirmed *in E. coli* where expression of a reporter gene driven from the *groESL1* or *groEL2* promoter is suppressed by co-expression of the *Anabaena hrcA* gene (Rajaram and Apte 2010). These results support the conclusion that CIRCE and HrcA constitute a regulatory mechanism that is involved in negative regulation of the *groESL* transcription in cyanobacteria.

A Novel Positive Regulation by K-Box

The negative regulation mediated by the CIRCE/HrcA system cannot be the only mechanism for regulation of cyanobacterial *groEL* gene expression because the *hrcA* mutant of *Synechocystis* sp. PCC6803 still responds to heat shock (Kojima and Nakamoto 2007, see Fig. 9.9). In order to study this *hrcA*-independent regulatory mechanism, we searched conserved sequences present in upstream regions of *groEL*s of various cyanobacterial species. Further upstream from CIRCE, there was a highly conserved sequence, which we called K-box (Fig. 9.10). In order to evaluate whether K-box is involved in the *groESL1* transcription, various upstream regions of the operon were fused (individually) to a reporter gene and inserted into a neutral site of the *Synechocystis* chromosome. Transcription of the *groESL1* operon

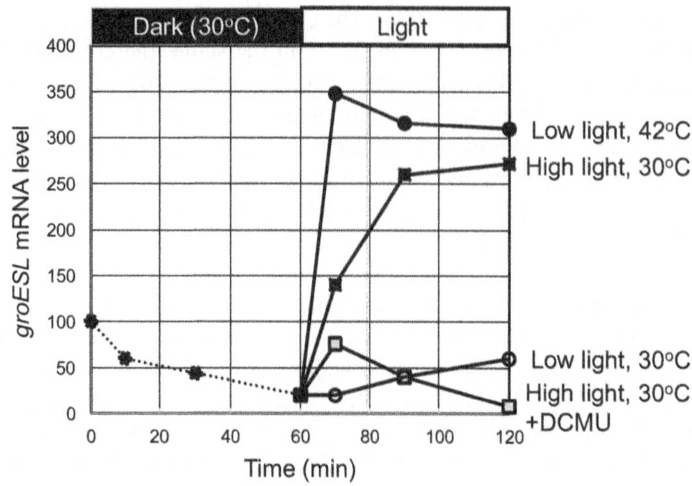

Fig. 9.9 Time course of the *groESL1* mRNA level of *Synechocystis hrcA* mutant under various conditions. Even in the absence of the HrcA repressor, the *groESL1* mRNA level in a cell decreases in the dark and increases upon heat shock. This novel heat shock induction is dependent on light and also on the photosynthetic electron transport. Adapted from Nakamoto et al. (2003)

under various conditions can be quantified as the reporter activity. The reporter assays showed that K-box is involved in the heat and/or light (see below) induction of *groESL1* transcription in *Synechocystis* sp. PCC6803. In the absence of K-box, *groESL1* promoter activity was completely abolished, indicating that K-box is essential not only for the induction but also for the basal expression of the *groESL1* operon (Kojima and Nakamoto 2007). An *Anabaena hrcA* null mutant, which expresses GroEL1 and GroEL2 constitutively, also shows a further increase in GroEL1 (but no increase in GroEL2) during temperature upshift (Rajaram and Apte 2010).

Transcription of the *groESL1* gene is induced by light and heat in the *hrcA* mutant of *Synechocystis* sp. PCC 6803 (Fig. 9.9). As described above, reporter assays demonstrated that K-box is involved in this light activation of *groESL1* transcription. Light modulates the heat shock response in the wild type of *Synechocystis* sp. PCC 6803: Transcription of *groESL1* is much more rapid and intense when cells are heat-shocked in the light than in darkness (Glatz et al. 1997; Asadulghani and Nakamoto 2003). Light appears to "activate" the gene expression through photosynthetic electron transport as DCMU, an inhibitor of the electron transport, abolishes the light activation totally (Kojima and Nakamoto 2007, see Fig. 9.9). Like CIRCE, K-box is highly conserved in the *groESL1* operons from various cyanobacterial species except *Gloeobacter violaceus* although the *groESL1s* of this cyanobacterium contain CIRCE. The regulatory mechanism for the *groESL1* operon in cyanobacteria is schematically depicted in Fig. 9.10. From genome-wide search, it appears that upstream regulatory regions of *groEL2* are much more diversified than those of *groEL1* (Nakamoto and Kojima 2017, see Fig. 9.11). For example, the *groEL2* genes

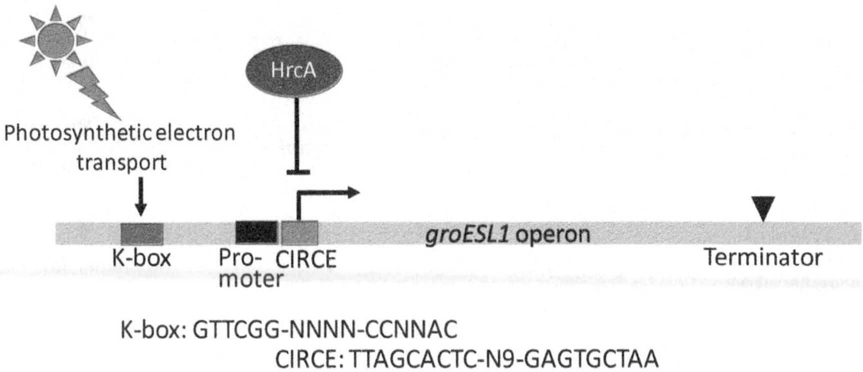

Fig. 9.10 CIRCE and K-box mediated regulation of the *groESL1* transcription in cyanobacteria. In cyanobacteria, the CIRCE/HrcA system (Fig. 9.8b) is involved in the regulation. Furthermore, the K-box element plays a major role in the regulation. Light/the photosynthetic electron transport exerts its effect through the K-box

from *S. elongatus* PCC 7942 and *T. elongatus* lack both CIRCE and K-box. Nevertheless, their transcription is induced by heat and/or light (Kojima and Nakamoto 2007), indicating that the heat- and/or light-mediated induction of the *groESL1* and *groEL2* expression is coordinated by different mechanisms.

K-box is also conserved in the upstream regions of *dnaK2* from various cyanobacterial species. There are three homologs of *dnaK* in cyanobacteria including *Synechocystis* sp. PCC 6803 and *Synechococcus elongatus* PCC 7942. The *dnaK2* gene from *S. elongatus* PCC 7942 is heat- and/or light-induced (Sato et al. 2007). Their detailed transcriptional analysis also showed that K-box is essential not only for the stress induction but also for the basal expression of *dnaK2*. Thus, K-box is a regulatory element of genes encoding major molecular chaperones in cyanobacteria whose expressions are modulated by heat and/or light. Besides CIRCE and K-box, species-specific regulatory sequences/motifs such as N-box (Nakamoto and Kojima 2017) and H-box (Rajaram and Apte 2010) have been reported to be involved in the *groESL1* transcription.

As described above (Sect. 9.3.3.2), the cellular level of free GroEL is thought to negatively regulate the transcription of the *B. subtilis groESL* operon in the CIRCE/HrcA system. Free GroEL is titrated by denatured proteins in a cell. The more denatured proteins are present, the less free GroEL is. A similar feedback mechanism may account for regulation of transcription of the *groE* genes in cyanobacteria although no evidence has been provided yet. It is reasonable that protein denaturation acts as a primary trigger for the expression of molecular chaperones. Furthermore, changes in lipid membrane's physical/structural properties may initiate the heat shock response, which is known as the membrane sensor hypothesis (Vígh et al. 2007). In vitro studies demonstrated that *E. coli* GroEL associates with model lipid membranes. The GroEL binding increases the molecular order of lipid bilayers (the lipid order in the liquid crystalline state), suggesting that GroEL rigidifies/stabilizes

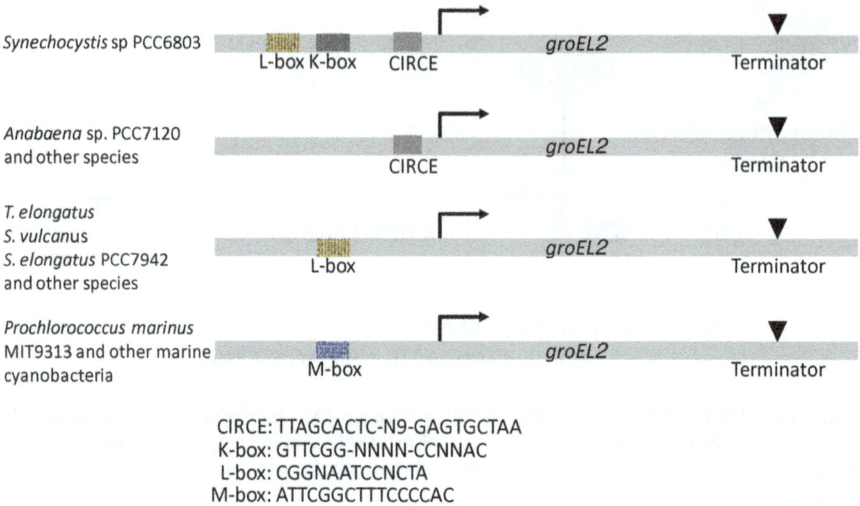

Fig. 9.11 Upstream regulatory regions in the *groEL2* genes from various cyanobacteria. L-box and M-box are putative regulatory motifs identified in the upstream regions of various *groEL2* genes. For details, see Nakamoto and Kojima (2017)

lipid membranes during heat stress (Török et al. 1997). Thus, GroEL may suppress/shut off the transcription of heat shock genes including the *groE* genes by binding membrane to restore its physical order.

9.3.3.4 The Evolution of Regulatory Mechanisms in Cyanobacterial groEL Paralogs

All cyanobacterial genomes except that of *Gloeobacter violaceus* encode at least one *groESL* operon and a single (two in rare cases) monocistronic *groEL* gene (Lund 2009; Weissenbach et al. 2017, see Figs. 9.6 and 9.7). *Gloeobacter violaceus* has two *groESL* operons, but no *groEL2*. In terms of regulatory elements, both of these *groESL* operons contain CIRCE (Nakamoto and Kojima 2017). It also has a gene coding for HrcA. However, there is no K-box upstream of any of the two *Gloeobacter groESL* operons. *Gloeobacter violaceus* lacks thylakoid membranes (Rippka et al. 1974) and is thought to be a member of an early branching lineage (Honda et al. 1999). Based on this, I propose that the cyanobacterial ancestor may have had only one *groESL* operon, which duplicated to yield two operons. These *groESL* paralogs had the CIRCE/HrcA system to regulate the heat shock response. During evolution, one of the paralogs lost the *groES* gene, resulting in the *groEL2* gene. On the other hand, the other one (*groESL1*) retained it. Furthermore, *groESL1* acquired K-box, whereas some *groEL2*s lost CIRCE, but have acquired new regulatory elements other than K-box (Fig. 9.12). The acquisition of K-box may be related to that of thylakoid membranes. The photosystems located in thylakoids regulate the expression of the *groESL1* operon via K-box. The diversification of the regulatory sequences suggests that the *groEL2* paralog is an outcome of

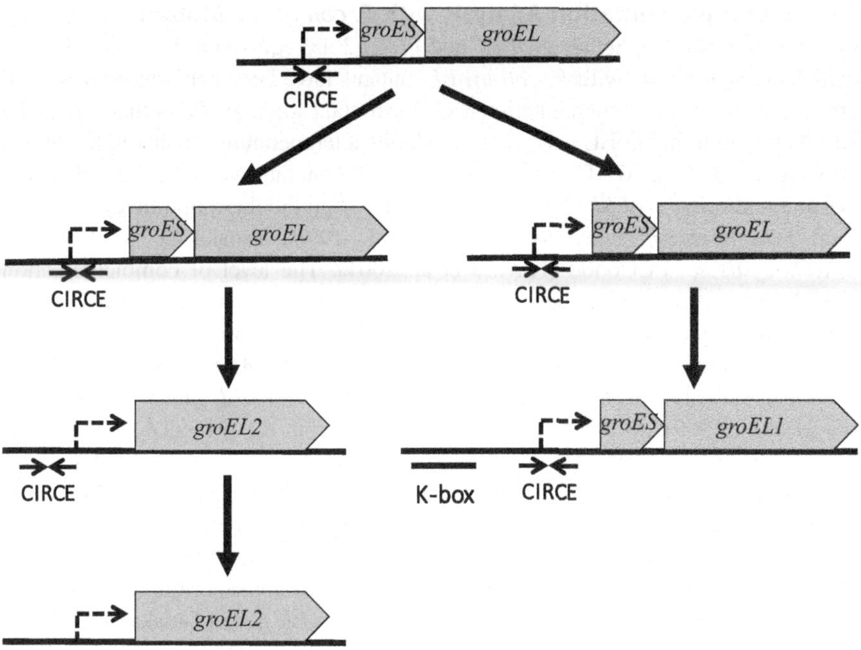

Fig. 9.12 Hypothetical model for evolution of *groESL1* and *groEL2* in cyanobacteria. Ancient cyanobacteria have two *groESL* operons with CIRCE around their promoters like *Gloeobacter violaceus* PCC7421. *G. violaceus* PCC7421 is thought to be diverged very early during the evolution of cyanobacteria. During evolution, one of the duplicated genes has lost *groES* to become the *groEL2* gene. Furthermore, *groEL2* in some cyanobacterial species has lost CIRCE and/or acquire a new regulatory element(s). The other *groESL* operon has conserved the *groES* gene, CIRCE, and further acquired K-box

neofunctionalization. If it were an outcome of subfunctionalization, then the expression of *groEL1* and *groEL2* would have to be regulated by the same mechanism.

9.4 Structure and Function of GroEL Paralogs in Cyanobacteria

9.4.1 Function of GroEL1 and GroEL2

The amino acid sequences of the two GroELs from *S. elongatus* PCC7942 or *Synechocystis* sp. PCC 6803 are ~60% identical. Amino acid sequence comparison shows that *E. coli* GroEL is equally similar to GroEL1 and GroEL2. *E. coli* GroEL, GroEL1, and GroEL2 are all acidic proteins with sizes around 58 kDa. These data do not show a significant difference between the GroELs; however, various studies have shown that GroEL1 and GroEL2 are mutually different as described below.

9.4.1.1 Complementation Analysis with *E. coli groEL* Mutants

In order to evaluate whether *groEL1* and *groEL2* are equivalent to *E. coli groEL*, complementation tests with *E. coli groEL* mutants have been performed by several groups. In the early complementation tests, the mutant strain *groEL44* that carries the E191G mutation in GroEL was used. It exhibits a temperature-sensitive phenotype, growing at 30 °C or 37 °C, but not at 42 °C. The mutant was transformed with a plasmid harboring *groESL1*, gro*EL1*, or *groEL2* from the thermophilic cyanobacterium *Synechococcus vulcanus* (Furuki et al. 1996; Tanaka et al. 1997) or *Synechocystis* sp. PCC6803 (Kovács et al. 2001). The level of complementation was evaluated by assessing the recovery of thermotolerance. In more recent complementation assays, the mutant strain MGM100 (McLennan and Masters 1998) was used. The expression of the native *groESL* operon in MGM100 is controlled by an arabinose-inducible pBAD promoter. The *groES* and *groEL* genes are essential (Fayet et al. 1989), so the strain can be kept viable in the presence of arabinose. Growth/survival of MGM100 without arabinose takes place if *groES/groEL* introduced by a plasmid can complement the native *groESL* gene in the MGM100 strain. Results of complementation tests up to now are summarized in Table 9.3.

The table shows that the *groEL1* gene from the thermophilic unicellular cyanobacterium *S. vulcanus* or the mesophilic unicellular cyanobacterium *Synechocystis* sp. PCC6803 can complement the native *E. coli groEL* gene (Tanaka et al. 1997; Kovács et al. 2001). Furthermore, the *groEL1* or *groEL1.2* gene expressed with the *groES1* gene from *C. fritschii* PCC6912 (see Fig. 9.7) can complement the native *E. coli groESL* operon (Weissenbach et al. 2017). On the other hand, the *groEL2* genes from *S. vulcanus* and *C. fritschii* PCC 6912 are unable to do so under normal or heat-stressed conditions regardless of the presence or absence of a *groES* gene (Furuki et al. 1996; Weissenbach et al. 2017). The *groEL2* gene from *Synechocystis* sp. PCC6803 can complement the native *E. coli groEL* gene, but only partially (Kovács et al. 2001). In contrast to the abovementioned studies, the *groEL1* or *groEL2* gene from *Anabaena* sp. L-31 can complement the native *groESL* gene in the MGM100 strain without co-expression of the *groES* gene at a high temperature (Potnis et al. 2016). Although there may be exceptions, I infer that GroEL1 is equivalent to the *E. coli* GroEL, whereas GroEL2 is not.

9.4.1.2 Function of GroEL1 and GroEL2 in Cyanobacteria

If GroEL1 is equivalent to the *E. coli* GroEL, then GroEL1 is expected to play an essential role under normal and stress conditions in cyanobacterial cells. As described below, there is evidence that GroEL1 is essential, whereas GroEL2 is nonessential in cyanobacteria.

There is a report that it is impossible to disrupt the *groEL1* or *groEL2* gene in all the genome copies of *S. elongatus* PCC7942 (Sato et al. 2007), indicating that these genes are essential. Some cyanobacterial species like *S. elongatus* PCC7942 and *T. elongatus* contain multiple complete copies of their single chromosome.

However, we were able to disrupt the *groEL2* gene by insertion of a chloramphenicol resistance gene cassette into all the genome copies of the thermophilic cyanobacterium *T. elongatus*, indicating that *groEL2* is nonessential under normal

Table 9.3 Complementation analysis with *E. coli groE* mutants

Species (reference)	*E. coli* host	Gene introduced by a plasmid	Induction	Growth/survival at 30 or 42 °C	Growth medium
Synechococcus vulcanus (Furuki et al. 1996; Tanaka et al. 1997)	*groEL44* (a strain generated by missense mutation)	*groESL1*	Heat shock	+++ (42 °C)	Agar plate
		groEL2		– (42 °C)	
Synechocystis sp. PCC6803 (Kovács et al. 2001)		*groEL1*		++++ (42 °C)	Agar plate
		groEL2		+ (42 °C)	
Anabaena sp. L-31 (Potnis et al. 2016)	MGM100 (No expression of the native *groESL* operon in the absence of arabinose)	*groESL1*	Light	++++ (42 °C)	Liquid medium
		groEL2	Light	++++ (42 °C)	
Chlorogloeopsis fritschii PCC 6912 (Weissenbach et al. 2017)		*groES1/groEL1*	Anhydrotetracycline	++++ (30 °C)	Agar plate
		groES1/groEL1.2		++++ (30 °C)	
		groEL1		– (30 °C)	
		groEL1.2		– (30 °C)	
		groEL2		– (30 °C)	
		groES1/groEL2		– (30 °C)	
		groES1.2/groEL2		– (30 °C)	

++++, normal growth; +++, +, reduced growth; –, no growth

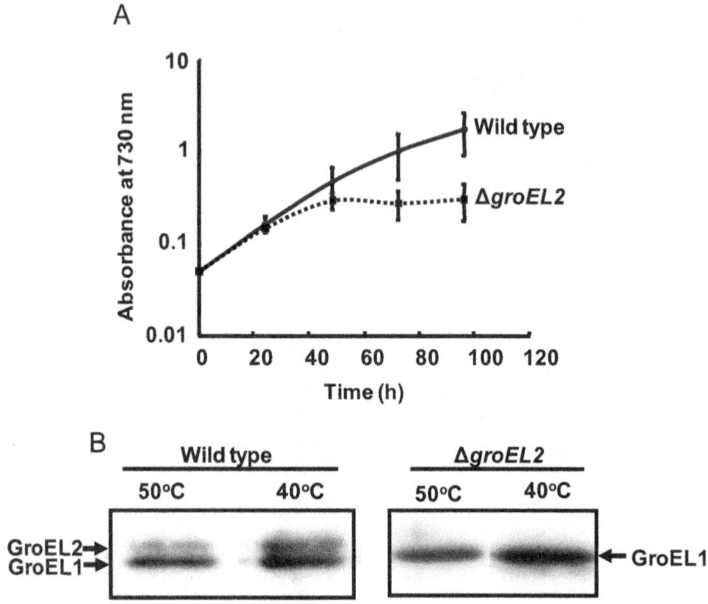

Fig. 9.13 Growth at 40 °C (**a**) and cellular levels of GroEL1 and GroEL2 (**b**) of the wild-type *Thermosynechococcus elongatus* and its *groEL2* mutant. Adapted from Sato et al. (2008)

growth conditions at 50 °C (Sato et al. 2008). This was the first evidence that one of the *groEL* genes in cyanobacteria is not essential. The *groEL2* mutant strain and the wild-type strain grow similarly at 50 °C. However, at a high temperature of 62 °C or a low temperature of 40 °C the mutant is unable to grow (Sato et al. 2008, see Fig. 9.13), indicating high- and low-temperature sensitivity of this mutant. Thus, GroEL2 plays a crucial role under both high and low temperatures. Consistent with the GroEL2 function under the temperature stresses, the wild type induces the *groEL2* gene at both 40 °C and 63 °C.

The difference in the essentiality between the *groEL1* and the *groEL2* genes is not due to difference in preferred (growth) temperature range of the species, that is, between the mesophilic cyanobacterium *S. elongatus* PCC7942 and the thermophilic cyanobacterium *T. elongatus*. Recent studies (Rubin et al. 2015) have analyzed the complete set of genomic regions necessary for survival in *S. elongatus* PCC7942. They report that *groEL1* is essential, whereas *groEL2* is nonessential. Thus, I conclude that GroEL1 is essential, whereas GroEL2 is nonessential in both mesophilic and thermophilic cyanobacteria.

The difference in the essentiality indicates that GroEL1 plays essential roles which GroEL2 is unable to substitute for. Conversely, can GroEL1 substitute for GroEL2 regarding the acquisition of low- and high-temperature tolerance? GroEL1 appears not to replace GroEL2 under low-temperature stress in *T. elongatus*. This is indicated by the fact that GroEL1 is present at the same or even higher level at the low temperature in the *groEL2* mutant cell (Sato et al. 2008, see Fig. 9.13). However, the cells are cold-sensitive, suggesting that GroEL1 is unable to substitute for the function of GroEL2. Furthermore, GroEL2 is induced at 40 °C, but GroEL1 is not in the wild type. This expression pattern is also consistent with the assumption that GroEL2 has a particularly important role at this temperature. I hypothesize that GroEL2 has evolved to confer stress tolerances, such as cold tolerance, to cells.

E. coli is cold-sensitive: Growth is impaired at temperatures below 20 °C, and no growth occurs below 8 °C (Ferrer et al. 2003). However, *E. coli* acquires great cold tolerance by heterologous expression of GroES (Cpn10) and GroEL (Cpn60) of *Oleispira antarctica*, a psychrophilic bacterium isolated from Antarctic seawater (Ferrer et al. 2003). *E. coli* expressing the chaperonin/co-chaperonin can grow even below 4 °C. It may be surmised that *T. elongatus* has an additional GroEL (GroEL2) that has evolved independently from GroEL1 in order to confer cold tolerance on cells.

Not only cyanobacteria, other bacteria like *Mycobacterium smegmatis,* have multiple *groEL* paralogs as well. *M. smegmatis* is used as a model for tuberculosis because it is a nonpathogenic cousin of *M. tuberculosis*. *M. smegmatis* contains three *groEL* genes. It has been suggested that two of them result from gene duplication. One of the two (*groEL1* or cpn60.1) is nonessential, whereas the other one (*groEL2* or *cpn60*.2) is essential. The nonessential *groEL* gene is required for biofilm formation (Ojha et al. 2005). Biofilms are microbial communities that are encased in an extracellular matrix that is composed of proteins, polysaccharides, DNAs, and lipids. They increase the resistance of microorganisms to antimicrobial agents, indicating that the GroEL protects cells from the agents by aiding biofilm formation. GroEL1 is also reported to be associated with nucleoids (Basu et al. 2009), suggesting the functional diversity of this nonessential GroEL.

9.4.2 Oligomers of GroEL

If the two GroELs have different functions, their structures are expected to be different. It is well known that *E. coli* forms a 14 mer or, more precisely, a double ring of heptamers (Fig. 9.2), which is essential for GroEL to assist non-native protein to fold (Fig. 9.3). So, the functional difference between GroEL1 and GroEL2 may be reflected in the oligomeric state.

Kovács et al. (2001) isolated GroEL1/GroEL2 (a mixture of the two GroELs) from *Synechocystis* sp. PCC6803 by sucrose gradient and gel filtration column chromatography. When the purified chaperonins were subjected to native PAGE, they observed only tetradecamers in the presence of glycerol, whereas in the absence of glycerol, predominantly the monomer form was detected, although a small amount of heptamer was also present. It should be noted that their sample for native PAGE contained both GroEL1 and GroEL2 whose sizes are almost the same. Thus, their data indicate that the two GroELs behave in a similar way. Alternatively, it is suggested that they form a hetero-oligomer.

We evaluated oligomeric states of highly purified recombinant GroEL1 and GroEL2 from *S. elongatus* by various methods (Huq et al. 2010). Analysis by native

PAGE showed that GroEL1 and GroEL2 do not form a 14 mer, whereas the *E. coli* GroEL gives a sharp (monodisperse) band of the 14 mer under the same conditions. Different sized oligomers ranging from pentamer to dimer of GroEL1 were detected. On the other hand, GroEL2 always forms a dimer. Glycerol and MgATP, which stabilize a 14 mer of *Synechocystis* GroELs, do not affect the oligomeric state of GroEL1 and GroEL2. Co-existence of the two GroELs does not affect it either, suggesting that GroEL1 and GroEL2 do not form a hetero-oligomer. Oligomeric forms of GroELs in cell extracts were also evaluated by native PAGE/Western blot analysis. Cell extracts of both the wild-type *S. elongatus* and its Δ*groEL2* mutant were analyzed in order to discriminate GroEL1 from GroEL2 (Huq et al. 2010). GroEL in *E. coli* cell extract was used as a 14-mer control. A 14-mer band was detected on a gel with both the wild-type cell extracts and the *groEL2* mutant cell extracts, indicating that GroEL1 forms a 14 mer. Two bands were detected between 67 kDa and 140 kDa size markers only with the wild-type cell extracts, indicating that GroEL2 exists as a monomer and/or dimer.

High protein concentration often stabilizes protein oligomers. We prepared 50 μM GroEL1 and GroEL2 solutions and analyzed the oligomeric state of GroELs by gel filtration chromatography. It was found that GroEL1 can form a 14 mer at a high concentration, but the largest oligomer of GroEL2 is a heptamer under the same conditions. Potnis et al. (2016) showed that *Anabaena* GroEL1 produces two peaks in gel filtration chromatography. The size of the bigger one is calculated as >700 and the other one is 61.7 kDa, corresponding to a high oligomer (>12 mer) and a monomer, respectively.

The above results are summarized in Table 9.4, from which it is clear that the oligomeric state of cyanobacterial GroEL1 and GroEL2 varies depending on cyanobacterial species, analytical methods, and experimental conditions. We are the only group that has analyzed the oligomeric state of GroEL2 and never observed a GroEL2 14 mer (Table 9.4).

Recently, bacterial two-hybrid analysis was employed in order to study interactions between GroEL and GroES from *C. fritschii* (Weissenbach et al. 2017). The genome of *C. fritschii* contains two *groESL1* operons and one *groEL2* gene (Fig. 9.7). As summarized in Table 9.5, the analysis indicated that GroEL2 does not interact with either itself or with GroEL1/GroEL1.2 (the other GroEL1 paralog, see Fig. 9.7), whereas GroEL1 or GroEL1.2 can interact with itself. It appears that GroEL1 or GroEL1.2 forms homo-oligomer, whereas GroEL2 is a monomer. In the two-hybrid analysis, interaction between fusion proteins is evaluated. Further biochemical analysis with isolated GroELs and GroESs is necessary to confirm these physical interactions and to determine the oligomeric state of GroELs.

Taking all the data into consideration, I conclude that GroEL1 is able to form a 14 mer like *E. coli* GroEL. On the other hand, GroEL2 is unable to form a 14 mer.

Table 9.4 Oligomeric states of GroEL1 and GroEL2

	Synechocystis sp. PCC6803 (Kovács et al. 2001)	*Synechococcus elongatus* PCC7942 (Huq et al. 2010)	*Anabaena* sp. L-31 (Potnis et al. 2016)
Native PAGE	[a]GroEL1/GroEL2: 14 mer in the presence of glycerol	GroEL1: 5 mer to 2 mer (several bands) GroEL2: 2 mer	
		[a]GroEL1: 14 mer (and other minor bands) [a]GroEL2: 2 mer and/or monomer	
Gel filtration chromatography		GroEL1: 14 mer to tetramer (polydisperse) GroEL2: 7 mer and 2 mer	GroEL1: >12 mer and monomer
Analytical ultracentrifugation		GroEL1: 14 mer to monomer (polydisperse) GroEL2: Monomer	

[a]Native GroEL proteins. Other GroELs are recombinant His-tagged proteins

Table 9.5 Interaction of GroESs and GroELs from *Chlorogloeopsis fritschii* as analyzed by bacterial two-hybrid method. Adapted and modified from Weissenbach et al. (2017)

	GroES1	GroEL1	GroES1.2	GroEL1.2	GroEL2
GroES1	+				
GroEL1	+	+			
GroES1.2	+	+	+		
GroEL1.2	+	−	−	+	
GroEL2	−	−	−	−	−

+, interaction detected; −, no interaction

9.4.3 Interaction of GroEL1 and GroEL2 with GroES

E. coli GroEL cooperates with GroES to assist folding of non-native protein (Figs. 9.3 and 9.14). *Synechocystis* sp. PCC6803 GroEL1 (and/or GroEL2) co-migrates with GroES during sucrose gradient centrifugation and gel filtration column chromatography (Kovács et al. 2001), indicating that GroES interacts with GroEL1 and/or GroEL2. In addition, *Chlorogloeopsis* GroEL1 or GroEL1.2 interacts with GroES1 (one of the two GroES), which has been shown by bacterial two-hybrid analysis (Weissenbach et al. 2017, see Table 9.5). In contrast, GroEL2 interacts with none of the GroESs. The "functional" interaction of *Chlorogloeopsis* GroEL1 and GroEL1.2 with GroES1 is revealed by complementation analysis, which shows that the *groEL1* or *groEL1.2* can complement the native *groEL* in MGM100 only when the *groES1* gene is co-expressed (Weissenbach et al. 2017, Table 9.3). Results mentioned above indicate that GroEL1 cooperates with GroES like *E. coli* GroEL, whereas GroEL2 does not.

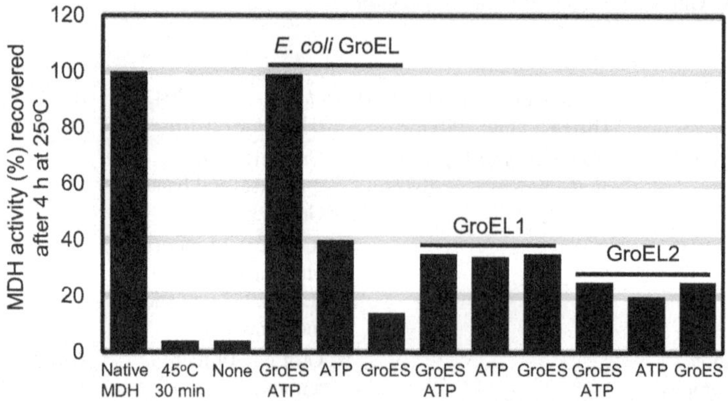

Fig. 9.14 Refolding of heat-denatured MDH with assistance of GroEL1, GroEL2, and *E. coli* GroEL. Chemically denatured Rubisco, heat-denatured MDH, and other non-native substrate polypeptides have been shown to reactivate with assistance of *E. coli* GroEL in a GroES- and ATP-dependent way. In contrast, GroES and/or ATP does not affect refolding of heat-denatured MDH and other non-native substrate polypeptides with assistance of cyanobacterial GroEL1 and GroEL2. Data were adapted and modified from Huq et al. (2010)

9.4.4 In Vitro Chaperone Function of GroEL1 and GroEL2

9.4.4.1 Anti-Aggregation Activity of GroEL1 and GroEL2

The capacity to prevent protein aggregation is characteristic among evolutionarily conserved molecular chaperones including GroEL. Non-native/denatured proteins are specifically recognized and captured by molecular chaperones, which usually results in suppression of aggregation of the proteins. Prevention of protein aggregation forms a first line of cellular defense under stress. This anti-aggregation activity can be quantified by measuring the apparent absorbance (turbidity or light scattering) increase in a solution where denatured proteins form their aggregates.

Both *Synechococcus* GroEL1 and GroEL2 suppress aggregation of heat (45 °C)-denatured malate dehydrogenase (MDH), just like *E. coli* GroEL does, at pH 8.0 (Huq et al. 2010). *Anabaena* GroEL1 also suppresses aggregation of heat (55 °C)-denatured MDH at pH 7.4 (Potnis et al. 2016). These results do not show any difference in anti-aggregation activity between GroEL1, GroEL2, and *E. coli* GroEL. However, we have observed differential effects of pH on the anti-aggregation activity of each GroEL (Akter and Nakamoto 2021). When pH decreases from 8.5 to 7.0, GroEL1 loses its activity to suppress aggregation of heat-denatured MDH most sharply, whereas the activity of *E. coli* GroEL is most resistant to pH changes. Compared with GroEL1, GroEL2 shows a modest response to the change.

9.4.4.2 ATPase Activity of GroEL1 and GroEL2

ATP binding and hydrolysis are essential in GroEL-mediated protein folding (Fig. 9.3). The k_{cat} values for ATP hydrolysis by *Synechococcus* GroEL1 and GroEL2 are 17% and 4% of that of *E. coli* GroEL, respectively (Huq et al. 2010). The ATPase activity of *Anabaena* GroEL1 is ~60% of that of *E. coli* GroEL (Potnis et al. 2016). The differences in the ATPase activity of GroEL1 may be due to differences in species and/or assay methods. The lower activity of cyanobacterial GroELs compared to *E. coli* GroEL may be related to the stability of the GroEL 14 mer: It is known that the *E. coli* GroEL monomer shows only about one-seventh the ATPase activity of the 14 mer (Ybarra and Horowitz 1995).

The ATPase activities of *Anabaena* GroEL1, *Synechococcus* GroEL1, and GroEL2 are enhanced by GroES, whereas that of *E. coli* GroEL is inhibited by GroES. This unusual property of cyanobacterial GroELs is reminiscent of a single-ring mutant of *E. coli* GroEL (Kovács et al. 2010). I infer that the low ATPase activity and enhancement of ATPase activity by GroES are due to the instability and inexistence of the 14 mer of GroEL1 and GroEL2, respectively.

9.4.4.3 Refolding of Non-native Protein with the Assistance of GroEL1 and GroEL2

E. coli GroEL assists the folding of non-native protein in a way that is both ATP- and GroES-dependent (Goloubinoff et al. 1989). As shown in Fig. 9.14, dimeric Rubisco composed of only large subunits from *Rhodospirillum rubrum* that is completely denatured in guanidine-HCl reactivates in the presence of *E. coli* GroEL, GroES, and ATP. We analyzed refolding of heat-denatured MDH assisted by *E. coli* GroEL (used as a positive control) and *Synechococcus* GroEL1 and GroEL2 (Huq et al. 2010). MDH denatured at 45 °C for 30 min in the presence of *E. coli* GroEL is fully reactivated in the presence of GroES and ATP within 2 h (Fig. 9.14). This reactivation is largely dependent on GroES and ATP. On the other hand, when cyanobacterial GroEL1 and GroEL2 are used, MDH recovers much less activity (Fig. 9.14). This activation is totally independent of GroES and ATP. However, much less, significantly higher activities than those in the presence of a negative control protein (BSA instead of GroEL) are recovered. Like *Synechococcus* GroELs, *Anabaena* GroEL1 and GroEL2 also assist refolding of different protein substrates including heat-denatured MDH in a GroES- and ATP-independent way (Potnis et al. 2016).

The GroES- and ATP-independent chaperone function of cyanobacterial GroEL1 and GroEL2 suggests that cyanobacterial GroELs are fundamentally different from the *E. coli* GroEL. However, both *Chlorogloeopsis* GroEL1 and GroEL1.2 require GroES in order to play the essential role of the native GroEL and GroES in *E. coli* cells as described above (Weissenbach et al. 2017, see Table 9.3). Thus, conditions for in vitro refolding experiments are not appropriate to detect a GroES- and ATP-dependent chaperone function of cyanobacterial GroELs. Otherwise, cyanobacterial GroELs may assist refolding of denatured proteins in a nonclassical way. The paradigm for the chaperone mechanism (Fig. 9.3) has been established in *E. coli* GroEL. However, there are a considerable number of reports that GroEL assists folding of a non-native protein in a GroES-independent fashion (Schmidt et al. 1994, and references therein). ATP (or even nonhydrolyzable ATP analogs and ADP) thus seems sufficient for the chaperone function (Mizobata et al. 1992). In addition to the protein substrate used for in vitro folding reactions, it has been shown that folding environments matter when it comes to whether GroES and/or ATP are required. In a "permissive" environment, where unassisted, spontaneous folding could occur, GroES is not mandatory (Schmidt et al. 1994). Further studies are necessary to determine whether cyanobacterial GroEL1 can function without GroES and/or ATP.

9.5 Concluding Remarks

Table 9.6 compares the characteristics of cyanobacterial GroEL1 and GroEL2 with those of *E. coli* GroEL. Clearly, GroEL2 is different from GroEL1 and *E. coli* GroEL. I conclude that the *groEL2* gene is the outcome of neofunctionalization (Fig. 9.5). It has acquired a novel, beneficial structure, and function, with the *groEL1* gene retaining the original function. GroEL2 may play an essential function under stress conditions as we proved for GroEL2 from *T. elongatus* (Fig. 9.15), and thus was preserved by natural selection.

Table 9.6 Comparison of *E. coli* GroEL with cyanobacterial GroEL1 and GroEL2

	E. coli GroEL	GroEL1	GroEL2
Essential?	Yes	Yes	No
Stress inducible?	Yes	Yes	Yes
Complements an *E. coli groEL* mutant	Yes	Yes	No
Suppresses aggregation of denatured proteins?	Yes	Yes	Yes
Forms an oligomer of 14 subunits?	Yes	Yes (unstable)	No
Assists folding of a protein in a GroES/ATP-dependent way?	Yes	Not detected yet	Not detected yet

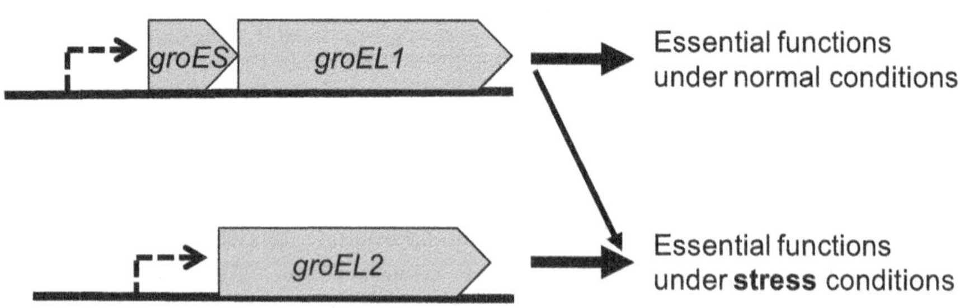

Fig. 9.15 GroEL2 has acquired a novel structure and function, which are beneficial to cyanobacterial survival under selective pressure, with GroEL1 performing the essential function

Acknowledgement This work was supported in part by Grants-in-Aid for Scientific Research (C) (No. 18 K05407, to H.N.) from the Ministry of Education, Culture, Sports, Science and Technology of Japan.

References

Akter T, Nakamoto H (2021) pH-mediated control of anti-aggregation activities of cyanobacterial and *E. coli* chaperonin GroELs. J Biochem 169(3):351–361.

Anfinsen CB (1973) Principles that govern the folding of protein chains. Science 181:223–230

Asadulghani SY, Nakamoto H (2003) Light plays a key role in the modulation of heat shock response in the cyanobacterium *Synechocystis* sp PCC 6803. Biochem Biophys Res Commun 306:872–879

Barraclough R, Ellis RJ (1980) Protein synthesis in chloroplasts. IX. Assembly of newly-synthesized large subunits into ribulose bisphosphate carboxylase in isolated intact pea chloroplasts. Biochim Biophys Acta 608:19–31

Basu D, Khare G, Singh S, Tyagi A, Khosla S, Mande SC (2009) A novel nucleoid-associated protein of *Mycobacterium tuberculosis* is a sequence homolog of GroEL. Nucleic Acids Res 37:4944–4954

Chaudhry C, Farr GW, Todd MJ, Rye HS, Brunger AT, Adams PD, Horwich AL, Sigler PB (2003) Role of the gamma-phosphate of ATP in triggering protein folding by GroEL-GroES: function, structure and energetics. EMBO J 22:4877–4887

Ellis RJ (2001) Macromolecular crowding: obvious but underappreciated. Trends Biochem Sci 26:597–604

Fayet O, Ziegelhoffer T, Georgopoulos C (1989) The *groES* and *groEL* heat shock gene products of *Escherichia coli* are essential for bacterial growth at all temperatures. J Bacteriol 171:1379–1385

Ferrer M, Chernikova TN, Yakimov MM, Golyshin PN, Timmis KN (2003) Chaperonins govern growth of *Escherichia coli* at low temperatures. Nat Biotechnol 21:1266–1267

Furuki M, Tanaka N, Hiyama T, Nakamoto H (1996) Cloning, characterization and functional analysis of *groEL*-like gene from thermophilic cyanobacterium *Synechococcus vulcanus*, which does not form an operon with *groES*. Biochim Biophys Acta 1294:106–110

Glatz A, Horváth I, Varvasovszki V, Kovács E, Török Z, Vigh L (1997) Chaperonin genes of the *Synechocystis* PCC 6803 are differentially regulated under light-dark transition during heat stress. Biochem Biophys Res Commun 239:291–297

Goloubinoff P, Christeller JT, Gatenby AA, Lorimer GH (1989) Reconstitution of active dimeric ribulose bisphosphate carboxylase from an unfolded state depends on two chaperonin proteins and Mg-ATP. Nature 342:884–889

Guisbert E, Yura T, Rhodius VA, Gross CA (2008) Convergence of molecular, modeling, and systems approaches for an understanding of the *Escherichia coli* heat shock response. Microbiol Mol Biol Rev 72:545–554

Hayer-Hartl M, Bracher A, Hartl FU (2016) The GroEL-GroES chaperonin machine: a nano-cage for protein folding. Trends Biochem Sci 41:62–76

Honda D, Yokota A, Sugiyama J (1999) Detection of seven major evolutionary lineages in cyanobacteria based on the 16S rRNA gene sequence analysis with new sequences of five marine *Synechococcus* strains. J Mol Evol 48:723–739

Horwich AL, Fenton WA, Chapman E, Farr GW (2007) Two families of chaperonin: physiology and mechanism. Annu Rev Cell Dev Biol 23:115–145

Huq S, Sueoka K, Narumi S, Arisaka F, Nakamoto H (2010) Comparative biochemical characterization of two GroEL homologs from the cyanobacterium *Synechococcus elongatus* PCC 7942. Biosci Biotechnol Biochem 74:2273–2280

Kojima K, Nakamoto H (2007) A novel light- and heat-responsive regulation of the *groE* transcription in the absence of HrcA or CIRCE in cyanobacteria. FEBS Lett 581:1871–1880

Kovács E, van der Vies SM, Glatz A, Török Z, Varvasovszki V, Horváth I, Vígh L (2001) The chaperonins of *Synechocystis* PCC 6803 differ in heat inducibility and chaperone activity. Biochem Biophys Res Commun 289:908–915

Kovács E, Sun Z, Liu H, Scott DJ, Karsisiotis AI, Clarke AR, Burston SG, Lund PA (2010) Characterisation of a GroEL single-ring mutant that supports growth of *Escherichia coli* and has GroES-dependent ATPase activity. J Mol Biol 396:1271–1283

Lehel C, Los D, Wada H, Györgyei J, Horváth I, Kovács E, Murata N, Vigh L (1993) A second *groEL*-like gene, organized in a *groESL* operon is present in the genome of *Synechocystis* sp. PCC 6803. J Biol Chem 268:1799–1804

Lund PA (2009) Multiple chaperonins in bacteria—why so many? FEMS Microbiol Rev 33:785–800

Lynch M, Conery JS (2000) The evolutionary fate and consequences of duplicate genes. Science 290:1151–1155

McLennan N, Masters M (1998) GroE is vital for cell-wall synthesis. Nature 392:139

Mizobata T, Akiyama Y, Ito K, Yumoto N, Kawata Y (1992) Effects of the chaperonin GroE on the refolding of tryptophanase from *Escherichia coli*. J Biol Chem 267:17773–17779

Nakamoto H (2013) Molecular chaperones and stress tolerance in cyanobacteria. In: Srivastava AK, Rai AN, Neilan BA (eds) Stress biology of cyanobacteria. CRC Press, New York, pp 113–144

Nakamoto H, Kojima K (2017) Non-housekeeping, non-essential GroEL (chaperonin) has acquired novel structure and function beneficial under stress in cyanobacteria. Physiol Plant 161:296–310

Nakamoto H, Suzuki M, Kojima K (2003) Targeted inactivation of the *hrcA* repressor gene in cyanobacteria. FEBS Lett 549:57–62

Ojha A, Anand M, Bhatt A, Kremer L, Jacobs WR Jr, Hatfull GF (2005) GroEL1: a dedicated chaperone involved in mycolic acid biosynthesis during biofilm formation in mycobacteria. Cell 123:861–873

Potnis AA, Rajaram H, Apte SK (2016) GroEL of the nitrogen-fixing cyanobacterium *Anabaena* sp. strain L-31 exhibits GroES and ATP-independent refolding activity. J Biochem 159:295–304

Rajaram H, Apte SK (2010) Differential regulation of *groESL* operon expression in response to heat and light in Anabaena. Arch Microbiol 192:729–738

Rensing SA (2014) Gene duplication as a driver of plant morphogenetic evolution. Curr Opin Plant Biol 17:43–48

Rippka R, Waterbury J, Cohen-Bazire G (1974) A cyanobacterium which lacks thylakoids. Arch Microbiol 100:419–436

Rubin BE, Wetmore KM, Price MN, Diamond S, Shultzaberger RK, Lowe LC, Curtin G, Arkin AP, Deutschbauer A, Golden SS (2015) The essential gene set of a photosynthetic organism. Proc Natl Acad Sci U S A 112:E6634–E6643

Saito M, Watanabe S, Nimura-Matsune K, Yoshikawa H, Nakamoto H (2020) Regulation of the *groESL1* transcription by the HrcA repressor and a novel transcription factor Orf7.5 in the cyanobacterium *Synechococcus elongatus* PCC7942. J Gen Appl Microbiol 66:85–92

Sato M, Nimura-Matsune K, Watanabe S, Chibazakura T, Yoshikawa H (2007) Expression analysis of multiple *dnaK* genes in the cyanobacterium *Synechococcus elongatus* PCC 7942. J Bacteriol 189:3751–3758

Sato S, Ikeuchi M, Nakamoto H (2008) Expression and function of a *groEL* paralog in the thermophilic cyanobacterium *Thermosynechococcus elongatus* under heat and cold stress. FEBS Lett 582:3389–3395

Schmidt M, Buchner J, Todd MJ, Lorimer GH, Viitanen PV (1994) On the role of groES in the chaperonin-assisted folding reaction. J Biol Chem 269:10304–10311

Schumann W (2016) Regulation of bacterial heat shock stimulons. Cell Stress Chaperones 21:959–968

Tanaka N, Hiyama T, Nakamoto H (1997) Cloning, characterization and functional analysis of *groESL* operon from thermophilic cyanobacterium *Synechococcus vulcanus*. Biochim Biophys Acta 1343:335–348

Török Z, Horváth I, Goloubinoff P, Kovács E, Glatz A, Balogh G, Vígh L (1997) Evidence for a lipochaperonin: association of active protein-folding GroESL oligomers with lipids can stabilize membranes under heat shock conditions. Proc Natl Acad Sci U S A 94:2192–2197

Vígh L, Török Z, Balogh G, Glatz A, Piotto S, Horváth I (2007) Membrane-regulated stress response: a theoretical and practical approach. Adv Exp Med Biol 594:114–131

Weissenbach J, Ilhan J, Bogumil D, Hülter N, Stucken K, Dagan T (2017) Evolution of chaperonin gene duplication in Stigonematalean cyanobacteria (subsection V). Genome Biol Evol 9:241–252

Xu Z, Horwich AL, Sigler PB (1997) The crystal structure of the asymmetric GroEL–GroES–$(ADP)_7$ chaperonin complex. Nature 388:741–750

Ybarra J, Horowitz PM (1995) Inactive GroEL monomers can be isolated and reassembled to functional tetradecamers that contain few bound peptides. J Biol Chem 270:22962–22967

Chromatic Acclimation in Cyanobacteria: Photomorphogenesis in Response to Light Quality

10

Pankaj K. Maurya, Vinod Kumar, Soumila Mondal, and Shailendra P. Singh

Abstract

Cyanobacteria are found in different habitats where they are exposed to fluctuating light conditions. The quality and quantity of light vary significantly in low and high light environments, which also change at diurnal and seasonal bases. Thus, cyanobacteria are always challenged by the ambient light environment, which affects the ecologically important function of photosynthesis and developmental processes in these organisms. The developmental process regulated by light signals is known as photomorphogenesis, and in this chapter, we present one of the well-known light-mediated developmental processes in cyanobacteria called chromatic acclimation (CA). CA allows cyanobacteria to utilize different quality and quantity of light conditions prevailing in their habitats to support the photosynthesis. This is achieved by altering the composition and size of their light-harvesting systems. Here, we briefly discuss the history of CA and present different types of CA known so far. We also present a molecular mechanism of well-known and studied type 3 CA and describe other aspects of type 3 CA that take place at a cellular level, besides, to change in light-harvesting machinery.

Keywords

Carboxysome · Cytoskeleton · Morphogenesis · Oxidative stress · ROS

P. K. Maurya · V. Kumar · S. Mondal · S. P. Singh (✉)
Department of Botany, Centre of Advanced Study in Botany, Institute of Science, Banaras Hindu University, Varanasi, India
e-mail: spsingh@bhu.ac.in

R. P. Rastogi (ed.), *Ecophysiology and Biochemistry of Cyanobacteria*,
https://doi.org/10.1007/978-981-16-4873-1_10

10.1 Introduction

Cyanobacteria are exposed to fluctuating environmental conditions due to their photoautotrophic mode of nutrition. These organisms are Gram-negative bacteria that possess plant-type photosynthetic machinery and produce oxygen during fixation of carbon dioxide (CO_2). Cyanobacteria are subjected to seasonal and diurnal change in light quality and quantity, temperature, nutrient availability, CO_2 levels, salinity, and pH that affect their fitness by affecting several cellular processes (Singh et al. 2010; Singh and Montgomery 2013a; Singh and Montgomery 2013b; Singh et al. 2014). Cyanobacteria are well known for their role in global CO_2 and N_2 fixation, acted as a model organism to understand several important biological processes such as photosynthesis, iron homeostasis, redox homeostasis, and N_2 fixation. These organisms are also a source of myriads of economically important compounds and have shown their potential to be utilized in the synthesis of nanoparticles and production of carbon-neutral bioenergy (Rajneesh et al. 2017; Pathak et al. 2018; Pathak et al. 2019).

However, the growth of cyanobacteria in their natural habitat and in outdoor and indoor artificial growth systems is affected by abovementioned abiotic stressors that ultimately affect productivity and therefore limit their utilization for various purposes (Rajneesh et al. 2017). Cyanobacteria have evolved several mitigation strategies to counter abiotic stressors and reshuffle their photosynthetic machinery to maximally absorb available radiant energy that drives photochemistry (Kehoe and Gutu 2006; Singh et al. 2010). Several cyanobacteria can reshuffle the pigment composition of their light-harvesting complex called phycobilisomes (PBSs) in response to available quality and quantity of light, which is known for a long time as complementary chromatic adaptation/acclimation (CCA) (Bennett and Bogorad 1973; Kehoe and Gutu 2006). Thus, CCA provides an advantage to cyanobacteria over other photosynthetic organisms to efficiently harvest available photons of light in their ecological niches.

CCA has emerged as an ecologically important phenomenon in recent years, which can provide an advantage in artificial growth systems to maximize the productivity by efficiently utilizing the available wavelengths of lights especially when cultures are dense. In this chapter, we provide a brief history of CCA, the emergence of different terminology, and discuss various chromatic acclimation processes that have emerged and redefined recently. We also include signal transduction mechanism of well-established type 3 chromatic acclimation, the significance of chromatic acclimation, and briefly present other aspects of type 3 chromatic acclimation that have emerged recently.

10.2 History and Concept of Complementary Chromatic Adaptation Vs. Acclimation

The complementary chromatic adaptation was first described in filamentous cyanobacteria more than 100 years ago when these organisms reported changing their color according to the quality of light they were receiving for photoautotrophic growth (Engelmann 1883; Engelmann 1902; Fujita and Hattori 1962). The term complementary chromatic adaptation was first used by Gaidukov in 1902 (Gaidukov 1902) when a filamentous cyanobacterium *Oscillatoria sancta* showed the ability to change its color in complementation with the ambient light quality. *Oscillatoria sancta* were red-colored in green light (GL) and appeared blue-green when grown in red light (RL) condition (Gaidukov 1902). This change in color of the organism was later found to be the result of a change in levels of phycobiliproteins (PBPs), which are the part of a light-harvesting complex in cyanobacteria called phycobilisomes (PBSs) (Boresch 1922; Bennett and Bogorad 1973; Tandeau de Marsac 1977). The phenomenon of complementary chromatic adaptation has been long known in cyanobacteria primarily for their response to the relative abundance of red or green wavelengths of light in a reversible manner (Bennett and Bogorad 1973; Tandeau de Marsac 1977).

From genetic perspectives, the term "adaptation" basically involves the fundamental changes in the genetic composition of the organisms, which allow their growth in prevailing environmental conditions. However, alteration in pigment composition in response to a spectrum of light is achieved by a change in the expression of the associated genes rather than a change in genetic composition. Thus, change in pigment composition is a reversible process dictated by the quality of light, while adaptation is accompanied by a change in genetic composition and renders a fixed phenotype. Therefore, "complementary chromatic adaptation" has been redefined as an "acclimation" response (Kehoe and Gutu 2006), and from here onwards, CCA represents "complementary chromatic acclimation" in this chapter. However, "adaptation" term can still be used for those organisms, which are unable to change their pigment composition in response to a prevailing light condition as they are genetically fixed for particular pigment composition. Similarly, removal of "complementary" has been recommended recently as several other types of chromatic acclimations are known where the color of the cell is not complementary to the light color that was originally noticed in the Gaidukov phenomenon (Gaidukov 1902; Sanfilippo et al. 2019). Therefore, for a broader perspective on complementary chromatic acclimation, from here, we refer it to chromatic acclimation (CA) as suggested recently (Sanfilippo et al. 2019).

10.3 Structural Components of PBS and CA

Phycobilisomes are light-harvesting complexes located on the outer surface of thylakoid membrane where they absorb light energy and transmit it to the reaction centers of both photosystems I and II (Liu et al. 2013; Watanabe et al. 2014). PBSs

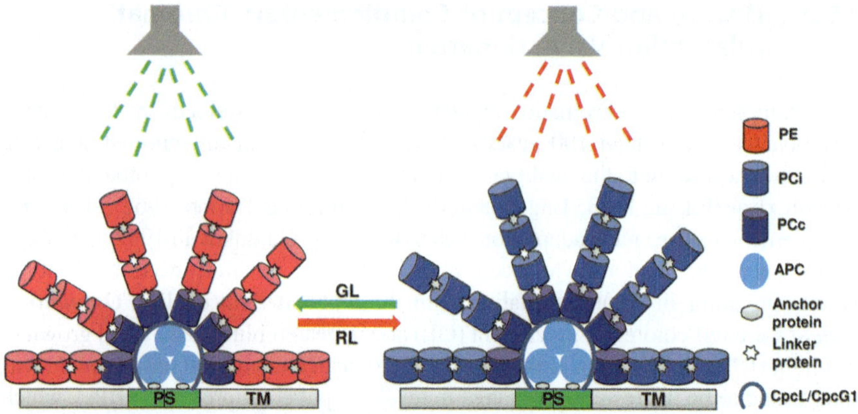

Fig. 10.1 Structural composition of phycobilisome and its reconstruction in response to green light (GL) and red light (RL) signals. *PS* photosystem, *TM* thylakoid membrane, *PE* phycoerythrin, *PCc* constitutive phycocyanin, *PCi* inducible phycocyanin, *APC* allophycocyanin

are composed of different types of PBPs, which are water-soluble pigmented proteins having α and β apoprotein subunits and a bilin chromophore (open chain tetrapyrrole) attached to a conserved cysteine residue (Anderson and Toole 1998; Samsonoff and MacColl 2001). The covalent attachment of chromophores to specific cysteines of α or β subunit of a particular phycobiliprotein is catalyzed by phycobilin lyases, which are required for stable heterodimer formation (Scheer and Zhao 2008; Wiethaus et al. 2010). The heterodimers of PBPs assemble into hexamers (αβ)6 and form hexameric disk, which are hollow in their centers and cylindrical in shape. The hexamer that makes up the core of the PBS is assembled in pairs with the hexamer of rods radiating from the core (Fig. 10.1). The hexamer units of PBPs are joined together by linker polypeptide to form a hemidiscoidal shape PBS, which can be divided into two parts, i.e., inner core part that attaches PBS to photosystem, and outer rod part that remain attached to the core part and does the job of light harvesting (Fig. 10.1) (Bennett and Bogorad 1973; Bogorad 1975; Bryant et al. 1979). The anchor polypeptides attach the PBSs to the thylakoid membrane at the site of reaction centers (Robinson and Miller 1970; Bryant et al. 1979; Glazer 1982; Beguin et al. 1985). Thus, light harvested by the rod part is transferred to the core part and ultimately reaches the photosystems to carry out the photochemistry.

The core part of the PBS always consists of the PBPs, allophycocyanin (APC; $\lambda_{max} = 650$ nm), which attaches the PBSs to the thylakoid membrane at the site of reaction centers. However, the rod part of PBS, which radiates from the core part, is composed of either phycoerythrin (PE; $\lambda_{max} = 565$ nm) or phycocyanin (PC; $\lambda_{max} = 620$ nm) depending upon the prevailing light conditions (Fig. 10.1). Notably, the inner part of the rod is always composed of the constitutive phycocyanin (PCc), irrespective of the light quality to efficiently channel the absorbed radiant energy from rod parts of PBS to the core part. However, the outer part of the rod is either composed of inducible phycocyanin (PCi) or PE depending on the light quality

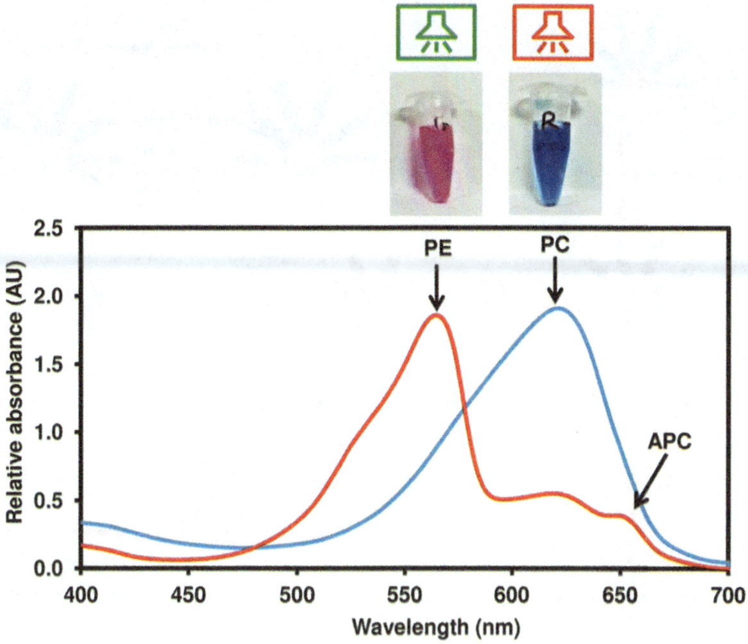

Fig. 10.2 Absorption spectra of phycoerythrin (PE; λ_{max} = 565 nm), phycocyanin (PC; λ_{max} = 620 nm), and allophycocyanin (APC; λ_{max} = 650 nm) extracted from culture of *Fremyella diplosiphon* grown under green and red light

received by the organism (Bennett and Bogorad 1973; Bogorad 1975; Bryant et al. 1979). For example, in *Fremyella diplosiphon*, RL-enriched environment promotes the incorporation of RL-absorbing PC, whereas GL-enriched low light environment promotes the incorporation of GL-absorbing PE in the outer part of rods (Bennett and Bogorad 1973) (Fig. 10.2). These GL- and RL-dependent changes in PE and PC levels have been considered as a characteristic feature of CA; however, further advancement in our knowledge on CA has led to the identification of six types of CAs (Sanfilippo et al. 2019). In the following sections, different types of CAs have been presented according to their responsive light quality (Fig. 10.3).

10.3.1 Green/Red Responsive CA

10.3.1.1 Type 1

The green-red light-responsive type 1 CA was recently identified in *Synechocystis* sp. PCC 6803, which is characterized by a change in rod linker protein rather than a change in PBP of PBS as in the case of type 2 and type 3 CAs (Tandeau de Marsac 1977; Kondo et al. 2005). Type 1 CA is carried out by a photoreceptor CcaS and its cognate response regulator CcaR. CcaS is a GL- and RL-responsive photoreceptor

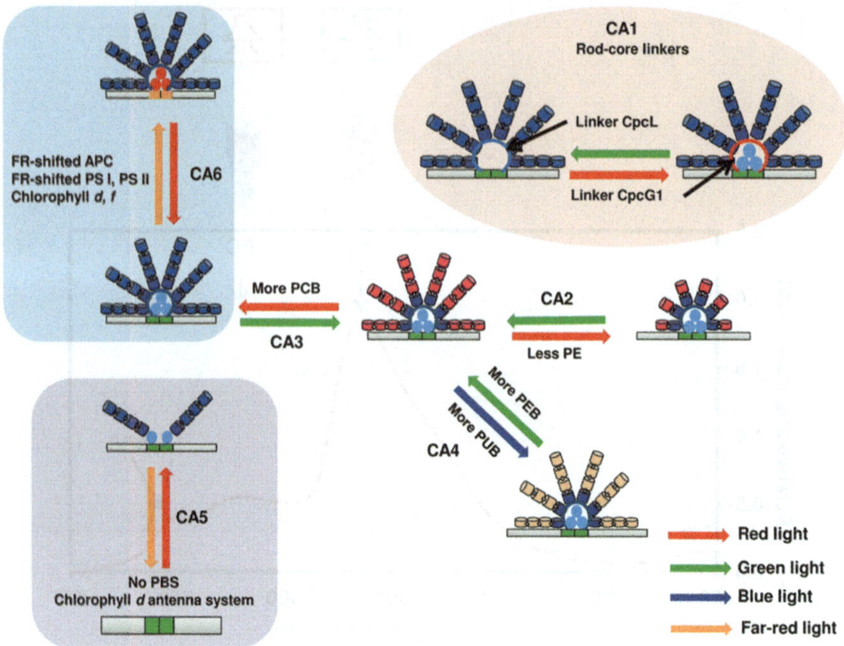

Fig. 10.3 Different types of chromatic acclimations (CAs) reported in cyanobacteria that maximize their performance and fitness in their ecological niches subjected to fluctuating light conditions. *FR* far-red, *PCB* phycocyanobilin, *PE* phycoerythrin, *PEB* phycoerythrobilin, *PS* photosystem, *PUB* phycourobilin

and control transcript levels of rod linker proteins encoding genes *cpcL* and *cpcG1* via CcaR (Watanabe et al. 2014). In RL condition, CpcG1 connects rod part of PBS to core part; however, in GL condition, CpcL rod linker protein directly attaches rod part to photosystem without AP core part of PBS (Fig. 10.3) (Kondo et al. 2005; Kondo et al. 2007; Hirose et al. 2008; Deng et al. 2012). Thus, in type 1 CA, PBS is without core part under GL condition and it is still not clear how CpcL attaches rod parts to photosystem without core part (Fig. 10.3).

10.3.1.2 Type 2

In this type of CA, PBSs contain longer rods composed of PE under GL condition, while there is no change in the level of PC under RL condition (Fig. 10.3). However, there is a significant reduction in PE-containing rods under RL condition (Tandeau de Marsac 1977; Hirose et al. 2010). The molecular mechanism of CA2 has been well studied in *Nostoc punctiforme* PCC 73102 where photoreceptor CcaS and response regulator CcaR regulate expression of genes involved in PE and linker CpcL biosynthesis (Hirose et al. 2010).

10.3.1.3 Type 3

Type 3 CA is found in a cyanobacterium *Fremyella diplosiphon* (also known as *Calothrix* sp. PCC 7601 or *Tolypothrix* sp. PCC 7601), which has acted as a model system to study various aspects of CA3 (Kehoe and Gutu 2006; Gutu and Kehoe 2012). In *F. diplosiphon*, GL promotes PE color-containing PBSs, while RL promotes PC-containing PBSs (Fig. 10.3). Thus, in CA3, the color of organism changes under both light conditions imparting brick red color under GL due to higher accumulation of PE and bluish-green color under RL growth condition due to higher accumulation of PC in rods of PBSs (Fig. 10.2) (Kehoe and Gutu 2006; Gutu and Kehoe 2012). Notably, chromophores phycocyanobilin (PCB) and phycoerythrobilin (PEB) are also altered in addition to PBPs in CA3. This light-dependent differential accumulation of PC or PE in rods part of PBSs is regulated by a photosensor RcaE, which perceives GL and RL signals, and control expression of PE and PC encoding genes via its cognate response regulators RcaF and RcaC (Kehoe and Grossman 1996; Kehoe and Grossman 1997).

10.3.2 Blue/Green Responsive CA

10.3.2.1 Type 4

This is a blue light (BL) and GL-responsive phenomenon where a change in only chromophores takes place without major change at the level of PBS proteins (Palenik 2001; Everroad et al. 2006). CA4 is exhibited by marine *Synechococcus,* which significantly contribute to global oceanic CO_2 fixation (Flombaum et al. 2013). In marine *Synechococcus*, BL and GL regulate the levels of phycourobilin (PUB) and phycoerythrobilin (PEB) chromophores in outer rod part of PBSs, which is primarily composed of PE. BL promotes PUB, while GL promotes higher levels of PEB in the PE (Fig. 10.3) (Palenik 2001; Everroad et al. 2006). Thus, in CA4, the ratio of PUB: PEB in PE is altered in response to BL and GL signals (Fig. 10.3).

10.3.3 Red/Far-Red Responsive CA

10.3.3.1 Type 5

Acaryochloris marina exhibit type 5 CA, which is responsive to far-red light condition. In far-red light condition, *A. marina* possesses chlorophyll *d*-containing light-harvesting antenna system, which is replaced by the rod-shaped PC-containing antenna system under RL condition (Gloag et al. 2007; Chen et al. 2009; Duxbury et al. 2009). Notably, cyanobacteria possess only chlorophyll *a;* however, *A. marina* is known to possess chlorophyll *a, d,* and *f* molecules.

10.3.3.2 Type 6

Similar to CA5, CA6 is also responsive to far-red and RL condition; however, CA6 is physiologically different from CA5 and also known as FaRLiP (far-red light photoacclimation) (Gan et al. 2014). CA6 has been well studied in the number of organisms growing in various ecological niches where it maximizes the utilization of far-red light to support the photosynthesis in *Leptolyngbya* sp. JSC-1, *Chlorogloeopsis fritschii* sp. PCC 9212, *Chroococcidiopsis thermalis* sp. PCC 7203, and *Synechococcus* sp. PCC 7335 (Brown et al. 2010; Gan et al. 2014; Behrendt et al. 2015; Gan et al. 2015; Ho et al. 2017; Averina et al. 2018). CA6 includes a change in photosynthetic apparatus and light-harvesting antenna system in the presence of far-red light, which is characterized by the presence of far-red light-absorbing APC, chlorophyll *d*, and chlorophyll *f*, and different photosystem proteins (Fig. 10.3) (Xu et al. 2016).

10.4 Molecular Mechanism of CA

Although six types of CAs are known, CA3 is the oldest and widely used as a representative of CA. Molecular mechanism and signal transduction pathway of CA3 are well-studied in comparison with other CAs that are still emerging. Therefore, we have included the molecular mechanism of type 3 CA in this section, which is well established and understood in *F. diplosiphon*.

The regulator for CA (Rca) is a two-component regulatory system, which involves phytochrome-related photoreceptor and its cognate response regulators (Sobczyk et al. 1993; Parkinson 1995; Kehoe and Grossman 1996; Terauchi et al. 2004). The first component of the Rca pathway is a photoreceptor RcaE, which perceives GL and RL signals and accordingly dictates the molecular composition of PBSs (Kehoe and Grossman 1996). RcaE is a 74 KDa histidine kinase, which possesses N-terminal chromophore binding domain and C-terminal kinase domain (Taylor and Zhulin 1999). In the absence of RcaE, *F. diplosiphon* cannot differentiate between RL and GL signals and behaves like a color blind. Therefore, RcaE mutant accumulates intermediate levels of both PE and PC irrespective of the light condition and appears black (Kehoe and Grossman 1996; Seib and Kehoe 2002; Alvey et al. 2003; Terauchi et al. 2004). Several studies suggest that RcaE can sense light quality and promote PE under GL, and PC under RL condition in the rods of PBSs via its response regulators RcaF and RcaC (Kehoe and Grossman 1996; Kehoe and Grossman 1997; Seib and Kehoe 2002; Alvey et al. 2003; Terauchi et al. 2004; Hirose et al. 2013).

These studies also suggest that RcaE acts as a kinase in RL condition and phosphatase in GL condition, and control activity of its RRs RcaF and RcaC by phosphorylation and dephosphorylation (Fig. 10.4) (Kehoe and Grossman 1996; Kehoe and Grossman 1997; Seib and Kehoe 2002; Alvey et al. 2003; Terauchi et al. 2004; Hirose et al. 2013). RcaF has a single receiver domain that consists of 124 amino acids and found next to RcaE in the signal transduction cascade (Kehoe and Grossman 1997). RcaC is a second response regulator that works next to RcaF in the Rca pathway. It has two receiver domains, one histidine phosphotransfer domain and one DNA-binding domain (DBD) (Chiang et al. 1992; Kehoe and Grossman 1997). DBD helps in interaction of RcaC with a genome of organisms to control light-dependent transcription of genes associated with PC

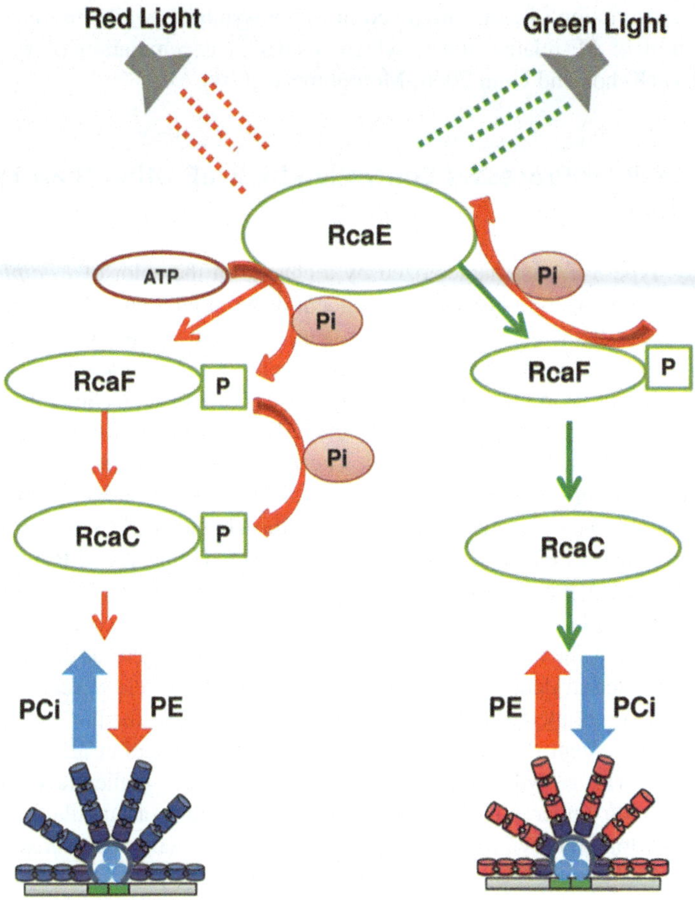

Fig. 10.4 Molecular mechanism of signal transduction cascade controlled by photosensor RcaE that leads to reconstruction of phycobilisomes in type 3 chromatic acclimation in response to green light and red light signals. Upward and downward arrows in blue and red colors indicate induction and inhibition of inducible phycocyanin and phycoerythrin

and PE biosynthesis. Several mutational analyses revealed that conserved aspartate residues at amino acid positions 51 and 576 of receiver modules are sites of reversible phosphorylation and are essential for the regulation of CA (Li and Kehoe 2005).

In RL condition, the kinase activity of RcaE leads to phosphorylation of RcaF using adenosine triphosphate (ATP), which further activates RcaC through phosphorylation. Phosphorylated RcaC is associated with activation of transcription of PC biosynthetic genes and simultaneously represses transcription of PE biosynthetic genes (Fig. 10.4). Conversely, under GL condition, the phosphatase activity of RcaE removes phosphate group from the RcaF. RcaF is inactive in unphosphorylated state, which leads to accumulation of RcaC in an inactive form. Therefore, RcaC cannot activate transcription of genes involved in PC biosynthesis while no longer inhibit transcription of PE-related genes, which results in accumulation of PE in rods (Fig. 10.4) (Kehoe and Gutu 2006; Montgomery 2008).

10.5 Cellular Processes Controlled by RcaE Other than Type 3 CA

Generally, type 3 CA is characterized by a change in the color of *F. diplosiphon,* which is caused due to light quality-dependent differential accumulation of PBPs in outer rods of PBSs (Fig. 10.2). However, studies conducted recently revealed that several other changes take place at a cellular level in *F. diplosiphon* during CA. For example, in addition to the change in color, filament length and cellular morphology are also altered reversibly under GL and RL growth conditions (Fig. 10.5) (Montgomery 2008; Singh et al. 2013). The filaments are longer with rectangular morphology of cells under green-enriched low light environment, while shorter filaments with spherical cell morphology are the characteristic feature of growth under RL-enriched high light environment (Fig. 10.5) (Bennett and Bogorad 1973; Bordowitz and Montgomery 2008; Singh and Montgomery 2011; Singh and Montgomery 2014).

In recent years, the mechanistic insight of cellular morphogenesis during CA3 has been investigated in detail and it was found that cellular levels of reactive oxygen species (ROS) are altered during CA, which is responsible for the change in cell shape to some extent (Singh and Montgomery 2012; Walters et al. 2013; Kumar et al. 2019). The growth under RL environment induces higher levels of ROS, whereas growth under GL environment results in lower accumulation of ROS (Singh and Montgomery 2012). Interestingly, this light quality-dependent differential accumulation of ROS is associated with a change in morphology of filaments and cell shape, and furthermore, RcaE plays important role in maintaining cellular ROS

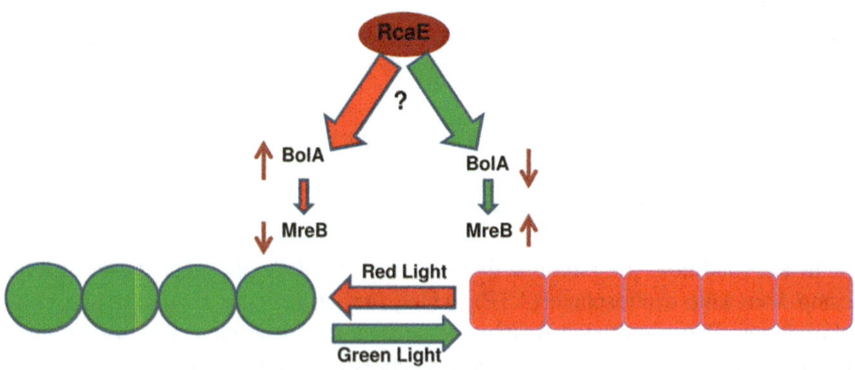

Fig. 10.5 Model showing mechanistic insight of morphological changes that take place in *Fremyella diplosiphon* during type 3 chromatic acclimation in response to green light and red light signals

levels in *F. diplosiphon* (Singh and Montgomery 2012). Thus, this was the first time when the role of photosensor RcaE in controlling ROS levels was established (Singh and Montgomery 2012).

RcaE was found to regulate cellular morphology and ROS levels in *F. diplosiphon* by regulating the level of BolA protein in a light quality-dependent manner via a still unknown mechanism (Singh and Montgomery 2014; Singh and Montgomery 2015). BolA is accumulated at a higher level in RL condition and at a lower level under GL growth condition, and this light-dependent differential accumulation of BolA protein is regulated at the transcription level in RcaE-dependent manner but independent of RcaF and RcaC (Singh and Montgomery 2014; Singh and Montgomery 2015). Thus, RcaE regulates the levels of BolA protein independent of the signal cascade, which operates in type 3 CA.

BolA is an oxidoreductase protein, which was found to regulate the ROS levels in *F. diplosiphon* as low levels of BolA are associated with higher levels of ROS (Singh and Montgomery 2014; Singh and Montgomery 2015). However, higher accumulation of BolA due to overexpression of *bolA* gene results in low levels of ROS and long filaments in *F. diplosiphon* (Singh and Montgomery 2014; Singh and Montgomery 2015). BolA also acts as a transcription factor and negatively controls transcription of gene encoding rod-shaped determining cytoskeleton protein MreB by binding to its promoter region (Singh and Montgomery 2014; Singh and Montgomery 2015). Thus, RcaE-dependent higher accumulation of BolA due to higher transcription of its encoding gene under RL downregulates MreB protein and thereby imparting spherical morphology to cells (Fig. 10.5). However, lower accumulation of BolA under GL due to lower transcription of *bolA* gene cannot inhibit the accumulation of MreB protein and therefore results in rod shape morphology of cells (Singh and Montgomery 2014; Singh and Montgomery 2015).

We have recently found the role of RcaE in controlling number and morphology of carboxysome, which is an essential component of carbon concentrating mechanism in *F. diplosiphon* (Rohnke et al. 2018). Thus, recent studies have shown that RcaE impacts several other cellular processes in addition to its well-established function in the reshuffling of PBS composition and therefore plays a critical role in determining the fitness of organism under fluctuating light conditions.

10.6 Significance of CA

The quality and quantity of light change significantly at different depths in aquatic ecosystems; RL-enriched high light environment is found at the surface of water column, whereas GL-enriched low light environment is found at higher depths. Thus, different cyanobacteria compete with other photosynthetic organisms to harvest photons to drive photosynthesis, and CA gives them a competitive advantage over other photosynthetic organisms in maximally utilizing available photons to drive photosynthesis. Thus, by incorporating PE in rods of PBS under low light environment, cyanobacteria can utilize GL to fuel photosynthesis, which cannot be utilized by other photosynthetic organisms except members of Rhodophyceae and

Chrysophyceae. Similarly, incorporation of far-red light-absorbing pigments or BL-, GL-, and RL-absorbing chromophores increases the fitness of cyanobacteria over other organisms in low light conditions existing in terrestrial and aquatic ecosystems.

The morphological changes that take place during CA benefit cyanobacteria by extending surface area for efficient absorption of available photons. For example, in a low light environment, rod shape morphology of cell and longer filament length in *F. diplosiphon* could provide increased surface area to efficiently absorb radiant energy. Further, rod shape morphology provides more space to accumulate more PBPs in the low light environment in comparison with a high light environment (Montgomery 2008; Singh and Montgomery 2011). Thus, CA increases the fitness of cyanobacteria and provides a competitive advantage over other photosynthetic organisms and supports the ecologically important process of photosynthesis in low light environments.

Acknowledgments PKM and VK are thankful to the UGC, New Delhi, India (3616/NET-DEC2014), and Council of Scientific and Industrial Research, New Delhi, India (09/013(0716)/2017-EMR-I), respectively, for the grants in the form of senior research fellowships. SPS acknowledges the UGC for start-up grant (F.30-370/2017; BSR) and the SERB for early-career research award (ECR/2016/000578). SM is thankful to UGC and BHU for providing UGC research fellowship. We thank Anjali Gupta for critically reading the MS and suggestions.

References

Alvey RM, Karty JA, Roos E, Reilly JP, Kehoe DM (2003) Lesions in phycoerythrin chromophore biosynthesis in *Fremyella diplosiphon* reveal coordinated light regulation of apoprotein and pigment biosynthetic enzyme gene expression. Plant Cell 15(10):2448–2463

Anderson LK, Toole CM (1998) A model for early events in the assembly pathway of cyanobacterial phycobilisomes. Mol Microbiol 30(3):467–474

Averina S, Velichko N, Senatskaya E, Pinevich A (2018) Far-red light photoadaptations in aquatic cyanobacteria. Hydrobiologia 813(1):1–17

Beguin S, Guglielmi G, Rippka R, Cohen-Bazire G (1985) Chromatic adaptation in a mutant of *Fremyella diplosiphon* in capable of phycoerythrin synthesis. Biochemist 67(1):109–117

Behrendt L, Brejnrod A, Schliep M, Sorensen SJ, Larkum AWD, Kuhl M (2015) Chlorophyll *f*-driven photosynthesis in a cavernous cyanobacterium. ISME J 9(9):2108–2111

Bennett A, Bogorad L (1973) Complementary chromatic adaptation in a filamentous blue-green alga. J Cell Biol 58(2):419–435

Bogorad L (1975) Phycobiliproteins and complementary chromatic adaptation. Annu Rev Plant Physiol 26(1):369–401

Bordowitz JR, Montgomery BL (2008) Photoregulation of cellular morphology during complementary chromatic adaptation requires sensor-kinase-class protein RcaE in *Fremyella diplosiphon*. J Bacteriol 190(11):4069–4074

Boresch K (1922) Die komplementäre chromatic adaptation. Arch Protistenkd 44:1–70

Brown II, Bryant DA, Casamatta D, Thomas-Keprta KL, Sarkisova SA, Shen G, Graham JE, Boyd ES, Peters JW, Garrison DH, McKay DS (2010) Polyphasic characterization of a thermotolerant siderophilic filamentous cyanobacterium that produces intracellular iron deposits. Appl Environ Microbiol 76(19):6664–6672

Bryant DA, Guglielmi G, Tandeau de Marsac N, Castets AM, Cohen-Bazire G (1979) The structure of cyanobacterial phycobilisomes: a model. Arch Microbiol 123(2):113–127

Chen M, Floetenmeyer M, Bibby TS (2009) Supramolecular organization of phycobiliproteins in the chlorophyll d-containing cyanobacterium *Acaryochloris marina*. FEBS Lett 583 (15):2535–2539

Chiang GG, Schaefer MR, Grossman AR (1992) Complementation of a red-light-indifferent cyanobacterial mutant. Proc Natl Acad Sci U S A 89(20):9415–9419

Deng GP, Liu F, Liu XW, Zhao JD (2012) Significant energy transfer from CpcG2 phycobilisomes to photosystem I in the cyanobacterium *Synechococcus* sp. PCC 7002 in the absence of ApcD-dependent state transitions. FEBS Lett 586(16):2342–2345

Duxbury Z, Schliep M, Ritchie RJ, Larkum AWD, Chen M (2009) Chromatic photoacclimation extends utilisable photosynthetically active radiation in the chlorophyll *d*-containing cyanobacterium, *Acaryochloris marina*. Photosynth Res 101(1):69–75

Engelmann TW (1883) Farbe und assimilation. Assimilation findet nur in den farbstoffhaltigen plasmathielchen statt II Näherer zusamennhang zwischen lichtabsorption und assimilation. Bot Z 41:1–13

Engelmann TW (1902) Untersuchungen über die qualitativen beziehungen zwieschen absorbtion des lichtes und assimilation in pflanzenzellen. I. Das mikrospectraphotometer, ein apparat zur qualitativen mikrospectralanalyse. II. Experimentelle grundlangen zur ermittelung der quantitativen beziehungen zwieschen assimilationsenergie und absorptiongrösse. III. Bestimmung der vertheilung der energie im spectrum von sonnenlicht mittels bacterien-methode und quantitativen mikrospectralanalyse. Bot Z 42:81–105

Everroad C, Six C, Partensky F, Thomas JC, Holtzendorff J, Wood AM (2006) Biochemical bases of type IV chromatic adaptation in marine *Synechococcus* spp. J Bacteriol 188(9):3345–3356

Flombaum P, Gallegos JL, Gordillo RA, Rincón J, Zabala LL, Jiao N, Karl DM, Li WKW, Lomas MW, Veneziano D, Vera CS, Vrugt JA, Martiny AC (2013) Present and future global distributions of the marine cyanobacteria *Prochlorococcus* and *Synechococcus*. Proc Natl Acad Sci U S A 110(24):9824–9829

Fujita Y, Hattori A (1962) Photochemical interconversion between precursors of phycobilin chromoproteins in *Tolypothrix tenuis*. Plant Cell Physiol 3(3):209–220

Gaidukov N (1902) Über den einfluss farbigen lichts auf die färbung lebender Oscillarien. Abh Preuss Akad Wiss 5:1–36

Gan F, Zhang S, Rockwell NC, Martin SS, Lagarias JC, Bryant DA (2014) Extensive remodeling of a cyanobacterial photosynthetic apparatus in far-red light. Science 345(6202):1312–1317

Gan F, Shen G, Bryant DA (2015) Occurrence of far-red light photoacclimation (FaRLiP) in diverse cyanobacteria. Life 5(1):4–24

Glazer AN (1982) Phycobilisomes: structure and dynamics. Annu Rev Microbiol 36(1):173–198

Gloag RS, Ritchie RJ, Chen M, Larkum AWD, Quinnell RG (2007) Chromatic photoacclimation, photosynthetic electron transport and oxygen evolution in the chlorophyll *d*-containing oxyphotobacterium *Acaryochloris marina*. Biochem Biophys Acta Bioenergetics 1767 (2):127–135

Gutu A, Kehoe DM (2012) Emerging perspectives on the mechanisms, regulation, and distribution of light color acclimation in cyanobacteria. Mol Plant 5(1):1–13

Hirose Y, Shimada T, Narikawa R, Katayama M, Ikeuchi M (2008) Cyanobacteriochrome CcaS is the green light receptor that induces the expression of phycobilisome linker protein. Proc Natl Acad Sci U S A 105(28):9528–9533

Hirose Y, Narikawa R, Katyama M, Ikeuchi M (2010) Cyanobacteriochrome CcaS regulates phycoerythrin accumulation in *Nostoc punctiforme*, a group II chromatic adaptor. Proc Natl Acad Sci U S A 107(19):8854–8859

Hirose Y, Rockwell NC, Nishiyama K, Narikawa R, Ukaji Y, Inomata K, Lagarias JC, Ikeuchi M (2013) Green/red cyanobacteriochromes regulate complementary chromatic acclimation *via* a protochromic photocycle. Proc Natl Acad Sci U S A 110(13):4974–4979

Ho MY, Gan F, Shen G, Zhao C, Bryant DA (2017) Far-red light photoacclimation (FaRLiP) in *Synechococcus* sp. PCC 7335: I. regulation of FaRLiP gene expression. Photosyn Res 131 (2):173–186

Kehoe DM, Grossman AR (1996) Similarity of a chromatic adaptation sensor to phytochrome and ethylene receptors. Science 273(5280):1409–1412

Kehoe DM, Grossman AR (1997) New classes of mutants in complementary chromatic adaptation provide evidence for a novel four-step phosphorelay system. J Bacteriol 179(12):3914–3921

Kehoe DM, Gutu A (2006) Responding to color: the regulation of complementary chromatic adaptation. Annu Rev Plant Biol 57:127–150

Kondo K, Geng XX, Katayama M, Ikeuchi M (2005) Distinct roles of CpcG1 and CpcG2 in phycobilisome assembly in the cyanobacterium *Synechocystis* sp. PCC 6803. Photosynth Res 84(1–3):269–273

Kondo K, Ochiai Y, Atayama M, Ikeuchi M (2007) The membrane-associated CpcG2-phycobilisome in *Synechocystis*: a new photosystem I antenna. Plant Physiol 144(2):1200–1210

Kumar V, Maurya PK, Mondal S, Sinha RP, Singh SP (2019) Photomorphogenesis in the cyanobacterium *Fremyella diplosiphon* improves photosynthetic efficiency. In: Mishra AK, Tiwari DN, Rai AN (eds) Cyanobacteria: from basic science to applications. Academic Press, Cambridge, MA, pp 131–143

Li L, Kehoe DM (2005) In vivo analysis of the roles of conserved aspartate and histidine residues within a complex response regulator. Mol Microbiol 55(5):1538–1552

Liu H, Zhang H, Niedzwiedzki DM, Prado M, He G, Gross ML, Blankenship RE (2013) Phycobilisomes supply excitations to both photosystems in a megacomplex in cyanobacteria. Science 342(6162):1104–1107

Montgomery BL (2008) Shedding new light on the regulation of complementary chromatic adaptation. Open Life Sci 3(4):351–358

Palenik B (2001) Chromatic adaptation in marine *Synechococcus* strains. Appl Environ Microbiol 67(2):991–994

Parkinson JS (1995) Genetic approaches for signaling pathways and proteins. In: Hoch J, Silhavy T (eds) Two-component signal transduction. ASM Press/American Society of Microbiology, Washington, DC, pp 9–23

Pathak J, Maurya PK, Singh SP, Häder DP, Sinha RP (2018) Cyanobacterial farming for environment friendly sustainable agriculture practices: innovations and perspectives. Front Environ Sci 6:7

Pathak J, Ahmed H, Singh DK, Pandey A, Singh SP, Sinha RP (2019) Recent developments in green synthesis of metal nanoparticles utilizing cyanobacterial cell factories. In: Tripathi DK, Ahmad P, Sharma S, Chauhan DK, Dubey NK (eds) Nanomaterials in plants algae and microorganisms: concepts and controversies, vol 2. Elsevier Inc., San Diego, CA, pp 237–265

Rajneesh R, Singh SP, Pathak J, Sinha RP (2017) Cyanobacterial factories for the production of green energy and value-added products: an integrated approach for economic viability. Renew Sust Energ Rev 69:578–595

Robinson BL, Miller JH (1970) Photomorphogenesis in the blue-green alga *Nostoc commune* 584. Physiol Plant 23(3):461–472

Rohnke BA, Singh SP, Pattanaik B, Montgomery BL (2018) RcaE-dependent regulation of carboxysome structural proteins has a central role in environmental determination of carboxysome morphology and abundance in *Fremyella diplosiphon*. mSphere 3(1):1–19

Samsonoff WA, MacColl R (2001) Biliproteins and phycobilisomes from cyanobacteria and red algae at the extremes of habitat. Arch Microbiol 176(6):400–405

Sanfilippo JE, Garczarek L, Partensky F, Kehoe DM (2019) Chromatic acclimation in cyanobacteria: a diverse and widespread process for optimizing photosynthesis. Annu Rev Microbiol 73:407–433

Scheer H, Zhao KH (2008) Biliprotein maturation: the chromophore attachment. Mol Microbiol 68 (2):263–276

Seib LO, Kehoe DM (2002) A turquoise mutant genetically separates expression of genes encoding phycoerythrin and its associated linker peptides. J Bacteriol 184(4):962–970

Singh SP, Montgomery BL (2011) Determining cell shape: adaptive regulation of cyanobacterial cellular differentiation and morphology. Trends Microbiol 19(6):278–285

Singh SP, Montgomery BL (2012) Reactive oxygen species are involved in the morphology-determining mechanism of *Fremyella diplosiphon* cells during complementary chromatic adaptation. Microbiology 158(9):2235–2245

Singh SP, Montgomery BL (2013a) Distinct salt-dependent effects impair *Fremyella diplosiphon* pigmentation and cellular shape. Plant Signal Behav 8(7):e24713

Singh SP, Montgomery BL (2013b) Salinity impacts photosynthetic pigmentation and cellular morphology changes by distinct mechanisms in *Fremyella diplosiphon*. Biochem Biophys Res Commun 433(1):84–89

Singh SP, Montgomery BL (2014) Morphogenes bolA and mreB mediate the photoregulation of cellular morphology during complementary chromatic acclimation in *Fremyella diplosiphon*. Mol Microbiol 93(1):167–182

Singh SP, Montgomery BL (2015) Regulation of BolA abundance mediates morphogenesis in *Fremyella diplosiphon*. Front Microbiol 6:1215

Singh SP, Häder D-P, Sinha RP (2010) Cyanobacteria and ultraviolet radiation (UVR) stress: mitigation strategies. Ageing Res Rev 9(2):79–90

Singh SP, Miller HL, Montgomery BL (2013) Temporal dynamics of changes in reactive oxygen species (ROS) levels and cellular morphology are coordinated during complementary chromatic acclimation in *Fremyella diplosiphon*. Photosynth Res 118(1–2):95–104

Singh SP, Rastogi RP, Häder DP, Sinha RP (2014) Temporal dynamics of ROS biogenesis under simulated solar radiation in the cyanobacterium *Anabaena variabilis* PCC 7937. Protoplasma 251(5):1223–1230

Sobczyk A, Schyns G, Tandeau de Marsac N, Houmard J (1993) Transduction of the light signal during complementary chromatic adaptation in the cyanobacterium *Calothrix* sp. PCC 7601: DNA-binding proteins and modulation by phosphorylation. EMBO J 12(3):997–1004

Tandeau de Marsac N (1977) Occurrence and nature of chromatic adaptation in cyanobacteria. J Bacteriol 130(1):82–91

Taylor BL, Zhulin IB (1999) PAS domains: internal sensors of oxygen, redox potential, and light. Microbiol Mol Biol Rev 63(2):479–506

Terauchi K, Montgomery BL, Grossman AR, Lagarias JC, Kehoe DM (2004) RcaE is a complementary chromatic adaptation photoreceptor required for green and red light responsiveness. Mol Microbiol 51(2):567–577

Walters KJ, Whitaker MJ, Singh SP, Montgomery BL (2013) Light intensity and reactive oxygen species are centrally involved in photoregulatory responses during complementary chromatic adaptation in *Fremyella diplosiphon*. Commun Integr Biol 6(5):e25005

Watanabe M, Semchonok DA, Webber Birungi MT, Ehira S, Kondo K, Narikawa R, Ikeuchi M (2014) Attachment of phycobilisomes in an antenna-photosystem I supercomplex of cyanobacteria. Proc Natl Acad Sci U S A 111(7):2512–2517

Wiethaus J, Busch AWU, Dammeyer T, Frankenberg-Dinkel N (2010) Phycobiliproteins in *Prochlorococcus marinus*: biosynthesis of pigments and their assembly into proteins. Eur J Cell Biol 89(12):1005–1010

Xu QZ, Han JX, Tang QY, Ding WL, Miao D, Zhou M, Scheer H, Zhao KH (2016) Far-red light photoacclimation: Chromophorylation of FR induced α- and β-subunits of allophycocyanin from *Chroococcidiopsis thermalis* sp. PCC7203. Biochim Biophys Acta Bioenerg 1857 (9):1607–1616

Phenomenon of Allelopathy in Cyanobacteria

11

Sylwia Śliwińska-Wilczewska, Kinga A. Wiśniewska, Gracjana Budzałek, and Zofia Konarzewska

Abstract

The phenomenon of cyanobacterial allelopathy is widespread and occurs in almost all aquatic habitats. Production of active allelopathic compounds is an essential adaptation performed by some cyanobacteria to get a competitive advantage over the other primary producers. Some studies showed that the target organisms can be completely eliminated, inhibited, and sometimes even stimulated by allelopathic compounds secreted by cyanobacteria. That is why, due to potential selective stimulation or growth inhibition of individual species, allelopathy may act as an effective strategy, promoting succession of some phytoplankton species in the water bodies and thus, contributing to massive blooms in many aquatic habitats. The recurring presence of certain bloom-forming species contributes to the emergence of many ecological and economic problems. That is why allelopathic effects among cyanobacteria in aquatic ecosystems have been intensively studied. In the past, several world-renowned books and papers regarding the allelopathic effects of aqueous photoautotrophs have been presented. Furthermore, in recent years, the number of reports on the cyanobacterial allelopathy phenomenon in aquatic ecosystems have increased significantly. This chapter compiles the current knowledge regarding the allelopathy phenomenon of cyanobacteria and their allelochemicals affecting species

S. Śliwińska-Wilczewska (✉) · G. Budzałek · Z. Konarzewska
Division of Marine Ecosystems Functioning, Institute of Oceanography, University of Gdańsk, Gdynia, Poland
e-mail: ocessl@ug.edu.pl

K. A. Wiśniewska
Division of Marine Chemistry and Environmental Protection, Institute of Oceanography, University of Gdańsk, Gdynia, Poland

R. P. Rastogi (ed.), *Ecophysiology and Biochemistry of Cyanobacteria*,
https://doi.org/10.1007/978-981-16-4873-1_11

variability in aquatic ecosystems. Thus, allelopathy can be an effective strategy that causes massive blooms in many aquatic habitats.

Keywords

Allelopathy · Allelochemicals · Cyanobacterial bloom · Species interactions

11.1 Introduction

Allelopathy is any process that refers to organisms that can produce biologically active metabolites that influence the variability of other plant and animal species (Legrand et al. 2003). It is believed that allelopathy is a unique strategy of organisms that discourage or eliminate competitors and predators living in the same ecosystem (Sarkar et al. 2006; Granéli et al. 2008). Metabolites that are produced and released into the adjacent environment by various cyanobacteria have been named allelopathic compounds or allelochemicals. This definition includes cyanobacterial toxins and cyanobacterial secondary metabolites (Leflaive and Ten-Hage 2007). Examples of cyanobacteria in which allelopathic activity against other cyanobacteria and microalgae was confirmed are shown in Fig. 11.1.

In the aquatic environment, there are various types of allelopathic mechanisms affecting species variability (Leflaive and Ten-Hage 2007). Allelopathy is thought to be both antagonistic and synergistic, and the allelopathic effect depends on various environmental factors, being more complex than the effect observed in laboratory experiments (Suikkanen et al. 2004, 2005). The level of allelopathic impact on aquatic ecosystems depends on the production and secretion of active allelopathic compounds and their effective spread to target organisms, as well as the sensitivity of target species to release allelochemicals (Lewis Jr 1986). Moreover, the effect of allelochemicals depends on the nature of the interaction between donor and target organisms and the activity of the chemical compounds responsible for this interaction and can often be highly strain-specific.

In the past, several world-renowned books and papers regarding the allelopathic effects of aqueous photoautotrophs have been presented (Rice 1979; Lewis Jr 1986; Rizvi and Rizvi 1992; Rizvi et al. 1992; Gopal and Goel 1993; Dakshini 1994; Inderjit and Dakshini 1994; Gross 2003; Legrand et al. 2003; Granéli and Turner 2006; Reigosa et al. 2006). Previous review papers described in detail the phenomenon of allelopathy in freshwater cyanobacteria (Leão et al. 2009a). Furthermore, in recent years, the number of reports on the allelopathy phenomenon of cyanobacteria in aquatic ecosystems have increased significantly (Poulin et al. 2018a, b; Barreiro Felpeto et al. 2019; Corcoran et al. 2019; Śliwińska-Wilczewska et al. 2019; Zhu et al. 2021). The newly carried out studies demonstrated allelopathic activity of completely new species, described new methods of research on the allelopathy phenomenon, showed new factors that can affect the cyanobacterial allelopathy, and also exhibited new modes of action of cyanobacterial allelochemicals. According to the author's best knowledge, in recent years there has been no review

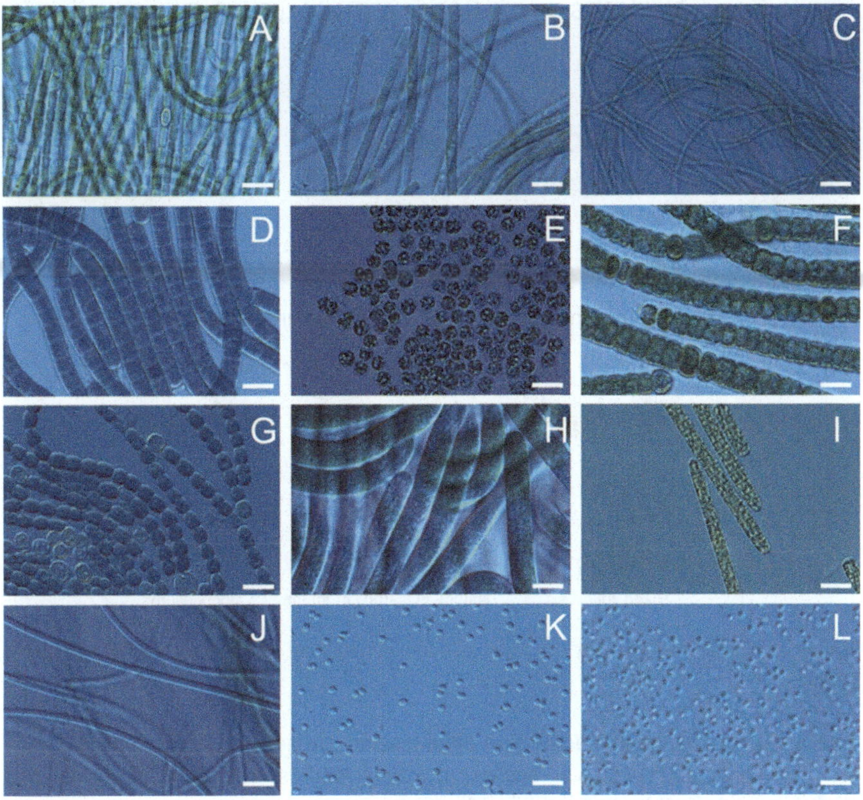

Fig. 11.1 Examples of allelopathic cyanobacteria: *Anabaena* sp. (**a**), *Aphanizomenon* sp. (**b**), *Geitlerinema* sp. (**c**), *Lyngbya* sp. (**d**), *Microcystis* sp. (**e**), *Nodularia spumigena* (**f**), *Nostoc* sp. (**g**), *Phormidium* sp. (**h**), *Planktothrix* sp. (**i**), *Pseudanabaena* sp. (**j**), *Synechocystis* sp. (**k**), *Synechococcus* sp. (**l**); scale bars = 10 μm. Photographs by Śliwińska-Wilczewska

paper presenting the whole issue such extensively. Therefore, the main goal of this chapter is to review the current knowledge of the cyanobacterial allelopathy phenomenon and allelochemicals affecting other cyanobacteria and microalgae variability in aquatic ecosystems. Furthermore, this section discusses the taxonomy of the allelopathy cyanobacteria and the factors affecting cyanobacterial allelopathy.

11.2 Methods of Cyanobacterial Allelopathy Examination

To detect the allelopathy phenomenon in aquatic ecosystems, it is necessary to use many different methods, from classical culture studies to advanced physiological and chemical analysis (Fig. 11.2). Although the first reports of allelopathy come from environmental observations (Akehurst 1931; Keating 1977), a detailed laboratory study is strongly recommended or even required to be carried out in order to

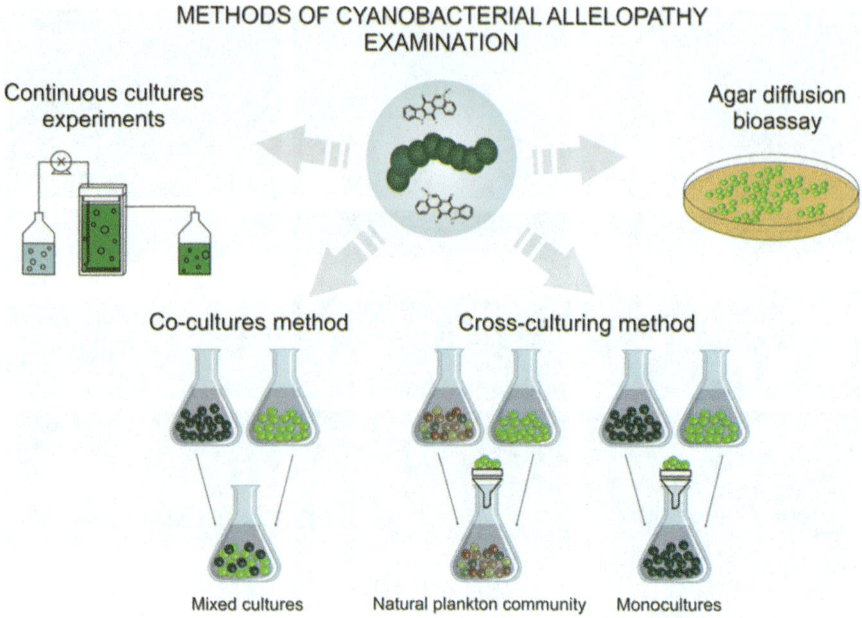

Fig. 11.2 The most commonly used methods for investigating the allelopathy phenomenon of cyanobacteria

determine the allelopathic activity of cyanobacteria in aquatic ecosystems (Leflaive and Ten-Hage 2007). This is because the field studies on cyanobacteria are still slightly constrained by technological limitations.

The classic but still widely used approach to study the allelopathy phenomenon are the algal tests carried out on solid media (Schagerl et al. 2002; Paz-Yepes et al. 2013; Brilisauer et al. 2019). Chan et al. (1980), as one of the first researchers, suggested that algal plating was a sufficient tool for investigating allelopathy among marine microalgae. Schagerl et al. (2002) observed the allelopathic activity of six strains of cyanobacteria isolated from Lake Neusiedlersee using plate diffusion assays. Recent research also uses agar tests to detect allelopathic effects. Paz-Yepes et al. (2013) used plate assays to determine whether allelopathic interactions occur between three strains of marine picocyanobacteria *Synechococcus* sp. In addition, Brilisauer et al. (2019) using agar-diffusion plate assay showed that *Synechococcus elongatus* supernatant extract inhibited the growth of other cyanobacteria *Anabaena variabilis*.

However, the most common method for studying the allelopathy phenomenon is the "cross-culturing" method. In this method, the cell-free filtrate obtained from the donor culture with nutrient-enriched media is added into the media with the target organism examined the effect of allelopathic compounds naturally released into the environment (Legrand et al. 2003). The use of a cell-free filtrate to determine allelopathic interactions is a preferred method as it excludes direct contact of donor and target cells in which the examined organisms could compete for nutrients

(Suikkanen et al. 2004). This method is useful in allelopathic experiments including both monocultures (Suikkanen et al. 2004; Śliwińska-Wilczewska et al. 2016, 2017a; Barreiro Felpeto et al. 2018; Konarzewska et al. 2020) and natural plankton community (Suikkanen et al. 2005; Śliwińska-Wilczewska et al. 2017b; Bubak et al. 2020). However, in the experiments where a single filtrate of the donor is added, the allelopathic effect may sometimes disappear during a few days of exposure due to degradation of allelopathic compounds, or activation of defense mechanisms in target organisms (Suikkanen et al. 2004). Recent studies have shown that the repeated filtrate addition has a more significant effect than the single addition of filtrate on target organisms (Barreiro Felpeto et al. 2018; Śliwińska-Wilczewska et al. 2019). All these reports indicate that some of the allelopathic compounds produced by cyanobacteria are not persistent thus, a single filtrate addition may not be representative of the allelopathic effects examined in the aquatic habitats and, as a result, data obtained may be underestimated (Granéli et al. 2008; Barreiro Felpeto and Vasconcelos 2014). In aquatic ecosystems, allelopathic compounds are released continuously, so experiments including repeated filtrate additions create a situation similar to the natural environment (Śliwińska-Wilczewska et al. 2019). Experiments with a single filtrate addition reveal the presence of the allelopathic phenomenon of cyanobacteria while the multiple additions of the filtrate make the observed effect stronger and close enough to the natural environment allelopathic effects (Suikkanen et al. 2004). Therefore, it is believed that the effects of single and repeated filtrate addition should be examined and compared. Barreiro Felpeto et al. (2018) showed initial and final concentrations of the macronutrients in the cell-free filtrates experiments and demonstrated that the effects of major nutrients limitations in the control culture and allelochemical treatments with cell-free filtrate could be excluded. Moreover, Schmidt and Hansen (2001) have shown in their laboratory experiment that in a nutrient-rich mineral medium usually less than 10–15% of major compounds are consumed by photoautotrophs. That study also proves that factors other than nutrient limitation are responsible for the observed growth inhibition of target species.

Another often used method for studying allelopathic effects is "mixed cultures", "bi-cultures", or "co-cultures". In this method, potentially allelopathic species grow together with target organisms in the medium that ensures the active growth of these species (Żak et al. 2012; Barreiro Felpeto et al. 2018; Śliwińska-Wilczewska et al. 2018). This method most often determines the abundance of donor and target species by counting them with a light microscope (Żak et al. 2012; Barreiro Felpeto et al. 2018; Śliwińska-Wilczewska et al. 2018) or using a flow cytometer (Śliwińska-Wilczewska et al. 2018).

Another approach to the study of the allelopathy phenomenon is to conduct experiments in continuous cultures. Barreiro Felpeto et al. (2017) demonstrated the allelopathic interaction between cyanobacteria and green algae by performing long-term competition experiments in nitrate-limited continuous cultures, and by describing the population dynamics using a mechanistic model, which was the first experimental confirmation that allelopathy can alter the predicted outcome of inter-specific competition in a nutrient-limited environment.

11.3 Taxonomic Position of the Allelopathic Cyanobacteria and Their Effect on Coexisting Phytoplankton Species

The phenomenon of allelopathy is recorded in many different species of cyanobacteria. In this chapter, we presented the number of donor cyanobacteria against other cyanobacteria and microalgae published in scientific articles based on studies of allelopathy contained in the title or keywords (until the year 2020). Additionally, we analyzed six papers (Lam and Silvester 1979; Bagchi et al. 1993; Issa 1999; Valdor and Aboal 2007; Żak and Kosakowska 2015; García-Espín et al. 2017) that examined the activity of cyanobacteria on other cyanobacteria and microalgae but in which the word allelopathy did not occur. We have compiled the information about the allelopathic ability in 25 different genera of cyanobacteria (belonging to four orders - Nostocales, Oscillatoriales, Synechococcales, Chroococcales) based on 46 literature reports. The allelopathic activity has been studied most often in cyanobacteria belonging to the genera *Microcystis*, *Synechococcus*, *Nostoc*, and *Cylindrospermopsis* (Fig. 11.3, Table 11.1). The least numerous studies for allelopathic ability were conducted for organisms belonging to the genera *Anabaenopsis*, *Chrysosporum*, *Cyanobium*, *Cylindrospermum*, *Geitlerinema*, *Lyngbya*, *Pseudanabaena*, *Synechocystis*, *Tolypothrix*, and *Trichormus*. It should also be noted that among these genera, different species and even strains belonging to the same species exhibit allelopathic activity (Table 11.1).

The frequency of cyanobacterial allelopathy research largely varies worldwide (Fig. 11.4). The greatest number of microorganisms used for allelopathic research are isolated in Europe, and significantly less in North and South America. In Asia, on the other hand, the highest proportion of isolated cyanobacteria is reported in China, while no studies have been reported using organisms isolated from other countries, among others. Interestingly, the number of research conducted in Australia and Oceania is scarce, despite many studies regarding other aspects of cyanobacteria

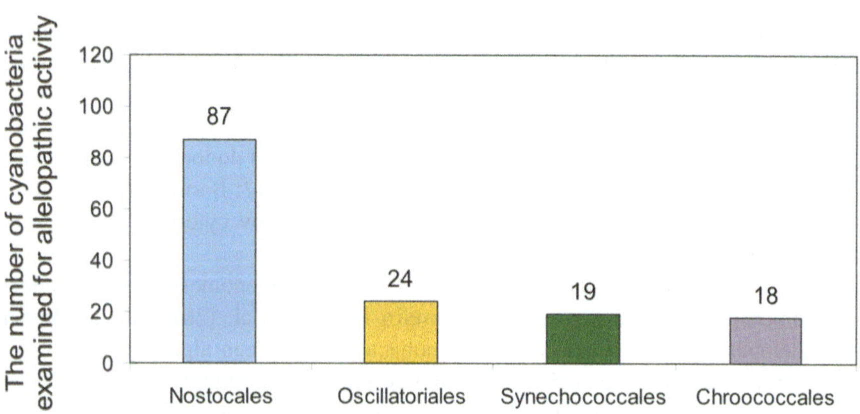

Fig. 11.3 The number of donor cyanobacteria divided into orders used in the allelopathic studies based on 46 literature reports

Table 11.1 Examples of allelopathic activity of cyanobacteria against other cyanobacteria and microalgae. Specific cyanobacterial strains are shown in brackets

Donor cyanobacteria	Target cyanobacteria	Effect	References
Anabaena cylindrica (ASW 01035)	*Anabaena cylindrica*, *Microcystis flos-aquae*	0	Schagerl et al. (2002)
Anabena lemmermannii (KAC 16)	*Anabaena* sp., *Aphanizomenon* sp., *Nodularia spumigena*, *Pseudanabaena* sp., *Snowella* sp.	0/+	Suikkanen et al. (2005)
Anabaena oscillarioides	*Microcystis aeruginosa*	–	Lam and Silvester (1979)
Anabaena torulosa (ASW 01028)	*Anabaena cylindrica*, *Microcystis flos-aquae*	–	Schagerl et al. (2002)
Anabaenopsis elenkinii (ASW 01027)	*Anabaena cylindrica*, *Microcystis flos-aquae*	0	Schagerl et al. (2002)
Aphanizomenon flexuosum (ASW 01033)	*Anabaena cylindrica*, *Microcystis flos-aquae*	0	Schagerl et al. (2002)
Aphanizomenon flos-aquae (Tr183)	*Anabaena* sp., *Aphanizomenon* sp., *Nodularia spumigena*, *Pseudanabaena* sp., *Snowella* sp.	0/+/–	Suikkanen et al. (2005)
Calothrix parietina	*Anabaena spiroides*, *Calothrix parietina*, *Microcystis aeruginosa*, *Nostoc muscorum*, *Oscillatoria angustissima*, *Phormidium mölle*, *Scytonema hofmanii*, *Synechococcus* sp.	0/–	Issa (1999)
Calothrix sp. (CAN 95/2, CAN 95/3, WA 96/8)	*Anabaena circinalis*, *Microcytis aeruginosa*, *Nodularia spumigena*	0/–	Schlegel et al. (1999)
Chrysosporum ovalisporum (CFWA01007)	*Microcystis panniformis*	–	Zhang et al. (2016)
Cylindrospermopsis raciborskii (LS117, LS118, LS123, LS124)	*Microcystis aeruginosa*	–	Figueredo et al. (2007), Rzymski et al. (2014)
Cylindrospermum sp. (ASW 01016)	*Anabaena cylindrica*, *Microcystis flos-aquae*	–	Schagerl et al. (2002)
Fischerella sp. (CAN 96/12, CAN 96/13, JAVA 94/20, LOM 95/3, LOM 95/9, LOM 95/17, NEP 95/1, NT 97/5, 52–1, VIET 97/2)	*Anabaena circinalis*, *Anabaena doliolum*, *Lyngbya* sp., *Microcystis aeruginosa*, *Nodularia spumigena*, *Nostoc* sp., *Pseudanabaena* sp., *Scytonema* sp.	0/–	Schlegel et al. (1999), Gantar et al. (2008)

(continued)

Table 11.1 (continued)

Donor cyanobacteria	Target cyanobacteria	Effect	References
Geitlerinema splendidum	*Nostoc* sp., *Pseudocapsa* sp., *Scytonema* sp.	0/–	Valdor and Aboal (2007)
Lyngbya sp. (15–2)	*Fischerella* sp., *Nostoc* sp., *Pseudanabaena* sp., *Scytonema* sp.	0/–	Gantar et al. (2008)
Microcystis aeruginosa (ASW 01002)	*Anabaena cylindrica*, *Anabaena oscillarioides*, *Anabaena* sp., *Cylindrospermopsis raciborskii*, *Microcystis flos-aquae*, *Oscillatoria angutissima*	0/+/–	Lam and Silvester (1979), Schagerl et al. (2002), El-Sheekh et al. (2010), Rzymski et al. (2014)
Microcystis flos-aquae (ASW 01004)	*Anabaena cylindrica*	0	Schagerl et al. (2002)
Microcystis panniformis (CFWA01028)	*Chrysosporum ovalisporum*	+	Zhang et al. (2016)
Nodularia harveyana (44.85)	*Arthrospira laxissima*, *Chroococcus minutus*, *Nostoc carneum*, *Nostoc insulare*, *Synechocystis aquatilis*	–	Volk and Furkert (2006)
Nodularia spumigena (CCBA15, KAC 13)	*Anabaena* sp., *Aphanizomenon* sp., *Nodularia spumigena*, *Pseudanabaena* sp., *Snowella* sp., *Synechococcus* sp.	0/+	Suikkanen et al. (2005), Barreiro Felpeto et al. (2018)
Nostoc insulare (54.79)	*Arthrospira laxissima*, *Chroococcus minutus*, *Nostoc carneum*, *Nostoc insulare*, *Synechocystis aquatilis*	–	Volk and Furkert (2006)
Nostoc muscorum (ASW 01011)	*Anabaena cylindrica*, *Microcystis flos-aquae*	–	Schagerl et al. (2002)
Nostoc sp. (37, 58–2, ASW 01010, ASW 01020, 23–2, Ev-1, NSW 95/10, WA 96/19)	*Anabaena circinalis*, *Anabaena cylindrica*, *Anabaena doliolum*, *Fischerella* sp., *Lyngbya* sp., *Microcystis aeruginosa*, *Microcystis flos-aquae*, *Nodularia spumigena*, *Nostoc* sp., *Pseudanabaena* sp., *Scytonema* sp.	0/–/+	Schlegel et al. (1999), Schagerl et al. (2002), Gantar et al. (2008)
Oscillatoria angustissima	*Anabaena spiroides*, *Calothrix parietina*, *Microcystis aeruginosa*, *Nostoc muscorum*, *Oscillatoria angustissima*, *Phormidium mölle*, *Scytonema hofmanii*, *Synechococcus* sp.	0/–	Issa (1999)

(continued)

Table 11.1 (continued)

Donor cyanobacteria	Target cyanobacteria	Effect	References
Oscillatoria sp.	*Anacystis nidulans*, *Microcystis* sp., *Nostoc muscorum*, *Nostoc* sp., *Oscillatoria* sp., *Phormidium uncinatum*, *Phormidium* sp., *Plectonema boryanum*, *Pseudocapsa* sp., *Scytonema* sp., *Synechococcus* sp.	–	Bagchi et al. (1993), Valdor and Aboal (2007)
Phormidium sp.	*Nostoc* sp., *Pseudocapsa* sp., *Scytonema* sp.	0/–	Valdor and Aboal (2007)
Phormidium sp. (*Oscillatoria* sp.) (LEGE 05292)	*Aphanizomenon* sp., *Limnothrix* sp., *Merismopedia* sp., *Microcystis aeruginosa*, *Microcystis* sp., *Synechococcus* sp.	0/–/+	Leão et al. (2012), Dias et al. (2017)
Planktothrix rubescens (BC 9307, TCC 29–1, TCC 69–6, TCC 69–7)	*Anabaena cylindrica*, *Microcystis flos-aquae*, *Mougeotia gracillima*, *Planktothrix agardhii*	0/–	Schagerl et al. (2002), Oberhaus et al. (2008)
Pseudanabaena sp. (21–9-3)	*Fischerella* sp., *Lyngbya* sp., *Nostoc* sp., *Scytonema* sp.	0	Gantar et al. (2008)
Rivularia biasolettiana	*Nostoc* sp., *Pseudocapsa* sp., *Scytonema* sp.	0	Valdor and Aboal (2007)
Rivularia haematites	*Nostoc* sp., *Pseudocapsa* sp., *Scytonema* sp.	0/–	Valdor and Aboal (2007)
Scytonema myochrous	*Nostoc* sp., *Pseudocapsa* sp., *Scytonema* sp.	0/–	Valdor and Aboal (2007)
Scytonema sp. (26–1)	*Fischerella* sp., *Lyngbya* sp., *Nostoc* sp. *Pseudanabaena* sp.	0/–	Gantar et al. (2008)
Synechococcus elongatus (PCC 7942)	*Anabaena variabilis*	–	Brilisauer et al. (2019)
Synechococcus sp. (CC9311, CCBA AR258, CC9605, CCBA120, CCBA124, CCBA132, WH8102)	*Anabaena* sp., *Aphanizomenon flos-aquae*, *Aphanizomenon* sp., *Aphanocapsa* sp., *Aphanothece* sp., *Chroococcus* sp., *Cyanodictyon* sp., *Gloeocapsa* sp., *Lemmermanniella* sp., *Limnothrix* sp., *Microcystis* sp., *Nodularia spumigena*, *Nostoc* sp., *Phormidium* sp.,	0/–/+	Paz-Yepes et al. (2013), Śliwińska-Wilczewska et al. (2017a, b), Barreiro Felpeto et al. (2018), Bubak et al. (2020), Konarzewska et al. (2020)

(continued)

Table 11.1 (continued)

Donor cyanobacteria	Target cyanobacteria	Effect	References
	Planktolyngbya sp., *Pseudanabaena* sp., *Rivularia* sp., *Snowella* sp., *Synechococcus* sp., *Synechocystis* sp., *Woronichinia* sp.		
Tolypothrix distorta	*Nostoc* sp., *Pseudocapsa* sp., *Scytonema* sp.	0/−	Valdor and Aboal (2007)
Trichormus doliolum	*Anabaena variabilis*, *Anabaena* sp., *Microcystis* sp.	−	von Elert and Jüttner (1997)
Donor cyanobacteria	**Target chlorophyta**	**Effect**	**References**
Anabaena cylindrica (ASW 01035)	*Pandorina morum*, *Pandorina* sp., *Scenedesmus acutus*, *Scenedesmus armatus* var. *maior*	0/−	Schagerl et al. (2002)
Anabena lemmermannii (KAC 16)	*Oocystis* sp., *Planktonema lauterbornii*	0/+	Suikkanen et al. (2005)
Anabaena oscillarioides	*Chlorella* sp.	−	Lam and Silvester (1979)
Anabaena torulosa (ASW 01028, SAG 26.79)	*Pandorina morum*, *Pandorina* sp., *Scenedesmus acutus*, *Scenedesmus armatus* var. *maior*	−	Schagerl et al. (2002)
Anabaena variabilis (29413)	*Chlorella vulgaris*	−	Żak et al. (2012)
Anabaena sp. (J1, J20, J46)	*Ankistrodesmus falcatus*, *Chlorella vulgaris*	0/+/−	Leão et al. (2009b)
Anabaenopsis elenkinii (ASW 01027)	*Pandorina morum*, *Pandorina* sp., *Scenedesmus acutus*, *Scenedesmus armatus* var. *maior*	0	Schagerl et al. (2002)
Aphanizomenon flexuosum (ASW 01033)	*Pandorina morum*, *Pandorina* sp., *Scenedesmus acutus*, *Scenedesmus armatus* var. *maior*	0	Schagerl et al. (2002)
Aphanizomenon flos-aquae (ACT 9605, CCAP 1401/3, J74, Tr183)	*Ankistrodesmus falcatus*, *Chlorella vulgaris*, *Oocystis* sp., *Planktonema lauterbornii*, *Scenedesmus quadricauda*	0/+/−	Suikkanen et al. (2005), Leão et al. (2009b), Żak and Kosakowska (2015), Kovács et al. (2018)
Aphanizomenon issatchenkoi (ACT 9602, J52)	*Ankistrodesmus falcatus*, *Chlorella vulgaris*, *Scenedesmus quadricauda*	−	Leão et al. (2009b), Kovács et al. (2018)

(continued)

Table 11.1 (continued)

Donor cyanobacteria	Target cyanobacteria	Effect	References
Aphanizomenon ovalisporum (APH OVAL)	*Ankistrodesmus falcatus*, *Chlorella vulgaris*	0	Leão et al. (2009b)
Calothrix parietina	*Ankistrodesmus falcatus*, *Chlorella fusca*, *Scenedesmus obliquus*	–	Issa (1999)
Calothrix sp. (CAN 95/2, CAN 95/3, WA 96/8)	*Coelastrum microporum*, *Monoraphidium convolutum*, *Scenedesmus acutus*	0/–	Schlegel et al. (1999)
Cyanobium gracile (ACT 9701)	*Scenedesmus quadricauda*	–	Kovács et al. (2018)
Cylindrospermopsis raciborskii (4799, 4899, ACT 9502, ACT 9505, AQS, CAIA, LEGE 99043, LS117, LS118, LS123, LS124, MARAU)	*Ankistrodesmus falcatus*, *Chlorella vulgaris*, *Coelastrum sphaericum*, *Scenedesmus quadricauda*	0/+/–	Figueredo et al. (2007), Leão et al. (2009b), Antunes et al. (2012), Kovács et al. (2018)
Cylindrospermum sp. (ASW 01016)	*Pandorina morum*, *Pandorina* sp., *Scenedesmus acutus*, *Scenedesmus armatus* var. *maior*	0/–	Schagerl et al. (2002)
Fischerella sp. (52–1, CAN 96/12, CAN 96/13, JAVA 94/20, LOM 95/3, LOM 95/9, LOM 95/17, LOM 96/37, NEP 95/1, NT 97/5, VIET 97/2)	*Ankistrodesmus* sp., *Chlamydomonas* sp., *Chlorella* sp., *Coelastrum microporum*, *Excentrosphaera* sp., *Monoraphidium convolutum*, *Rhizoclonium* sp., *Scenedesmus acutus*, *Selenastrum* sp.	0/–	Schlegel et al. (1999), Gantar et al. (2008)
Geitlerinema splendidum	*Klebsormidium* sp.	0	Valdor and Aboal (2007)
Lyngbya sp. (15–2)	*Ankistrodesmus* sp., *Chlamydomonas* sp., *Chlorella* sp., *Excentrosphaera* sp., *Rhizoclonium* sp., *Selenastrum* sp.	0/–	Gantar et al. (2008)
Microcystis aeruginosa (SAG 1450–1, SAG 14.58, FACHB-905, ASW 01002, FACHB-905, FACHB-469, BCCUSP232)	*Chlorella pyrenoidosa*, *Chlorella vulgaris*, *Monoraphidium convolutum*, *Pandorina morum*, *Pandorina* sp., *Scenedesmus acuminatus*, *Scenedesmus acutus*, *Scenedesmus armatus* var. *maior*, *Scenedesmus quadricauda*	0/+/–	Schagerl et al. (2002), Bittencourt-Oliveira et al. (2015), Ma et al. (2015), Song et al. (2017), Wang et al. (2017), Kovács et al. (2018)

(continued)

Table 11.1 (continued)

Donor cyanobacteria	Target cyanobacteria	Effect	References
Microcystis aeruginosa (AM VIVO 11, M1 23, M6)	*Ankistrodesmus falcatus*, *Chlorella vulgaris*, *Chlorella* sp., *Oocystis marsonii*, *Scenedesmus obliquus*,	0/+/−	Lam and Silvester (1979), Leão et al. (2009b), El-Sheekh et al. (2010), Dunker et al. (2013)
Microcystis flos-aquae (ASW 01004)	*Pandorina morum*, *Pandorina* sp., *Scenedesmus acutus*, *Scenedesmus armatus* var. *maior*	0	Schagerl et al. (2002)
Microcystis panniformis (BCCUSP200)	*Monoraphidium convolutum*, *Scenedesmus acuminatus*	−	Bittencourt-Oliveira et al. (2015)
Nodularia spumigena (ASW 01037, KAC13, ZGNS1)	*Chlorella vulgaris*, *Oocystis* sp., *Planktonema lauterbornii*, *Scenedesmus acutus*, *Scenedesmus armatus* var. *maior*	0/+/−	Schagerl et al. (2002), Suikkanen et al. (2005), Żak et al. (2012)
Nostoc muscorum (ASW 01011)	*Pandorina morum*, *Pandorina* sp., *Scenedesmus acutus*, *Scenedesmus armatus* var. *maior*	0/−	Schagerl et al. (2002)
Nostoc sp. (23-2, 37, 58-2, ASW 01010, ASW 01020, Ev-1, LOM 95/16, LOM 95/18, LOM 96/36, NSW 95/10, WA 96/15, WA 96/19, WA 96/22)	*Ankistrodesmus* sp., *Chlamydomonas* sp., *Chlorella* sp., *Coelastrum microporum*, *Excentrosphaera* sp., *Monoraphidium convolutum*, *Pandorina morum*, *Pandorina* sp., *Rhizoclonium* sp., *Scenedesmus acutus*, *Scenedesmus armatus* var. *maior*, *Selenastrum* sp.	0/−/+	Schlegel et al. (1999), Schagerl et al. (2002), Gantar et al. (2008)
Oscillatoria angustissima	*Ankistrodesmus falcatus*, *Chlorella fusca*, *Scenedesmus obliquus*	−	Issa (1999)
Oscillatoria sp. (4899 OSC, INFOUT OSC V, LEGE 05292, MELAH OSC, OSC AP 1, PCC 6506, SITE BIG 4)	*Ankistrodesmus falcatus*, *Chlorella pyrenoidosa*, *Chlorella vulgaris*, *Chlamydomonas reinhardtii*, *Klebsormidium* sp., *Scenedesmus quadricauda*, *Selenastrum capricornutum*	0/−/+	Bagchi et al. (1993) Valdor and Aboal (2007), Leão et al. (2009b), Barreiro and Vasconcelos (2014), Barreiro Felpeto et al. (2017), Kovács et al. (2018)
Phormidium sp. (EDAH OSCI, EDAH OSCI II)	*Ankistrodesmus falcatus*, *Chlorella vulgaris*, *Klebsormidium* sp.	−/+	Valdor and Aboal (2007), Leão et al. (2009b)

(continued)

Table 11.1 (continued)

Donor cyanobacteria	Target cyanobacteria	Effect	References
Planktothrix agardhii (SCCAP K-0546)	*Chlorella vulgaris*	0/−/+	Żak and Kosakowska (2015)
Planktothrix rubescens (BC 9307)	*Pandorina morum*, *Pandorina* sp., *Scenedesmus acutus*, *Scenedesmus armatus* var. *maior*	0	Schagerl et al. (2002)
Planktothrix sp. (PP)	*Ankistrodesmus falcatus*, *Chlorella vulgaris*	−/+	Leão et al. (2009b)
Pseudanabaena sp. (21–9-3)	*Ankistrodesmus* sp., *Chlamydomonas* sp., *Chlorella* sp., *Excentrosphaera* sp., *Rhizoclonium* sp., *Selenastrum* sp.	0	Gantar et al. (2008)
Rivularia biasolettiana	*Klebsormidium* sp.	0	Valdor and Aboal (2007)
Rivularia haematites	*Klebsormidium* sp.	0	Valdor and Aboal (2007)
Scytonema myochrous	*Klebsormidium* sp.	0	Valdor and Aboal (2007)
Scytonema sp. (26–1)	*Ankistrodesmus* sp., *Chlamydomonas* sp., *Chlorella* sp., *Excentrosphaera* sp., *Rhizoclonium* sp., *Selenastrum* sp.	0/−	Gantar et al. (2008)
Synechococcus sp. (CCBA AR258, CCBA120, CCBA124, CCBA132)	*Ankistrodesmus* sp., *Chlorella fusca*, *Chlorella vulgaris*, *Chlorella* sp., *Coelastrum* sp., *Coenocystis* sp., *Crucigenia* sp., *Crucigeniella* sp., *Cylindrocystis* sp., *Desmodesmus* sp., *Dictyosphaerium* sp., *Kirchneriella obesa*, *Koliella longiseta* cf. *longiseta*, *Monoraphidium convolutum* var. *pseudosabulosum*, *Monoraphidium* sp., *Oocystis submarina*, *Pediastrum* sp., *Phacotus* sp., *Planctonema* sp., *Scenedesmus* sp., *Sphaerocystis* sp., *Stichococcus bacillaris*, *Stichococcus* sp., *Tetraëdron* sp., *Tetrastrum* sp.	0/−/+	Śliwińska-Wilczewska et al. (2017b, 2018), Śliwińska-Wilczewska and Latała (2018), Bubak et al. (2020), Konarzewska et al. (2020)

(continued)

Table 11.1 (continued)

Donor cyanobacteria	Target cyanobacteria	Effect	References
Synechocystis sp. (CCBA MA-01)	*Chlorella vulgaris*	–	Barreiro Felpeto et al. (2019)
Tolypothrix distorta	*Klebsormidium* sp.	0	Valdor and Aboal (2007)
Donor cyanobacteria	**Target charophyta**	**Effect**	**References**
Anabaena cylindrica (ASW 01035)	*Cosmarium* sp., *Staurastrum crenulatum*	0	Schagerl et al. (2002)
Anabaena torulosa (ASW 01028)	*Cosmarium* sp., *Staurastrum crenulatum*	–	Schagerl et al. (2002)
Anabaenopsis elenkinii (ASW 01027)	*Cosmarium* sp., *Staurastrum crenulatum*	0	Schagerl et al. (2002)
Aphanizomenon flexuosum (ASW 01033)	*Cosmarium* sp., *Staurastrum crenulatum*	0/–	Schagerl et al. (2002)
Cylindrospermum sp. (ASW 01016)	*Cosmarium* sp., *Staurastrum crenulatum*	0	Schagerl et al. (2002)
Microcystis aeruginosa (ASW 01002)	*Cosmarium* sp., *Staurastrum crenulatum*	0	Schagerl et al. (2002)
Microcystis flos-aquae (ASW 01004)	*Cosmarium* sp., *Staurastrum crenulatum*	0	Schagerl et al. (2002)
Nostoc muscorum (ASW 01011)	*Cosmarium* sp., *Staurastrum crenulatum*	0/–	Schagerl et al. (2002)
Nostoc sp. (ASW 01010, ASW 01020)	*Cosmarium* sp., *Staurastrum crenulatum*	0/–	Schagerl et al. (2002)
Planktothrix rubescens (BC 9307)	*Cosmarium* sp., *Staurastrum crenulatum*	0	Schagerl et al. (2002)
Synechococcus sp. (CCBA AR258)	*Cosmarium subcostatum*	0/+/–	Bubak et al. (2020)
Donor cyanobacteria	**Target bacillariophyta**	**Effect**	**References**
Anabaena cylindrica (ASW 01035)	*Fragilaria* sp.	0	Schagerl et al. (2002)
Anabena lemmermannii (KAC 16)	*Chaetoceros* sp., *Thalassiosira weissflogii*	0/+/–	Suikkanen et al. (2004, 2005)
Anabaena torulosa (ASW 01028)	*Fragilaria* sp.	–	Schagerl et al. (2002)
Anabaenopsis elenkinii (ASW 01027)	*Fragilaria* sp.	0	Schagerl et al. (2002)
Aphanizomenon flos-aquae (Tr183)	*Chaetoceros* sp., *Thalassiosira weissflogii*	0/–	Suikkanen et al. (2004, 2005)
Aphanizomenon flexuosum (ASW 01033)	*Fragilaria* sp.	0	Schagerl et al. (2002)
Cylindrospermum sp. (ASW 01016)	*Fragilaria* sp.	0	Schagerl et al. (2002)
Microcystis aeruginosa (FACHB-905)	*Cyclotella meneghiniana*	+/–	Wang et al. (2017)
Microcystis aeruginosa (ASW 01002)	*Fragilaria* sp.	0	Schagerl et al. (2002)
Microcystis flos-aquae (ASW 01004)	*Fragilaria* sp.	0	Schagerl et al. (2002)

(continued)

Table 11.1 (continued)

Donor cyanobacteria	Target cyanobacteria	Effect	References
Nodularia spumigena (ASW 01037, KAC13)	*Chaetoceros* sp., *Fragilaria* sp., *Thalassiosira weissflogii*	0/–	Schagerl et al. (2002), Suikkanen et al. (2004, 2005)
Nostoc muscorum (ASW 01011)	*Fragilaria* sp.	0	Schagerl et al. (2002)
Nostoc sp. (ASW 01010, ASW 01020)	*Fragilaria* sp.	–	Schagerl et al. (2002)
Planktothrix rubescens (BC 9307)	*Fragilaria* sp.	0	Schagerl et al. (2002)
Rivularia haematites	*Stephanodiscus minutulus*, *Gomphonema parvulum*, *Fistulifera pelliculosa*, *Nitzschia frustulum*	–	García-Espín et al. (2017)
Rivularia biasolettiana	*Stephanodiscus minutulus*, *Gomphonema parvulum*, *Fistulifera pelliculosa*, *Nitzschia frustulum*	–	García-Espín et al. (2017)
Synechococcus sp. (CCBA AR258, CCBA120, CCBA124, CCBA132)	*Achnanthes* sp., *Amphora coffeaeformis*, *Amphora* sp., *Aulacoseira granulata granulata*, *Bacillaria paxillifer*, *Bacillaria* sp., *Chaetoceros* sp., *Coscinodiscus* sp., *Cyclotella meneghiniana*, *Cyclotella* sp., *Diploneis* sp., *Fistulifera saprophila*, *Fragilaria* sp., *Gomphonema* sp., *Grammatophora* sp., *Navicula perminuta*, *Navicula* sp., *Nitzschia dissipata*, *Nitzschia fonticola*, *Nitzschia* sp., *Odontella* sp., *Pinnularia* sp., *Rhoicosphenia* sp., *Skeletonema marinoi*, *Stauroneis* sp., *Stephanodiscus* sp., *Ulnaria* sp.	0/+/–	Śliwińska-Wilczewska et al. (2016, 2017b, 2018, 2019), Śliwińska-Wilczewska and Latała (2018), Bubak et al. (2020), Konarzewska et al. (2020)
Synechocystis sp. (CCBA MA-01)	*Fistulifera* sp.	–	Barreiro Felpeto et al. (2019)
Donor cyanobacteria	**Target Miozoa**	**Effect**	**References**
Anabena cf. *lemmermannii* (KAC 16)	*Amphidinium* sp., *Dinophysis norvegica*, *Paulsenella* sp.	0/+/–	Suikkanen et al. (2005)
Aphanizomenon flos-aquae (Tr183)	*Amphidinium* sp., *Dinophysis norvegica*, *Paulsenella* sp.	0/+/–	Suikkanen et al. (2005)
Microcystis sp. (MG, MB)	*Peridinium gatunense*	–	Sukenik et al. (2002)

(continued)

Table 11.1 (continued)

Donor cyanobacteria	Target cyanobacteria	Effect	References
Nodularia spumigena (KAC13)	*Amphidinium* sp., *Dinophysis norvegica*, *Paulsenella* sp.	0/+	Suikkanen et al. (2005)
Synechococcus sp. (CCBA124)	*Gymnodinium* sp., *Peridinium* sp.	0	Śliwińska-Wilczewska et al. (2017b)
Donor cyanobacteria	**Target cryptophyta**	**Effect**	**References**
Anabena lemmermannii (KAC 16)	*Rhodomonas* sp.	–	Suikkanen et al. (2004)
Aphanizomenon flos-aquae (Tr183)	*Rhodomonas* sp.	–	Suikkanen et al. (2004, 2006)
Dolichospermum sp. (BIR250A, BIR256, BIR257)	*Rhodomonas nottbecki*	–	Brutemark et al. (2015)
Microcystis aeruginosa (BGSD 243)	*Cryptomonas ovata*	–	Viktoria et al. (2012)
Nodularia spumigena (AV1, KAC13)	*Rhodomonas* sp.	–	Suikkanen et al. (2004, 2006)
Synechococcus sp. (CCBA AR258)	*Plagioselmis* sp.	+	Bubak et al. (2020)
Donor cyanobacteria	**Target haptophyta**	**Effect**	**References**
Anabena lemmermannii (KAC 16)	*Prymnesium parvum*	0	Suikkanen et al. (2004)
Aphanizomenon flos-aquae (Tr183)	*Prymnesium parvum*	0	Suikkanen et al. (2004)
Nodularia spumigena (KAC13)	*Prymnesium parvum*	0	Suikkanen et al. (2004)
Synechococcus sp. (CCBA124)	*Prymnesium parvum*	+	Śliwińska-Wilczewska et al. (2018)
Donor cyanobacteria	**Target ochrophyta**	**Effect**	**References**
Synechococcus sp. (CCBA AR258)	*Dinobryon divergens*	0	Bubak et al. (2020)
Donor cyanobacteria	**Target euglenozoa**	**Effect**	**References**
Synechococcus sp. (CCBA AR258)	*Trachelomonas volvocina*, *Trachelomonas hispida*	0	Bubak et al. (2020)
Donor cyanobacteria	**Target rhodophyta (unicellular)**	**Effect**	**References**
Rivularia haematites	*Chroothece richteriana*	–	García-Espín et al. (2017)
Rivularia biasolettiana	*Chroothece richteriana*	–	García-Espín et al. (2017)
Synechococcus sp. (CCBA124)	*Porphyridium purpureum*	–	Śliwińska-Wilczewska et al. (2018)
Synechocystis sp. (CCBA MA-01)	*Porphyridium purpureum*	–	Barreiro Felpeto et al. (2019)

where: – means inhibiting effect, + means stimulating effect, 0 means lack of effect

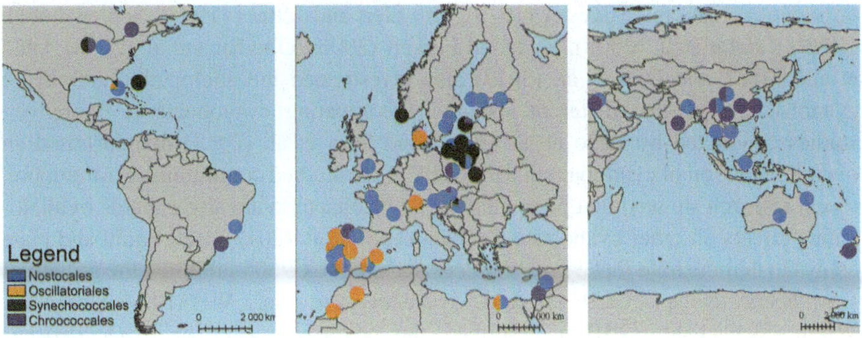

Fig. 11.4 Share of allelopathic cyanobacteria isolated in selected regions found in literature, divided into taxa (ArcMap 10.6.1)

(as studies of bloom formation). Furthermore, the only studies regarding allelopathy in Africa were conducted in Morocco and Egypt. On a global scale, it can be seen that mainly Nostocales are subject to experiments. Much of the research on Synechococcales is carried out by organisms from the Baltic Sea. Oscillatoriales, however, are isolated mostly in Portugal and Chlorococcales in China.

11.3.1 The Allelopathic Interaction Between Cyanobacteria

Literature data indicated that some cyanobacteria could produce the allelopathic compounds that affect the growth of other cyanobacterial species (Table 11.1). Bagchi et al. (1993) showed that *Oscillatoria* sp. produced allelopathic compounds that affected *Microcystis aeruginosa*. Issa (1999) also studied the effects of allelochemicals produced by *Oscillatoria angustissima* and *Calothrix parietina* on cyanobacteria. Based on those results, the author noted that cyanobacteria from *Oscillatoria*, *Calothrix*, *Nostoc*, and *Anabaena* genera were the most insensitive species to the allelopathic compounds produced and released by analyzed donor cyanobacteria. Furthermore, Schagerl et al. (2002) noted that *Anabaena torulosa* and *Nostoc* sp. strongly reduced the growth of *A. cylindrica* and *M. flos-aquae*. Valdor and Aboal (2007) also showed that the extracts obtained from the cyanobacteria *Oscillatoria* sp., *Rivularia biasolettiana*, *Rivularia haematites*, *Geitlerinema splendidum*, *Phormidium* sp., *Tolypothrix distorta*, and *Scytonema myochrous* had an inhibitory effect on the growth of *Nostoc* sp., *Pseudocapsa* sp., and *Scytonema* sp. After 4 days of exposure, the growth of *Scytonema* sp. was inhibited by all cyanobacterial extracts except *R. biasolettiana*. *Pseudocapsa* sp. was hindered by the extract obtained from *R. haematites*, *G. splendidum*, *Phormidium* sp., and *Oscillatoria* sp. Additionally, it was found that *Pseudocapsa* sp. and *Nostoc* sp. were the most allelochemical-sensitive species and their growth was completely inhibited by all cyanobacterial extracts (Valdor and Aboal 2007). Figueredo et al. (2007) and Rzymski et al. (2014) examined that *Cylindrospermopsis raciborskii* decreased the growth and negatively influenced the metabolism of *Microcystis aeruginosa*. Lam and Silvester (1979), von Elert and Jüttner (1997), Schlegel et al. (1999), Gantar et al. (2008), Volk and Furkert (2006), Oberhaus et al. (2008), Leão et al. (2012), and Dias et al. (2017) also investigated the allelopathic activity of cyanobacteria against different species of competing cyanobacteria. Moreover, studies conducted by Leão et al. (2012) and Dias et al. (2017) demonstrated an overall reduction of cyanobacterial diversity of the studied community. Furthermore, recent research showed that picoplanktonic cyanobacteria are also capable of allelopathic effects on other cyanobacteria. Paz-Yepes et al. (2013) used liquid and plate assays to demonstrate that *Synechococcus* sp. strain CC9605 inhibited the growth of *Synechococcus* sp. CC9311 and *Synechococcus* sp. WH8102. Śliwińska-Wilczewska et al. (2017a) described the adverse impact of *Synechococcus* sp. filtrate against *Nostoc* sp. and *Phormidium* sp. Moreover, the authors showed that the addition of picocyanobacterial filtrate stimulated the growth of *A. flos-aquae* and had no allelopathic effects on *Rivularia* sp. Also, Barreiro Felpeto et al. (2018) demonstrated that *Synechococcus* sp. had a strong inhibitory effect on *N. spumigena*, and there was no target organism reciprocal effect. Also, Brilisauer et al. (2019), Bubak et al. (2020), and Konarzewska et al. (2020) showed allelopathic activity of *Synechococcus* sp. on other cyanobacteria species. On the other hand, Suikkanen et al. (2005) showed that the Baltic cyanobacteria *Nodularia spumigena*, *Anabaena* cf. *lemmermannii*, and *Aphanizomenon* sp. had different effects on the natural phytoplankton community, especially on other cyanobacteria. These authors have noted the stimulatory effect of donor organisms on selected species of cyanobacteria, while other microalgae were significantly inhibited. Furthermore, in the study the cell numbers of *Snowella* sp. and *Pseudanabaena* sp. were considerably higher in all experiments with cyanobacterial filtrate addition than in control treatments. It was found that additions of *N. spumigena* filtrate significantly increase the abundance of *N. spumigena* and *Anabaena* sp. Moreover, the filtrate from *Aphanizomenon* sp. results in a 50-fold increase in the number of cells of *Aphanizomenon* sp. itself when compared to filtrate-free culture. Zhang et al. (2016) also showed that coculturing *C. ovalisporum* with *M. panniformis* caused a strong inhibition of *M. panniformis* growth but stimulation of *C. ovalisporum*. Suikkanen et al. (2005) have shown that cyanobacteria can affect the natural phytoplankton community differently, depending on the coexisting species. Śliwińska-Wilczewska et al. (2017b) also indicated that the degree of inhibition was different for each species, causing a change in the phytoplankton abundance and dominance during the time of the experiment. The authors demonstrated that picocyanobacterium *Synechococcus* sp. filtrate generally had an inhibitory effect on all phytoplankton community except the cyanobacteria *N. spumigena* and *Gloeocapsa* sp., in which the number of cells increased in the filtrate treatment.

Nevertheless, it is still not understood precisely why cyanobacteria produce compounds that perform the stimulatory activity. Some researchers believe that cyanobacteria are capable of secreting some autostimulators that accelerate the development of the same species in the environment (Suikkanen et al. 2004, 2005). Moreover, it is commonly known that in laboratory experiments using monocultures, generally, cyanobacteria inhibit the growth of other cyanobacteria; however, in natural assemblies, many co-occurring species could have developed some protective mechanisms against cyanobacterial metabolites and even benefit from them (Śliwińska-Wilczewska et al. 2018). These observations indicate that some groups of organisms may show tolerance for allelopathic compounds, which may be the result of coevolution during their coexistence in the aquatic ecosystem (Suikkanen et al. 2004). It is believed that allelopathy in the aquatic habitats is one of the most competitive cyanobacterial strategies wherein analyzed organisms can affect other cyanobacterial species. Suikkanen et al. (2004) and Figueredo et al. (2007) suggested that the ecological role of allelopathic compounds produced by studied cyanobacteria may cause their dominance in the environment, during their bloom. In conclusion, the allelopathic effects of cyanobacteria might be a pivotal contributor to the formation of monospecific blooms of these organisms in freshwater, marine, and brackish ecosystems (Ma et al. 2015).

11.3.2 The Allelopathic Effect Between Cyanobacteria and Green Algae

Studies have shown that cyanobacteria can inhibit the growth of some green algae species (Table 11.1). Żak et al. (2012) noted that the Baltic cyanobacteria *Anabaena variabilis* and *N. spumigena* showed allelopathic activity on the growth of *Chlorella vulgaris* in both the coculture and cell-free filtrate experiments. It was found that the filtrate obtained from *A. variabilis* inhibited *C. vulgaris*. Moreover, the strong allelopathic effect of *A. variabilis* was observed in cocultures. Furthermore, Żak et al. (2012) noted the allelopathic effect of *N. spumigena* on *C. vulgaris* in cocultures and filtrate additions. It was found that the donor cyanobacterium stimulated the growth of these green algae. The authors observed an adverse effect and lack of allelopathic impact of filtrate on *C. vulgaris* as well. They suggested that different responses of green algae to the filtrate obtained from *N. spumigena* may have been caused by the change of bioactive compounds associated with the concentration of cyanobacteria. Three years later, Żak and Kosakowska (2015) demonstrated that the allelopathic compounds obtained from *Planktothrix agardhii* affected the growth of green algae *C. vulgaris* positively, after the addition of a small amount of filtrate and negatively, with a higher volume of cell-free filtrate added. Wang et al. (2017) noted that exudates of *Microcystis aeruginosa* both inhibited and facilitated the growth of *Scenedesmus quadricauda* and *Chlorella pyrenoidosa* depending on the growth phase of donor cyanobacteria. Ma et al. (2015) demonstrated that the growth of *Chlorella vulgaris*, cocultured with the toxic or nontoxic *Microcystis aeruginosa* strains, was increased and decreased depending on temperature, with a decrease of chlorophyll *a* concentration noted at 15 °C and an increase in higher temperatures (20, 25, 30 °C). Schlegel et al. (1999) studied the allelopathic activity of cyanobacteria *Fischerella* sp., *Nostoc* sp., and *Calothrix* sp. on the growth of selected green algae *Coelastrum microporum, Monoraphidium convolutum*, and *Scenedesmus acutus*. Authors reported that strains of *Fischerella* sp. and *Calothrix* sp. inhibited the growth of all analyzed green algae. Conversely, the allelopathic compounds produced by *Nostoc* sp. did not affect the growth of any tested organism. The different effect of allelopathic compounds may be the indicator of a different level of cell membrane permeability of the analyzed green algae. Issa (1999) reported the allelopathic effect of cyanobacteria *Oscillatoria angustissima* and *Calothrix parietina* on different green algae species (*Ankistrodesmus falcatus, Scenedesmus obliquus, Chlorella fusca*); it was found that, using antibiotic treatments, all analyzed green algae showed a decrease in growth after addition of allelopathic compounds, produced by donor cyanobacteria. Song et al. (2017) described that *Microcystis aeruginosa* inhibited the growth of *Chlorella vulgaris*. Lam and Silvester (1979) also demonstrated that *A. oscillarioides* and *M. aeruginosa* significantly inhibited the growth of *Chlorella* sp. Schagerl et al. (2002) showed that compounds produced by *Nostoc muscorum* strongly inhibited the growth of *Scenedesmus acutus* and *Pandorina morum*. On the other hand, authors noted weak allelopathic effects only of *Aphanizomenon flexuosum* and *Anabaena torulosa* on the tested green algae *Cosmarium* sp., *Scenedesmus armatus* var. *maior* and *P. morum*. Valdor and Aboal (2007) showed that *Klebsormidium* sp. was not susceptible to the extracts of cyanobacteria *Oscillatoria* sp., *Rivularia biasolettiana*, *Geitlerinema splendidum*, and *Phormidium* sp. The allelopathic activity of cyanobacteria on target green algae was also demonstrated by Bagchi et al. (1993), Suikkanen et al. (2005), Figueredo et al. (2007), Leão et al. (2009b), Antunes et al. (2012), Barreiro Felpeto and Vasconcelos (2014), Bittencourt-Oliveira et al. (2015), Barreiro Felpeto et al. (2017) (Table 11.1). Śliwińska-Wilczewska et al. (2018) demonstrated that both the addition of *Synechococcus* sp. cell-free filtrate and coculture inhibited the growth of *Stichococcus bacillaris*. Moreover, Śliwińska-Wilczewska and Latała (2018) noted that *Synechococcus* sp. inhibited also the growth of *C. vulgaris*. Contrary to that, *Synechococcus* sp. filtrate had no allelopathic effects on *O. submarina*. Recently, Kovács et al. (2018) demonstrated that the freshwater picocyanobacterium *Cyanobium gracile* had a substantial negative impact on the coexisting *Scenedesmus quadricauda*. Also, Śliwińska-Wilczewska et al. (2017b), Barreiro Felpeto et al. (2019), Bubak et al. (2020), and Konarzewska et al. (2020) showed allelopathic activity of *Synechococcus* sp. on different green algae (Table 11.1). Those results may indicate that picocyanobacteria are also capable of producing more than one bioactive compound that affects different processes in target organisms, as well as different sensitivity of the target organisms to secreted allelochemicals. The allelopathic activity of cyanobacteria against green algae could also constitute an interesting concept in terms of evolution. Allelopathic effects recognized in cyanobacteria may play an important role in deterrence target organisms from colonization of cyanobacteria filaments (Gantar et al. 2008). Moreover, Schlegel et al. (1999) suggested that allelopathic compounds secreted by cyanobacteria may be responsible for the natural selection and their ecological succession by inhibiting co-occurring competitive green algae species.

11.3.3 The Allelopathic Effect Between Cyanobacteria and Diatoms

Diatoms seem to be highly sensitive to allelopathic compounds. Table 11.1 presents some studies that documented the allelopathic effect of cyanobacteria on selected diatoms. Keating (1977) demonstrated for the first time, the inhibition of diatoms growth by the addition of filtrate from the Linsley Pond (North Branford, United States) where cyanobacteria dominated. First detailed data on the allelopathic effects of cyanobacteria isolated from the Baltic Sea on diatoms were provided by Suikkanen et al. (2004). In that study, the allelopathic effects of *N. spumigena*, *Anabena lemmermannii*, and *Aphanizomenon flos-aquae* on the diatom *Thalassiosira weissflogii* were demonstrated. It was the first observation of allelopathic properties of *N. spumigena* in the Baltic Sea. Moreover, *A. flos-aquae* caused a 57% decrease in *T. weissflogii* abundance during the first day, but later on, the cells were able to grow again. According to that research, diatoms demonstrated some tolerance for a single filtrate addition, but their growth was significantly inhibited when the filtrate addition was repeated. Wang et al. (2017) reported that the filtrate of *Microcystis aeruginosa* from the exponential growth phase and the stationary phase significantly inhibited the growth of *Cyclotella meneghiniana*, whereas the filtrate from the decline phase increased the diatom growth. Schagerl et al. (2002) and Barreiro Felpeto et al. (2019) described the allelopathic activity of cyanobacterium on *Fistulifera* sp. García-Espín et al. (2017) showed an inhibitory effect of *Rivularia haematites* and *Rivularia biasolettiana* on selected diatoms. Śliwińska-Wilczewska et al. (2016) showed that the picocyanobacterium *Synechococcus* sp. affected coexisting diatom *N. perminuta* negatively and it was the first such report in the literature. One year later, Śliwińska-Wilczewska et al. (2017b) examined the influence of allelopathic compounds on the growth, total abundance, and composition of the phytoplankton community by adding the cell-free filtrate of *Synechococcus* sp. into the medium. That study pointed to the diatoms of the genera *Navicula*, *Chaetoceros*, *Amphora*, *Coscinodiscus*, *Grammatophora*, and *Nitzschia* as the most allelochemicals-sensitive organisms. Moreover, Śliwińska-Wilczewska and Latała (2018) and Śliwińska-Wilczewska et al. (2019) demonstrated that the addition of *Synechococcus* sp. filtrate inhibited the growth of *Skeletonema marinoi* and *Bacillaria paxillifer*. Konarzewska et al. (2020) also demonstrated that three different *Synechococcus* sp. phenotypes had a significant allelopathic effect on the selected species of diatoms (*Cyclotella meneghiniana*, *Amphora coffeaeformis*, *Navicula perminuta*, *Nitzschia fonticola*, *Fistulifera saprophila*, *Skeletonema marinoi*). In contrast, Śliwińska-Wilczewska et al. (2018) showed that *N. dissipata* was not affected by the picocyanobacterial filtrate or coculture. Furthermore, Suikkanen et al. (2005) and Bubak et al. (2020) described that the cyanobacteria showed a diversified allelopathic effect on diatoms residing in the natural phytoplankton assemblages. The susceptibility of target diatoms to allelochemicals may depend on the nature of allelopathic compounds to which they are exposed, because the same target organisms may respond differently to the filtrate obtained from different donor organisms (Konarzewska et al. 2020). Additionally, some coevolutionary aspects may contribute to the observed results (Suikkanen et al. 2004). In the natural environment, diatoms generally do not have the opportunity to develop any defense mechanism for the allelopathic compounds secreted by cyanobacteria due to the lack of long-term interactions in natural settings. That could be a reason for the cyanobacterial allelopathic effect-driven significant inhibition of diatoms growth (Suikkanen et al. 2004).

11.3.4 The Allelopathic Effect Between Cyanobacteria and Other Microalgae

Literature reports are indicating that cyanobacteria can produce and release secondary metabolites that affect the growth of other microalgae species belonging to Charophyta, Miozoa, Cryptophyta, Euglenozoa, Ochrophyta, Haptophyta, and the unicellular Rhodophyta phyla (Table 11.1). Schagerl et al. (2002) and Bubak et al. (2020) described that cyanobacteria may affect the growth of *Cosmarium* sp. and *Staurastrum* sp. (Charophyta). In turn, the influence of cyanobacteria on Miozoa was investigated by Suikkanen et al. (2005) and Śliwińska-Wilczewska et al. (2017b). Other studies have shown that cyanobacteria, in general, have a negative effect on single-celled Rhodophyta (García-Espín et al. 2017; Śliwińska-Wilczewska et al. 2018; Barreiro Felpeto et al. 2019), as well as Cryptophyta (Suikkanen et al. 2004, 2006; Brutemark et al. 2015) (Table 11.1). Only Bubak et al. (2020) showed the stimulating effect of cyanobacteria on the Cryptophyta *Plagioselmis* sp. Moreover, cyanobacteria have been shown to have no effect (Suikkanen et al. 2004), or even stimulate (Śliwińska-Wilczewska et al. 2018) the growth of *Prymnesium parvum* (Haptopyta). Moreover, Bubak et al. (2020) examined that the freshwater picocyanobacterium *Synechococcus* sp. showed no allelopathic effect on *Dinobryon divergens* (Ochrophyta) as well as *Trachelomonas volvocina* and *Trachelomonas hispida* (Euglenozoa). However, to fully understand the allelopathic effects in aquatic environments, which may depend on the specificity of the donor and target group, studies on many different phytoplankton species are still needed to be performed.

11.4 Factors Affecting Cyanobacterial Allelopathy and Modes of Action of Cyanobacterial Allelochemicals

As it has been already pronounced, allelopathic compounds produced by cyanobacteria can affect the surrounding ecosystem and cause a variety of responses of the functioning of target organisms. However, it should be noted that the different mechanism of interaction, i.e., abiotic and biotic factors affecting allelopathy phenomenon, is not fully recognized. Despite that, Granéli and Hansen (2006) and Reigosa et al. (2006) strongly suggest that abiotic and biotic factors may affect the production of allelopathic compounds directly. Furthermore, Tang et al. (1995) and Reigosa et al. (1999) claim the same factors may also affect the sensitivity of the target organism to the allelopathic compounds consequently strengthening or weakening the allelopathy phenomenon. Thus, environmental factors that increase the intensity of allelopathic effects may also alter the ratio between the abundance of donor and target organisms that occur in the same aquatic ecosystem. Some selected interactions between the environment and the allelopathy phenomenon that were already observed and reported have been gathered and described below.

It was noted that light intensity is an indispensable factor in the production of secondary metabolites by some species of cyanobacteria (Antunes et al. 2012; Śliwińska-Wilczewska et al. 2016; Barreiro Felpeto et al. 2018). In their conclusions, the authors demonstrated that variation in light intensity should be considered while estimating the allelopathic effects of bloom-forming cyanobacteria in aquatic environments; the effect should be considered for both issues: the influence of the environment on the allelopathic compounds production and the influence of the environment on the allelopathic impact of one organism on another (Barreiro Felpeto et al. 2018). These few observations indicate a possible significant effect of various light ntensities on the production of allelopathic compounds, which may be especially crucial for planktonic species.

The production of secondary metabolites by donor cyanobacteria may also depend on temperature. The increase in the allelopathic activity of different species of cyanobacteria under high temperature was reported by Antunes et al. (2012), Ma et al. (2015), Śliwińska-Wilczewska et al. (2016), and Barreiro Felpeto et al. (2019). Moreover, more studies proved that low temperature caused damage to photosynthetic activity due to photosystem II (PSII) degradation (Sakamoto and Bryant 1999). Thus, the predicted increase in temperature caused by the greenhouse effect may increase the production of allelochemicals which may encourage the formation of massive cyanobacterial blooms (O'Neil et al. 2012; Ma et al. 2015).

Some reports indicate that the concentration of nutrients may have a significant effect on the allelopathic activity of different cyanobacteria species (Barreiro Felpeto and Hairston Jr 2013; Śliwińska-Wilczewska and Latała 2018). Eutrophication is a leading threat for many freshwaters and coastal marine ecosystems in the world. Therefore, determining the influence of nutrients availability on the allelopathic activity should be a core issue for marine and freshwater research (Thornton et al. 2013). There are few studies that describe some cyanobacteria allelopathic effects in nutrient excess conditions (von Elert and Jüttner 1997; Antunes et al. 2012; Barreiro Felpeto et al. 2017; Śliwińska-Wilczewska and Latała 2018). It was demonstrated that nutrients sufficiency level influenced the picocyanobacterium *Synechococcus* sp. functioning by affecting its production of allelochemicals, i.e., it increased this production rate (Śliwińska-Wilczewska and Latała 2018). On the other hand, the cyanobacterium *C. raciborskii* uniquely exhibits the greatest allelopathic activity under phosphorus deficiency, which coincides with optimal conditions for its growth (Antunes et al. 2012). This relationship may explain that allelopathic activity may be the strongest under the optimal growth environmental factors for some cyanobacteria and microalgae. It has also been reported that nutrient deficiency caused an increase of allelopathic activity of the donor organisms. It was found that a single extracellular extract obtained from cyanobacterium *Trichormus doliolum* cultures that were grown under phosphate limitation, strongly inhibited the growth of target *Anabaena variabilis* (von Elert and Jüttner 1997). Furthermore, the same studies noted that *T. doliolum* that was cultured in phosphorus deficiency produced and released about 200 times more allelopathic compounds per biomass unit than in cultures with phosphorus surplus. In a more recent study, Barreiro Felpeto et al. (2017) examined the allelopathic interaction between *Ankistrodesmus falcatus* and *Oscillatoria* sp., which competed for nitrate as a single limiting nutrient. The authors performed long-term competition experiments in nitrate-limited continuous cultures and confirmed that allelopathy can alter the predicted outcome of interspecific competition in a nutrient-limited environment. These results suggest that the availability of nutrients is an essential factor in the regulation of allelopathic compounds production. These observations may also indicate that physiological stress, caused by nutrient limitation, may result in an increase in productivity of secondary metabolites for some donor cyanobacteria.

Salinity is another abiotic factor that can influence the production of secondary metabolites by some species of cyanobacteria. However, according to the author's best knowledge, only two studies indicate that salinity in the Baltic Sea region can be an important factor affecting cyanobacterial allelopathy (Brutemark et al. 2015; Śliwińska-Wilczewska et al. 2016). The ability to produce allelopathic compounds by some cyanobacteria can give them an advantage over other phytoplankton species especially in waters with variable salinity.

An essential biotic factor influencing allelopathic interactions is the growth phase of the donor organism. Suikkanen et al. (2004) found that the exudates collected during the exponential growth phase from the *Nodularia spumigena* culture had a negative allelopathic effect on *Thalassiosira weissflogii* and *Rhodomonas* sp. while the stationary growth phase did not show any significant impact on the same examined species. Similarly, the *Synechococcus* sp. and *N. spumigena* were allelopathic only in the exponential growth phase whereas the filtrate from the stationary phase did not have an adverse allelopathic effect on *B. paxillifer*, *C. vulgaris*, *O. submarina*, and *S. marinoi* (Śliwińska-Wilczewska et al. 2019). Moreover, further studies concluded that cultures in the exponential growth phase would exhibit greater allelopathic effects than cultures in the stationary phase (Schmidt and Hansen 2001; Antunes et al. 2012).

In the natural environment, the functional principle of allelopathic compounds is highly diverse and donor species can affect target organisms in many different ways. The mode of action of allelopathic compounds depends on the nature of the interaction between donor and target organisms and the allelopathic compounds activity itself (Barreiro Felpeto et al. 2018). Cyanobacterial secondary metabolites contain different active compounds that have an allelopathic effect on coexisting species. Although allelochemicals sometimes have a stimulatory effect, most of the studies mainly demonstrated the inhibitory activity of cyanobacteria on target organisms (Table 11.1). Inhibition of growth and high mortality of the target organism as a consequence of a negative influence of allelopathic compounds are widespread in previous studies; this is the most frequently described mode of action of donor cyanobacteria (Suikkanen et al. 2004, 2005, 2006; Antunes et al. 2012; Żak and Kosakowska 2015; Wang et al. 2017; Barreiro Felpeto et al. 2018; Śliwińska-Wilczewska et al. 2018, 2019; Bubak et al. 2020; Konarzewska et al. 2020; Table 11.1).

Allelopathic compounds secreted by cyanobacteria can also inhibit fluorescence and photosynthesis process (Figueredo et al. 2007; Gantar et al. 2008; Śliwińska-Wilczewska et al. 2016, 2017a, 2018, 2019; Barreiro Felpeto et al. 2019; Konarzewska et al. 2020), reduce pigment content (Suikkanen et al. 2006; Antunes et al. 2012; Śliwińska-Wilczewska et al. 2017a; Barreiro Felpeto et al. 2018; Konarzewska et al. 2020), and negatively affect cell morphology of target species (Valdor and Aboal 2007; Gantar et al. 2008; Barreiro Felpeto et al. 2018; Śliwińska-Wilczewska et al. 2019). Cyanobacteria and microalgae produce bioactive substances that can inhibit the photosynthesis of coexisting phytoplankton species. It was noted that the photosynthesis of target microalgae was significantly inhibited after exposure to the cyanobacterial filtrate (Issa 1999; Śliwińska-Wilczewska et al. 2019). The author concluded that the inhibition of the photosynthesis process may result in lower primary production, and consequently in reducing the growth rates of coexisting species (Figueredo et al. 2007). Chlorophyll fluorescence measurements can also point to changes in PSII activity. The results obtained from these measurements may indicate the condition of the tested organism (Suresh Kumar et al. 2014; Machado et al. 2015) because chlorophyll fluorescence is a measure of the effectiveness of the photosynthetic apparatus (Qian et al. 2010). This is also a highly sensitive method for examining cyanobacterial and algal responses to stress (Suresh Kumar et al. 2014; Machado et al. 2015; Song et al. 2017; Śliwińska-Wilczewska et al. 2017a, 2018; Konarzewska et al. 2020). Different photochemical parameters reflect the physiological condition of an organism but the maximum quantum yield of PSII (F_v/F_m) is correlated with photosynthesis and plant viability the most (Kolber et al. 1988). The high values of the F_v/F_m indicate the relatively high potential of PSII in tested organisms. On the other hand, low levels of these values stand for some disturbance in the photosynthesis process (Suresh Kumar et al. 2014). Sukenik et al. (2002) investigated the allelopathic effects of cyanobacterium *Microcystis* sp. on *Peridinium* sp. using the PAM method; that was one of the first researches of that kind. Due to the sensitivity of the method, it is believed that the measurement of PSII performance may provide for allelopathic effect observations when target organisms are the least susceptible to allelopathic compounds. The reduction of pigmentation as well as the destruction of cell membranes is another possible mode of action of allelopathic compounds. These alterations would contribute to physiological changes, as suggested by Sedmak and Elersek (2005). However, in most cases, the allelopathy mechanism of action is still insufficiently understood. Thus, more research is needed to characterize and understand the impact of allelochemicals modes of action on target organisms in an improved and complete manner.

11.5 Conclusions

Allelopathic interactions between cyanobacteria and other phytoplankton species by secretion of secondary metabolites play a significant role in aquatic environments. Production of active allelopathic compounds is an important adaptation performed by some cyanobacteria to get a competitive advantage over the other primary producers. It is proven that changes in the composition and structure of phytoplankton due to the variety of cyanobacterial allelopathic compounds affecting different target organisms are apparent in the environment. Some studies showed that target organisms can be completely eliminated, inhibited, and sometimes even stimulated by allelopathic compounds secreted by donor's cyanobacteria to the adjacent environment. It is believed that the selective stimulation or growth inhibition of individual species may contribute to the succession of some cyanobacteria in the water bodies. That is why for a decade allelopathic effects among cyanobacteria in aquatic ecosystems have been intensively studied.

Various environmental factors influence the allelopathic effects of cyanobacteria. However, the impact of abiotic and biotic factors on allelopathy is still not well understood. Based on already carried studies, the main abiotic factors influencing the allelopathy phenomenon of cyanobacteria are light intensity, temperature, availability of nutrients (especially nitrates and phosphates), and salinity. Moreover, the phase of growth of analyzed organisms is claimed to be a significant biotic factor influencing cyanobacterial allelopathy in aquatic ecosystems. Allelopathic compounds are released into the environment, which affects the growth of target organisms. Cyanobacterial allelochemicals may also reduce pigment content and inhibit the photosynthesis process as well as negatively affect the cell morphology of target species. Therefore, it is crucial to examine allelopathic interactions in controlled laboratory conditions to investigate the nature of released substances and their effect on target organisms.

Acknowledgments This work has been funded by the Polish National Science Centre project (contract no. 2019/33/N/ST10/00585). This study was also supported by BMN grant, Poland, No. 539-O160-B432-20.

References

Akehurst SC (1931) Observations on pond life, with special reference to the possible causation of swarming of phytoplankton. J R Microsc Soc 9:1–48

Antunes JT, Leão PN, Vasconcelos VM (2012) Influence of biotic and abiotic factors on the allelopathic activity of the cyanobacterium *Cylindrospermopsis raciborskii* strain LEGE 99043. Microb Ecol 64:584–592

Bagchi SN, Chauhan VS, Marwaii JB (1993) Effect of an antibiotic from *Oscillatoria late-virens* on growth, photosynthesis, and toxicity of *Microcystis aeruginosa*. Curr Microbiol 26:223–228

Barreiro Felpeto A, Hairston NG Jr (2013) The influence of resource limitation on the allelopathic effect of *Chlamydomonas reinhardtii* on other unicellular freshwater planktonic organisms. J Plankton Res 35:1339–1344

Barreiro Felpeto A, Vasconcelos VM (2014) Interactions between allelopathic properties and growth kynetics in four freshwater phytoplankton species studied by model simulations. Aquat Ecol 48:191–205

Barreiro Felpeto A, Roy S, Vasconcelos VM (2017) Allelopathy prevents competitive exclusion and promotes phytoplankton biodiversity. Oikos 127:85–98

Barreiro Felpeto A, Śliwińska-Wilczewska S, Złoch I, Vasconcelos V (2018) Light-dependent cytolysis in the allelopathic interaction between picoplanktic and filamentous cyanobacteria. J Plankton Res 40:165–177

Barreiro Felpeto A, Śliwińska-Wilczewska S, Klin M, Konarzewska Z, Vasconcelos V (2019) Temperature-dependent impacts of allelopathy on growth, pigment, and lipid content between a subpolar strain of *Synechocystis* sp. CCBA MA-01 and coexisting microalgae. Hydrobiologia 835(1):117–128

Bittencourt-Oliveira M, Chia MA, de Oliveira HSB, Araújo MKC, Molica RJR, Dias CTS (2015) Allelopathic interactions between microcystin-producing and non-microcystin-producing cyanobacteria and green microalgae: implications for microcystins production. J Appl Phycol 27:275–284

Brilisauer K, Rapp J, Rath P, Schöllhorn A, Bleul L, Weiß E, Stahl M, Grond S, Forchhammer K (2019) Cyanobacterial antimetabolite 7-deoxy-sedoheptulose blocks the shikimate pathway to inhibit the growth of prototrophic organisms. Nat Commun 10:545

Brutemark A, Vandelannoote A, Engström-Öst J, Suikkanen S (2015) A less saline Baltic Sea promotes cyanobacterial growth, hampers intracellular microcystin production, and leads to strain-specific differences in allelopathy. PLoS One 10:e0128904

Bubak I, Śliwińska-Wilczewska S, Głowacka P, Szczerba A, Możdżeń K (2020) The importance of Allelopathic Picocyanobacterium *Synechococcus* sp. on the abundance, biomass formation, and structure of phytoplankton assemblages in three Freshwater Lakes. Toxins 12:–259

Chan AT, Andersen RJ, Le Blanc MJ, Harrison PJ (1980) Algal plating as a tool for investigating allelopathy among marine microalgae. Mar Biol 59:7–13

Corcoran AA, Seger M, Niu R, Nirmalakhandan N, Lammers PJ, Holguin FO, Boeing WJ (2019) Evidence for induced allelopathy in an isolate of *Coelastrella* following co-culture with *Chlorella sorokiniana*. Algal Res 41:101535

Dakshini KMM (1994) Algal allelopathy. Bot Rev 60:182–196

Dias F, Antunes JT, Ribeiro T, Azevedo J, Vasconcelos V, Leão PN (2017) Cyanobacterial allelochemicals but not cyanobacterial cells markedly reduce microbial community diversity. Front Microbiol 8:1495

Dunker S, Jakob T, Wilhelm C (2013) Contrasting effects of the cyanobacterium Microcystis aeruginosa on the growth and physiology of two green algae, *Oocystis marsonii* and *Scenedesmus obliquus*, revealed by flow cytometry. Freshw Biol 58:1573–1587

El-Sheekh MM, Khairy HM, El-Shenody RA (2010) Allelopathic effects of cyanobacterium *Microcystis aeruginosa* Kützing on the growth and photosynthetic pigments of some algal species. Allelopathy J 26(2):275–290

Figueredo CC, Giani A, Bird DF (2007) Does allelopathy contribute to *Cylindrospermopsis raciborskii* (cyanobacteria) bloom occurrence and geographic expansion? J Phycol 43:256–265

Gantar M, Berry JP, Thomas S, Wang M, Perez R, Rein KS, King G (2008) Allelopathic activity among cyanobacteria and microalgae isolated from Florida freshwater habitats. FEMS Microbiol Lett 64:55–64

García-Espín L, Cantoral EA, Asencio AD, Aboal M (2017) Microcystins and cyanophyte extracts inhibit or promote the photosynthesis of fluvial algae. Ecological and management implications. Ecotoxicology 26:658–666

Gopal B, Goel U (1993) Competition and allelopathy in aquatic plant communities. Bot Rev 59:155–210

Granéli E, Hansen PJ (2006) Allelopathy in harmful algae: a mechanism to compete for resources? In: Granéli E, Turner J (eds) Ecology of harmful algae, series: ecological studies 189. Springer, Heidelberg, pp 189–201

Granéli E, Turner JT (eds) (2006) Ecology of harmful algae, ecological studies, vol 189. Springer, Heidelberg

Granéli E, Salomon PS, Fistarol GO (2008) The role of allelopathy for harmful algal bloom formation. In: Evangelista V, Barsanti L, Frassanito A, Passarelli V, Gualtieri P (eds) Algal toxins: nature, occurrence, effect and detection, NATO Science for Peace and Security Series A: Chemistry and Biology. Springer, Netherlands, pp 159–178

Gross EM (2003) Allelopathy of aquatic autotrophs. Crit Rev Plant Sci 22:313–339

Inderjit KM, Dakshini M (1994) Algal allelopathy. Bot Rev 60:182–197

Issa AA (1999) Antibiotic production by the cyanobacteria *Oscillatoria angustissima* and *Calothrix parietina*. Environ Toxicol Pharmacol 8:33–37

Keating KI (1977) Allelopathic influence on blue-green bloom sequence in a Eutrophic Lake. Science 196:885–887

Kolber ZS, Zehr J, Falkowski PG (1988) Effects of growth irradiance and nitrogen limitation on photosynthetic energy conversion in photosystem II. Plant Physiol 88:72–79

Konarzewska Z, Śliwińska-Wilczewska S, Felpeto AB, Vasconcelos V, Latała A (2020) Assessment of the allelochemical activity and biochemical profile of different phenotypes of picocyanobacteria from the genus *Synechococcus*. Mar Drugs 18:179

Kovács AW, Tóth VR, Pálffy K (2018) The effects of interspecific interactions between bloom forming cyanobacteria and *Scenedesmus quadricauda* (chlorophyta) on their photophysiology. Acta Biol Hung 69:210–223

Kumar KS, Dahms HU, Lee JS, Kim HC, Lee WC, Shin KH (2014) Algal photosynthetic responses to toxic metals and herbicides assessed by chlorophyll a fluorescence. Ecotoxicol Environ Safety 104:51–71

Lam CWY, Silvester WB (1979) Growth interactions among blue-green (*Anabaena oscillarioides, Microcystis aeruginosa*) and green (*Chlorella* sp.) algae. Hydrobiologia 63:135–143

Leão PN, Vasconcelos MTS, Vasconcelos VM (2009a) Allelopathy in freshwater cyanobacteria. Crit Rev Microbiol 35:271–282

Leão PN, Vasconcelos MTS, Vasconcelos VM (2009b) Allelopathic activity of cyanobacteria on green microalgae at low cell densities. Eur J Phycol 44:347–355

Leão PN, Engene N, Antunes A, Gerwick WH, Vasconcelos V (2012) The chemical ecology of cyanobacteria. Nat Prod Rep 29:372–391

Leflaive J, Ten-Hage L (2007) Algal and cyanobacterial secondary metabolites in freshwaters: a comparison of allelopathic compounds and toxins. Freshw Biol 52:199–214

Legrand C, Rengefors K, Fistarol GO, Granéli E (2003) Allelopathy in phytoplankton – biochemical, ecological and evolutionary aspects. Phycologia 42:406–419

Lewis WM Jr (1986) Evolutionary interpretation of allelochemical interactions in phytoplankton algae. Am Nat 127:184–194

Ma ZL, Fang TX, Thring RW, Li YB, Yu HG, Zhou Q, Zhao M (2015) Toxic and non-toxic strains of *Microcystis aeruginosa* induce temperature dependent allelopathy toward growth and photosynthesis of *Chlorella vulgaris*. Harmful Algae 48:21–29

Machado MD, Lopes AR, Soares EV (2015) Responses of the alga *Pseudokirchneriella subcapitata* to long-term exposure to metal stress. J Hazard Mater 296:82–92

O'Neil JM, Davis TW, Burford MA, Gobler CJ (2012) The rise of harmful cyanobacteria blooms: the potential roles of eutrophication and climate change. Harmful Algae 14:313–334

Oberhaus L, Briand JF, Humbert JF (2008) Allelopathic growth inhibition by the toxic, bloom-forming cyanobacterium *Planktothrix rubescens*. FEMS Microbiol Ecol 66:243–249

Paz-Yepes J, Brahamsha B, Palenik B (2013) Role of a Microcin-C–like biosynthetic gene cluster in allelopathic interactions in marine *Synechococcus*. Proc Natl Acad Sci 110(29):12030–12035

Poulin RX, Poulson-Ellestad KL, Roy JS, Kubanek J (2018a) Variable allelopathy among phytoplankton reflected in red tide metabolome. Harmful Algae 71:50–56

Poulin RX, Hogan S, Poulson-Ellestad KL, Brown E, Fernández FM, Kubanek J (2018b) Karenia brevis allelopathy compromises the lipidome, membrane integrity, and photosynthesis of competitors. Sci Rep 8:9572

Qian HF, Yu SQ, Sun ZQ, Xie XC, Liu WP, Fu ZW (2010) Effects of copper sulfate, hydrogen peroxide and N-phenyl-2-naphthylamine on oxidative stress and the expression of genes involved photosynthesis and microcystin disposition in *Microcystis aeruginosa*. Aquat Toxicol 99:405–412

Reigosa MJ, Sanchez-Moreiras A, Gonzalez L (1999) Ecophysiological approach in allelopathy. Crit Rev Plant Sci 18:577–608

Reigosa MJ, Pedrol N, González L (2006) Allelopathy: a physiological process with ecological implications. In: Reigosa MJ, Pedrol N, González L (eds) Springer science & business media. Netherlands, Dordrecht, p 637

Rice EL (1979) Allelopathy-an update. Bot Rev 45:15–109

Rizvi SJH, Rizvi V (1992) Allelopathy: basic and applied aspects. In: Rizvi SJH, Rizvi V (eds) Chapman and hall. UK, London, p 480

Rizvi SJH, Haque H, Singh VK, Rizvi V (1992) A discipline called allelopathy. In: Allelopathy. Springer, Dordrecht, pp 1–10

Rzymski P, Poniedziałek B, Kokociński M, Jurczak T, Lipski D, Wiktorowicz K (2014) Interspecific allelopathy in cyanobacteria: Cylindrospermopsin and *Cylindrospermopsis raciborskii* effect on the growth and metabolism of *Microcystis aeruginosa*. Harmful Algae 35:1–8

Sakamoto T, Bryant DA (1999) Nitrate transport and not photoinhibition limits growth the freshwater cyanobacterium *Synechococcus* species PCC 6301 at low temperature. Plant Physiol 119:785–794

Sarkar RR, Petrovskii SV, Biswas M, Gupta A, Chattopadhyay J (2006) An ecological study of a marine plankton community based on the field data collected from Bay of Bengal. Ecol Model 193:589–601

Schagerl M, Unterrieder I, Angeler DG (2002) Allelopathy among cyanoprokaryota and other algae originating from Lake Neusiedlersee (Austria). Int Rev Hydrobiol 87:365–374

Schlegel I, Doan NT, de Chazal N, Smith GD (1999) Antibiotic activity of new cyanobacterial isolates from Australia and Asia against green algae and cyanobacteria. J Appl Phycol 10:471–479

Schmidt LE, Hansen PJ (2001) Allelopathy in the prymnesiophyte *Chrysochromulina polylepis*: effect of cell concentration, growth phase and pH. Mar Ecol Prog Ser 216:67–81

Sedmak B, Elersek T (2005) Microcystins indice morphological and physiological changes in selected representative phytoplanktons. Microb Ecol 50:298–305

Śliwińska-Wilczewska S, Latała A (2018) Allelopathic activity of the bloom-forming picocyanobacterium *Synechococcus* sp. on the coexisting microalgae: the role of eutrophication. Int Rev Hydrobiol 103:37–47

Śliwińska-Wilczewska S, Pniewski F, Latała A (2016) Allelopathic activity of the picocyanobacterium *Synechococcus* sp. under varied light, temperature and salinity conditions. Int Rev Hydrobiol 101:69–77

Śliwińska-Wilczewska S, Maculewicz J, Barreiro Felpeto A, Vasconcelos V, Latała A (2017a) Allelopathic activity of the picocyanobacterium *Synechococcus* sp. on filamentous cyanobacteria. J Exp Mar Bio Ecol 496:16–21

Śliwińska-Wilczewska S, Maculewicz J, Tuszer J, Dobosz K, Kalusa D, Latała A (2017b) First record of allelopathic activity of the picocyanobacterium *Synechococcus* sp. on a natural plankton community. Ecohydrol Hydrobiol 17:227–234

Śliwińska-Wilczewska S, Barreiro Felpeto A, Maculewicz J, Sobczyk A, Vasconcelos V, Latała A (2018) Allelopathic activity of the picocyanobacterium *Synechococcus* sp. on unicellular eukaryote planktonic microalgae. Mar. Freshwater Res 69:1472–1479

Śliwińska-Wilczewska S, Barreiro Felpeto A, Możdżeń K, Vasconcelos V, Latała A (2019) Physiological effects on coexisting microalgae of the allelochemicals produced by the bloom-forming cyanobacteria *Synechococcus* sp. and *Nodularia spumigena*. Toxins 11(12):712

Song H, Lavoie M, Fan X, Tan H, Liu G, Xu P, Fu Z, Paerl HW, Qian H (2017) Allelopathic interactions of linoleic acid and nitric oxide increase the competitive ability of *Microcystis aeruginosa*. ISME J 11:1865–1876

Suikkanen S, Fistarol GO, Granéli E (2004) Allelopathic effects of the Baltic cyanobacteria *Nodularia spumigena*, *Aphanizomenon flos-aquae* and *Anabaena lemmermannii* on algal monocultures. J Exp Mar Bio Ecol 308:85–101

Suikkanen S, Fistarol GO, Granéli E (2005) Effects of cyanobacterial allelochemicals on a natural plankton community. Mar Ecol Prog Ser 287:1–9

Suikkanen S, Engström-Öst J, Jokela J, Sivonen K, Viitasalo M (2006) Allelopathy of Baltic Sea cyanobacteria: no evidence for the role of nodularin. J Plankton Res 28:543–550

Sukenik A, Eskhol R, Livne A, Hadas O, Rom M, Tchernov D, Vardi A, Kaplan A (2002) Inhibition of growth and photosynthesis of the dinoflagellate *Peridinium gatunense* by *Microcystis* sp. (cyanobacteria): a novel Allelopathic mechanism. Limnol Oceanogr 47:1656–1663

Tang CS, Cai WF, Kohl K, Nishimoto RK (1995) Plant stress and allelopathy. In: Inderjit KM, Dakshini M, Einhellig FA (eds) Allelopathy: organisms, processes and applications, vol 582. American Chemical Society, Ames, IA, pp 142–157

Thornton JA, Harding WR, Dent M, Hart RC, Lin H, Rast CL, Sven-Olof R, Slawski TM (2013) Eutrophication as a 'wicked'problem. Lakes Reserv Res Manag 18:298–316

Valdor R, Aboal M (2007) Effects of living cyanobacteria, cyanobacterial extracts and pure microcystins on growth and ultrastructure of microalgae and bacteria. Toxicon 49:769–779

Viktoria B, Grigorszky I, Vasas G, Borics G, Várbíró G, Nagy SA, Borbèly G, Bácsi I (2012) The effects of *Microcystis aeruginosa* (cyanobacterium) on *Cryptomonas ovata* (Cryptophyta) in laboratory cultures: why these organisms do not coexist in steady-state assemblages? Hydrobiologia 691:97–107

Volk RB, Furkert FH (2006) Antialgal, antibacterial and antifungal activity of two metabolites produced and excreted by cyanobacteria during growth. Microbiol Res 161:180–186

von Elert E, Jüttner F (1997) Phosphorus limitation and not light controls the extracellular release of allelopathic compounds by *Trichormus doliolum* (cyanobacteria). Limnol Oceanogr 42:1796–1802

Wang L, Zi J, Xu R, Hilt S, Hou X, Chang X (2017) Allelopathic effects of *Microcystis aeruginosa* on green algae and a diatom: evidence from exudates addition and co-culturing. Harmful Algae 61:56–62

Żak A, Kosakowska A (2015) The influence of extracellular compounds produced by selected Baltic cyanobacteria, diatoms and dinoflagellates on growth of green algae *Chlorella vulgaris*. Estuar Coast Shelf Sci 167:113–118

Żak A, Musiewicz K, Kosakowska A (2012) Allelopathic activity of the Baltic cyanobacteria against microalgae. Estuar Coast Shelf Sci 112:4–10

Zhang W, Jeppesen E, Wang M, Xu X, Wang L (2016) Allelopathic effect boosts *Chrysosporum ovalisporum* dominance in summer at the expense of *Microcystis panniformis* in a shallow coastal water body. Environ Sci Pollut Res 24:4666–4675

Zhu X, Dao G, Tao Y, Zhan X, Hu H (2021) A review on control of harmful algal blooms by plant-derived allelochemicals. J Hazard Mater 401:123403

Nitrogen Metabolism in Cyanobacteria 12

Kunal Seth, Geetanjali Kumawat, Mukesh Kumar, Vishambhar Sangela, Nitika Singh, Amit Kumar Gupta, and Harish

Abstract

Cyanobacteria are known to have unique capability of nitrogen fixation in their specialized cell known as heterocyst. However, differentiation of vegetative cell toward heterocyst reduces competitive ability of cyanobacteria because it led to a shift of energy allocation from carbon to nitrogen metabolism. Therefore, heterocyst formation is regulated to avoid the differentiation commitment due to short-term nitrogen fluctuation. Once nitrogen deficiency signal is sensed by the cyanobacteria, pattern of heterocyst formation is determined that ensures equidistance formation of heterocyst cells with about one heterocyst per ten vegetative cells. After differentiation, heterocyst provides anaerobic condition that is prerequisite for the nitrogenase complex to fix the atmospheric dinitrogen. Microoxic condition inside the heterocyst is attained by elimination of oxygen-producing photosystem II activity, increasing respiration rate, and by formation of thick heterocyst-specific exopolysaccharide and glycolipid layer. Nitrogen-fixing machinery is assembled and activated during heterocyst differentiation. The nitrogenase complex is encoded by *nif* gene family. Many of these genes are interrupted in the vegetative cells by interruption elements and these are excised during differentiation of heterocyst by a site-specific recombinase, leading to the activation of genes. In this chapter, we have outlined the molecular circuit of heterocyst differentiation and discussed the assembly of nitrogen-fixing machinery and role of key enzymes in the nitrogen metabolism in the cyanobacteria.

K. Seth
Department of Botany, Government Science College, Valsad, Gujarat, India

G. Kumawat · M. Kumar · V. Sangela · A. K. Gupta · Harish (✉)
Department of Botany, Mohanlal Sukhadia University, Udaipur, Rajasthan, India

N. Singh
Department of Botany, Government College Bundi, Bundi, Rajasthan, India

R. P. Rastogi (ed.), *Ecophysiology and Biochemistry of Cyanobacteria*,
https://doi.org/10.1007/978-981-16-4873-1_12

Keywords

Biofertilizer · Cyanobacteria · Nitrogen fixation · Heterocyst · Nitrogenase

12.1 Introduction

Nitrogen metabolism is the group of reactions, which includes the conversion of atmospheric dinitrogen (N_2) into glutamate at the expense of energy. This conversion is necessary because of two reasons; one is the storage of ammonia that is quite toxic to a cell and another is the amino acid glutamate that acts as a precursor for many metabolic pathways like 5-aminolevulinate, phycobilin, and chlorophyll biosynthesis (Flores and Herrero 1994). Nitrogen fixation, which is characteristic feature of some prokaryotes, is an important phenomenon to sustain nutrient cycling. It occurs in symbiotic bacteria like *Rhizobium* or in cyanobacteria like *Nostoc* and *Anabaena*.

Cyanobacteria are prokaryotic autotrophs which play an imperative role in food chain and have a diverse ecological niche. Cyanobacteria are not confined to oceanic condition yet their assorted variety in biochemistry has enabled these groups of species to occupy almost any terrestrial and aquatic living space on earth (Schirrmeister et al. 2013). Cyanobacteria normally live in marine or fresh water, some cyanobacteria live in a place with earthly biological system and some even thrive in outrageous conditions like desert, the polar area or warm water (Muro-Pastor and Hess 2020). Cyanobacteria are not in every case free-living yet many are fundamental for organizing complex microbial networks in endolithic form for example, in stromatolites, microbial mats, coastal and desert biological soil, and as symbionts of certain higher plants and fungi (Muro-Pastor and Hess 2020). They are also called as diazotrophs because they have capability to fix the atmospheric nitrogen (Lee 2018). Diazotrophic cyanobacteria species contribute considerable quantity of combined nitrogen into the biosphere by changing dinitrogen into ammonia, a procedure known as biological nitrogen fixation (Muro-Pastor and Hess 2020).

These are the oldest multicellular organism on earth (Herrero et al. 2016). According to paleobotanical study, these organisms were thought to evolve about 2.7 billion years ago and have characteristic blue green color due to the principal pigment c-phycocyanin and c-phycoerythrin. The cells of cyanobacteria are covered with mucilaginous sheath called capsule. Cyanobacterial cell wall shows similarity with gram-negative bacteria, a peptidoglycan layer is present at the outer side of the cell membrane. This layer is composed of NAM (N-acetyl muramic acid), NAG (N-acetyl glucosamine), and tetrapeptides, which are further linked by amide bond. They have porin protein on outer membrane which is permeable for all macro- and micromolecules. But the plasma membrane acts as a permeability barrier for these biomolecules.

Cyanobacteria are for the most part described by their high protein content. In these, nitrogen metabolism is directed by a saved calibrating framework, which detects the cell balance among carbon and nitrogen level (Forchhammer and Lüddecke 2016). In *Synechocystis* PCC 6803, for example, photosynthetic rates are found to be positively correlated with amino acid and protein levels, but not with growth rates (Esteves-Ferreira et al. 2017). The growth of any cyanobacteria requires two interdependent cell types, viz., vegetative cells for oxygenic photosynthesis and heterocyst for dinitrogen fixation. The fix ratio of two macronutrients carbon and nitrogen (5:1) plays an important role in metabolic homeostasis. Vegetative cells supply reduced carbon to heterocyst, similarly heterocyst supply fixed nitrogen to vegetative cells and maintains the carbon-nitrogen pool. The balanced metabolism of C and N is essential for optimal growth. Heterocysts are connected with vegetative cells through microplasmodesmata or septosome for minerals and substrates, so it manifests the best example of cell-cell communication in cyanobacteria. Heterocyst itself is a modified vegetative cell, thick walled, pale yellow in color due the principal pigment carotenoid, lacks oxygen evolving PSII activity, and creates a microoxic environment for the key enzyme nitrogenase (Harish 2020).

Heterocysts develop from vegetative cells by decomposition of granular inclusions (carboxysomes and glycogen granules), disintegration of photosynthetic thylakoids, and formation of new membrane structures. They neither fix carbon dioxide nor produce oxygen, but have a high oxygen consumption rate via respiration, surrounded by thick layered laminated cell wall. A special system "Honey comb" is present close to heterocyst poles and has a role in respiration and photosynthesis. The differentiation of heterocyst is completed in two steps—first step is reversible in which the vegetative cell senses the nitrogen-deprived condition and converts it into proheterocyst and the next step is irreversible in which conversion of proheterocyst to heterocyst occurs and *nif* gene is activated. Proteins like NtcA, HetR, HetC, PatA, PatS, and PatB participate in heterocyst differentiation and pattern formation (Harish 2020). Nitrogen-deprived condition induces the vegetative filament for heterocyst differentiation, accumulation of 2-oxoglutarate (an intermediate of tricarboxylic acid (TCA) cycle), which acts as a signaling molecule for heterocyst differentiation and pattern formation (Esteves-Ferreira et al. 2018). The 2-OG provides a carbon framework for ammonia assimilation through GS-GOGAT cycle (Zhang et al. 2018; Forchhammer and Selim 2020). Heterocyst differentiation is the suitable example of remodeling and cell differentiation.

The key enzyme of biological nitrogen fixation is nitrogenase. There is a temporal and spatial separation in cyanobacteria to prevent denaturation of nitrogenase. Nitrogenase has two components, first is dinitrogenase reductase (a type of iron protein) and second is dinitrogenase (a type of molybdenum-iron protein) (Flores and Herrero 2005). Ammonia is the most preferable nitrogen source because it diffuses passively through the membrane. The ammonia is further converted into glutamate through GS-GOGAT pathway by succeeding reaction held by glutamine synthetase (GS) and glutamate synthase (GOGAT). The reaction catalyzed by GS is ATP-dependent and GOGAT is an amino-transferase which transfers amide group of glutamine to 2-OG resulting in formation of glutamate amino acid. Cyanobacteria can assimilate many organic and inorganic nitrogen-containing compounds other than atmospheric nitrogen, these may be nitrate, nitrite, ammonium, urea, cyanate, and amino acids such as arginine, glutamine, and glutamate, but ammonia is the favored nitrogen source (Esteves-Ferreira et al. 2017). The concentration of ammonia in a medium acts as a regulator (inducer or repressor) for the signal transduction pathway; this pathway is closely correlated with plants because the evolution of plastid is phylogenetically correlated with cyanobacteria by endosymbiosis theory. Nitrogen fixation is a metabolically expensive process because it involves 16 ATP for fixing each molecule of nitrogen. Like GS-GOGAT cycle, there are some amino acids like arginine and aspartate that combinedly form a nitrogen storage reservoir called cyanophycin. It is a nonribosomal synthesized protein like polymer which is arranged in a poly-aspartate form (Lee 2018).

12.2 Heterocyst Differentiation

Heterocyst differentiation is a quite complex mechanism and many proteins are involved in its regulation. Multiple layers of regulation ensure that cyanobacteria do not commit heterocyst formation due to short-term fluctuations in the soil nitrogen source content. When cyanobacterial filament receives a lasting signal for nitrogen depletion condition, it led to synthesis of 2-oxoglutarate. This is considered as first sensing signal for induction of heterocyst formation to overcome the nitrogen starvation condition. 2-oxoglutarate or α-ketoglutarate of the Krebs cycle and this metabolite connect C and N metabolism (Huergo and Dixon 2015). Increase in concentration of 2-oxoglutarate triggers the synthesis of NtcA protein also known as global nitrogen regulator, due to its role in overall regulation of nitrogen metabolism in the cyanobacterial filament. Further, in its downstream cascade signal perpetuation, synthesis of HetR protein occurs. This protein is known as specific master regulator, due to its specific role in heterocyst differentiation. HetR itself is regulated by Pkn22 Kinase, cyABrB1, some other genes like *asl1930, alr3234,* and *alr2902.* Multiple layers of regulation ensure fine tuning of development mechanism for heterocyst differentiation. Further, HetR itself has autocatalytic activity and phosphorylation of serine residue present at 130 positions in HetR protein is essential for its activity. HetR protein then interacts with HetP and HetZ proteins and developmental signal is passed to these proteins. HetP protein led to irreversible commitment toward heterocyst formation.

Number of heterocyst cells with respect to vegetative cells are regulated by pattern determination. Because too many heterocyst cells will incur huge energy cost in terms of entire filament and will reduce the competitive ability of the filament. Heterocyst differentiation is energy intensive phenomenon for the cyanobacteria. Therefore, equal distribution of heterocyst cells throughout the filament ensures the equal distribution of fixed nitrogen compound to neighboring vegetative cells, and also conserves the energy for carbon metabolism in the filament. Pattern distribution is therefore equally important aspect when considering the heterocyst differentiation. Proteins of *Pat* gene family regulate this aspect in the cyanobacteria. HetR protein is involved in synthesis of PatS, which is an inhibitor of the HetR protein itself, thereby controlling the number of heterocyst cells in the filament. PatS is processed to short peptide that acts as concentration-dependent manner to inhibit the heterocyst formation. HetC protein is known for its role in transport of short peptides of PatS. The concentration ratio of PatS and HetR determines the development of heterocyst and position of the heterocyst cell in the filament. PatX is another protein that inhibits heterocyst formation and there is functional overlap between PatS and PatX. PatC protein ultimately selects the cell for differentiation into heterocyst and thereby governs the spatial pattern determination of heterocyst in cyanobacterial filament. Other proteins are also identified that play role in regulation of heterocyst frequency like PatD, PatN, and PknH.

Cell wall of the heterocyst is thick to keep the oxygen concentration minimum in the interior of the cell. Therefore, entire remodeling of the cell wall is done during differentiation of vegetative cell to the heterocyst cell. Cyanobacteria are gram-negative as far as their cell wall organization is concerned, but the thickness of peptidoglycan layer is intermediate (15–35 nm) between gram-positive bacteria and gram-negative bacteria and cross-linking is also higher in cyanobacterial cell wall in comparison to gram-negative bacteria. During heterocyst differentiation, two additional layers are developed. The external polysaccharide layer is known as hep layer and internal glycolipid layer is known as hgl layer. HepA protein is involved in synthesis of hep layer and gene of this protein is regulated by HetR and HepK proteins. Some more genes involved in synthesis of hep layer are *hepA, hepK, hepN, hepS, henR, murB, murC, hcwA, amiC1, amiC2, pbp6, sepJ (fraG), fraC, fraD, and sjcF1*. Additionally, hgl layer prevents the entry of oxygen inside the cell and therefore ensures low oxygen concentration for functioning of nitrogenase complex. Some genes, which regulated the formation of hgl layer, have been identified like *hgdB, hgdC, devBCA* operon, *devH,* and *hglE* (Table 12.1).

The availability of ammonia in a medium, acts as a regulator (inducer or repressor) for the signal transduction pathway. Where global nitrogen regulator gene *ntcA* and signal transducer P-II (which is encoded by *glnB*) control the activity of many genes like *henA*, *hetR*, *hetC*, *patA*, *patB*, and *patC* which are responsible for the heterocyst differentiation and pattern formation. NtcA is a bacterial transcription factor which is a member of catabolic repressor protein. NtcA can inactivate GS-activity by coding inhibitory polypeptides (IF-7 and IF-17) by protein – protein interaction (Muro-Pastor and Florencio 2003). By this, cyanobacteria maintain the metabolic homeostasis. In nitrogen-starved condition, storage level of 2-OG is very high and NtcA self-regulates the expression of *hetR* for heterocyst differentiation (Muro-Pastor et al. 2001). HetR is a kind of serine type protease and also a DNA-binding protein. In in vivo condition, HetR performs as a homodimer and this homodimer is essential for DNA-binding activity and heterocyst differentiation (Huang et al. 2004). Another gene *patS* inhibits this DNA-binding affinity (Huang et al. 2004). The nitrogen regulatory protein PII (PII) interacts with 2-OG and brings conformational changes of PII leading to the release of the PII interacting protein X (PipX). PipX interacts with the nitrogen control factor (NtcA) of cyanobacteria.

Table 12.1 Role of different protein identified during heterocyst differentiation and nitrogen metabolism in cyanobacteria

Protein/gene	Function
NtcA	An autocatalytic protein acts as a transcriptional regulator of many genes of nitrogen metabolism
HetR	A homodimer DNA-binding protein (Ser type protease) regulates heterocyst differentiation
Pkn22	A hanks-type kinase which phosphorylates HetR protein at 130th position of Ser residue
cyABrB1	Negatively controls heterocyst formation
asl1930	Negatively controls HetR protein by holding up the commitment of heterocyst differentiation
alr2902	Negatively regulatory proteins which obstruct the development of heterocyst
alr3234	Delays heterocyst differentiation by inhibiting HetR
HetP	Interacts with HetR and down-regulates heterocyst differentiation
HetZ	Interconnects with HetZ and alr2902 and acts as a regulatory protein of pattern formation phase of cellular differentiation
PatS	Negatively controls heterocyst differentiation by post translational degradation of HetR
PatD	Controls heterocyst frequency
PatN	Plays an important role in pattern formation
PatA	Helps in pattern formation
PatC	Plays a key role in pattern formation by selecting cells for differentiation
PatX	Negatively controls heterocyst differentiation as PatS
PatB (CnfR)	Activates *Nif-B* gene
HetN	Helps in fate determination
PknH	Maintaining the heterocyst system by connecting the vegetative cell and heterocyst.
HepA	Plays a key role in the formation of hep layers
HepK	Controls the expression of HepA and functions as a bacterial two component regulatory system
HepN, HepS	Subsidiary genes which are essential for Hep layer formation
HenR	Subsidiary gene which is essential for Hep layer formation
MurB	Transforms UDP-N-acetylglucosamine enolpyruvate to N-acetyl muramic acid
MurC	Attaches pentapeptide chains
HcwA	Involved in rearrangement of the peptidoglycan layer
AmiC1	Amidase encoding gene, forms nanopore at septal junction and is also involved in rearrangement of peptidoglycan layer
AmiC2	Amidase encoding gene, forms nanopore at septal junction and is also involved in rearrangement of peptidoglycan layer
Pbp6	A penicillin-binding protein functions as origination of the peptidoglycan layer and maturing it
Yfr1 (sRNA)	Reduces heterocyst differentiation along with ten different m-RNA
SepJ (FraG)	Maintains the number of nanopores
FraC	Maintains the number of nanopores
FraD	Maintains the number of nanopores

(continued)

Table 12.1 (continued)

Protein/gene	Function
SjcF1	Regulates the size of nanopore diameter
HgdB	A kind of membrane fusion protein which helps in to find out the correct configuration and ratio of glycolipids in heterocyst
HgdC	A type of permease which helps in to find out the correct configuration and ratio of glycolipids in heterocyst
DevBCA	Exports glycolipid along with ABC exporter
DevH	Positively regulates the gene expression of HglE
HglE	Helps in synthesis of hgl
NifH1	Main nitrogen fixation gene
NifD	Plays a key role in coding the alpha subunit of nitrogenase
HupL	Helps in nitrogen fixation
Primase P4	Helps in nitrogen fixation
NifB	Essential for the formation of Fe-Mo cofactor
NifS	Essential for the formation of Fe-Mo cofactor
NifU	Essential for the formation of Fe-Mo cofactor
FdxN	Encodes ferredoxin
NifH	Encodes dinitrogenase complex
NifK	Plays a key role in coding the beta subunit of nitrogenase
NifB	Activated by Cnfr
Cox2	A respiratory oxidase which helps in enhancing the respiration rate in heterocyst
Cox3	A respiratory oxidase which helps in enhancing the respiration rate in heterocyst
Flv1B	Decreases oxygen concentration to form water solely in heterocyst
Flv3B	Decreases oxygen concentration to form water solely in heterocyst
PetH	Encodes ferredoxin-NADP oxidoreductase
GlnA	Encodes glutamine synthetase
CphA1	Encodes cyanophycin synthetase which synthesizes cyanophycin granules
CphB1	Encodes cyanophycinase which forms cyanophycin granules

12.3 Nitrogenase and Alternate Nitrogenase

In order to enter the biogeochemical cycle, atmospheric N_2 must be first reduced to a form that can be readily assimilated by organisms in a process known as nitrogen fixation. In cyanobacteria and other N_2-fixing prokaryotes, molecular dinitrogen (N_2) is reduced in multiple electron transfer reactions requiring 16 ATPs per N_2 fixed, resulting in the synthesis of ammonia and the release of hydrogen as a by-product. H_2 generated during the N_2 fixation process may be oxidized by a hydrogenase in a subsequent step (Esteves-Ferreira et al. 2018).

$$N_2 + 8H^+ + 8e^- + 16ATP \rightarrow 2NH_3 + 16(ADP + P_i) + H_2$$

The reduction of molecular nitrogen to ammonium is catalyzed in all nitrogen-fixing organisms via the nitrogenase enzyme complex. Nitrogenase can also reduce many other substances, such as acetylene, hydrogen azide, hydrogen cyanide, or nitrous oxide. Of these, acetylene reduction to ethylene can be monitored because both acetylene and ethylene can be detected easily by gas chromatography (Fay 1992). Based on the type of metal center, there are three well-known types of nitrogenases: iron and molybdenum (Fe/Mo) nitrogenase, iron and vanadium (Fe/V) nitrogenase, and iron only (Fe) nitrogenase. The Fe/Mo-type is the most commonly found in cyanobacteria and rhizobia. The Fe-only and V-nitrogenases are referred as alternative nitrogenases and are considered as "backup" enzymes when Mo is limiting (McRose et al. 2017). The Fe/Mo-nitrogenase is encoded by nitrogen fixation genes (*nifHDK*), the V-nitrogenase by vanadium-dependent nitrogen fixation genes (*vnfHDK*), and the Fe-nitrogenase by alternative nitrogen fixation genes (*anfHDK*) (McRose et al. 2017).

In nonheterocystous cyanobacteria, nitrogenase enzyme is present in all vegetative cells, while in heterocystous form it is localized only in heterocysts. The enzyme nitrogenase that is expressed in heterocyst is Mo-dependent nitrogenase (Nif-1), which has two components—a Mo-Fe protein (molybdoferredoxin or dinitrogenase) and Fe protein (azoferredoxin or dinitrogenase reductase). The dinitrogenase (Mo-Fe protein) is an $\alpha_2\beta_2$ tetramer and its subunits are encoded by *nifD* and *nifK* genes, respectively. The other component, dinitrogenase reductase (Fe protein) is a dimer of two identical subunits (γ) encoded by *nifH* gene. Fifteen nitrogen fixation-related genes are found clustered together in six transcriptional units: *nifB-fdxN-nifS-nifU*, *nifHDK*, *nifEN*, *nifX-orf2*, *nifW-hesA-hesB*, and *fdxH*. A gene-designated *glbN* is found positioned between *nifU* and *nifH,* which encodes monomeric hemoglobin called cyanoglobin. A second functional Mo-dependent nitrogenase Nif2 has been reported in *Anabaena variabilis* ATCC 29413 which is synthesized in the vegetative cells solely under anoxic conditions after the cells have been starved of nitrogen and long before heterocysts form (Schrautemeier et al. 1995; Thiel et al. 1997). Nif2 has also been observed in vegetative cells of nonheterocystous species (Berman-Frank et al. 2003).

Vanadium-containing nitrogenase was first reported in *Anabaena variabilis*, which significantly reduced acetylene (C_2H_2) to ethane (C_2H_6) under Mo deficiency and in the presence of vanadium (V). It was further identified that the V-nitrogenase is encoded by *vnf* genes cluster (*vnfDGKEN*) in *A. variabilis* (Thiel 1996). The V-dinitrogenase is actually encoded by *vnfDGK* gene cluster, while, *vnfEN* gene cluster located downstream of *vnfDGK* is found to be essential for V-nitrogenase activity. In addition to *vnfDGKEN* gene cluster, four other *vnfH* genes are located 23 kb downstream of *vnfN and* are responsible for encoding dinitrogenase reductase of V-nitrogenase.

The V-nitrogenase is a heterooctomer consisting of two α-subunit (VnfD), two β-subunit (VnfK), four δ-subunits (VnfG), and two Fe-V cofactors (Thiel and Pratte 2014). In comparison to Mo-nitrogenase, the V-nitrogenases have lower substrate-binding efficiency; therefore, it reduces less dinitrogen and produces three times more hydrogen than the Mo-nitrogenase (Thiel and Pratte 2014).

$$N_2 + 12H^+ + 12e^- + 24ATP \rightarrow 2NH_3 + 24(ADP + P_i) + 3H_2$$

Nitrogenase is extremely oxygen sensitive. The oxygen is kept far away from nitrogenase by biochemical pathways like the Mehler-reaction or by special oxygen scavenging molecules such as cyanoglobin that binds oxygen reversibly, with high affinity and noncooperatively (Thorsteinsson et al. 1996). In addition, cyanobacteria are diverse group of gram-negative bacteria which coordinate two mutually exclusive process; O_2-evolving photosynthesis and O_2-sensitive nitrogenase-dependent nitrogen fixation. Cyanobacteria have an efficient way to protect O_2-sensitive nitrogenase from O_2_evolved during photosynthesis. In cyanobacteria, these processes are either separated temporally (as in nonheterocystous form/unicellular cyanobacteria, where alternate cycles of nitrogen fixation and photosynthesis take place) or spatially (as in heterocystous forms). Interestingly, heterocyst lacks photosystem II activity; therefore, they do not evolve oxygen that inhibits nitrogen fixation.

Numerous nonheterocystous cyanobacterial strains can fix and reduce atmospheric N_2 to ammonium when confronting nitrogen hardship, for example, *Synechocystis* and *Arthrospira* (*Spirulina*) *maxima* (Esteves-Ferreira et al. 2017). N-fixation is a costly metabolic reaction catalyzed by nitrogenase, which is restrained by O_2 (Esteves-Ferreira et al. 2018). To shield nitrogenase from O_2, photosynthesis and N-fixation are transiently isolated. High nitrogenase activity peaks 12 h after the peak of photosynthesis, at the same time with higher respiratory rates. An alternate N-fixation methodology is solely seen in strains of the genera *Trichodesmium* (Bergman et al. 2013). In these genera, nitrogenase is situated in roughly 20% cells of the filament, and inquisitively these cells display high N_fixation rates at midday (Rodriguez and Ho 2014).

12.4 Uptake of Nitrogen Sources

The most commonly used nitrogen sources by cyanobacteria are nitrate, ammonium, urea, and dinitrogen. Ammonium is the most reduced inorganic form of nitrogen and preferred source of nitrogen for cyanobacteria. When present in the environment, a decrease in the abundance of nitrogen assimilatory enzymes and a reduced expression of nitrogen transport systems leads to a process referred as global nitrogen control (Esteves-Ferreira et al. 2018). Ammonium indirectly represses the expression of *nif* genes by blocking the transcription of NtcA. In natural environments, ammonium is generally present at low concentrations; therefore, specific permeases namely Amt1, Amt2, and Amt3 are required for efficient cellular uptake of ammonium (Esteves-Ferreira et al. 2018). It has been identified that Amt1 is the main permease for ammonium uptake in *Synechocystis* (Montesinos et al. 1998).

Nitrate and nitrite are the most frequent sources of nitrogen for cyanobacteria. In order to be assimilated by cyanobacteria, nitrate is reduced to ammonium via two sequential reactions catalyzed by enzymes nitrate reductase and nitrite reductase. The reductions of nitrate to nitrite and nitrite to ammonium are Fd-dependent and energetically costly (Flores and Herrero 2005). Nitrate uptake and nitrate reductase system are not found in heterocyst. Nitrate and nitrite are actively transported by the ABC-type NrtABCD transporter in freshwater cyanobacterial strains (Maeda et al. 2015). However, it has been reported that nitrate utilization by cyanobacteria in saline environments may be mediated by NapA (NrtP) rather than NrtABCD transporters. The genes for NrtABCD transporter (*nrtA, nrtB, nrtC, and nrtD*) are commonly present in the *nirA* operon (i.e., *nirA-nrtABCD-narB*). The *nirA* and *narB* genes encode the enzymes Fd-nitrite reductase (NirA) and Fd-nitrate reductase (NarB), respectively, which lead to the formation of ammonium. In *Synechocystis*, *nirA* has been found to be separated from *nrtABCD-narB* (Ohashi et al. 2011). Certain marine and saline water cyanobacterial strains have nitrite transporter of the formate/nitrite transporter (FNT) family, and the cyanate ABC-type transporter which transport nitrite with a much lower affinity than for cyanate (Maeda and Omata 2009; Maeda et al. 2015). A transporter, encoded by the gene *nrtP*, additionally displays high affinity for nitrate and nitrite and was distinguished in the genome of cyanobacterial strains from freshwater and marine conditions (Sakamoto et al. 1999; Bird and Wyman 2003; Maeda et al. 2015).

Many cyanobacteria have shown to import urea at concentrations as low as 0.1–0.6 mM (Mitamura et al. 2000). But before assimilation, urea needs to be hydrolyzed to ammonium and CO_2 catalyzed by a Ni^{2+}-dependent urease. The urease is typically a constitutive enzyme which is not regulated by nitrogen-containing compounds (Ludwig and Bryant 2012). However, a low urease action has been noted in some cyanobacterial strains in presence of ammonium (Singh 1992). These cyanobacteria, such as *Synechocystis*, have high-affinity urea ABC-type transporter responsible for urea uptake at concentrations lower than 1 mM (Esteves-Ferreira et al. 2018). The urea ABC-type transporter is encoded by *urt* genes *urtA, urtB, urtC, urtD,* and *urtE*. These genes are normally organized in an operon, although in *Synechocystis* they have been found to spread along the chromosome (Valladares et al. 2002).

12.5 Ammonium Incorporation into Carbon Skeletons

In cyanobacteria, ammonium is incorporated into carbon skeletons mainly through the glutamine synthetase/glutamine oxoglutarate aminotransferase (GS/GOGAT) cycle. GS catalyzes the ATP-dependent incorporation of ammonium into glutamate to form glutamine. In the following reaction, GOGAT (glutamate synthase) catalyzes the transfer of amide group from glutamine to 2-oxoglutarate (2-OG) to form two molecules of glutamate. Subsequently, aminotransferases can transfer the amino group from glutamate to other carbon skeletons to form other amino acids (Esteves-Ferreira et al. 2018).

In cyanobacteria, there is only one GS (GSI) which is encoded by the gene *glnA*. GSI activity is negatively regulated in presence of ammonium by protein-protein interaction of two inactivating factors (i.e., IF7 and IF17). In contrast, under nitrogen deficiency or in the presence of nitrogen sources other than ammonium, *glnA* expression is up-regulated (Esteves-Ferreira et al. 2018). Interestingly, in some cyanobacterial strains such as *Synechocystis*, *Synechococcus*, and *Gloeocapsa* sp. PCC 7428, a second type of GS encoded by *glnN*, referred as GS type III (GSIII) has also been observed (Reyes and Florencio 1994). It has been observed that *glnN* transcription is more sensitive to ammonium and nitrate than *glnA*, and maximal GSIII activity can reach at most 24% of GSI activity in cells under nitrogen starvation (Reyes and Florencio 1994). In *Pseudanabaena* sp. strain PCC 6903, only the GSIII has been found to be responsible for all ammonium assimilation. This indicates that the presence of an additional glutamine synthetase (GSIII) in some cyanobacterial strains indicates its possible role in increasing the efficiency of ammonium assimilation under nitrogen deficiency.

GOGAT (glutamine 2-oxoglutarate aminotransferase) is a constitutive expressed enzyme. Two different forms of GOGAT have been described in photosynthetic organisms, one is a Fd-GOGAT encoded by *glsF* (*gltS*), and, second is a complex of NADH-GOGAT encoded by *gltB* and *gltD gltB* (large subunit) and *gltD* (small subunit) (Muro-Pastor and Florencio 2003). All cyanobacteria encompass Fd-GOGAT (Muro-Pastor et al. 2005) though NADH-GOGAT has been also identified in *Synechocystis* and *Plectonema boryanum* (Wang et al. 2004). Although, both the forms of this enzyme are simultaneously active in these strains, but Fd-GOGAT is found to be more active and has a more prominent role in ammonium assimilation and growth.

Alternatively, the glutamate dehydrogenase (GDH) can catalyze the formation of glutamate directly from 2-oxoglutarate and ammonium in an NADPH-dependent reaction. However, since GDH has higher K_m (and consequently low affinity) for ammonium than GSI, hence, this enzyme is not likely to be intricated in main ammonium assimilation in cyanobacteria. It has been suggested that this enzyme is involved in the removal of excess cellular ammonium, so as to guard the proton gradient within the thylakoid and periplasmic spaces (Chávez et al. 1999).

12.6 Cyanobacteria as Biofertilizer

Due to ability of cyanobacteria to fix the atmospheric nitrogen, they are conventionally used as biofertilizers in agriculture field. Their importance as biofertilizer has increased mainly to avoid usage of synthetic fertilizers. Cyanobacteria not only increase the nitrogen content in the soil, but also improve the soil structure via released polysaccharides, increase the soil aeration by their filamentous structure, improve soil health by releasing growth hormones, decrease the soil salinity, and improve water holding capacity of the soil (Kuraganti et al. 2020). Because of their ecofriendly organic nature and economic feasibility, cyanobacteria are commonly used as biofertilizer mainly in the paddy field. Generally, *Anabaena*, *Nostoc*, *Aulosira*, and *Tolypothrix* are used as biofertilizer. However, use of *Azolla-Anabaena* symbiotic N_2-fixing complex has also been reported and it is found to have this symbiotic complex as more advantageous than free-living cyanobacteria from agronomic point of view.

12.7 Conclusions

Nitrogen-fixing capability of cyanobacteria has made this group of organisms very important for the research purpose. Cyanobacteria require two interdependent cell types for growth, i.e., vegetative cells for oxygenic photosynthesis and heterocyst for nitrogen fixation. Heterocyst differentiation process involves many proteins like NtcA, HetR, HetP, HetZ, PatS, PatX, PatC, PatD, PatN, PknH, HepA, HepK, etc. The first sensing signal for induction of heterocyst formation is synthesis of 2-oxoglutarate which is a part of Kreb cycle as alpha ketoglutarate. Increase in concentration of 2-oxoglutarate triggers the synthesis of NtcA protein. NtcA further activates the other genes involved in heterocyst differentiation. Development of heterocyst and their spatial distribution both are equally important. Proteins of *het* gene family are involved in differentiation and proteins of *Pat* gene family regulate distribution pattern.

In this chapter, it is discussed that most commonly used nitrogen sources by cyanobacteria are nitrate, ammonium, urea, and dinitrogen, but ammonium is most preferred nitrogen source. In nitrogen fixation, dinitrogen is reduced to ammonia in multiple electron transfer reactions requiring 16 ATPs per N_2 fixed. In nitrogen assimilation by cyanobacteria, nitrate is reduced to ammonium via two sequential reactions catalyzed by enzymes nitrate reductase and nitrite reductase. Nitrate and nitrite are actively transported by the ABC-type NrtABCD transporter in freshwater cyanobacterial strains, and in saline environments may be mediated by NapA (NrtP). Urea is hydrolyzed to ammonium and CO_2 by a Ni^{2+}-dependent urease. The urea ABC-type transporter is encoded by *urt* genes.

In cyanobacteria, ammonium is combined into carbon skeletons mostly through the GS/GOGAT cycle. GS is encoded by the gene *glnA*. Two diverse forms of GOGAT have been described in photosynthetic organisms, a Fd-dependent GOGAT (*glsF*) and NADH-dependent GOGAT (*gltB* and *gltD*). Alternatively, GDH catalyzes the synthesis of glutamate from 2-oxoglutarate and ammonium in an NADPH-dependent reaction, albeit with low affinity.

Studies involving nitrogen metabolism led to understanding of molecular mechanism of nitrogen fixation on exposure to various nitrogen sources. Better understanding of the mechanisms involved in nitrogen metabolism would increase the biotechnological potential of these organisms (Zehr and Capone 2020).

Acknowledgments MK acknowledges financial assistance from University Grants Commission (UGC), New Delhi in a form of JRF-NET (NTA Ref. No. 191620001766). Contribution of VS to this study was financially supported by UGC, New Delhi, India in the form of BSR meritorious fellowship [F.25-a/2014-15(BSR)/7-125/2007(BSR)]. Financial support from Department of Science and Technology, New Delhi for laboratory infrastructure is also acknowledged (SERB File Number: EEQ/2020/000011). We are thankful to the Head, Department of Botany for providing necessary facility.

References

Bergman B, Sandh G, Lin S, Larsson J, Carpenter EJ (2013) Trichodesmium–a widespread marine cyanobacterium with unusual nitrogen fixation properties. FEMS Microbiol Rev 37(3):286–302

Berman-Frank I, Lundgren P, Falkowski P (2003) Nitrogen fixation and photosynthetic oxygen evolution in cyanobacteria. Res Microbiol 154(3):157–164

Bird C, Wyman M (2003) Nitrate/nitrite assimilation system of the marine picoplanktonic cyanobacterium *Synechococcus* sp. strain WH 8103: effect of nitrogen source and availability on gene expression. Appl Environ Microbiol 69(12):7009–7018

Chávez S, Lucena JM, Reyes JC, Florencio FJ, Candau P (1999) The presence of glutamate dehydrogenase is a selective advantage for the cyanobacterium *Synechocystis* sp. strain PCC 6803 under nonexponential growth conditions. J Bacteriol 181(3):808–813

Esteves-Ferreira AA, Cavalcanti JHF, Vaz MGMV, Alvarenga LV, Nunes-Nesi A, Araújo WL (2017) Cyanobacterial nitrogenases: phylogenetic diversity, regulation and functional predictions. Gen Mol Biol 40(1):261–275

Esteves-Ferreira AA, Inaba M, Fort A, Araújo WL, Sulpice R (2018) Nitrogen metabolism in cyanobacteria: metabolic and molecular control, growth consequences and biotechnological applications. Crit Rev Microbiol 44(5):541–560

Fay P (1992) Oxygen relations of nitrogen fixation in cyanobacteria. Microbiol Mol Biol Rev 56(2):340–373

Flores E, Herrero A (1994) Assimilatory nitrogen metabolism and its regulation. In: The molecular biology of cyanobacteria. Springer, Dordrecht, pp 487–517

Flores E, Herrero A (2005) Nitrogen assimilation and nitrogen control in cyanobacteria. Biochem Soc Trans 33(1):164–167

Forchhammer K, Lüddecke J (2016) Sensory properties of the PII signalling protein family. The FEBS J 283(3):425–437

Forchhammer K, Selim KA (2020) Carbon/nitrogen homeostasis control in cyanobacteria. FEMS Microbiol Rev 44(1):33–53

Harish SK (2020) Molecular circuit of heterocyst differentiation in cyanobacteria. J Basic Microbiol. https://doi.org/10.1002/jobm.202000266

Herrero A, Stavans J, Flores E (2016) The multicellular nature of filamentous heterocyst-forming cyanobacteria. FEMS Microbiol Rev 40(6):831–854

Huang X, Dong Y, Zhao J (2004) HetR homodimer is a DNA-binding protein required for heterocyst differentiation, and the DNA-binding activity is inhibited by PatS. PNAS USA 101(14):4848–4853

Huergo LF, Dixon R (2015) The emergence of 2-oxoglutarate as a master regulator metabolite. Microbiol Mol Biol Rev 79(4):419–435

Kuraganti G, Edla S, Pallaval VB (2020) Cyanobacteria as biofertilizers: current research, commercial aspects, and future challenges. In: Yadav A, Rastegari A, Yadav N, Kour D (eds) Advances in plant microbiome and sustainable agriculture. Microorganisms for sustainability, vol 20. Springer, Singapore. https://doi.org/10.1007/978-981-15-3204-7_11

Lee RE (2018) Phycology. Cambridge University Press, New York

Ludwig M, Bryant DA (2012) Acclimation of the global transcriptome of the cyanobacterium Synechococcus sp. strain PCC 7002 to nutrient limitations and different nitrogen sources. Front Microbiol 3:145

Maeda SI, Murakami A, Ito H, Tanaka A, Omata T (2015) Functional characterization of the FNT family nitrite transporter of marine picocyanobacteria. Life 5(1):432–446

Maeda SI, Omata T (2009) Nitrite transport activity of the ABC-type cyanate transporter of the cyanobacterium *Synechococcus elongatus*. J Bacteriol 191(10):3265–3272

McRose DL, Zhang X, Kraepiel AM, Morel FM (2017) Diversity and activity of alternative nitrogenases in sequenced genomes and coastal environments. Front Microbiol 8:267

Mitamura O, Kawashima M, Maeda H (2000) Urea degradation by picophytoplankton in the euphotic zone of Lake Biwa. Limnology 1(1):19–26

Montesinos ML, Muro-Pastor AM, Herrero A, Flores E (1998) Ammonium/methylammonium permeases of a cyanobacterium. Identification and analysis of three nitrogen-regulated amt genes in *Synechocystis* sp. PCC 6803. J Biol Chem 273(47):31463–31470

Muro-Pastor MI, Florencio FJ (2003) Regulation of ammonium assimilation in cyanobacteria. Plant Physiol Biochem 41(6–7):595–603

Muro-Pastor AM, Hess WR (2020) Regulatory RNA at the crossroads of carbon and nitrogen metabolism in photosynthetic cyanobacteria. BBA-Gene Regul Mech 1863(1):194477

Muro-Pastor MI, Reyes JC, Florencio FJ (2001) Cyanobacteria perceive nitrogen status by sensing intracellular 2-oxoglutarate levels. J Biol Chem 276(41):38320–38328

Muro-Pastor MI, Reyes JC, Florencio FJ (2005) Ammonium assimilation in cyanobacteria. Photosyn Res 83(2):135–150

Ohashi Y, Shi W, Takatani N, Aichi M, Maeda SI, Watanabe S, Omata T (2011) Regulation of nitrate assimilation in cyanobacteria. J Exp Bot 62(4):1411–1424

Reyes JC, Florencio FJ (1994) A new type of glutamine synthetase in cyanobacteria: the protein encoded by the glnN gene supports nitrogen assimilation in *Synechocystis* sp. strain PCC 6803. J Bacteriol 176(5):1260–1267

Rodriguez IB, Ho TY (2014) Diel nitrogen fixation pattern of *Trichodesmium*: the interactive control of light and Ni. Sci Rep 4:4445

Sakamoto T, Inoue-Sakamoto K, Bryant DA (1999) A novel nitrate/nitrite permease in the marine cyanobacterium *Synechococcus* sp. strain PCC 7002. J Bacteriol 181(23):7363–7372

Schirrmeister BE, de Vos JM, Antonelli A, Bagheri HC (2013) Evolution of multicellularity coincided with increased diversification of cyanobacteria and the great oxidation event. PNAS USA 110(5):1791–1796

Schrautemeier B, Neveling U, Schmitz S (1995) Distinct and differently regulated Mo- dependent nitrogen-fixing systems evolved for heterocysts and vegetative cells of *Anabaena variabilis* ATCC 29413: characterization of the fdxH1/2 gene regions as part of the nif1/2 gene clusters. Mol Microbiol 18(2):357–369

Singh S (1992) Regulation of urease cellular levels in the cyanobacteria *Anacystis nidulans* and *Nostoc muscorum*. Biochem Physiol Pflanz 188(1):33–38

Thiel T (1996) Isolation and characterization of the VnfEN genes of the cyanobacterium *Anabaena variabilis*. J Bacteriol 178(15):4493–4499

Thiel T, Pratte BS (2014) Regulation of three nitrogenase gene clusters in the cyanobacterium *Anabaena variabilis* ATCC 29413. Life 4(4):944–967

Thiel T, Lyons EM, Erker JC (1997) Characterization of genes for a second Mo-dependent nitrogenase in the cyanobacterium *Anabaena variabilis*. J Bacteriol 179(16):5222–5225

Thorsteinsson MV, Bevan DR, Ebel RE, Weber RE, Potts M (1996) Spectroscopical and functional characterization of the hemoglobin of *Nostoc commune* UTEX 584 (cyanobacteria). BBA Protein Struct Mol Enzym 1292(1):133–139

Valladares A, Montesinos ML, Herrero A, Flores E (2002) An ABC type, high affinity urea permease identified in cyanobacteria. Mol Microbiol 43(3):703–715

Wang HL, Postier BL, Burnap RL (2004) Alterations in global patterns of gene expression in *Synechocystis* sp. PCC 6803 in response to inorganic carbon limitation and the inactivation of ndhR, a LysR family regulator. J Biol Chem 279(7):5739–5751

Zehr JP, Capone DG (2020) Changing perspectives in marine nitrogen fixation. Science 368(6492): eaay9514. https://doi.org/10.1126/science.aay9514

Zhang CC, Zhou CZ, Burnap RL, Peng L (2018) Carbon/nitrogen metabolic balance: lessons from cyanobacteria. Trends Plant Sci 23(12):1116–1130

Phycoremediation of Wastewater 13

Simranjeet Singh, Vijay Kumar, Shweta Shekar, Dhriti Kapoor, Deepika Bhatia, Daljeet Singh Dhanjal, Praveen C. Ramamurthy, and Joginder Singh

Abstract

In the past few decades, the world has encountered the climatic hitches, depletion of fossil fuel reserves, and the global rise in temperature and water pollution, which have been of utmost concern lately. These issues have raised various environmental, economic, and geopolitical concerns, also threatening global security and influencing the society in distinct ways. Despite the availability of substitutes for fossil fuels, they pose certain limitations which restrain their application in the global market. This shortcoming has intrigued the researchers worldwide and shifted the research focus toward the development, production, and commercialization of renewable and sustainable biofuels as an effective alternative for conventional energy sources. Cultivation of microalgal organisms with the inherent potential of phycoremediation can help curb the alarming global

S. Singh · P. C. Ramamurthy (✉)
Interdisciplinary Centre for Water Research (ICWaR), Indian Institute of Sciences, Bangalore, India

V. Kumar
Regional Ayurveda Research Institute for Drug Development, Gwalior, Madhya Pradesh, India

S. Shekar
Department of Material Engineering, Indian Institute of Sciences, Bangalore, India

D. Kapoor
Department of Botany, Lovely Professional University, Phagwara, Punjab, India

D. Bhatia
Department of Biotechnology and Medical Sciences, Baba Farid College, Bathinda, Punjab, India

D. S. Dhanjal
Department of Biotechnology, Lovely Professional University, Phagwara, Punjab, India

J. Singh (✉)
Department of Microbiology, Lovely Professional University, Phagwara, Punjab, India
e-mail: joginder.15005@lpu.co.in

R. P. Rastogi (ed.), *Ecophysiology and Biochemistry of Cyanobacteria*,
https://doi.org/10.1007/978-981-16-4873-1_13

issues keeping into consideration the economics and sustainability of the process. The process of phycoremediation involves the employment of macro- or microalgal organisms that can reduce or biotransform various nutrients, toxic chemicals, and pollutants from wastewaters of diverse origins. This book chapter addresses different properties of algae contributing to wastewater treatment, more specifically the use of cyanobacteria and its consortium for effective removal of numerous pollutants from waste effluents.

Keywords

Environment · Contaminants · Phycoremediation · Cyanobacterial species

13.1 Introduction

The escalating rates of population, urbanization, industrialization, and indirect hampering of ecosystem services have resulted in making concerns related to environment pollution more critical (Sood et al. 2012). Consequently, the scientific community faces this grievous challenge of rectifying the problems like enormous formation of sewage/wastewater, the inclusion of this unprocessed sewage into water resources, an outbreak of waterborne diseases, and global warming at large, so that our planet is able to adequately sustain the needs of the present as well as the coming generations. Every aspect of environment pollution comes with its own set of problematic issues and needs specific skills and technical proficiency to conquer this challenge completely. Out of all, the addition of this unrefined wastewater in freshwater bodies is becoming an impending hazard which is putting the stability of nations at stake (Renuka et al. 2014). The major reason behind it is the direct or indirect dependency of people residing in developing countries who use the fresh-water resources to get their routine activities done.

Moreover, the dearth of freshwater arising due to the ever-increasing population in developing countries like India has made this scenario more gruesome and largely contributing to the generation of wastewater (Badr et al. 2010). Apart from this, rapid industrialization, unmindful, and superfluous usage of fertilizers and pesticides are also defiling our on-hand water resources owing to the incorporation of untreated wastewater in them (Ghosh et al. 2012). Further, increased subsistence of surplus nutrients (Nitrogen and Phosphorus) in the wastewater has aggravated the problems like algal blooms, eutrophication, unrestricted expansion of some emergent hydrophytes, reduction in oxygen content, and vanishing of major plant and animal species causing complete and permanent deterioration of water bodies (Khan and Ansari 2005). Hereof, spotting out of economic, environment-friendly technologies based on the simple framework, inputs and minimal expertise with ease of use even for a layman is necessitated more than ever. Applicability of these technologies at a small-scale level with possible utilization at the commercial level in the near future is a requisite feature (Ji et al. 2013).

On the basis of type and magnitude of pollution, three procedures namely physical, chemical, and biological, either alone or in combination, are adopted for reforming the contaminated water (Hongyang et al. 2011). The course of actions for individual wastewater treatments is further grouped into primary, secondary, and tertiary methods so as to attain maximum elimination of contaminants. Expensive nature of physical and chemical methods along with upsurge in normal pH value, conductivity, and piling up of dissolved substances in water by chemical methods particularly, mar their efficacy. Thence, the biological treatment of sewage stands out as a favorable option. Activated sludge method either by itself (Radjenovic et al. 2009) or together with algae (Su et al. 2012) is a popular practice in biological treatment. But, generation of sludge, its removal, and then disposal are the concerns which force to search out for better alternatives (Kim et al. 2010).

Phytoremediation, a propitious choice now a day, is the application of corrective measures for purifying sewage and removing contaminants of emerging concern like heavy metals from it with the help of plants especially lower ones such as algae and related microflora (Franchino et al. 2013). The usage of microalgae for sewage water treatment is specifically termed as Phycoremediation by John (2000), as proposed by Souza et al. (2012). The main highlight of using phycoremediation lies in its potentiality to manage multiple issues together. Employment of microalgae in combating this problem has several advantages, for instance, (i) forbearance of drastic conditions (Makandar and Bhatnagar 2010), (ii) greater photosynthetic efficiency in comparison to higher plants (Bhatnagar et al. 2011), (iii) inherent ability to accumulate inorganic nutrients such as Nitrogen and Phosphorus from the wastewater resulting in their trouble-free disposal (Mata et al. 2012), (iv) capability to absorb and utilize CO_2 release from power plants with the aid of solar energy accounting for their minimum energy requirements (Razzak et al. 2013), (v) vast implementations of harvested biomass (Gupta et al. 2013), and (vi) potential to curb the emission of greenhouse gases (Bhola et al. 2014). All in all, microalgae are a whole lot better than higher aquatic macrophytes in decreasing the harmful qualities of wastewater.

Growing of algae in wastewater serves many useful purposes like purification of wastewater, abatement of greenhouse gases in addition to the production of algal biomass (Pittman et al. 2011). This biomass can be put to use variously—in augmentation of protein and other nutrients in animal and human feed, as a potential source of bioenergy like biogas and other biofuels, as bio-ore for replenishing precious heavy metals, cosmetics, pharmaceuticals, and other beneficial chemicals (Gupta et al. 2013).

13.2 Cyanobacterial Diversity in Wastewater

Photoautotrophic eukaryotic microalgae and prokaryotic cyanobacteria, dwelling in fresh and marine water places, likewise, come under the broader term Microalgae with a display of great variations in their thallus organization and habitat (Lee 2008). The huge biodiversity of microalgae with an estimated number of about

200,000–800,000 species is yet to be explored fully as only 50,000 species are described so far (Starckx 2012). Its wide range of diversity and ubiquitous distribution have resulted in scrutinizing and identifying favorable strains/species/genera so as to evolve best microalgae-based practices for treating foul water (Fouilland 2012).

Overview of relevant literature in context to various researches carried out pertaining to the occurrence of microalgal heterogeneity in different types of wastewater divulge that analysis of phytoplankton diversity in sewage water and drain water mixture was performed by Hussein and Gharib (2012), in which they detected the presence of 152 taxa in total, including Bacillariophyceae, Chlorophyceae, Cyanophyceae, Euglenophyceae, and Dinophyceae with Bacillariophyta being in the pioneering position, constituting approximately 40% of inclusive diversity in the drain. In contrast, such a study was undertaken in case of an open sewage-contaminated channel by Renuka et al. (2013a, b) discerning the dominance of Cyanophycean members (58%) like species of *Chroococcus*, *Lyngbya*, *Phormidium*, *Limnothrix*, *Oscillatoria*, and *Planktothrix*, followed by members of Chlorophyceae (25%) and Bacillariophyceae (17%). Further, an effort to divulge information regarding alteration in microalgal population in batch reactors being used during municipal sewage water treatment containing wastewater from dairies as well was made by Bernal et al. (2008), and findings revealed the presence of microalgae belonging to Cyanophyta, Chlorophyta, and Euglenophyta groups in all the phases of the treatment process with *Arthrospira jenneri* and *Coccomonas* sp. of most common occurrence.

Wastewater bodies have been commonly utilized as an easily accessible and cost-effective substrate for growth, biomass formation, and nutrient exclusion and have been considered as the only efficient and sustainable source of nutrients for blue-green algae biomass production (Arias et al. 2020a). Some animal wastewaters hold greater amounts of NH_3 and although cyanobacteria can develop in a high salt concentration, wastewaters must first be diluted prior to practice them as cultivation culture, particularly for pH level greater than 8.0, due to the reason that the free form of NH_3 is noxious for blue-green algae development (Markou and Georgakakis (2011). In wastewater treatment plants of agri-food industry, *Lyngbya* and *Oscillatoria* were the major species occurred whereas most frequent cyanobacteria moieties found in municipal wastewater of Portugal were *Planktothrix mougeotii*, *Microcystis aeruginosa*, and *Pseudanabaena mucicola* subsequently, these cyanobacterial species responsible for the total phytoplankton amount with highest values of 99.8% and 100% (Martins et al. 2011). In Morocco, *Synechocystis* sp. and *Pseudanabaena galeata* exist in wastewater treatment plants; whereas in Egypt, *Oscillatoria* spp. were the chief species, contributing total phytoplankton volume of 97.8% (Badr et al. 2010; Martins et al. 2011). In Brazil, two nonheterocystous forms, two unicellular forms, and one filamentous heterocystous form were extracted from a municipal wastewater which leads to the occurrence of six cyanobacterial genera (*Synechocystis* sp., *Aphanocapsa* sp., *Merismopedia punctata*, *Merismopedia tenuissima*, *Lyngbya* sp., *Phormidium* sp., and *Pseudoanabaena* sp.). Due to the elevation in the sunlight availability and temperature, maximum range of blue-green algae occurs in spring and summer time. For instance, *Planktothrix mougeotii* exists at greater densities of 6.6 × 106 cells ml^{-1} during April and May, in comparison to *Microcystis aeruginosa* which had the maximum density range of 3.2 × 106 cells ml^{-1} in month of July in municipal wastewater treatment plants (Martins et al. 2011). Diversity of blue-green algae is 91.7% in summer and 96.4% in autumn which is greater than the diversity of algae, i.e., 8.3% in summer and 3.6% in autumn, exhibiting the greater involvement of these microbes to phytoplankton groups (Furtado et al. 2009).

Besides this, a number of reports are also out there in which distribution and diversity of microalgae found in industrial discharges are studied instead of sewage water treatment plants (Dubey et al. 2011). Incidences of the presence of cyanobacterial species maximum in sugar mill effluents (93%) followed by dye effluents (91%), paper mill (76%), and pharmaceutical effluents (50%) were observed by Vijayakumar et al. (2007). Cyanobacterial genus *Oscillatoria* is predominant in its occurrence in all these effluents, and then followed by *Phormidium, Lyngbya, Microcystis,* and *Synechococcus*. A total of 25 cyanobacterial members in paper mill and pharmaceutical effluents were recorded by Dubey et al. (2011) in which species like *Microcystis aeruginosa, Oscillatoria curviceps, O. princeps, Phormidium ambiguum,* and *P. corium* were common to both the effluents with Oscillatoria topping them all. Presence of cyanobacterial communities that belong to Oscillatoriales and Chroococcales in the effluents of pulp and paper secondary waste treatment systems of Brazil, Canada, New Zealand, and the USA was also listed by Kirkwood et al. (2001) where *Phormidium, Geitlerinema, Pseudanabaena,* and *Chroococcus* as the dominant genera. Diversity of microalgal communities in processed and unprocessed carpet mill effluent was assessed by Chinnasamy et al. (2010a, b). They reported equivalent diversity of both Cyanophycean and Chlorophycean members in processed wastewater during spring season and dominance of Chlorophyta in unprocessed wastewater during all the seasons were found (Rawat et al. 2011).

It has been substantiated so far that algae are a major biotic component of aquatic ecosystems with a great range of diversity in morphology and systematics and get extensively affected by amount and type of different pollutants present in their habitats (Lim et al. 2010). Even though a plethora of information regarding environmental factors, range of residing areas of algae, numerous types of wastewater are available, yet it is quite cumbersome to zero in upon generalizations about qualitative and quantitative aspects of diversity in microalgal communities (Min et al. 2011).

13.3 Use of Cyanobacterial Monocultures in Nutrient Sequestration and Biomass Production

Various research efforts regarding usage of microalgae in remedial measures undertaken to purify the municipal wastewater have now been decades-old (Singh et al. 2017). A number of microalgae are reported to be quite adept in eliminating inorganic nutrients like nitrogen and phosphorus from wastewater and then using them for their growth and proliferation. The degree of efficiency in nutrient sequestering or the foul water rectification differs greatly across genera, species, and strains of these algal species, which is evaluated further in different microalgal monocultures in terms of nutrient removal compared with biomass production in wastewaters of various types (Abou-Shanab et al. 2013). Again, a generalization is hard to be reached at in context to quality and quantity of generated biomass due to varying extent of different physicochemical attributes of wastewater from separate sources, having a direct or indirect influence on the growth of algal communities. So, the worthiness of such disparate wastewater forms for growing monocultures of microalgae is to be ascertained first and then utilized at large scale on biotechnological levels for further downstream processing. Scrutiny of individual cultures of microalgae (or microalgal monocultures) is of great help in getting to know the potentiality of individual strain versus elimination of a particular or several nutrients altogether (Frampton et al. 2013).

About 84–96% & 72–87% of ammonium nitrogen and phosphorus, were sequestered by *Arthospira* sp. from piggery wastewater, respectively (Deviram et al. 2020), whereas nitrate recovery was 76% in *Spirulina* sp. which was grown in oxygen-deficient digested pig waste, and improved progress with $NaHCO_3$ amount around 16.8 g L^{-1} (Markou and Georgakakis 2011). *Phormidium valderianum* BDU20041, a marine cyanobacterium, sequestered available nitrate by 66.35% and Phosphorus by 35.66% and also decreased the calcium chloride and magnesium content upon treatment by 16.34% and 23.07% (Dineshbabu et al. 2017). *Anabaena ambigua* significantly decreased the chloride, nitrate, and phosphate concentrations and reported from biomass formation kinetics that alga can develop even on 100% textile wastewater (Brar et al. 2019). Soil cyanobacteria resulted in greater hydraulic holding period and small nutrients load that attained the maximum exclusion efficacies in total N > 95%, total P 35–78%, total organic carbon >93%, and total inorganic carbon >82%, and a biomass production of 0.05–0.07 mg L^{-1} d^{-1} with abundance of filamentous N-fixing blue-green alga. Otherwise, smaller hydraulic holding period and large nutrients load escalate carbon diminution which results in reduced Nitrogen and Phosphorus sequestration, and small biomass generation (Arias et al. 2020b). Among seven cyanobacterial strains, i.e., *Synechocystis* PCC 6803, *Synechococcus* PCC 7942, *Nostoc muscorum, Oscillatoria* sp., *Anabaena cylindrica, Lyngbya* sp., and *Phormidium* sp., highest precise growth proportion, biomass, and lipid yield were found in *Synechocystis* PCC 6803 (Patel et al. 2018). Cyanobacterial biofilms made from cyanobacterial mats which are dominated by *Phormidium* and *Oscillatoria* spp. significantly removed PO_4^{3-} -Phosphate from domestic sewage and other nutrient-enriched wastewaters (Rai et al. 2016). Markou and Georgakakis (2011) reported complete and 69% of ammonia and phosphorous removal, by using cyanobacteria *Phormidium bohneri* of the family *Oscillatoriaceae* with an exclusion amount of 5.9 mg N-NH3/Ld and 2.9 mg P-PO_4^{3-}/Ld, respectively.

13.4 Significance and Promise of Cyanobacterial Consortial Approach in the Remediation of Wastewater

The challenges with excessive generation and release of untreated wastewaters into natural streams have been a severe issue worldwide in the current time (Bhatia et al. 2017; Bhatia et al. 2018). From an eco-friendly perception, the cyanobacteria and their consortium have been mainly recognized for biological treatment of effluents these days (Perera et al. 2018). For efficient use of cyanobacteria and its consortium in effluent treatment, a thorough understanding of their structural properties, behavior, and interaction is necessary. Cyanobacteria are pervasive and belong to an oxygenic photosynthetic class of microorganisms and are usually associated symbiotically with bacteria to form a consortium. During the wastewater remediation, this encouraging symbiotic association can sustain low-cost oxygenation, minimize the environmental repercussion related with other mechanical and chemical treatment processes, and elevate nutrient recovery as well (Patel et al. 2019). Additionally, remediation of elements such as nitrogen, carbon, and phosphorous by employing the consortium of bacteria and cyanobacteria and microalgae instead of individual microorganisms is extensively more efficient. The interactions between the sustainability and among partners of non-native algal bodies are the major point of intention. In consortium, one strain could be effective in removing a single contaminant while the strain in remediation of another. Therefore, the development of consortium for remediation using strains also proves more effectiveness and synergistic in remediation of the wastewater. Such an association also diminishes the greenhouse effect involved in effluent treatment and enhances the growth of microalgae which is an imperative factor for the biological application of the consortium (Sood et al. 2015).

13.5 Wastewater Treatment Using Cyanobacterial Consortia

Recently, phycoremediation has gained the significant and emerged as a beneficial technique to improve the physiochemical parameters of wastewater for being eco friendly, and harmless approach in contrast to other conventional approaches. *Botryococcus, Chlorella, Chlamydomonas, Cosmarium, Nitzschia, Pediastrum,* and *Scenedesmus* are the algal species that are used for wastewater treatment. Following are the reports about the different types of wastewater treated by microalgae and their consortium.

13.5.1 Municipal Wastewater

The combination or mixture of water discharge from household activities and from industrial facilities and institutes is designated as municipal wastewater. Untreated effluent generally comprises a high concentration of organic matter, several harmful microorganisms' pathogenic compounds. The partially treated and untreated municipal effluents represent an ideal medium for the growth of various microalgal species as it is rich in all the vital nutrients such as carbon, ammoniacal nitrogen, nitrate nitrogen, and phosphorus. In addition, cyanobacteria offer an economical and feasible approach to remediate municipal wastewater during the second and tertiary stages of water treatment. The major benefits of using algae lie in the fact that they can be reused for the generation of value-added products. They also offer several advantages such as fast growth rate, photosynthesis efficiency, active adaptive capability, and CO_2 sequestration (Biglari Quchan Atigh et al. 2020). Symbiotic association is commonly seen in bacterial–microalgal interaction. During the process of photosynthesis, algal bodies release certain organic compounds that are utilized by the bacteria as energy source and carbon, oxygen which used for oxidation of organic matter. On the other hand, bacterial population releases carbon dioxide which is used for photosynthetic reactions in plants. Many species of algae have been reported to be appropriate for municipal effluent treatments such as *Chlorella* sp., *Micratinium* sp., *Desmodesmus* sp., and *Scenedesmus* sp. (Gani et al. 2016); *Chlorella* sp. (Abdel-Raouf et al. 2012); and *S. dimorphus*, *A. variabilis*, *Plectonema* sp., *Nostoc* sp., *Phormidium* sp., *Oscillatoria* sp. (Dewangan 2015). Consequently, phycoremediation of municipal wastewater presents massive scope for the better and efficient recycling of wastewater and effective biomass production as well. The algal biomass thus generated has vast industrial applications where value-added products can be synthesized. Phycoremediation of municipal effluent was collected from different stages of the pretreatment and was used for the cultivating *Chlorella* sp. (Li et al. 2011). This study indicated the ability of *Chlorella* sp. in reducing chemical oxygen demand from 2390 to 230 mg/L within 2 days of incubation and by the end of the experimentation, 90.8% of chemical oxygen demand, 93% of ammoniacal nitrogen, 89% of total nitrogen, and 79% of total phosphorous were removed.

13.5.2 Industrial Wastewaters

The presence of detergents, phenols, and heavy metals in municipal wastewater with nutrients has become the matter of concern, but microalgal species have emerged as a solution as they can sequester these compounds. Usually, these heavy metals obstruct the uptake of macronutrients, but when they are present in trace amounts, these metals act as micronutrients which aid growth of these microalgae. Hence, comprehensive information related to wastewater is required for effective proliferation of microalgae before using them as cultivation media. Other than this, culturing of microalgal species in wastewater contaminated with heavy metals creates the problem for the use of biomass as food or related purpose in food as well as pharmaceutical industries. However, heavy metal-containing biomass can be used for the above-said purpose after removing the heavy metal from it. Rai et al. (2020) conducted phycoremediation study to eradicate pollutants from coke-oven wastewater. They reported that application of cyanobacteria consortia (*Leptolyngbya* sp. and *Planktothrix* sp.) to remediate ammoniacal-N, phenolic compounds, and nitrates from coke- oven wastewater as a part of tertiary treatment. A one factor at a time approach (OTA) was used to determine optimum experimental conditions for removal by varying pH in the range of 8–10; the size of inoculum (5–10%); and initial concentration of phenolic compounds (2-3 mgL^{-1}), ammoniacal-N (150–200 mg/L), and nitrate (30–40 mg/L). Response surface methods were used to achieve the optimum operating condition for the removal of the pollutants and production of biomass. The results obtained from both approaches (one factor at a time approach and response surface methods) were satisfactory for real-time treatment of effluent, nevertheless from an economic point of view, OTA analysis results were more promising.

The potential of *C. vulgaris* has been explored for remediation of leather industry effluents which mainly consists of inorganics (heavy metal), casein pigment, and few another chemical (Solanki et al. 2019). The results indicated the considerable reduction, viz., 22% in biological oxygen demand; 38% in chemical oxygen demand, 80% in free ammonia, 89% in nitrite, 63% in calcium, and 50% in magnesium concentration. In a similar way, the application of various algae sp., i.e., *Chlorella* sp., *Chlamydomonas* sp., *Gloeocystis* sp., and *Cyanobacteria* was identified for phycoremediating effluents generated from carpet mills (Chinnasamy et al. 2010a, b). Whilst *Nostoc* sp. was used for phycoremediation of effluents discharged from the dairy industry (Brar et al. 2017). Algal species are also reported to bisphenols, chlorophenols, nitrophenols, nitrates, phosphates, dye removal, etc. with good removal efficiency. It was also reported that *Chlorella pyrenoidosa* is efficient in 62% reduction in nitrates and 87% reduction in phosphates and is capable of remediating textile dye industry effluents (Pathak et al. 2014).

Phycoremediation capabilities of different green algal bodies like *Chlorella vulgaris, Chlorella pyrenoidosa, Scenedesmus obliquus, Chlamydomonas reinhardtii* are also reported to remove organic compounds such as personal care and pharmaceutical compounds, endocrine disrupting chemicals, etc. (Zhou et al. 2014).

13.5.3 Heavy Metal Removal by Cyanobacteria

Sewage wastewater is a well-known source of amelioration of lot many nutrients associated with the growth of plants so that it can be used for irrigation in agricultural practices. On the other hand, if the sewage wastewater gets used for agricultural activities without any pretreatment, it will pass the toxic and pathogenic contaminants. It could include inorganic (heavy metals) or organic (pesticide residues) into the soil and finally transform crops and other plants. Prevailing issues, related with accumulation of these heavy metals in soil, bioremediation is recognized as the efficient technique to eradicate pollutants from sewage wastewater. Employing the use of algae for bioremediation is termed as "phycoremediation approach". It is algae-based remediation of the polluted site, in which various species of algae are studied to remove particular noxious waste (inorganic and organic) from contaminated soil and water.

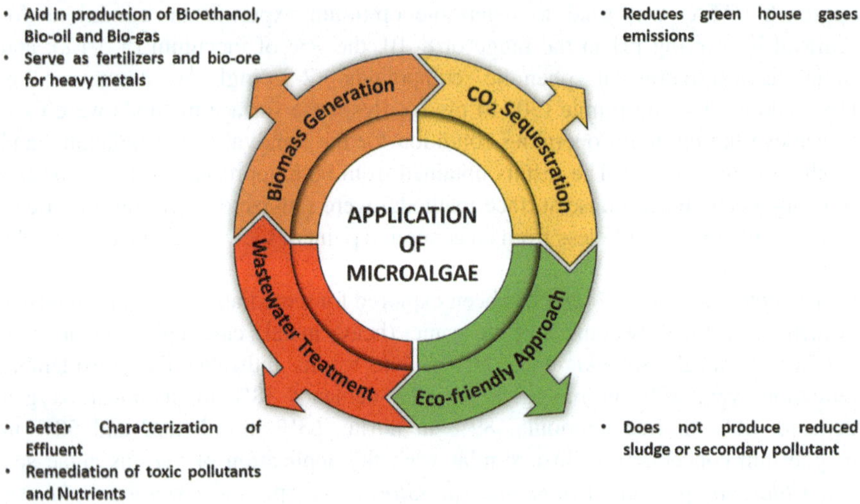

Fig. 13.1 Outline of phycoremediation in effluent treatment and its application

Additionally, to the recuperation of land, algal biomass is also formed by fixation of carbon (Solanki et al. 2019). This algal biomass further can be employed for the preparation of biofuel (biodiesel). The outline of phycoremediation in effluent treatment and its application is shown in Fig. 13.1.

For the bioremediation of sewage wastewater, various cyanobacterial sp. are being used. The algae come under the category of autotrophic microorganism which requires the maximum amount of phosphorous and nitrogen to synthesize protein for metabolic activities. Researchers have reported various strains of algae for absorbing a considerable amount of toxic contaminants, viz., heavy metals and other organic contaminants as well (Abdel-Razek et al. (2019).

Warjri and Syiem (2018) isolated a cyanobacterium, identified as *Nostoc* sp. from the water samples collected from colliery in the hilly areas of the Meghalayan region, India. Their study revealed the tolerance capacity of *Nostoc* sp. in the presence of chromium and reported its ability to grow in the presence of 15 mg/L of chromium, that is 30 times more than the concentration present in that site. Most favorable conditions for algae were recorded at pH in the range of 5–6 with 3 μg mL^{-1} biomass. Maximum adsorption capacity showed by *Nostoc* sp. for Cr was 20 mg/g of algal biomass. Dixit and Singh (2013) reported the impact of different factors on the adsorption of lead and cadmium by *N. muscorum*. They displayed that this strain showed maximum biosorptions of cadmium and lead were 85.2 and 93.3%, respectively within 30 and 15 minutes at 60 and 80 μg ml^{-1} concentration of metal whereas, with an increase in biosorbent dosage lead to increase in the rate of biosorption. The most favorable pH for the absorption of lead was 5 and for the cadmium was 6, and the most favorable temperature was recorded as 40 °C. Sibi (2014) studied the bioremediation potential of few algal strains *Chlorella, Spirogyra, Oscillatioria, Pandorina,* and *Scenedesmus* found in freshwater against As(III and V). The outcome of this study showed that these algal species were As-resistant, the maximum concentration of As adsorption was found 0.8 gL^{-1} at 32 °C temperature with pH 4. The uptaking of metals by these microalgae species for arsenic (III) was much higher than the arsenic (V). All five algal sp. displayed variable growth rates in the presence of As. The growth phase showed by *Oscillatioria* was of 8–10 days, 11 days by *Pandorina,* 13 days by *Chlorella and Scenedesmus,* and 5 days by *Spirogyra.*

Kumar and Oommen (2012) studied the sorption capacity of *S. hyalina* for the remediation of multielements, viz., cobalt, mercury, lead, cadmium, and arsenic present in industrial effluent. The outcome of this study indicated that lead and cobalt adsorbed at 80 mgL^{-1} whereas, cadmium, mercury, and arsenic were adsorbed maximum at an initial concentration of 40 mgL^{-1}. The metal uptake order was found to be mercury>lead>cadmium>arsenic>cobalt by died algal biomass. Kumar et al. (2015) investigated the efficiency of mobilized cells of marine microalga *C. marina* to remediate diverse types of wastewaters, viz., household activities, industrial, and from aquaculture streams. The study resulted in the reduction of heavy metals. The concentration of chromium was reduced by 89% and lead by 87% in the aquaculture wastewater stream. The mobilized cells of *C. marina* were found efficient in reducing 85–88% of nitrite, 70–75% of nitrate, 60–64% of ammonia, and 51% of phosphorous. The concentration of biosorbent (cells of *C. marina*) was also found to increase from 3×10^6 to 1.5×10^7 cells/ml after an incubation period of 7 days. The percentage removal of heavy metal by different algal species in various wastewater, viz., municipal, urban, and industrial, is displayed in Table 13.1.

13.5.3.1 Mechanism of Heavy Metal Removal by Cyanobacteria

Absorption of heavy metals by algal strains and cyanobacteria is directly through the attachment by a cellular surface which is called as physical adsorption. The route through which a particular pollutant enters into the cytoplasm and finally degraded by the enzymatic system to breaking them into microelements is well-known as chemisorption (Zinicovscaia and Cepoi 2016). The cyanobacteria are capable of releasing thick polysaccharides of anionic nature in the culture medium, making these thick polymers available for biological applications. The mechanisms behind the removal of heavy metals by cyanobacteria are based on two principles via active uptake by cells (known as bioaccumulation) and by passive absorption of the metals to the cell surface or by released polymers (ion-exchange) (Mota et al. 2016).

A unicellular and marine cyanobacterium, *Cyanothece* sp. CCY 0110, earlier reported to be very efficient in secreting polysaccharides, and capable of removing copper, lead, and cadmium most commonly found in industrial wastewater. Basically, this polysaccharide polymer was the key player in metal removal, revealing that phycoremediation is mainly occurring by the mechanisms of biosorption (Mota et al. 2016).

The strains of algae are one of the most successful and efficient classes of microorganisms to remediate heavy metals from industrial wastewater as they can survive the high concentration of these contaminants (Ahmad et al. 2020). In

Table 13.1 The removal of heavy metal by different algal species in a different type of wastewater viz., (municipal, urban, and industrial)

Wastewater type	Microalgae sp.	Heavy metals	Per cent removal	Reference
Iron-mining industry	*Oscillatoria* sp. *Leptolyngbya* sp.	Iron, Chromium, Copper, Lead, and Nickel	–	Biglari Quchan Atigh et al. (2020)
Urban wastewater and agricultural drainage	*Chlorella vulgaris, Scenedesmus Quadricauda, Spirulina platensis*	Cadmium, Nickel, Lead	–	Abdel-Razek et al. (2019)
Industrial wastewater	*Botryococus brurauni*	Lead Cadmium Copper	93% 89% 82%	Uddin and Lall (2019)
Coal industry	*Nostoc* sp.	Chromium	–	Warjri and Syiem (2018)
Wet market wastewater	*Scenedesmus* sp.	Cadmium Chromium Iron Zinc	93.06% 91.5% 92.47% 92.40%	Apandi et al. (2018)
Industrial wastewater	*Chlorella* sp. *Scenedesmus* sp.	Calcium Calcium Magnesium Magnesium	56% 56% 59% 29%	Raikova et al. (2016)
Petrochemical wastewater	Consortium (*A. nodosum, F. spiralis, L. hyperborea*, and *P. canaliculata*)	Zinc Nickel Copper	93% 94% 94%	Cechinel et al. (2016)
Domestic sewage and industrial waste	*Botryococcus* sp.	Aluminum	41%	Ab Razak et al. (2016)
Industrial waste	*Cyanothece* sp. CCY 0110	Copper, Lead, cadmium	–	Mota et al. (2016)
Municipal wastewater	*Spirulina* sp.	Copper Calcium	91% 98%	Al-Homaidan et al. (2014)
Municipal wastewater	*Chlorella minutissima*	Zinc Manganese Cadmium Copper	62% 84% 74% 84%	Yang et al. (2015)
Domestic, industrial effluents, and aquaculture streams	*Chlorella marina*	Chromium Lead	89% 87%	Kumar et al. (2015)
Acid mine drainages	*Oedogonium* sp.	Copper Nickel Zinc Cobalt	46% 34% 48% 50%	Bakatula et al. (2014)

(continued)

Table 13.1 (continued)

Wastewater type	Microalgae sp.	Heavy metals	Per cent removal	Reference
Freshwater	*Chlorella, Spirogyra, Oscillatioria, Pandorina,* and *Scenedesmus*	As (III & V)		Sibi (2014))
Synthetic solution of heavy metals	*Nostoc muscorum*	Lead Cadmium	85.2% 93.3%	Dixit and Singh (2013)
Oil sands tailings ponds	*Cladophora fracta*	Copper Zinc	99% 85%	Mahdavi et al. (2012)
Industrial wastewater	*Spirogyra hyalina*	Cobalt, Mercury, Lead, Cadmium, and arsenic	–	Kumar and Oommen (2012)
Tannery wastewater	*Oscillatoria tenuis*	Chromium	–	Dwivedi et al. (2010)

addition to that, they possess a large surface area to absorb a significant quantity of toxins from wastewater. They are capable of growing either in autotrophic mode or in heterotrophic mode. Algae also have the potential for genetic manipulation (Jais et al. 2017). The peptide chains present on the outer surface tend to form organometallic complex after binding with heavy metal which in turn incorporates into the vacuoles to regulate the amount of heavy metals in the cytoplasm. In this manner, algal cells keep on checking the noxious impact of heavy metals. The peptide chains present on the algal surface are known as metallothioneins and phytochelatins (Balzano et al. 2020). Genes code the metallothioneins, and phytochelatins are synthesized by enzymatic activities. The phytochelatins are referred to as class-III metallothioneins. Class-II and class-III metallothioneins are present in algae, but class-I metallothioneins are absent (Sinaei et al. 2018). Synthesis of class-III metallothioneins can be induced by the action of heavy metals like cadmium, silver, zinc, mercury, and lead. Class-III metallothioneins are very crucial peptide molecules in algae as their presence makes the algal cells capable of withstanding a high amount of heavy metals (Balzano et al. 2020). Moreover, the synthesis of class-III metallothioneins is directly proportional to the degree of pollution.

13.5.4 Water Quality Improvement by Cyanobacteria

El-Bestawy (2008) explored the ability of *A. variabilis*, *T. ceylonica,* and *A. oryzae* in the enhancement of superiority of water originating from household and industrial activities. Maximum decline in biological oxygen demand (89%) and dissolved solids (39%) was recorded with *A. variabilis*, while the maximum decline in chemical oxygen demand (74%) with *A. oryzae* and 64% reduction in suspended solids with *T. ceylonica* in 7 days. The phases of growth and industrial effluent treatment potential of *S. platensis* were studied in swine wastewater in the absence of air by Cheunbarn and Peerapornpisal (2010). The highest decline of 45% in biological oxygen demand and 23% in chemical oxygen demand was noted on twelfth day with 10% dilution with sodium bicarbonate and sodium nitrate at $8.0 gL^{-1}$ and $1.5 gL^{-1}$, respectively. In another similar kind of study, *S. platensis* was found to remove 81.02% of chemical oxygen demand from effluent diluted with Zarrouk medium (13%) in 28 days (Magro et al. 2012).

Pandi et al. (2009) conducted a study to improve water quality in Retan chrome liquor by using *S. fusiformis*. The cyanobacterium was capable of removing 17–22.6% of total solids, 18–22.5% of dissolved solids, 15–23% of suspended solids, 73.8–78.9% of biological oxygen demand, 82.4–88.5% of chemical oxygen demand, and 93–99% of chromium (VI) in retan chrome liquor with varying chromium concentrations in the range of 100–300 mg/L. One more cyanobacterium, *N. muscorum* was found efficient in lowering the biological oxygen demand and chemical oxygen demand up to 54% and 69%, respectively from distillery wastewater in a month (Selvam et al. 2011).

Shankar et al. (2013) reported that *O. annae* efficiently eradicated 36% concentration of salts and biological oxygen demand from tannery wastewater in 15 days. However, after immobilized on coir pith as a media, displayed more efficient in reducing salt concentration and biological oxygen demand by 55 and 63%. Renuka et al. (2013a, b) revealed 99% reduction in chemical oxygen demand and 89% reduction in biological oxygen demand of sewage effluent by using cyanobacterium consortia comprising indigenous strains (*Phormidium* sp., *Anabaena* sp., *Westiellopsis* sp., *Limnothrix* sp., *Spirogyra* sp., and *Fischerella* sp.).

13.5.5 CO_2 Sequestration

The industrial revolution, globalization, and elevating needs for transportation have greatly increased the concentrations of greenhouse gases in the environment, mainly carbon dioxide, which have resulted in the rise of atmospheric temperature. In this perspective, sequestration of carbon dioxide is extremely important and recently one of the most researched fields around the world. Though various physio-chemical methods have been suggested, the biological sequestration remains efficient and promising one. Many telluric plants alleviate enormous amounts of carbon dioxide from the atmosphere; however, due to less percentage of carbon dioxide (0.04%) in the environment, the use of terrestrial plants is no more efficient. Moreover, the concentration of carbon dioxide present in the exhaust air released by industries is much higher than that already present in the atmosphere (10 to 20%). Consequently, there is a dire need to develop such strategies based on industrial emissions aforementioned. Therefore, carbon dioxide biofixation has appeared as a propitious option. The biochemical analyses have revealed that microalgal biomass comprises of 40–50% carbon, which suggests that approximately 1.5–2 kg of carbon dioxide is requisite to produce 1 kg of algal biomass (Sobczuk et al. 2000).

Kassim and Meng (2017) reported the carbon dioxide fixation by using two algae sp., i.e., *T. suecica* and *Chlorella* sp. under elevated carbon dioxide concentration.

The impact of altered concentration of CO_2 on these two algae sp. growth kinetics, biofixation, and its chemical composition were recorded using 0.04, 5, 15, and 30% carbon dioxide. The alteration in initial pH and its correlation on carbon dioxide concentration toward growth medium were also studied. The results obtained from this study resulted in assessing different tolerance mechanisms of both microalgae sp. toward carbon dioxide concentration. The maximum biomass production and biofixation for *Chlorella* sp. of 0.64 g/L and 96.89 mg/L/d were obtained when the cultivation was carried out using 5% and 15% carbon dioxide, respectively. In the contrary, the maximum biomass production and carbon dioxide biofixation for *T. suecica* of 0.72 g/L and 111.26 mg/L/d were obtained from cultivation using 15% and 5% carbon dioxide. The optimum pH value for the cultivation medium using carbon dioxide was in the range of 7.5 and 9, which is favorable for microalgal growth. In another study, biofixation of carbon dioxide was reported by cultivating microalgae *Chlorella* sp. at carbon dioxide concentrations (at 1.75% and 9.45% v/v) and gas flow rates (at 30, 50, and 70 ml/min) in a bubble column reactor (Pourjamshidian et al. 2019). The utmost specific growth rate of *Chlorella* sp. was obtained for the carbon dioxide concentration of 1.75% and the flow rate of 50 mL/min. The maximum biomass productivity rate (at 0.17 g/L/day) was for a sample with 1.75% carbon dioxide concentration and at a gas flow rate of 70 ml/min. Furthermore, the results have also indicated a direct relationship between the rate of growth and carbon dioxide sequestration with culturing of *Chlorella* sp. Consequently, the microalgae *Chlorella* sp. has a vast potential for the production of biofuel and carbon dioxide capturing so as to lessen the harmful impacts of greenhouse gases and global warming.

13.6 Conclusion

Literature review and studies from diverse sources indicated phycoremediation technology is proven as a green technology, eco-friendly, cost-effective, and easy to use approach. It produces no hazardous secondary by-products, and the remains of the process can be a source of biofuel production. The available data clearly states that cyanobacteria are imperative elements of the marine ecosystem and commonly found in contaminated habitat, as they have enormous potential to survive in a eutrophic environment. This book chapter is an attempt to bridge the existing literature highlighting the potential of cyanobacteria and in the form of the consortium to remediate contaminants present in municipal and industrial wastewater and their possible mechanism. Various researchers reported the efficacy of cyanobacteria for improving the water quality and fixation of carbon dioxide as well. Though, these scientific fields still require more depth knowledge and understanding. The basic dissimilarities in the features and composition of different effluents are the chief factors accountable for the different response of cyanobacteria for their potential for phycoremediation.

13.7 Future Perspectives

Phycoremediation approach is quite easy in the remediation of contaminants and employs re-use of water. It is a promising technique which could be used in small-scale plants and large- scale industrial units, by agriculturist, in villages which are lacking the municipal sewage treatment plants. By means of recombinant DNA technology and genetic manipulation of different algal species, this method can be made more beneficial for use. These future outlooks are as follows: (a) Classification of the wastewater which stimulates the synthesis of lipid in microalgae for better understanding of biofuel production, (b) Isolation of new microalgal species or development of mutated microalgae for effective treatment of toxic wastewater, (c) Optimization of growth parameter to develop an integrated approach for biomass production and nutrient removal from wastewater, (d) Optimization of parameters for biogas production from defatted microalgae biomass, (e) Identification or development of single microalgal strain for biodiesel/ethanol production. Moreover, genetically engineered algal strains with altered morphological and physiological characteristics by inserting catabolic genes into cyanobacteria and algae would be one of the encouraging tools for future scientific research to boost the aforementioned features in different species of algae. The development of transgenic strains and combination with existing technology will open new avenues for research. Additionally, it will also help in understanding the full potential as well as the implication of algae in treating wastewater. Hence, new and more species of cyanobacteria need to be explored for their remediation potential, which could ascertain their growth in harsh habitats, and also helpful in scaling up of such technologies in the future. The achievement in the prevalent use of phycoremediation technology will be dependent on a combined attempt by researchers and academicians.

Acknowledgments Dr. Simranjeet Singh is thankful to the Interdisciplinary Center for Water Research (IcWaR), Indian Institute of Sciences, Bangalore for the financial assistance, IOE-IIsc Fellowship (Sr. No: IE/REAC-20-0134), and providing laboratory and library facilities.

References

Ab Razak AR, Sunar NM, Alias NA, Gani P, Subramaniam M (2016) Physiochemicals and heavy metal removal from domestic wastewater via phycoremediation. In: MATEC Web of Conferences (Vol. 47, p. 05003). EDP Sciences

Abdel-Raouf N, Al-Homaidan AA, Ibraheem IBM (2012) Microalgae and wastewater treatment. Saudi J Biol Sci 19(3):257–275

Abdel-Razek MA, Abozeid AM, Eltholth MM, Abouelenien FA, El-Midany SA, Moustafa NY, Mohamed RA (2019) Bioremediation of a pesticide and selected heavy metals in wastewater from various sources using a consortium of microalgae and cyanobacteria. Slov Vet 56(Suppl 22):61–73

Abou-Shanab RAI, Ji M, Kim H, Paeng K, Jeon B (2013) Microalgal species growing on piggery wastewater as a valuable candidate for nutrient removal and biodiesel production. J Environ Manag 115:257–264

Ahmad S, Pandey A, Pathak VV, Tyagi VV, Kothari R (2020) Phycoremediation: algae as eco-friendly tools for the removal of heavy metals from wastewaters. In: Bioremediation of industrial waste for environmental safety. Springer, Singapore, pp 53–76

Al-Homaidan AA, Al-Houri HJ, Al-Hazzani AA, Elgaaly G, Moubayed NM (2014) Biosorption of copper ions from aqueous solutions by *Spirulina platensis* biomass. Arab J Chem 7(1):57–62

Apandi N, Mohamed RMSR, Al-Gheethi A, Latiffi A, Arifin SNH, Gani P (2018) Phycoremediation of heavy metals in wet market wastewater. IOP Conf Ser Earth Environ Sci 140(1):012017

Arias DM, García J, Uggetti E (2020a) Production of polymers by cyanobacteria grown in wastewater: current status, challenges and future perspectives. New Biotechnol 55:46–57

Arias DM, Uggetti E, García J (2020b) Assessing the potential of soil cyanobacteria for simultaneous wastewater treatment and carbohydrate-enriched biomass production. Algal Res 51: 102042

Badr SA, Ghazy ME, Moghazy RM (2010) Toxicity assessment of cyanobacteria in a wastewater plant, Egypt. J Appl Sci Res 6:1511–1516

Bakatula EN, Cukrowska EM, Weiersbye IM, Mihaly-Cozmuta L, Peter A, Tutu H (2014) Biosorption of trace elements from aqueous systems in gold mining sites by the filamentous green algae (*Oedogonium* sp.). J Geochem Explor 144:492–503

Balzano S, Sardo A, Blasio M, Chahine TB, Dell'Anno F, Sansone C, Brunet C (2020) Microalgae metallothioneins and phytochelatins and their potential use in bioremediation. Front Microbiol 11:517

Bernal CB, Vazquez G, Quintal IB, Bussy AN (2008) Microalgal dynamics in batch reactors for municipal wastewater treatment containing dairy sewage water. Water Air Soil Pollut 190:259–270

Bhatia D, Sharma NR, Kanwar R, Singh J (2018) Physicochemical assessment of industrial textile effluents of Punjab (India). Appl Water Sci 8(3):83

Bhatia D, Sharma NR, Singh J, Kanwar RS (2017) Biological methods for textile dye removal from wastewater: a review. Crit Rev Environ Sci Technol 47(19):1836–1876

Bhatnagar A, Chinnasamy S, Singh M, Das KC (2011) Renewable biomass production by mixotrophic algae in the presence of various carbon sources and wastewaters. Appl Energy 88:3425–3431

Bhola V, Swalaha F, Kumar RR, Singh M, Bux F (2014) Overview of the potential of microalgae for CO_2 sequestration. Int J Environ Sci Technol 11:2103–2118

Brar A, Kumar M, Vivekanand V, Pareek N (2017) Photoautotrophic microorganisms and bioremediation of industrial effluents: current status and future prospects. 3 Biotech 7(1):18

Brar A, Kumar M, Vivekanand V, Pareek N (2019) Phycoremediation of textile effluent-contaminated water bodies employing microalgae: nutrient sequestration and biomass production studies. Int J Environ Sci Technol 16(12):7757–7768

Cechinel MA, Mayer DA, Pozdniakova TA, Mazur LP, Boaventura RA, de Souza AAU, Vilar VJ (2016) Removal of metal ions from a petrochemical wastewater using brown macro-algae as natural cation-exchangers. Chem Eng J 286:1–15

Cheunbarn S, Peerapornpisal Y (2010) Cultivation of *Spirulina platensis* using anaerobically swine wastewater treatment effluent. Int J Agric Biol 12(4):586–590

Chinnasamy S, Bhatnagar A, Claxton R, Das KC (2010a) Biomass and bioenergy production potential of microalgae consortium in open and closed bioreactors using untreated carpet industry effluent as growth medium. Bioresour Technol 101:6751–6760

Chinnasamy S, Bhatnagar A, Hunt RW, Das KC (2010b) Microalgae cultivation in a wastewater dominated by carpet mill effluents for biofuel applications. Bioresour Technol 101(9): 3097–3105

Deviram G, Mathimani T, Anto S, Ahamed TS, Ananth DA, Pugazhendhi A (2020) Applications of microalgal and cyanobacterial biomass on a way to safe, cleaner and a sustainable environment. J Clean Prod 253:119770

Dewangan N (2015) Wastewater treatment using inverse fluidization unit by algae (Doctoral dissertation)

Dineshbabu G, Uma VS, Mathimani T, Deviram G, Ananth DA, Prabaharan D, Uma L (2017) On-site concurrent carbon dioxide sequestration from flue gas and calcite formation in ossein effluent by a marine cyanobacterium Phormidium valderianum BDU 20041. Energy Convers Manag 141:315–324

Dixit S, Singh DP (2013) Phycoremediation of lead and cadmium by employing *Nostoc muscorum* as biosorbent and optimization of its biosorption potential. Int J Phytoremediation 15(8): 801–813

Dubey SK, Dubey J, Viswas AJ, Tiwali P (2011) Studies on cyanobacterial biodiversity in paper mill and pharmaceutical industrial effluents. Biotechnol J Int 1:61–67

Dwivedi S, Srivastava S, Mishra S, Kumar A, Tripathi RD, Rai UN, Trivedi PK (2010) Characterization of native microalgal strains for their chromium bioaccumulation potential: phytoplankton response in polluted habitats. J Hazard Mater 173(1–3):95–101

El-Bestawy E (2008) Treatment of mixed domestic–industrial wastewater using cyanobacteria. J Ind Microbiol Biotechnol 35(11):1503–1516

Frampton DMF, Gurney RH, Dunstan GA, Clementson LA, Toifl MC, Pollard CB, Burn S, Jameson LD, Blackburn SI (2013) Evaluation of growth, nutrient utilization and production of bioproducts by a wastewater-isolated microalga. Bioresour Technol 130:261–268

Franchino M, Comino E, Bona F, Riggio VA (2013) Growth of three microalgae strains and nutrient removal from an agro-zootechnical digestate. Chemosphere 92:738–744

Furtado ALFF, do Carmo Calijuri M, Lorenzi AS, Honda RY, Genuário DB, Fiore MF (2009) Morphological and molecular characterization of cyanobacteria from a Brazilian facultative wastewater stabilization pond and evaluation of microcystin production. Hydrobiologia 627(1): 195–209

Fouilland E (2012) Biodiversity as a tool for waste phycoremediation and biomass production. Rev Environ Sci Biotechnol 11(1):1–4

Gani P, Mohamed N, Matias-Peralta H, Latiff AAA (2016) Application of phycoremediation technology in the treatment of food processing wastewater by freshwater microalgae Botryococcus sp. J Eng Appl Sci 11(11):7288–7292

Ghosh S, Barinova S, Keshri JP (2012) Diversity and seasonal variation of phytoplankton community in the Santragachi lake, West Bengal, India. QScience Connect 2012:3

Gupta V, Ratha SK, Sood A, Chaudhary V, Prasanna R (2013) New insights into the biodiversity and applications of cyanobacteria (blue–green algae)—prospects and challenges. Algal Res 2: 79–97

Biglari Quchan Atigh Z, Heidari A, Sepehr A, Bahreini M, Mahbub KR (2020) Bioremediation of heavy metal contaminated soils originated from Iron ore mine by bio-augmentation with native cyanobacteria. Iran J Energy Environ 11(2):89–96

Hongyang S, Yalei Z, Chunmin Z, Xuefei Z, Jinpeng L (2011) Cultivation of *Chlorella pyrenoidosa* in soybean processing wastewater. Bioresour Technol 102:9884–9890

Hussein NR, Gharib SM (2012) Studies on spatio-temporal dynamics of phytoplankton in El-Umum drain in west of Alexandria. Egypt J Biol 33:101–105

Jais NM, Mohamed RMSR, Al-Gheethi AA, Hashim MA (2017) The dual roles of phycoremediation of wet market wastewater for nutrients and heavy metals removal and microalgae biomass production. Clean Techn Environ Policy 19(1):37–52

Ji M, Abou-Shanab RAI, Kim S, Salama E, Lee S, Kabra AN, Lee Y, Hong S, Jeon B (2013) Cultivation of microalgae species in tertiary municipal wastewater supplemented with CO_2 for nutrient removal and biomass production. Ecol Eng 58:142–148

John J (2000) A self-sustainable remediation system for acidic mine voids. In: 4th International conference of diffuse pollution, pp. 506–511

Kassim MA, Meng TK (2017) Carbon dioxide (CO_2) biofixation by microalgae and its potential for biorefinery and biofuel production. Sci Total Environ 584:1121–1129

Khan FA, Ansari AA (2005) Eutrophication: an ecological vision. Bot Rev 71:449–482

Kim J, Lingaraju BP, Rheaume R, Lee J, Siddiqui KF (2010) Removal of ammonia from wastewater effluent by *Chlorella vulgaris*. Tsinghua Sci Technol 4:391–396

Kirkwood AE, Nalewajko C, Fulthorpe RR (2001) The occurrence of cyanobacteria in pulp and paper waste-treatment systems. Can J Microbiol 47:761–766

Kumar D, Santhanam SP, Jayalakshmi T, Nandakumar R, Ananth S, Devi AS, Prasath BB (2015) Excessive nutrients and heavy metals removal from diverse wastewaters using marine microalga *Chlorella marina* (butcher). India J Geo-Mar Sci 44(1):97–103

Kumar JN, Oommen C (2012) Removal of heavy metals by biosorption using freshwater alga *Spirogyra hyalina*. J Environ Biol 33(1):27

Lee RE (2008) Phycology, 4th edn. Cambridge University Press, New York

Li Y, Chen YF, Chen P, Min M, Zhou W, Martinez B, Ruan R (2011) Characterization of a microalga Chlorella sp. well adapted to highly concentrated municipal wastewater for nutrient removal and biodiesel production. Bioresour Technol 102(8):5138–5144

Lim S, Chu W, Phang S (2010) Use of *Chlorella vulgaris* for bioremediation of textile wastewater. Bioresour Technol 101:7314–7322

Magro CD, Deon MC, Rossi AD, Reinehr CO, Hemkemeier M, Colla LM (2012) Chromium (VI) biosorption and removal of chemical oxygen demand by *Spirulina platensis* from wastewater-supplemented culture medium. J Environ Sci Health A 47(12):1818–1824

Mahdavi H, Ulrich AC, Liu Y (2012) Metal removal from oil sands tailings pond water by indigenous micro-alga. Chemosphere 89(3):350–354

Makandar MB, Bhatnagar A (2010) Morphotypic diversity of microalgae in arid zones of Rajasthan. J Algal Biomass Utln 1:74–92

Markou G, Georgakakis D (2011) Cultivation of filamentous cyanobacteria (blue-green algae) in agro-industrial wastes and wastewaters: a review. Appl Energy 88(10):3389–3401

Martins J, Peixe L, Vasconcelos VM (2011) Unraveling cyanobacteria ecology in wastewater treatment plants (WWTP). Microb Ecol 62(2):241–256

Mata TM, Melo AC, Simoes M, Caetano NS (2012) Parametric study of a brewery effluent treatment by microalgae *Scenedesmus obliquus*. Bioresour Technol 107:151–158

Min M, Wang L, Li Y, Mohr MJ, Hu B, Zhou W, Chen P, Ruan R (2011) Cultivating Chlorella sp. in a pilot-scale photobioreactor using centrate wastewater for microalgae biomass production and wastewater nutrient removal. Appl Biochem Biotechnol 165:123–137

Mota R, Rossi F, Andrenelli L, Pereira SB, De Philippis R, Tamagnini P (2016) Released polysaccharides (RPS) from *Cyanothece* sp. CCY 0110 as biosorbent for heavy metals bioremediation: interactions between metals and RPS binding sites. Appl Microbiol Biotechnol 100(17):7765–7775

Pandi M, Shashirekha V, Swamy M (2009) Bioabsorption of chromium from retan chrome liquor by cyanobacteria. Microbiol Res 164(4):420–428

Patel A, Matsakas L, Rova U, Christakopoulos P (2019) A perspective on biotechnological applications of thermophilic microalgae and cyanobacteria. Bioresour Technol 278:424–434

Patel VK, Sundaram S, Patel AK, Kalra A (2018) Characterization of seven species of cyanobacteria for high-quality biomass production. Arab J Sci Eng 43(1):109–121

Pathak VV, Singh DP, Kothari R, Chopra AK (2014) Phycoremediation of textile wastewater by unicellular microalga *Chlorella pyrenoidosa*. Cell Mol Biol 60(5):35–40

Perera I, Subashchandrabose SR, Venkateswarlu K, Naidu R, Megharaj M (2018) Consortia of cyanobacteria/microalgae and bacteria in desert soils: an underexplored microbiota. Appl Microbiol Biotechnol 102(17):7351–7363

Pittman JK, Dean AP, Osundeko O (2011) The potential of sustainable algal biofuel production using wastewater resources. Bioresour Technol 102:17–25

Pourjamshidian R, Abolghasemi H, Esmaili M, Amrei HD, Parsa M, Rezaei S (2019) Carbon dioxide biofixation by *Chlorella* sp. in a bubble column reactor at different flow rates and CO2 concentrations. Braz J Chem Eng 36(2):639–645

Radjenovic J, Petrovic M, Barcelo D (2009) Fate and distribution of pharmaceuticals in wastewater and sewage sludge of the conventional activated sludge (CAS) and advanced membrane bioreactor (MBR) treatment. Water Res 43:831–841

Rai A, Wadhwa GK, Chakrabarty J, Dutta S (2020) Application of cyanobacterial consortium to remove ammoniacal-N, phenol, and nitrate from synthetic coke-oven wastewater as tertiary treatment. J Environ Eng 146(7):04020062

Rai J, Kumar D, Pandey LK, Yadav A, Gaur JP (2016) Potential of cyanobacterial biofilms in phosphate removal and biomass production. J Environ Manag 177:138–144

Raikova S, Smith-Baedorf H, Bransgrove R, Barlow O, Santomauro F, Wagner JL, Chuck CJ (2016) Assessing hydrothermal liquefaction for the production of bio-oil and enhanced metal recovery from microalgae cultivated on acid mine drainage. Fuel Process Technol 142:219–227

Rawat I, Kumar RR, Mutanda T, Bux F (2011) Dual role of microalgae: phycoremediation of domestic wastewater and biomass production for sustainable biofuels production. Appl Energy 88:3411–3424

Razzak SA, Hossain MM, Lucky RA, Bassi AS, de Lasa H (2013) Integrated CO_2 capture, wastewater treatment and biofuel production by microalgae culturing—a review. Renew Sust Energ Rev 27:622–653

Renuka N, Sood A, Prasanna R, Ahluwalia AS (2014) Influence of seasonal variation in water quality on the microalgal diversity of sewage wastewater. S Afr J Bot 90:137–145

Renuka N, Sood A, Ratha SK, Prasanna R, Ahluwalia AS (2013a) Nutrient sequestration, biomass production by microalgae and phytoremediation of sewage water. Int J Phytoremediation 15: 789–800

Renuka N, Sood A, Ratha SK, Prasanna R, Ahluwalia AS (2013b) Evaluation of microalgal consortia for treatment of primary treated sewage effluent and biomass production. J Appl Phycol 25(5):1529–1537

Selvam GG, Baskaran R, Mohan PM (2011) Microbial diversity and bioremediation of distilleries effluent. J Res Biol 1(3):153–162

Shankar AM, Henciya S, Malliga P (2013) Bioremediation of tannery effluent using fresh water cyanobacterium *Oscillatoria annae* with coir pith. Int J Environ Sci 3(6):1881–1890

Sibi G (2014) Biosorption of arsenic by living and dried biomass of fresh water microalgae-potentials and equilibrium studies. J Bioremed Biodegr 5(6):249

Sinaei M, Loghmani M, Bolouki M (2018) Application of biomarkers in brown algae (Cystoseria indica) to assess heavy metals (Cd, Cu, Zn, Pb, Hg, Ni, Cr) pollution in the northern coasts of the Gulf of Oman. Ecotoxicol Environ Saf 164:675–680

Singh AK, Sharma N, Farooqi H, Abdin MZ, Mock T, Kumar S (2017) Phycoremediation of municipal wastewater by microalgae to produce biofuel. Int J Phytoremediation 19(9):805–812

Sobczuk TM, Camacho FG, Rubio FC, Fernández FA, Grima EM (2000) Carbon dioxide uptake efficiency by outdoor microalgal cultures in tubular airlift photobioreactors. Biotechnol Bioeng 67(4):465–475

Solanki P, Dotaniya ML, Khanna N, Udayakumar S, Dotaniya CK, Meena SS, Srivastava RK (2019) Phycoremediation of industrial effluents contaminated soils. In: New and future developments in microbial biotechnology and bioengineering. Elsevier, Netherlands, pp 245–258

Sood A, Uniyal PL, Prasanna R, Ahluwalia AS (2012) Phytoremediation potential of aquatic macrophyte, Azolla. Ambio 41:122–137

Sood A, Renuka N, Prasanna R, Ahluwalia AS (2015) Cyanobacteria as potential options for wastewater treatment. In: Phytoremediation. Springer, Cham, pp 83–93

Souza PO, Ferreira LR, Pires NRX, Filho PJS, Duarte FA, Pereira CMP, Mesko MF (2012) Algae of economic importance that accumulate cadmium and lead: a review. Braz J Pharmacognosy 22:825–837

Starckx S (2012) A place in the sun—algae is the crop of the future, according to researchers in Geel, Flanders today. http://www.flanderstoday.eu/content/place-sun

Su Y, Mennerich A, Urbana B (2012) Synergistic cooperation between wastewater-born algae and activated sludge for wastewater treatment: influence of algae and sludge inoculation ratios. Bioresour Technol 105:67–73

Uddin A, Lall AM (2019) Phycoremediation of heavy metals by *Botryococus brurauni* from wastewater. Biosci Biotechnol Res Asia 16(1):129–133

Vijayakumar S, Thajuddin N, Manoharan C (2007) Biodiversity of cyanobacteria in industrial effluents. Acta Bot Malacitana 32:27–34

Warjri SM, Syiem MB (2018) Analysis of biosorption parameters, equilibrium isotherms and thermodynamic studies of chromium (VI) uptake by a *Nostoc* sp. isolated from a coal mining site in Meghalaya, India. Mine Water Environ 37(4):713–723

Yang J, Cao J, Xing G, Yuan H (2015) Lipid production combined with biosorption and bioaccumulation of cadmium, copper, manganese and zinc by oleaginous microalgae *Chlorella minutissima* UTEX2341. Bioresour Technol 175:537–544

Zhou GJ, Ying GG, Liu S, Zhou LJ, Chen ZF, Peng FQ (2014) Simultaneous removal of inorganic and organic compounds in wastewater by freshwater green microalgae. Environ Sci: Processes Impacts 16(8):2018–2027

Zinicovscaia I, Cepoi L (eds) (2016) Cyanobacteria for bioremediation of wastewaters. Springer International Publishing, Berlin

Environmental Resilience and Circular Agronomy Using Cyanobacteria Grown in Wastewater and Supplemented with Industrial Flue Gas Mitigation

14

Vivek Dalvi, Krutika Patil, Harshita Nigam, Rahul Jain, Sunil Pabbi, and Anushree Malik

Abstract

Soaring levels of wastewater discharge into the local water bodies and flue gas emissions into the atmospheric environment have emerged as a global challenge in the past few decades. Cyanobacteria, commonly known as blue-green algae, uphold flair ability for the reclamation of these environmental matters of global concern. These microscopic photosynthetic microalgae can utilize the contaminants present in the wastewaters such as nitrogen, phosphorus for their own growth/metabolism and play a pivotal role in the removal of contaminants of emerging concern such as heavy metals, fertilizers, pharmaceutical wastes, and personal care products from a variety of wastewaters. These cyanobacteria also have wide applicability to mitigate industrial flue gases through its photo-CO_2-sequestration process.

The biomass generated through the application of cyanobacterial phycoremediation process for wastewaters and flue gas mitigation has important agroecological characteristics for application as soil conditioner/biofertilizer. As cyanobacteria possess properties such as atmospheric N_2 fixation, production of extra polymeric, and growth-promoting substances, their use as bioinoculant helps to improve soil fertility and productivity. Their ubiquitous nature and short regeneration span make these an ideal option for environmental resilience and circular agronomy. This chapter presents an overview of the tangible cyanobacteria-based phycoremediation technologies (implemented in the last

V. Dalvi · H. Nigam · R. Jain · A. Malik (✉)
Applied Microbiology Laboratory, Center for Rural Development & Technology, Indian Institute of Technology, Delhi, Hauz Khas, New Delhi, India

K. Patil · S. Pabbi
Division of Microbiology, Centre for Conservation and Utilisation of Blue Green Algae, ICAR—Indian Agricultural Research Institute, New Delhi, India

R. P. Rastogi (ed.), *Ecophysiology and Biochemistry of Cyanobacteria*,
https://doi.org/10.1007/978-981-16-4873-1_14

10 years), producing a large amount of biomass as a byproduct, and it is an on-field application for agriculture.

Keywords

Cyanobacteria · Phycoremediation · Environmental resilience · Circular agronomy · Biofertilizer · Soil conditioner

14.1 Introduction

The twentieth century is the century of rapid urbanization and industrialization with resource-intensive end-user centric product manufacturing processes (Lin et al. 2018). The adopted technologies are solely dependent upon fossil-derived fuels, such as coal for electricity, natural oil, and gas for transportation and cooking fuel. The advanced process used in energy production, petroleum refining, fertilizer, cement production, textile, and fast-moving consumer goods production industries contributes to the environmental deterioration by unattended emissions of wastewater into the natural waterways and flue gases into the atmosphere (Mezynska et al. 2018; Ebenstein et al. 2012).

These emissions into natural waterways include pollutants such as carbon (in the form, chemical oxygen demand, COD), nitrogen (ammoniacal nitrogen, nitrate, nitrite), phosphorus (in the form of phosphates), heavy metals, and pharmaceutical and personal care products (PPCPs). These pollutants are the sole cause of freshwater eutrophication and exhibit negative impacts on aquatic living systems (Bystrzejewska-Piotrowska et al. 2009; Guerra et al. 2014). The emissions into the atmospheric environment contained within flue gases include nitrogen dioxide (NOx), Sulfur oxides (SOx), particulate matters (PM), and greenhouse gases (GHG) such as carbon dioxide (CO_2) and methane (CH_4) (Thitakamol et al. 2012). There are increasing concerns for global climate change and thus heightened interest worldwide for reducing the emissions of GHG, mainly CO_2.

Cyanobacteria-based microbial processes possess the ability to treat the pollutants emitted into the environment by photosynthesis. Through photosynthesis, cyanobacteria consume organic and inorganic pollutants (carbon, nitrogen, and phosphorus) from wastewaters for their growth (Pacheco et al. 2015; Erb and Zarzycki 2018). Further, the cyanobacterial photosynthetic processes have shown potential in the removal of heavy metals and PPCPs (Cuellar-Bermudez et al., 2017). Also, cyanobacterial biomass generated during (wastewater and flue gas) remediation favors further application in agriculture as biofertilizer (Suleiman et al. 2020; Castro et al. 2020). Their ability as biofertilizer to fix atmospheric nitrogen, to carry out oxygenic photosynthesis, production of phytohormones, bioactive compounds, improve soil structure and fertility, and thereby increase crop productivity (Pabbi 2008; Singh et al. 2016a, b; Mogor et al. 2018; Renuka et al. 2018; Ronga et al. 2019; Deviram et al. 2020). Cyanobacteria are of significant stature due to their

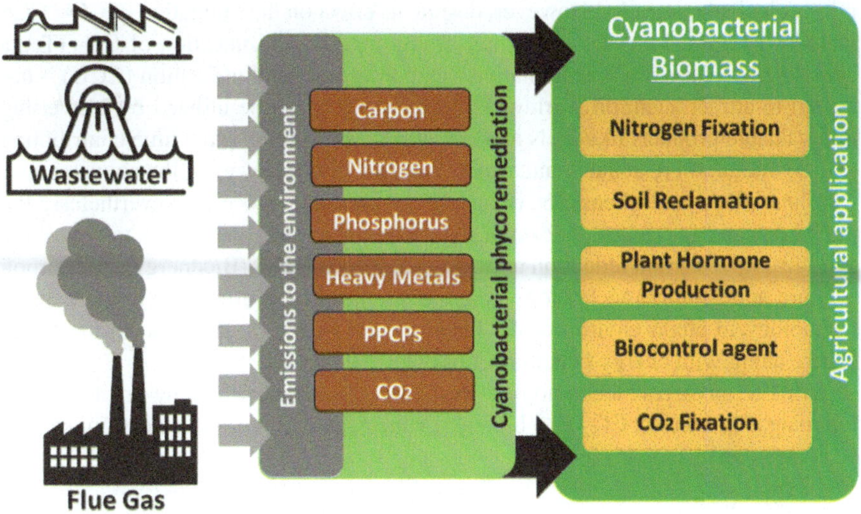

Fig. 14.1 Multifaceted role of cyanobacteria for wastewater remediation, flue gas mitigation, and sustainable agriculture

multifaceted role in wastewater remediation, flue gas mitigation, and agriculture as shown in Fig. 14.1.

14.2 Global Challenge of Wastewater Disposal and Flue Gas Mitigation

Around the globe, 80% of wastewater is discharged into the water streams without proper treatment. The significant difference is observed in wastewater treatment capacity in developed (around 70%), developing (around 35%), and undeveloped countries (around 30%). As untreated wastewater disposal has a direct or indirect impact on the environment, health, and economy, it is the need of the hour to take into consideration wastewater treatment using environmentally resilient technologies (WWAP (United Nations World Water Assessment Programme 2017).

Like this, the case with flue gas emissions has been observed where special attention to CO_2 mitigation is needed. Various policies have been made and implemented, which include emission standards, carbon tax which is a cost-effective tool in achieving CO_2-free environment. However, these policies have their limitations. Emission standards only restrict the emissions of air pollutants, unable for complete removal of CO_2 globally. In the case of a carbon tax, identifying the amount of taxable carbon at an industrial scale is a tedious job. Apart from this, the carbon tax elevates the price of related products along with the cost of enterprise and enforces a negative effect on economic growth (Lin and Li 2011).

One of the potential technologies for CO_2 emission is CO_2 capture and storage (CCS). CCS includes capturing CO_2 from flue gases and transporting it for injection into geological reservoirs. However, despite its effect on the mitigation of CO_2, CCS includes additional costs of capturing, transportation, and injection of CO_2 (Tapia et al. 2018). Therefore, an alternative - carbon capture and utilization (CCU) – has started to attract attention worldwide. In CCU, CO_2 can be utilized by converting into valuable products like fuels and chemicals, while at the same time contributing to CO_2 mitigation (Cuéllar-Franca and Azapagic 2015). However, the conversion of CO_2 into fuels and chemicals is an energy-intensive process. Nevertheless, the global demand for chemicals cannot sequester enough CO_2 emissions to contribute significantly to carbon reduction targets. Therefore, BECCS (Bioenergy with Carbon Capture and Storage) technology came as the novel and effective carbon capture technology to attain environmental benefits with zero CO_2 emission to the atmosphere along with energy benefits (Choi et al. 2019).

BECCS involves technologies such as cyanobacteria (blue-green algae), microalgae to convert CO_2 into biomass, which can be used in further applications such as biofertilizers, biofuel.

14.3 Cyanobacterial Wastewater Treatment

14.3.1 Types of Cyanobacterial WWT Systems

Cyanobacterial wastewater treatment systems have been broadly divided into two categories: (a) Suspension reactors and (b) Biofilm reactors. Suspension systems consist of open cultivation systems and closed photobioreactors. All these reactor systems are elaborated below:

14.3.1.1 High Rate Algal Pond (HRAP)

It is an open pond wastewater treatment system equipped with paddle wheels for continuous algae circulation and homogeneous mixing. They have the simplest construction design with the lowest cost (Young et al. 2017). Sunlight is the primary source of photosynthetic active light delivery to cyanobacterial cells; therefore, the pond depths vary from 20 to 30 cm for optimal light penetration. Wastewater treatment time of 4–10 days is typically required to obtain the desired remediation levels (Arashiro et al. 2019). However, they are more efficient in nutrient removal than waste stabilization ponds (WSP) that typically take longer wastewater treatment time 30–60 days (Ragush et al. 2017). Harvesting of cyanobacteria biomass is often challenging and adds to the cultivation cost. Chemical flocculants or auto sedimentation due to gravity are commonly employed strategies for biomass harvesting (Mubarak et al. 2019). Due to low costs related to them, they have been demonstrated at a larger scale in New Zealand and the United States (Craggs et al. 2014, Sutherland et al. 2014).

14.3.1.2 Closed Photobioreactors (PBRs)

They are closed suspension cultures with a wide range of designs and more excellent operational controls. The growth parameters like culture pH, temperature, light intensity, O_2, and CO_2 concentrations can be regulated to give maximum biomass yield and more significant wastewater remediation (Acién et al. 2017; Trebuch et al. 2020). PBRs are effective in cultivating a pure cyanobacterial strain with fewer chances of microbial contamination. The wastewater is re-circulated in the reactor using a high energy pump to remove excess O_2 produced during photosynthesis by degassing. All these advanced controls lead to higher installation and operational costs (Kumar et al. 2011; Vo et al. 2019). PBRs are generally employed to produce valuable products from cyanobacterial biomass with the utilization of wastewater as a growth medium (Troschl et al. 2018). PBRs are broadly classified based on reactor geometry and operation mode. In terms of geometrical design, the major category of reactors includes (a) flat, (b) vertical, horizontal, inclined, (c) tubular, (d) serpentine, (e) floating, and (f) spiral. Operational classification of PBRs includes (a) aeration or without aeration and (b) single-phase systems (where the gas exchange takes place in different gas exchanger) or two-phase systems (where both media and gas are present in the same vessel that helps in steady gas mass transfer) (Zittelli et al. 2013; Acién et al. 2017). Different types of PBRs have been designed based on the application area. Polyethylene bags or sleeves supported within a mesh frame and sparged by air are the most common cultivation systems used in hatcheries. The Near-Horizontal Tubular Reactor (NHTR) was developed to mitigate CO_2 emissions from the power plants by cultivating microalgae (Ugwu et al. 2008; Tredici et al. 2009). Similarly, vertically arranged manifold PBRs were constructed at an industrial scale to produce cosmetic, food, and pharmaceutical from microalgae (Torzillo et al. 2015). Annular column reactors that are made of two concentric glass cylinders have been used for oil production from microalgae. A "windy, wavy, and wiped" tubular photobioreactor (WWW-PBR) was developed to cultivate fragile and slow-growing algal species (Morweiser et al. 2010; Zittelli et al. 2013). Torous photobioreactor that enables light to be highly regulated in addition to very efficient mixing was used for H_2 production from microalgae (Pruvost et al. 2006). A UV-resistant transparent film PBR with particular coatings was used to cultivate genetically engineered cyanobacterium for the production of ethanol (Luo et al. 2010, Woods et al. 2010) . Various other configurations of the PBRs with their features are highlighted in Table 14.1.

14.3.1.3 Biofilm Reactors

The biofilm system consists of a support material on which cyanobacterial biofilm formation takes place. Broadly, two types of biofilm processes exist, one in which wastewater or medium flows over the growth substrate and the other in which the substrate is immersed in the nutrient medium for biofilm development (Hoh et al. 2016). Biomass cultivated is usually a consortium of mixed strains and forms a multilayer pattern. The harvesting is performed using mechanical scrapping and is relatively easy with low energy input (Choudhary et al. 2017). Recently, several works have shown better wastewater treatment and biomass production potential of biofilm reactors compared to suspended systems (Hodges et al. 2017; Tang et al. 2018). Various configurations of biofilm reactors have been constructed for wastewater treatment. These include algal turf scrubber (Mulbry et al. 2008), rotating

Table 14.1 Different types of closed photobioreactors (PBRs)

PBR configuration	Design types	Advantages	Drawbacks	Source
Vertical column	• Bubble column • Internal-loop (draft-tube) airlift • Split-column airlift • External-loop airlift	• High volumetric gas transfer coefficients • Homogeneous mixing with little shear stress	• Unequal distribution of light • Biofilm formation on reactor surface	Pawar (2016)
Flat panel (FP)	• Pump-drive FP-PBR • Airlift FP-PBR • Vertical alveolar panel PBR (VAP)	• Large illuminated surface to volume ratio	• Expensive • Deficiencies in culture flow control	Banerjee and Ramaswamy (2019)
Tubular	• Horizontal tubular • Vertical tubular • Helical tubular	• Most popular configurations of PBRs • High biomass productivity • Greater operational control • Flexibility in design	• Expensive temperature control systems • Water spraying: require large amount of water for cooling	García et al. (2018)
Plastic bag	• Polyethylene bag • Disposable plastic bag	• Low cost • Good sterility at start up	• Disposal of used plastic bags • Inadequate mixing	Chen et al. (2018)
Membrane PBR	• Membrane-sparged helical tubular PBR • Membrane-sparged horizontal tubular PBR	• Large surface areas facilitate gas/liquid mass transfer • Energy costs minimized • No shear stress to cells due to pumping or circulation	• Membranes need to be changed at regular intervals • Blocking of membranes	Chang et al. (2019)

biofilm reactor (Christenson and Sims 2012), attached biofilm reactor (Choudhary et al. 2017), flow lane biofilm reactor (Guzzon et al. 2019), vertical biofilm reactor (Zhang et al. 2018), polystyrene foam biofilm reactor (Ozkan et al. 2012), twin layer biofilm reactor (Goeres et al. 2020; Hoh et al. 2016), waveguide biofilm reactor (Genin et al. 2015), etc. Besides reactor designs, much research has been done to select effective support mediums for best cell attachment (Schnurr and Allen 2015). Properties like surface roughness, charge, material surface energy, pore size, surface area, and three-dimensional behavior of support material play an important role in support material selection and biofilm formation (Genin et al. 2014; de Assis et al. 2019).

14.3.2 Nutrients Removal

Cyanobacteria have been extensively studied for its wastewater treatment potential. Various studies highlight the cyanobacteria capability for the efficient removal of various contaminants from industrial, agricultural, and domestic wastewaters. Table 14.2 depicts some of the cyanobacterial strains and their wastewater treatment potential. Some of the significant nutrients consumed by cyanobacteria are discussed below:

Carbon: It is a crucial nutrient consumed by cyanobacteria in both the available forms: organic and inorganic. Inorganic carbon in the form of CO_2 and HCO_3^- is utilized by cyanobacteria through its CO_2 concentrating mechanism (Singh et al. 2016b). Organic carbon in the form of sugars, fatty, and amino acids is consumed by cyanobacterial strains, and it is species-dependent.

Nitrogen: It is an essential component required by cyanobacteria and consumed in various forms like NO_3^-, NO_2^- or NH_4^+, and N_2. NH_4^+ is the preferred form of nitrogen by cyanobacteria, followed by NO_3^- and N_2 (Andersen et al. 2020). If NH_4^+ is available in the water, cyanobacteria do not consume other forms of nitrogen even if they are present in abundance. The glutamine synthetase enzyme system is utilized for ammonium ions assimilation (Chawla et al. 2020). However, for the utilization of nitrate and nitrite, both are first reduced to ammonium using nitrate reductase and nitrite reductase enzymes, respectively. This intracellular reduction process is energy-intensive. The presence of an excess amount of ammonium ions has also proved toxic to cyanobacterial cells, and the threshold ammonium concentration is species-specific that varies with the cyanobacterial strain (Rossi et al. 2020).

Phosphorus: Although the cyanobacterial cell is composed of less than 1% phosphorus, but it is a significant growth-limiting factor. Cyanobacteria consume it in the form of orthophosphate (PO_4^{3-}) from the wastewater. Although the pentavalent form of phosphorus is dominant in water, it is hydrolyzed to PO_4^{3-} by extracellular enzymes. The uptake of phosphorus is energy-intensive and faster in light compared to dark (Gismondi et al. 2016; Kube et al. 2018).

Besides, these vital nutrients, cyanobacteria also consume other micro and macroelements from the surroundings for their growth and cell metabolism. Macroelements include calcium, potassium, and magnesium, whereas the micronutrients consist of iron, copper, zinc, cobalt, molybdenum, and boron (Fawzy and Mohamed 2017; Singh and Seneviratne 2017; Kulal et al. 2020).

Table 14.2 Nutrient removal using different cyanobacteria

Cyanobacteria	Wastewater Type	COD Removal	Nitrogen Removal	Phosphorus Removal	Source
Arthrospira platensis	Poultry litter wastewater	–	–s	99%	Markou et al. (2016)
Spirulina platensis	Olive-oil mill wastewater	73%	99%	99%	Markou et al. (2012)
Synechocystis salina	OECD test medium		52.3%	77%	Gonçalves et al. (2016a)
Scytonema hyalinum	Municipal wastewater	81.63%	90.64%	97.08%	Wu et al. (2020)
Oscillatoria species	Dairy wastewater	92.5%	75.2%	86%	Kabariya and Ramani (2018)
Phormidium species	Dairy wastewater	94.2%	81.7%	94%	Kabariya and Ramani (2018)
Microcystis aeruginosa	OECD test medium	–	73.9%	59.1	Gonçalves et al. (2016b)
Anabaena augstumalis	Simulated domestic wastewater	–	–	28%	Gismondi et al. (2016)
Calothrix sp.	Simulated domestic wastewater	–	–	43%	Gismondi et al. (2016)
Nostoc sp.	Simulated domestic wastewater	–	–	23%	Gismondi et al. (2016)
Consortium: *Chlorella and Phormidium*	Parboiled rice effluent	80.8%	99%	94.7%	Mukherjee et al. (2016)
Synechocystis sp	Wastewater dewatering process of anaerobically digested sludge	–	69%	–	Hughes et al. (2018)

14.3.3 Removal of Heavy Metals and PPCPs

Over the years, cyanobacteria have proved their potential in the removal of heavy metals, such as **copper** (Cu), **iron** (Fe), **zinc** (Zn), **mercury** (Hg), **lead** (Pb), **nickel** (Ni), **chromium** (Cr), **arsenic** (As), **cadmium** (Cd), and **silver** (Ag) (Hashim and Chu 2004; Pavasant et al. 2006; Saavedra et al. 2018) and contaminants of emerging concern used as pharmaceuticals, such as **endocrine-disrupting compounds**

(Estrone (E1), 17-Beta-estradiol (E2), 17-Alpha-ethinylestradiol (EE2), Bisphenol A, Octylphenol), **pesticides** (Atrazine, Diazinon, Benzothiazole, OH-Benzothiazole, Terbutrin, Triclosan), **personal care products** (Cashmeran, Celestolide, Galaxolide, Hydrocinnamic acid, Methyl dihydrojasmonate, Methylparaben, Oxybenzone), **pharmaceutical analgesics** (Carbamazepine, Codeine, Diclofenac, Ibuprofen, Phenazone, Naproxen), and **antibiotics** (Sulfathiazole, Sulfapyridine, Sulfamethazine, Sulfamethoxazole, Tetracycline, Oxytetracycline, Azithromycin) (Yu et al. 2006; Behera et al. 2011; Guerra et al. 2014; Ebele et al. 2017). The removal performance of these contaminants of emerging concern is listed in Table 14.3. The removal of heavy metals and pharmaceutical compounds occurs through mechanisms (Fig. 14.2), such as biosorption, bioaccumulation, photolysis, and intracellular and extracellular biodegradation, which are explained in detail below:

14.3.3.1 Biosorption

Biosorption is the key removal mechanism for heavy metals remediation shown by cyanobacteria (Chan et al. 2014; Escapa et al. 2017). The extracellular polymeric substances (EPS) released by cyanobacteria aid the process of metal absorption. Cyanobacterial EPS are compound heteropolysaccharides of anionic nature, consisting of different monosaccharides belonging to hexoses, pentoses, deoxyhexoses, and acidic hexoses. The anionic nature of EPS attracts cationic metals and pharmaceutical compounds for conjugate formation (Pandey 2017).

14.3.3.2 Bioaccumulation

The process of uptake of metals and PPCPs through metabolic activity is known as bioaccumulation. The bioaccumulation capability of the remediation process is dependent upon many factors, such as initial contaminants (metals and pharmaceuticals), residence time, pH, temperature, light irradiance, salinity, physiological state, and density of the cell in the system (Khan et al. 2017). The high concentration of these contaminants in wastewater is detrimental to the growth of cyanobacteria causing oxidative stress, which leads to reactive oxygen species formation (Pandey 2017).

14.3.3.3 Photolysis

Sunlight plays a vital role in the growth of cyanobacteria during wastewater remediation. The sunlight has a significant impact on the photolysis of pharmaceutical compounds. The photolysis of pharmaceutical compounds involves direct and indirect photolysis (Sui et al. 2015). Under direct photolysis, pharmaceutical compounds absorb direct sunlight and get broken down, while in indirect photolysis, exogenously present photosensitizers produce free radicals such as hydroxyl radicals (OH^-), peroxyl radicals (ROO^-), and singlet oxygen ($^1O_2$) under sunlight illumination. These exogenously produced free radicals to aid the process of breaking down of pharmaceutical compounds (Xiong et al. 2018; Vo et al. 2019).

Table 14.3 Removal of heavy metals and pharmaceutical compound using different cyanobacteria

	Cyanobacteria	Heavy metal removal	Removal mechanism	Source
Heavy metals	*Nostoc sphaeroides Kützing*	Cu, Cd, Cr, Pb, Ni, and Mn	Majorly adsorption, selective biosorption of Pb and Cr	Jiang et al. (2015)
	Chroococcus multicoloratus, Oscillatoria trichoides Szafer	Pb	Biosorption	Miranda et al. (2012)
	Nostoc ellipsosporum	As	Adsorption through EPS	Datta and Bhaduri (2020)
	Dolichospermum flos-aquae NTMS07	Cr	Bioaccumulation	Kumar et al. (2013a)
	Nostoc muscorum	Zn, Cu	Biosorption, influenced by pH, temp, inoculum age, and size	Diengdoh et al. (2017), Goswami et al. (2015)
	Synechococcus elongatus	^{234}U	Adsorption	Vijayaraghavan et al. (2018)
Pharmaceutical compounds	*Anabaena flos-aquae*	Tylosin, lincomycin, and trimethoprim	Photolysis, biodegradation	Guo et al. (2016)
	Anabaena sp., *Nostoc ellipsosporum*	Lindane, 4-chlorobenzoate, 4-iodobenzoate, 4-Hydroxybenzoate	Biodegradation, bioaccumulation	Kurtiz et al. (1995)
	Anabaena flos-aquae	Endocrine disruptors- phthalate esters	Photolysis, biodegradation	Babu and Wu (2010)
	Anabaena azotica	Organochlorine pesticide γ-hexachlorocyclohexane	Bioaccumulation, photolysis, biodegradation	(Zhang et al., 2012)
	Anabaena PD-1	Polychlorobiphenyl	Photolysis, biodegradation	Zhang et al. (2014)
	Nostoc sp. *and Anabaena* sp.	DDE (1,1-dichloro-2,2-bis (p- chlorophenyl) ethylene) and DDD (1,1-dichloro-2,2- bis L. Cepoi et al. 41 (p-chlorophenyl) ethane)	Photolysis, biosorption	Cepoi et al. (2016)
	Microcystis aeruginosa	Triclosan	Photolysis	Huang et al. (2016)

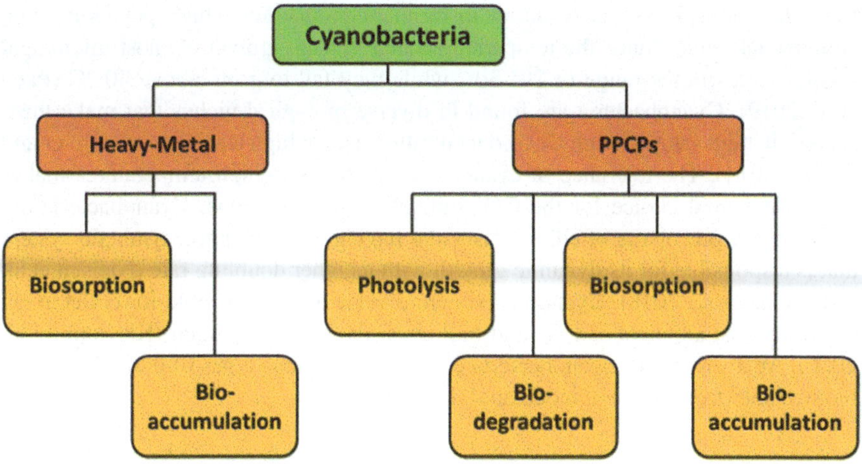

Fig. 14.2 Removal mechanism of heavy metals and pharmaceutical compounds by cyanobacteria

14.3.3.4 Biodegradation

These photosynthetic microorganisms have a complex enzyme system that plays a crucial role in the biodegradation of pharmaceutical compounds. These enzyme systems function in two phases, Phase-I (cytochrome450) and Phase-II (glutathione-S-transferases). The biodegradation process begins with the Phase-I enzyme, which makes pharmaceutical compounds hydrophilic by adding a hydroxyl group through reactions such as hydrolysis, oxidation, or reduction. Different enzymatic reactions such as hydroxylation, decarboxylation, dehydroxylation, carboxylation, hydrogenation, glycosylation, demethylation, ring cleavage, and bromination are also observed during biodegradation (Behera et al. 2011; Ebele et al. 2017; Vo et al. 2019). Phase-II enzymes carry out the conjugation reaction between electrophilic compounds and glutathione, which results in the opening of an epoxide ring for oxidative damage protection. Many enzymes such as pyruvate dehydrogenase, glutamyl-tRNA reductase, mono(di)oxygenase, soluble inorganic, pyrophosphatase, carboxylase/decarboxylase, dehydratase, alkaline and acid phosphatase, transferase, catalase, etc. play a key role in endogenous biodegradation (Evgenidou et al. 2015; Bai and Acharya 2017; Xiong et al. 2018). However, the exact role of these enzymes is not well established.

14.4 Cyanobacterial Flue Gas Mitigation

Among various techniques that are being projected for the sequestration of CO_2 and mitigate the outcome of greenhouse gases, the photosynthetic organisms are driving interests toward their efficient capability of sequestering CO_2 and coordinately mitigate the CO_2 level in the atmosphere. To meet this goal, researchers have been focusing on microalgae and cyanobacteria for proficiently reducing the CO_2 level from the atmosphere and resulting in biomass production which possesses high commercial value. Since the temperature range for the growth of most microalgal species varies in the range of 20°-30°, while they fail to grow above 40 °C (Patel et al. 2019). Cyanobacteria are found in diverse biological niches that make them survive in high salinity, high CO_2 concentration, and at high temperature (Oliver and Atsumi 2014). The tolerance of cyanobacterial strains to high temperatures makes them a potential choice for the CO_2 mitigation from flue gas. Cyanobacteria are evolutionary precursors of chloroplasts and having a simple photosynthetic system that is responsible for its dynamic growth with a higher doubling rate (Nielsen et al. 2016; Giannuzzi 2018). Cyanobacteria are diverse and responsible for collectively capturing vast amounts of CO_2 and biological fixation of nitrogen (Klawonn et al. 2016). Apart from this, cyanobacteria can provide various types of biofuels such as biomethane (from biogas), bioethanol, biohydrogen, etc.

14.4.1 CO_2 Sequestration by Cyanobacteria

The CO_2 sequestration by cyanobacteria is 10–15 times higher compared with terrestrial plants. Therefore, the utilization of these biological entities could be a practical approach in reducing the CO_2 concentration from the atmosphere and thus help in mitigating CO_2. It was observed that phytoplanktons were responsible for half of the global photosynthesis, and among that, 25% is accounted by marine cyanobacteria, dominantly *Synechococcus* and *Prochlorococcus*. Since the flue gas possesses a high concentration of CO_2 (up to 20%) and has a high temperature (~120 °C), the use of thermophilic cyanobacteria would be the most promising approach which is tolerant to temperature as well as high CO_2 concentration. However, the flue gas contains SOx, NOx, HCl, heavy metals, and other pollutants. However, thermophilic cyanobacteria can remove the impurities of SOx, NOx, etc. up to certain limits along with the CO_2 sequestration (Singh et al. 2016b). *Thermosynechococcus elongatus* TA-1, thermophilic cyanobacteria isolated from hot springs in Taiwan, showed enhanced growth in 10% or 20% CO_2 at a light intensity 6000 lux (Patel et al. 2019). Prominently the increased growth promotes the synthesis of phycocyanin (C-PC), an economically viable product, signifying that the CO_2 mitigation from industrial flue gas could be efficient in the production of high-value products. A study reported that *Thermosynechococcus elongatus* PKUAC-SCTE54 could grow at above 50 °C in 15% CO_2 and can resist at 200 ppm of NO and SO_2 which are components of flue gases (Liang et al. 2019). An overview of various applications of cyanobacteria is mentioned in Table 14.4.

Cyanobacteria have developed highly active carbon capture mechanisms (CCMs) that depend on inorganic carbon (IC) such as CO_2, CO_3^{-2},and HCO_3^- as a substrate for photosynthetic CO_2 fixation (Long et al. 2016). Cyanobacterial CCMs performed energy-dependent active transport of inorganic carbon (Ci) to enhance the intracellular concentration of CO_2 at the site of Rubisco (ribulose-1,5-bisphosphate carboxylase/oxygenase) to facilitate CO_2 fixation even in low concentration of carbon.

Table 14.4 CO_2 sequestration and applications of cyanobacteria

Cyanobacteria	Biotechnological application	References
Thermosynechococcus CL–1 (TCL–1)	CO_2 biofixation and bioethanol production	Su et al. (2017)
Phormidium valderianum BDU 2004	CO_2 sequestration	
Synechocystis salina and *Microcystis aeruginosa*	CO_2 biofixation	Gonçalves et al. (2016b)
Phormidium sp.	Bioethanol production	
Synechococcus aquatilis	CO_2 sequestration	Singh et al. (2016b)
Synechococcus lividus and *Mastigocladus laminosus*	CO_2 sequestration	Singh et al. (2016b)

14.4.2 Potential of Cyanobacterial Genetic Engineering

Genetic engineering of cyanobacteria is a promising tool for CO_2 mitigation in the atmosphere and saving Earth's energy crisis. The application of thermophilic cyanobacterial strains for capturing CO_2 and its conversion into biomass and production of metabolites via genetic engineering can be beneficial (Liang et al. 2019). The higher cultivation temperature would be advantageous in two terms. Firstly, a higher cultivation temperature environment would limit the contamination of other microbes. Secondly, thermophilic microbes can provide thermostable valuable products like phycocyanins, etc. The genome sequence of various cyanobacteria is now available, which will help reverse engineering tools like gene targeting, genetic transformation, selection of markers, etc. The presence of the genome sequence of cyanobacteria will disclose the metabolic pathways to project new genetically engineered biological entities for enhanced CO_2 sequestration, bioproducts synthesis, and biofuel production. Furthermore, thermophilic cyanobacteria are the source of the gene pool for thermostable enzymes. These enzymes have the potential to improve the plants in arid areas. Apart from this, genetically engineered cyanobacteria can produce different chemicals by carbon fixation such as ethylene (Xiong et al. 2015), isopropanol (Hirokawa et al. 2017), etc. Chwa et al. (2016) reported the production of photosynthetic acetone by continuous feeding of CO_2 under light and aerobic conditions by engineered *Synechococcus elongatus* PCC 7942. Moreover, the metabolic engineering of cyanobacteria is a powerful tool to engineer strains for CO_2 sequestration, enhanced biomass productivity, and value-added products.

14.5 On-field Challenges and Opportunities in Cyanobacteria-Based Remediation

In recent years, many different studies have been carried out at the pilot-scale (Dalvi et al. 2021; Hom-Diaz et al. 2017; Marazzi et al. 2019). Hom-Diaz et al. (2017) studied the performance of 1200 L outdoor PBR on toilet wastewater and observed significant removal of nutrients (COD >80%, $N\text{-}NH_4^+ > 99\%$, and TP > 40%) and PPCP removal (up to 45%) at HRT of 8 days. In another pilot study, Garcia-galan et al. (2018) conducted a pilot study on agricultural runoff using 8500 L hybrid-PBR during autumn and winter using a consortium of *Chlorella* sp., *Gloecapsa* sp., *Scenedesmus* sp., and *Pediastrum sp*. They observed up to 95% and 84% removal of total nitrogen during winter and autumn, respectively. Interestingly the increase in COD within the outlet of PBR during both winter and autumn was observed due to increased dissolved organic matter (DOM) production by microalgae.

Cyanobacterial cultivation could become more efficient that can contribute to bioenergy production, but there are certain technical as well as economic limitations on a large scale (Chew et al. 2017). The significant technical challenges include the cultivation and harvesting of cyanobacteria. Under outdoor cultivation, the performance of cyanobacterial systems is dependent upon environmental conditions. The performance gets significantly affected in winter due to lower solar irradiance and temperature (Gonçalves et al. 2017). In open pond cultivation of cyanobacteria, a limited number of species can survive successfully in controlled conditions. Closed system-based cultivation possesses high maintenance charge (Pawar 2016). Apart from this, harvesting techniques are an energy-intensive process that further adds expenses (Roselet et al. 2015). Furthermore, the utilization of cyanobacterial biomass in bioenergy products seems to be critical in terms of extraction processes. Based on extraction methods, the energy inputs for bioenergy products could probably surpass the output which is not the case for biofertilizer application as postprocess is not needed after harvesting.

Moreover, cyanobacterial cultivation also includes environmental concerns. Few cyanobacterial species release cyanotoxins as secondary metabolites such as cytotoxins, dermatoxins, hepatotoxins, and neurotoxins (Kumar et al. 2019). These cyanotoxins impose a severe threat to aquatic life, water quality, and health issues in the human body. Thus, this global issue of cyanobacterial blooms desires a sustainable solution that can maintain green ecology and water quality.

14.6 Cyanobacterial Biomass Application as Biofertilizer/Soil Conditioner

Biofertilizers are living microbial inoculants that help crop growth and development when applied to the field (Majumdar 2015). They are gaining significant popularity in sustainable agriculture as they are eco-friendly, economically feasible, and reduce environmental pollution. Different biofertilizers are available based on the kind of organisms to be used in a specific crop (Li et al. 2017). Cyanobacteria or blue-green algae (BGA) are of significant stature due to their multifaceted role in agriculture (Fig. 14.1) and explained as below:

14.6.1 Nitrogen and Phosphorus Contribution

Prokaryotic microorganisms are the only living creatures present on earth to carry out atmospheric nitrogen fixation. They constitute significant contributors of nitrogen present in the biosphere. It is estimated that two-thirds of the annual yield of nitrogen, which is about 100 to 200 metric tons, comes from biological nitrogen fixation. Cyanobacteria are the primitive group of microorganisms having an oxygenic photosynthetic ability. These organisms are ecologically diverse and found in almost every terrestrial, freshwater, and marine habitat, including extreme environmental conditions from oceans to deserts to Antarctic lakes to hot springs (Gupta et al. 2013). Cyanobacteria are morphologically diverse, and their structure is organized differently from unicellular, heterotrichous, filamentous heterocystous to nonheterocystous in nature (Table 14.5). Most of the cyanobacteria can fix atmospheric nitrogen, because of which, they form a readily available, efficient, and economically feasible source to be utilized as biofertilizers. Nitrogen-fixing cyanobacteria are found in abundance in paddy fields (Fig. 14.3). However, their existence varies according to the soil and other climatic conditions (Pabbi 2015). Their significant role in enhancing nitrogen levels in paddy fields was reported as early as the 1930s (De 1939).

Cyanobacteria can fix atmospheric nitrogen due to the presence of a special type of cell, known as heterocysts, which are formed by the modification in vegetative cells. They are larger and thicker than vegetative cellsthat act as a mechanical barrier against oxygen entry. Heterocysts contain the nitrogenase enzyme responsible for nitrogen fixation and photosystem I for ATP production (Singh et al. 2011). Principally, Photosystem II is absent in heterocyst to protect O_2-sensitive nitrogenase from oxygen. The enzyme is a complex; it catalyzes the molecular form of nitrogen (N_2) into reduced form like ammonia (NH_3). The fixed nitrogen is released in the form of free amino acids, polypeptides, vitamins, and auxin-like substances (Subramanian and Sundaram 1986). Several nonheterocystous and unicellular cyanobacteria are also capable of fixing atmospheric nitrogen though under microaerophilic conditions. Besides, symbiotically associated cyanobacteria can do it with counterparts like water fern *Azolla, Gunnera*, cycads, etc. Therefore, the cyanobacteria act as an excellent biofertilizer for enhancing soil fertility.

As stated, cyanobacteria play an essential role in the nitrogen ecosystem. These exploit the sun's energy captured during the process of oxygenic photosynthesis to

Table 14.5 Important genera of cyanobacteria belonging to different forms

Cyanobacterial forms	Members
Unicellular	*Synechococcus, Chroococcus, Aphanothece, Gleocapsa, Pleurocapsa, Dermacapsa, Xenococcus*
Filamentous, heterocystous	*Anabaena, Nostoc, Aulosira, Calothrix, Cylindrospermum, Tolypothrix, Stigonema, Scytonema, Westiellopsis, Gleotrichia, Hapalosiphon, Fischerella, Rivularia*
Filamentous, nonheterocystous	*Oscillatoria, Plectonema, Lingbya, Phormidium, Microcoleus, Trichodesmium, Pseudoanabaena, Schizothrix*

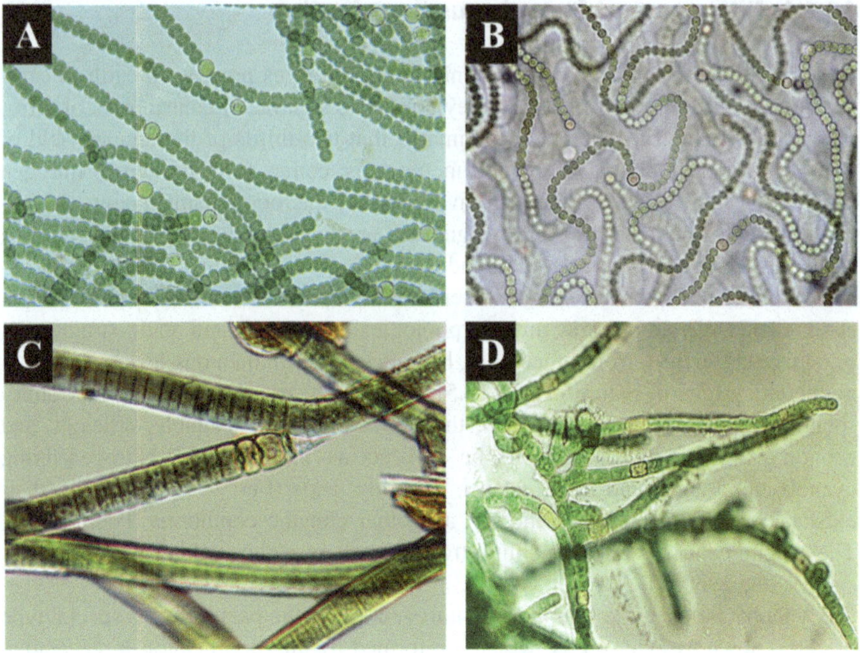

Fig. 14.3 Nitrogen-fixing cyanobacteria. (**a**) *Anabaena* sp., (**b**) *Nostoc* sp., (**c**) *Tolypothrix* sp., and (**d**) *Hapalosiphon* sp.

fix nitrogen and turn it into a form utilizable by the plants. Kalyansundaram et al. (2020) highlighted the importance of cyanobacteria as valuable resources in sustainable agriculture. Cyanobacteria play a significant role in contributing about 20–-30 kg N ha^{-1} per crop season and the organic matter to the soil (Issa et al. 2014). Several cyanobacterial species are found to be effective biofertilizers such as *Nostoc muscorum, Anabaena variabilis, Tolypothrix tenuis,* and *Aulosira fertilissima.* The nitrogen fixation process of cyanobacteria gets activated when the combined nitrogen level is less than the threshold level (~40 ppm), which is a kind of switch on & off mechanism. This permits algal biomass to provide significant nitrogen sources to the plant whenever the nitrogen level is reduced in the ecological system due to overexploitation and loss of fertilizers due to leaching and evaporation.

Several scientists have shown that cyanobacterial inoculation has resulted in a significant improvement in soil nitrogen content when applied to various crops (Venkataraman 1972; Rodgers et al. 1979; Tripathi et al. 2008; Renuka et al. 2016; Suleiman et al. 2020). This can save 25–40% of chemical nitrogen fertilizers and reduce the cost of cultivation significantly (Nisha et al. 2007; Nain et al. 2010; Prasanna et al. 2016a). A study conducted in the Indo-Gangetic plains region of India on comparative analysis of paddy yield, urea consumption, and farmers' income with or without BGA application (Bhooshan et al. 2018) revealed that farmers were able to harvest a 1% higher yield of paddy along with 3% higher

Table 14.6 Benefits of nitrogen-fixing cyanobacteria as biofertilizer in different crops

Cyanobacteria	Crops	Role	Reference
Anabaena iyengarii var. *tenuis*, *Nostoc commune*, *Nostoc linckia*, *Nostoc* sp. VICCRI, *Anabaena variabilis*	Rice	50% reduction in chemical fertilizers, improved grain yield, and quality	Pereira et al. (2009), Jha and Prasad (2006), Innok et al., (2009), Singh and Datta (2007)
Anabaena variabilis, *Nostoc muscorum*, *Tolypothrix tenuis*, *Aulosira fertilissima*	Rice	Improvement in soil enzymes activity	Mishra et al. (2005)
Anabaena sp. biofilm with *Mesorhizobium*	Wheat	Increased soil nitrogen content	Swarnalakshmi et al. (2013)
Nostoc entophytum, *Oscillatoria augustissima*	Pea	Increased nutritional value of seeds, 50% savings in chemical fertilizers	Osman et al. (2010)
Anabaena, *Nostoc* *Anabaena* + *Trichoderma* biofilm	Maize	Increased plant height and yield, increased soil nutrient	Prasanna et al. (2016a)
Anabaena + *Trichoderma*, *Anabaena* + *Azotobacter*	Chrysanthemum	Increased plant growth and soil nitrogen level	Prasanna et al. (2016b)
Anabaena doliolum HH-209, *Cylindrospermum sphaerica*	Pearl millet and wheat	Enhanced plant growth and yield; Improved carbon and nitrogen mineralization	Nisha et al. (2007)
Nostoc entophytum, *Oscillatoria angustissima*	Pea	Seed germination	Ismail and Ismail (2011)
Calothrix ghosei, *Hapalosiphon intricatus*, *Nostoc* sp.	Wheat	Increased plant height, dry weight, and yield	Karthikeyan et al. (2007, 2009)
Anabaena sp., *Calothrix* sp.	Okra	Increased nutrient availability, growth, and yield	Manjunatha et al. (2016)

income and 41.1% reduction in urea consumption. It is also shown that the use of nitrogen-fixing cyanobacteria as biofertilizers has resulted in increased soil fertility, grain yield, plant growth, and nutritional quality of seeds (Table 14.6).

Besides taking a significant part in the nitrogen cycle, cyanobacteria are also known to enhance the bioavailability of phosphorus to the plants. Phosphorus is considered as a major nutrient required by the plants beside nitrogen and potassium. In most of the aquatic ecosystems, P and N are the most limiting nutrients. Most Indian soils are deficient in P even though farmers apply enough phosphate fertilizers to the field. Phosphate often gets fixed in the soils and is not utilizable by the plants because of which plants often show P-deficiency symptoms like purple leaves and reduced oil quality of the seeds. Phosphate can be made available to the plants by increased activity of microorganisms in the soils having mineralization activity.

Some of the cyanobacteria act as phosphate solubilizers in the soil and can take out nutrients from their surrounding environment through structural changes, along with physiological and biochemical changes (Singh and Dhar 2007). With the help of phosphatase enzymes, cyanobacteria can solubilize and mobilize the insoluble organic phosphates present in the soil. They also can solubilize the insoluble form of $(Ca)_3(PO_4)_2$, $FePO_4$, $AlPO_4$, and hydroxyapatite $[Ca_5(PO_4)_3OH]$ in soils and sediments (Dorich et al., 1985; Wolf et al., 1985; Cameron and Julian 1988). Both laboratory and field studies have shown the phosphate solubilizing activity of many cyanobacteria (Yandigeri et al. 2010; Jaiswal et al. 2019). The significant sources of phosphate like Tricalcium Phosphate (TCP) and Mussoorie Rock Phosphate (MRP), which are mainly used for the development of affordable and low-cost phosphate fertilizers, are shown to be solubilized by heterocystous cyanobacteria *Westiellopsis prolifica* and *A. variabilis* (Yandigeri and Pabbi 2005; Yandegeri et al. 2011). The availability of the nutrients to the plants is mainly dependent on the decomposition rate of algal biomass in the soil. In algal biomass, phosphorus is accumulated as nucleic acids, phospholipids, polyphosphates, and proteins, which can be quickly converted into inorganic forms of phosphate upon degradation and efficiently utilized by the plant roots (Solevchenko et al. 2016). A study conducted by Schreiber et al. (2018) showed that P released from *Chlorella vulgaris* was transformed into plant-utilizable form in the soil.

14.6.2 Reclamation of Salt-affected Soils and Improvement in Soil Fertility

Salt-affected soils are basically classified into sodic, alkaline, and saline soils. These soils are very less productive and inflexible because of excessive salts in the upper layer of the soil surface. These soils are primarily impermeable to water, leading to water scarcity to the plant root system. Based on the salt content, these are categorized into saline and alkaline soils. The saline soils are described by the presence of high amounts of soluble salts (EC >4 dS cm^{-1}) which impart high osmotic pressure on plants for the uptake of water and nutrients from the soil. The alkaline soils are described by high pH, high exchangeable sodium, and high number of carbonates. They undergo extensive clay dispersion and deflocculation. The reduction in soil aeration and poor hydraulic conductivity make these soils unproductive (Pandey et al., 1992). Several physical and chemical methods are used for improvement of salt-affected soils like using excessive irrigation, the addition of gypsum, or sulfur, but these methods are not economically feasible and eco-friendly (Singh et al. 2016a; Li et al. 2019). Moreover, salinity has several side effects on the growth and other metabolic activities of plants and algae (Tang et al. 2007; Singh and Dhar 2010). Singh (1961) reported that the cyanobacteria might be utilized for the reclamation of Usar (Sodic/alkaline/saline) soils. He observed that the cyanobacteria form a thick covering on the soil surface, thereby maintain soil organic carbon, nitrogen, and phosphorus besides enhancing the water holding capacity. It is a well-established fact that the cyanobacteria improve soil permeability and aeration due to the addition of organic matter and nitrogen. Kaushik and Subhashini (1985) reported that the cyanobacteria are good candidates for the amendment of saline and alkali soils. They improve soil aggregation capacity by lowering the pH, electrical conductivity, and hydraulic conductivity contributing to the overall improvement in the physico-chemical quality. Apte et al. (1987) further elaborated that the cyanobacteria can curtail the influx of sodium ions. The outer surface network of the filaments/trichomes formed by the cyanobacteria on the soil binds the soil particles and, at the same time, entangles the soil particles at depth (Mazor et al. 1996; Nisha et al. 2007). The unique property of carbon and nitrogen fixation is also responsible for the improvement of the soil nutrient pool of organic carbon and nitrogen (Kolman et al. 2015; Li et al. 2015; Munoz-Rojas et al. 2018). Swapnil et al. (2015) observed enhanced salinity tolerance in the cyanobacteria is associated with proper maintenance of Ca^{2+}/Na^{+} ratio. Sheathed cyanobacteria are known to exert a mechanical effect on the soil and bind soil particles with the help of polysaccharides present in the sheath (MalamIssa et al. 1999, 2001). This polysaccharide sheath enhances soil aggregate size, thereby reducing soil compaction and preserving the nutrient status of the soil (Rogers and Burns 1994). Many cyanobacterial species excrete extracellular polymers substances (EPS). These EPS are mainly comprised of organic components like polysaccharides, nucleic acids, proteins, and inorganic components like carbonate, silica, etc. (Flemming and Wingender 2010). These are the main principal source of organic carbon in the soil (Mishra et al. 2011; Bondoc et al. 2016; Bhunia et al. 2018). EPS mainly help in improving soil surface stability and prevent it from erosion. They also play an essential role in the water table of the soil as their hygroscopic nature helps in improving the water retention ability of the soils. This, in turn, helps in better soil structure, infiltration rate, soil temperature, and aeration (Ibraheem, 2007; De Caire et al., 2000). EPS also help the plant in combating osmotic stress by improving antioxidant enzymes activity, phenolic compounds, and metabolites (Arroussi et al. 2018). Thus, cyanobacteria can be used as soil conditioners to improve soil structure, soil organic carbon, soil nitrogen, and soil fertility, thereby increasing plant productivity (Mahanty et al. 2017; Odjadjare et al. 2017). They are environment friendly, economical and provide safer ways to restore degraded soils (Pandey et al. 2005; Singh 2014; Nisha et al. 2018).

14.6.3 Plant Growth Promoters/Chemicals by Cyanobacteria

Cyanobacteria play a significant role in plant growth and development. They act as a biostimulant in crop production. They are known to elicit either direct or indirect effects on plant metabolism. They are involved directly via the production of phytohormones like auxins, cytokinins, gibberellic acid, jasmonic acid, etc.,

(Mazur et al. 2001; Stirk et al. 2002; Jadhav et al. 2018) and are involved indirectly via the production of siderophores, vitamins, amino acids, etc. (Singh and Trehan 1973; Ahmed and Holmstrom 2014; Jaiswal et al. 2018). Many reports are documented on the application of plant hormone-excreting cyanobacterial strains for *in vitro* and on-field studies of many agriculturally important crops (Table 14.7). Cyanobacterial species produce hormones either intracellular or extracellular in the growth medium and neighboring environmental conditions (Sergeeva et al. 2002; Lu and Xu et al. 2015; Romanenko et al. 2015), which help the plants against biotic and abiotic stresses (Rodreguez et al. 2006).

Siderophore production by cyanobacteria indirectly stimulates plant growth (Estep et al. 1975; Gademann and Portmann 2008). Siderophores are the organic acids which help in chelating iron when the iron is deficient in the surrounding environment (Ahmed and Holmstrom 2014). Many cyanobacteria are known to produce siderophores like *Anabaena cylindrica, Anabaena flos-aquae, Anabaena* sp., etc. which chelate iron along with other micronutrients such as copper (McKnight and Morel 1980; Goldman et al. 1983). Several studies are available on the use of cyanobacterial species along with eubacteria and other green algae for the biofortification of many essential food crops by enriching them with micronutrients like iron, copper, zinc, and manganese made available through BGA (Rana et al. 2012, 2015; Prasanna et al. 2015; Manjunath et al. 2016; Renuka et al. 2017). Exopolysaccharides (EPS) production by cyanobacteria is another crucial factor for plant development (De Caire et al. 2000). EPS increase the activity of agriculturally important microbes by colonizing them on the root system with beneficial biofilms (Weiss et al. 2012; Xiao and Zheng 2016). In the study performed by Priya et al. (2015), inoculation with *Calothrix elenkenii* increased the activity of microbiome in the rhizosphere of rice.

14.6.4 Biocontrol Agent

Inappropriate use of biocides (Pesticides, fungicides, nematicides, etc.) has posed a significant challenge to the ecosystem. They are incredibly toxic and dangerous to the environment as they leach out of the soil and water, leading to the accumulation of many harmful chemicals in the plant system. Therefore, considering the need for alternate ways of reducing pests and diseases, the use of biological sources offers an ethical, safe, and eco-friendly approach to combat these biotic stresses and improve plant growth (Spadaro and Gullino 2005). Cyanobacteria are considered as an excellent biocontrol agent against many pests and plant pathogens like fungi, bacteria, and nematodes (Table 14.8). They produce different kinds of bioactive compounds like benzoic acid, jasmonic acid, Ambigol A, majusculonic acid, etc. (Prasanna et al., 2008; Chaudhary et al. 2012). These antimicrobial compounds kill the pathogens by disrupting their cellular structures and cellular metabolism (Swain et al. 2017). Cyanobacteria also produce many hydrolytic enzymes like chitinase, β-1,3-endoglucanase, peroxidase, polyphenol oxidase, catalase, etc. (Kumar et al. 2013b; Babu et al. 2015; Priya et al. 2015), which facilitate in improving plant

Table 14.7 Plant growth-promoting activity of different cyanobacteria

Cyanobacteria	Plant growth-promoting substance	Effect	Reference
Anabaena fertilissima	Amino acids	Increased growth of rice seedlings	Singh and Trehan (1973)
Calothrix sp., *Phormidium animale*	Cytokinins, auxins	High growth rate and cytokinin-like activity in cucumber cotyledon expansion bioassay	Stirk et al. (2002)
Anabaena sp. MACC 643, *Leptolyngbya* sp. MACC 642	Auxin, cytokinin	Antherculture and regeneration in maize	Jager et al. (2010)
Chroococcidiopsis sp. Ck4, *Anabaena* sp. Ck1	Cytokinins, IAA	Increased seed germination, shoot length, tillering, number of lateral roots, spike length, and grain weight in wheat	Hussain and Hasnain (2011)
Chroococcidiopsis, Synechocystis, Leptolyngbya, Phormidium	Auxins	Increased growth in wheat	Mazar et al. (2013)
Aphanothece sp. MBDU 515	Indole-3-acetic acid	In vitro propagation of *Arachis hypogaea* and *Moringa oleifera*	Gayathri et al. (2015)
Nostoc sp.	Indole-3-acetic acid	Helps in plant symbiosis	Sergeeva et al. (2002)
Anabaena sp.	Indole acetic acid	Stimulates plant growth	Prasanna et al. (2010)
Anabaena vaginicola, Nostoc calicola	Indole-3-acetic acid, Indole butyric acid, Indole-3-propionic acid	Increased growth of vegetable crops	Hashtroudi et al. (2012)
Westillopsis sp. CCC554, *Nostoc* sp. CCC546, *Chroococcus minutus* CCC582, *Microcystis robusta* CCC568, *Anabaena* sp. CCC573	IAA, IBA	Increased seed germination, seedling root-shoot elongation in rice and wheat	Jadhav et al. (2018)
Anabaena variabilis, Nostoc muscorum, Tolypothrix tenuis, Aulosira fertilissima, Westelliopsis prolifica	Indole acetic acid, Siderophore	Increased seed germination, seedling root and shoot growth in rice	Jaiswal et al. (2018)

Table 14.8 Use of biocontrol capacity of cyanobacteria against different pests and pathogens

Cyanobacteria	Target organism (pest/pathogen)	Crop	Reference
Calothrix sp.	Damping-off	Vegetables	Manjunath et al. (2010)
Anabaena oscillarioides	*Pythium debaryanum, Pythium aphanidermatum, Fusarium oxysporum, Rhizoctonia solani*	Tomato	Dukare et al. (2011)
Anabaena, Calothrix, Nostoc, Oscillatoria, Nodularia	*Alternaria alternate, Rhizopus stolonifera, Botrytis cinerea*	Rice	Kim (2006)
Anabaena variabilis	*Fusarium moniliforme, Fusarium oxysporum lycopersici, Pythium debaryanum, Rhizoctonia solani*	Tomato	Chaudhary et al. (2012)
Oscillatoria chlorina	*Meloidogyne arenaria*	Tomato	Khan et al. (2007)
Aulosira fertilissima	*Meloidogyne triticoryzae*	Rice	Chandel (2009)

defense mechanisms. Many reports are available to prove the hydrolytic enzyme activity of these organisms. *Calothrix elenkenii* showed the activity of polyphenol oxidase, peroxidase, and ammonia-lyase in the rice root and shoot system (Priya et al. 2015). In the study performed by Prasanna et al. (2013), *A. variabilis* and *Anabaena laxa* were shown to suppress the growth of wilt-causing pathogen *Fusarium* sp. in tomato with the help of hydrolytic enzymes production.

14.7 Challenges and Opportunities of Cyanobacterial Biofertilizers

Biofertilizer is Mother Nature's way out for enhanced soil fertility. Cyanobacteria are one such example of biofertilizers, a kind of organic fertilizer containing living organisms that can live on a very minimum amount of innately available inputs like solar energy, water, and carbon dioxide (Woese 1987; Castenholz 2001) to ensure soil fertility and plant growth. These biofertilizers can be used by small-scale farmers to harvest more substantial and more sustainable yields and maintain healthier soils for future use (Higa and Wididana 1991). However, there are a few constraints in the extensive use of biofertilizers, and the major ones include their availability and quality. However, there are many technology interventions for improvement, and large-scale production of cyanobacterial biofertilizers, the available technologies that are economically viable and sustainable need to be promoted in a big way. There is an excellent opportunity in commercialization if one focuses on strain development, optimization of growth conditions (temperature, light, pH), suitable carrier material, etc., and proper management to avoid contamination. These optimization strategies shall also help in using cyanobacteria to a wide range of agriculturally important crops. The application of wastewater as a growth medium provides a low-cost solution. However, commercialization of wastewater grown algae imposes a great challenge as it contains many unwanted materials like harmful chemicals and pathogens. To meet the need of commercial farming, a massive quantity of nutrients is required and, in turn, produces the colossal amount of microbial biomass to meet these requirements. Anhydrous ammonia present in chemical fertilizers contains 82% N, whereas cyanobacterial biomass encompasses only 1–10% N (Cabanelas et al. 2013). Thus, cyanobacterial biomass is needed to be supplied 15 times more than that of chemical fertilizers to obtain that level of nutrients. However, the use of living inoculum provides an advantage over chemical fertilizers. They continuously multiply and provide nutrients to not only present crops but also to the subsequent cropping system. It is not easy to measure the nutrient content provided by cyanobacteria, as it is influenced by climatic conditions and other abiotic and biotic factors which may vary from location to location and season to season.

Genetic manipulation of organisms is gaining significant importance in achieving the needs of sustainable agriculture. However, the use of genetic engineering in the field of algae is in its initial phases. Many precautions need to be taken before conducting field studies involving genetically modified organisms, and the use of genetically modified cyanobacteria is of utmost concern before releasing into the environment. It is highly debatable in society as it is at the necessary level of the food chain (Andow and Zwahlen 2006). There is the foremost requirement of developing regulatory actions to monitor and analyze the significant risks involved in using genetically engineered cyanobacteria (Qin et al. 2012). Besides these challenges, the development of genetically modified cyanobacteria at the industrial level is another major problem due to the instability of developed mutants/modified strains.

14.8 Conclusion

The development of a sustainable agro-ecosystem is of utmost importance in this growing world. It maintains and preserves the diversity and complexity of nature. In this context, cyanobacteria play a multifaceted role for circular agronomy, from wastewater and flue gas remediation to agricultural field applications such as biofertilizer. Cyanobacterial use in remediation addresses major environmental concerns like pollution by the removal of nutrients, heavy metals, and pharmaceutical compounds of emerging concern from wastewater and CO_2 from flue gases. Following which generated biomass when applied in agriculture as biofertilizer, it plays a prominent role in improving soil fertility by means of its beneficial effects like nitrogen fixation, increase in soil carbon, and increasing bioavailability of phosphorus and other nutrients to the plant root system. The use of cyanobacteria as biofertilizer gives added advantage of continuous nutrient supply (macro and micronutrients), phytohormones, and essential metabolites to the growing plant. It acts as a soil conditioner as it improves soil structure, soil carbon, and nitrogen, which is not achievable with the use of chemical fertilizers alone. This multifaceted role played by the cyanobacteria in environmental resilience paves out the path toward the circular agronomy.

Acknowledgment The authors gratefully acknowledge IIT, Delhi for financial support (grant no. IIITD/IRD/MI02021/167142) (Faculty interdisciplinary research project - IITD in collaboration with IARI).

References

Acién FG, Molina E, Reis A, Torzillo G, Zittelli GC, Sepúlveda C, Masojídek J (2017) Photobioreactors for the production of microalgae. Microalgae-Based Biofuels and Bioproducts. https://doi.org/10.1016/B978-0-08-101023-5.00001-7

Ahmed E, Holmstrom SJ (2014) Siderophores in environmental research: roles and applications. Microb Biotechnol 7(3):196–208

Andersen IM, Williamson TJ, González MJ, Vanni MJ (2020) Nitrate, ammonium, and phosphorus drive seasonal nutrient limitation of chlorophytes, cyanobacteria, and diatoms in a hyper-eutrophic reservoir. Limnol Oceanogr 65:962–978. https://doi.org/10.1002/lno.11363

Andow DA, Zwahlen C (2006) Assessing environmental risks of transgenic plants. EcolLett 9(2):196–214

Apte SK, Reddy BR, Thomas J (1987) Relationship between sodium influx and salt tolerance of nitrogen-fixing cyanobacteria. Appl Environ Microbiol 53:1934–1939

Arashiro LT, Ferrer I, Rousseau DPL, Van Hulle SWH, Garfí M (2019) The effect of primary treatment of wastewater in high rate algal pond systems: Biomass and bioenergy recovery. Bioresour Technol 280:27–36

Arroussi HE, Benhima R, Elbaouchi A, Sijilmassi B, Mernissi NE, Aafsar A, Smouni A (2018) Dunaliella salina exopolysaccharides: a promising biostimulant for salt stress tolerance in tomato (Solanum lycopersicum). J Appl Phycol 30(5):2929–2941

Babu B, Wu JT (2010) Biodegradation of phthalate esters by cyanobacteria. J Phycol 46:1106–1113

Babu S, Bidyarani N, Chopra P, Monga D, Kumar R, Radha P, Kranthi S, Adak A, Saxena AK (2015) Evaluating microbe-plant interactions and varietal differences for enhancing biocontrol efficacy in root rot challenged cotton crop. Eur J Plant Pathol 142:345–362

Bai X, Acharya K (2017) Algae-mediated removal of selected pharmaceutical and personal care products (PPCPs) from Lake Mead water. Sci Total Environ 581–582:734–740

Banerjee S, Ramaswamy S (2019) Comparison of productivity and economic analysis of microalgae cultivation in open raceways and flat panel photobioreactor. Bioresour Technol Rep 8:100328

Behera SK, Kim HW, Oh JE, Park HS (2011) Occurrence and removal of antibiotics, hormones and several other pharmaceuticals in wastewater treatment plants of the largest industrial city of Korea. Sci Total Environ 409:4351–4360

Bhooshan N, Pabbi S, Singh A, Sharma A, ChetanJaiswal A, Kumar A (2018) Impact of blue green algae (BGA) technology: an empirical evidence from north western Indo-Gangetic Plains. 3 Biotech 8(8):324

Bhunia B, Prasad Uday US, Oinam G, Mondal A, Bandyopadhyay TK, Tiwari ON (2018) Characterization, genetic regulation and production of cyanobacterial exopolysaccharides and its applicability for heavy metal removal. Carbohydr Polym 179:228–243

Bondoc KGV, Heuschele J, Gillard J, Vyverman W, Pohnert G (2016) Selective silicate-directed motility in diatoms. Nat Commun 7:10540

Bystrzejewska-Piotrowska G, Golimowski J, Urban PL (2009) Nanoparticles: their potential toxicity, waste and environmental management. Waste Manag 29:2587–2595

Cabanelas ITD, Ruiz J, Arbib Z, Chinalia FA, Garrido-Perez C, Rogalla F, Nascimento IA, Perales JA (2013) Comparing the use of different domestic wastewaters for coupling microalgal production and nutrient removal. Bioresour Technol 131:429–436

Cameron HJ, Julian GR (1988) Utilisation of hydroxyl apatite by cyanobacteria as their sole source of phosphate and calcium. Plant Soil 109:123–124

Castenholz RW (2001) Phylum BX. Cyanobacteria. In: Boone DR, Castenholz RW (eds) Bergey's manual of systematic bacteriology, 2ndedn edn. Springer, New York, pp 473–599

Castro JD, Calijuri ML, Mattiello EM, Ribeiro VJ, Assemany PP (2020) Algal biomass from wastewater: Soil phosphorus bioavailability and plants productivity. Sci Total Enviro 711: 135088

Cepoi L, Donțu N, Şalaru V, Şalaru V (2016) Removal of organic pollutants from wastewater by cyanobacteria. Cyanobacteria Bioremediation Wastewaters 27:43

Chandel ST (2009) Nematicidal activity of the Cyanobacterium, *Aulosira fertilissima* on the hatch of *Meloidogyne triticoryzae* and *Meloidogyne incognita*. Arch Phytopathol Plant Protect 42(1):32–38

Chan A, Salsali H, McBean E (2014) Heavy metal removal (copper and zinc) in secondary effluent from wastewater treatment plants by microalgae. ACS Sustain Chem Eng 2:130–137

Chang H, Fu Q, Zhong N, Yang X, Quan X, Li S, Fu J, Xiao C (2019) Microalgal lipids production and nutrients recovery from landfill leachate using membrane photobioreactor. Bioresour Technol 277:18–26

Chaudhary V, Prasanna R, Nain L, Dubey SC, Gupta V, Singh R, Jaggi S, Bhatnagar AK (2012) Bioefficacy of novel cyanobacteria-amended formulations in suppressing damping off disease in tomato seedlings. World J Microbiol Biotechnol 28(12):3301–3310

Chawla P, Malik A, Sreekrishnan TR, Dalvi V, Gola D (2020) Selection of optimum combination via comprehensive comparison of multiple algal cultures for treatment of diverse wastewaters. Environ Technol Innov 18:100758

Chen CY, Nagarajan D, Cheah WY (2018) Eicosapentaenoic acid production from Nannochloropsis oceanica CY2 using deep sea water in outdoor plastic-bag type photobioreactors. Bioresour Technol 253:1–7

Chew KW, Yap JY, Show PL, Suan NH, Juan JC, Ling TC, Lee DJ, Chang JS (2017) Microalgae biorefinery: High value products perspectives. Bioresour Technol 229:53–62

Choi YY, Patel AK, Hong ME, Chang WS, Sim SJ (2019) Microalgae bioenergy with carbon capture and storage (BECCS): an emerging sustainable bioprocess for reduced CO_2 emission and biofuel production. Bioresour Technol Rep 7:100270

Choudhary P, Prajapati SK, Kumar P, Malik A, Pant KK (2017) Development and performance evaluation of an algal biofilm reactor for treatment of multiple wastewaters and characterization of biomass for diverse applications. Bioresour Technol 224:276–284

Christenson LB, Sims RC (2012) Rotating algal biofilm reactor and spool harvester for wastewater treatment with biofuels by-products. Biotechnol Bioeng 109:1674–1684

Chwa JW, Kim WJ, Sim SJ, Um Y, Woo HM (2016) Engineering of a modular and synthetic phosphoketolase pathway for photosynthetic production of acetone from CO_2 in *Synechococcus elongatus* PCC 7942 under light and aerobic condition. Plant Biotechnol J 14:1768–1776

Craggs R, Park J, Heubeck S, Sutherland D (2014) High rate algal pond systems for low-energy wastewater treatment, nutrient recovery and energy production. New Zeal J Bot 52:60–73

Cuellar-Bermudez SP, Aleman-Nava GS, Chandra R, Garcia-Perez JS, Contreras-Angulo JR, Markou G, Muylaert K, Rittmann BE, Parra-Saldivar R (2017) Nutrients utilization and contaminants removal. a review of two approaches of algae and cyanobacteria in wastewater. Algal Res 24:438–449

Cuéllar-Franca RM, Azapagic A (2015) Carbon capture, storage and utilisation technologies: A critical analysis and comparison of their life cycle environmental impacts. J CO2 Util 9:82–102

Dalvi V, Chawla P, Malik A (2021) Year-long performance assessment of an on-site pilot scale (100 L) photobioreactor on nutrient recovery and pathogen removal from urban wastewater using native microalgal consortium. Algal Res 55:102228

De Caire GZ, de Cano MS, Palma RM, de Mule CZ (2000) Changes in soil enzyme activities following additions of cyanobacterial biomass and exopolysaccharide. Soil BiolBiochem 32(13):1985–1987

De Assis LR, Calijuri ML, Assemany PP, Berg EC, Febroni LV, Bartolomeu TA (2019) Evaluation of the performance of different materials to support the attached growth of algal biomass. Algal Res 39:101440

De PK (1939) The role of blue-green algae in nitrogen fixation in rice fields. Proc R SocLond B 127(846):121–139

Deviram G, Mathimani T, Anto S, Ahamed TS, Ananth DA, Pugazhendhi A (2020) Applications of microalgal and cyanobacterial biomass on a way to safe, cleaner and a sustainable environment. J Clean Product 253:119770

Diengdoh OL, Syiem MB, Pakshirajan K, Rai AN (2017) Zn^{2+} sequestration by nostoc muscorum: Study of thermodynamics, equilibrium isotherms, and biosorption parameters for the metal. Environ Monitor Assess 189(7):314

Dorich RA, Nelson DW, Sommers LE (1985) Estimating algal available phosphorus in suspended sediments by chemical extraction. J Environ Qual 14:400–405

Dukare AS, Prasanna R, Chandra Dubey S, Nain L, Chaudhary V, Singh R, Saxena AK (2011) Evaluating novel microbe amended composts as biocontrol agents in tomato. Crop Prot 30(4):436–442

Dutta S, Bhadury P (2020) Effect of arsenic on exopolysaccharide production in a diazotrophic cyanobacterium. J Appl Phycol 32(5):2915–2926

Ebele AJ, Abou-Elwafa Abdallah M, Harrad S (2017) Pharmaceuticals and personal care products (PPCPs) in the freshwater aquatic environment. Emerg Contam 3:1–16

Ebenstein A (2012) The consequences of industrialization: evidence from water pollution and digestive cancers in China. Rev Econ Stat 94(1):186–201

Erb TJ, Zarzycki J (2018) A short history of RubisCO: the rise and fall (?) of Nature's predominant CO_2 fixing enzyme. Curr Opin Biotechnol 49:100–107

Estep M, Armstrong JE, Van Baalen C (1975) Evidence for the occurrence of specific iron (III)-binding compounds in near-shore marine ecosystems. Appl Microbiol 30:186–188

Escapa C, Coimbra RN, Nuevo C, Vega S, Paniagua S, García AI, Calvo LF, Otero M (2017) Valorization of microalgae biomass by its use for the removal of paracetamol from contaminated water. Water (Switzerland) 9:1–9

Evgenidou EN, Konstantinou IK, Lambropoulou DA (2015) Occurrence and removal of transformation products of PPCPs and illicit drugs in wastewaters: a review. Sci Total Environ 505: 905–926

Fawzy MA, Mohamed AKSH (2017) Bioremediation of heavy metals from municipal sewage by cyanobacteria and its effects on growth and some metabolites of Beta vulgaris. J Plant Nutr 40: 2550–2561

Flemming HC, Wingender J (2010) The biofilm matrix. Nat Rev Microbiol 8(9):623

Gademann K, Portmann C (2008) Secondary metabolites from cyanobacteria: complex structures and powerful bioactivities. Curr Org Chem 12:326–341

García J, Ortiz A, Álvarez E, Belohlav V, García-Galán MJ, Díez-Montero R, Álvarez JA, Uggetti E (2018) Nutrient removal from agricultural run-off in demonstrative full scale tubular photobioreactors for microalgae growth. Ecol Eng 120:513–521

García-Galán MJ, Gutiérrez R, Uggetti E, Matamoros V, García J, Ferrer I (2018) Use of full-scale hybrid horizontal tubular photobioreactors to process agricultural runoff. Biosyst Eng 166:138–149

Gayathri M, Kumar PS, Prabha AML, Muralitharan G (2015) In vitro regeneration of *Arachis hypogaea* L. and *Moringa oleifera* Lam. using extracellular phytohormones from *Aphanothece* sp. MBDU 515. Algal Res 7:100–105

Genin SN, Aitchison JS, Allen DG (2014) Design of algal film photobioreactors: material surface energy effects on algal film productivity, colonization and lipid content. Bioresour Technol 155: 136–143

Genin SN, Aitchison JS, Allen DG (2015) Novel waveguide reactor design for enhancing algal biofilm growth. Algal Res 12:529–538

Giannuzzi L (2018) Cyanobacteria growth kinetics. In: Algae. IntechOpen. https://doi.org/10.5772/intechopen.81545

Gismondi A, Di Pippo F, Bruno L, Antonaroli S, Congestri R (2016) Phosphorus removal coupled to bioenergy production by three cyanobacterial isolates in a biofilm dynamic growth system. Int J Phytoremediation 18:869–876

Goeres DM, Pedersen S, Warwood B, Walker DK, Parker AE, Mettler M, Sturman P (2020) Design and fabrication of biofilm reactors. In: Recent trends in biofilm science and technology 2020 Jan 1, pp 71–88

Goldman S, Lammers P, Berman M, Sanders-Loehr J (1983) Siderophore-mediated iron uptake in different strains of Anabaena sp. J Bacteriol 156(3):1144–1150

Gonçalves AL, Pires JCM, Simões M (2017) A review on the use of microalgal consortia for wastewater treatment. Algal Res 24:403–415

Gonçalves AL, Pires JCM, Simões M (2016a) Biotechnological potential of Synechocystis salina co-cultures with selected microalgae and cyanobacteria: nutrients removal, biomass and lipid production. Bioresour Technol 200:279–286

Gonçalves AL, Rodrigues CM, Pires JCM, Simões M (2016b) The effect of increasing CO_2 concentrations on its capture, biomass production and wastewater bioremediation by microalgae and cyanobacteria. Algal Res 14:127–136

Goswami S, Diengdoh OL, Syiem MB, Pakshirajan K, Kiran MG (2015) Zn(II) and Cu(II) removal by nostoc muscorum: a cyanobacterium isolated from a coal mining pit in Chiehruphi, Meghalaya, India. Can J Microbiol 61(3):209–215

Guerra P, Kim M, Shah A, Alaee M, Smyth SA (2014) Occurrence and fate of antibiotic, analgesic/anti-inflammatory, and antifungal compounds in five wastewater treatment processes. Sci Total Environ 473:235–243

Guo J, Selby K, Boxall AB (2016) Effects of antibiotics on the growth and physiology of chlorophytes, cyanobacteria, and a diatom. Arch Environ Con Tox 71(4):589–602

Gupta V, Ratha SK, Sood A, Chaudhary V, Prasanna R (2013) New insights into the biodiversity and applications of cyanobacteria (blue-green algae)—Prospects and challenges. Algal Res 2(2):79–97

Guzzon A, Di Pippo F, Congestri R (2019) Wastewater biofilm photosynthesis in photobioreactors. Microorganisms 7(8):252

Kabariya JH, Ramani VM (2018) Dairy wastewater treatment by cyanobacteria for removal of nutrients with extraction of high value compounds from biomass. Int J Curr Microbiol Appl Sci 7:1527–1538

Hashim MA, Chu KH (2004) Biosorption of cadmium by brown, green, and red seaweeds. Chem Eng J 97:249–255

Hashtroudi MS, Ghassempour A, Riahi H, Shariatmadari Z, Khanjir M (2012) Endogenous auxins in plant growth-promoting cyanobacteria—*Anabaena vaginicola* and *Nostoc calcicola*. J Appl Phycol 25(2):379–386

Higa T, Wididana GN (1991) Changes in the soil microflora induced by effective microorganisms. In: Parr JF, Hornick SB, Whitman CE (eds) Proceedings of the first international conference on kyusei nature farming. U.S. Department of Agriculture, Washington, DC, pp 153–162

Hirokawa Y, Dempo Y, Fukusaki E, Hanai T (2017) Metabolic engineering for isopropanol production by an engineered cyanobacterium, *Synechococcus elongatus* PCC 7942, under photosynthetic conditions. J Biosci Bioeng 123:39–45

Hodges A, Fica Z, Wanlass J, VanDarlin J, Sims R (2017) Nutrient and suspended solids removal from petrochemical wastewater via microalgal biofilm cultivation. Chemosphere 174:46–48

Hoh D, Watson S, Kan E (2016) Algal biofilm reactors for integrated wastewater treatment and biofuel production: a review. Chem Eng J 287:466–473

Hom-Diaz A, Jaén-Gil A, Bello-Laserna I, Rodríguez-Mozaz S, Vicent T, Barceló D, Blánquez P (2017) Performance of a microalgal photobioreactor treating toilet wastewater: Pharmaceutically active compound removal and biomass harvesting. Sci Total Environ 592: 1–11

Huang X, Tu Y, Song C, Li T, Lin J, Wu Y, Liu J, Wu C (2016) Interactions between the antimicrobial agent triclosan and the bloom-forming cyanobacteria *Microcystis aeruginosa*. Aquat Toxicol 172:103–110

Hughes AR, Sulesky A, Andersson B, Peers G (2018) Sulfate amendment improves the growth and bioremediation capacity of a cyanobacteria cultured on municipal wastewater centrate. Algal Res 32:30–37

Hussain A, Hasnain S (2011) Phytostimulation and biofertilization in wheat by cyanobacteria. J Ind Microbiol Biotechnol 38(1):85–92

Ibraheem IB (2007) Cyanobacteria as alternative biological conditioners for bioremediation of barren soil. Egypt J Phycol 8(100):99–117

Innok S, Chunleuchanon S, Boonkerd N, Teaumroong N (2009) Cyanobacterial akinete induction and its application as biofertilizer for rice cultivation. J Appl Phycol 21(6):737–744

Ismail AEA, Ismail MM (2011) Antagonistic activity of some fungi and cyanobacteria species against *Rhizoctonia solani*. Int J Plant Pathol 2(3):101–114

Issa AA, Abd-Alla MH, Ohyama T (2014) Nitrogen-fixing cyanobacteria: future prospect. In: Ohyama T (ed) Advances in Biology and Ecology of Nitrogen Fixation. InTech, Rijeka

Jadhav SD, Pabbi S, Gopal M, Chand S (2018) Physicochemical confirmatory evidences for cyanobacterial released plant growth hormones governing escalation of rice (*Oryza sativa* L.) and wheat (*Triticum* sp.) crop. Int J Res Bio Sci 7(3):1–17

Jager K, Bartok T, Ordog V, Barnabas B (2010) Improvement of maize (*Zea mays* L.) anther culture responses by algae-derived natural substances. S Afr J Bot 76(3):511–516

Jaiswal A, Das K, Koli DK, Pabbi S (2018) Characterization of cyanobacteria for IAA and siderophore production and their effect on rice seed germination. Int J Curr Microbiol App Sci (Special Issue-7):5212–5222

Jaiswal A, Mishra R, Koli DK, Sharma VK, Pabbi S (2019) Evaluation of growth, nitrogen fixation and P-solubilizing ability of diazotrophic cyanobacteria under mineral phosphorus sources. Indian J Agric Sci 89(3):420–425

Jha MN, Prasad AN (2006) Efficacy of new inexpensive cyanobacterial biofertilizer including its shelf-life. World J Microbiol Biotechnol 22(1):73–79

Jiang J, Zhang N, Yang X, Song L, Yang S (2015) Toxic metal biosorption by macrocolonies of cyanobacterium *Nostoc sphaeroides* Kützing. J Appl Phycol 28(4):2265–2277

Kalyanasundaram GT, Ramasamy A, Rakesh S, Subburamu K (2020) Microalgae and Cyanobacteria: role and applications in agriculture. In: Arumugam M, Kathiresan S et al (eds) Applied algal biotechnology. Nova Science Publishers, Inc., New York

Karthikeyan N, Prasanna R, Nain L, Kaushik BD (2007) Evaluating the potential of plant growth-promoting cyanobacteria as inoculants for wheat. Eur J Soil Biol 43(1):23–30

Karthikeyan N, Prasanna R, Sood A, Jaiswal P, Nayak S, Kaushik BD (2009) Physiological characterization and electron microscopic investigations of cyanobacteria associated with wheat rhizosphere. Folia Microbiol 54:43–51

Kaushik BD, Subhashini D (1985) Amelioration of salt-affected soils with blue-green algae: improvements in soil properties. Proc Ind Natl Sci Acad 51:380–389

Khan S, Shamshad I, Waqas M, Nawab J, Ming L (2017) Remediating industrial wastewater containing potentially toxic elements with four freshwater algae. Ecol Eng 102:536–541

Khan Z, Kim YH, Kim SG, Kim HW (2007) Observations on the suppression of root knot nematode (*Meloidogyne arenaria*) on tomato by incorporation of cyanobacterial powder (*Oscillatoria chlorina*) into potting field soil. Bioresour Technol 98(1):69–73

Kim JD (2006) Screening of cyanobacteria (blue-green algae) from rice paddy soil for antifungal activity against plant pathogenic fungi. Mycobiology 34(3):138–142

Klawonn I, Nahar N, Walve J, Andersson B, Olofsson M, Svedén JB, Littmann S, Whitehouse MJ, Kuypers MMM, Ploug H (2016) Cell-specific nitrogen- and carbon-fixation of cyanobacteria in a temperate marine system (Baltic Sea). Environ Microbiol 18:4596–4609

Kolman MA, Nishi CN, Perez-Cenci M, Salerno GL (2015) Sucrose in cyanobacteria: from a salt-response molecule to play a key role in nitrogen fixation. Life (Basel) 5:102–126

Kube M, Jefferson B, Fan L, Roddick F (2018) The impact of wastewater characteristics, algal species selection and immobilisation on simultaneous nitrogen and phosphorus removal. Algal Res 31:478–488

Kulal DK, Loni PC, Dcosta C, Some S, Kalambate PK (2020) Cyanobacteria: as a promising candidate for heavy-metals removal. In: Advances in cyanobacterial biology. Academic Press, London, pp 291–300

Kumar K, Dasgupta CN, Nayak B, Lindblad P, Das D (2011) Development of suitable photobioreactors for CO_2 sequestration addressing global warming using green algae and cyanobacteria. Bioresour Technol 102:4945–4953

Kumar MS, Praveenkumar R, Ilavarasi A, Rajeshwari K, Thajuddin N (2013a) Biochemical changes of fresh water Cyanobacteria Dolichospermum flos-aquae NTMS07 to chromium-induced stress with special reference to antioxidant enzymes and cellular fatty acids. Bull Environ Contam Toxicol 90(6):730–735. https://doi.org/10.1007/s00128-013-0984-9

Kumar M, Prasanna R, Bidyarani N, Babu S, Mishra BK, Kumar A, Adak A, Jauhari S, Yadav K, Singh R, Saxena AK (2013b) Evaluating the plant growth promoting ability of thermotolerant bacteria and cyanobacteria and their interactions with seed spice crops. Sci Hortic 164:94–101

Kumar P, Hegde K, Brar SK, Cledon M, Kermanshahi-pour A (2019) Potential of biological approaches for cyanotoxin removal from drinking water: a review. Ecotoxicol Environ Saf 172:488–503

Kuritz T, Wolk CP (1995) Use of filamentous cyanobacteria for biodegradation of organic pollutants. Appl Environ Microbiol 61(1):234–238

Li DJ, Wang L, Zhao QY, Wei W, Sun YH (2015) Improving high carbon dioxide tolerance and carbon dioxide fixation capability of *Chlorella* sp by adaptive laboratory evolution. Bioresour Technol 185:269–275

Li H, Zhao Q, Huang H (2019) Current states and challenges of salt-affected soil remediation by cyanobacteria. Sci Total Environ 669:258–272

Li R, Tao R, Ling N, Chu G (2017) Chemical, organic and bio-fertilizer management practices effect on soil physicochemical property and antagonistic bacteria abundance of a cotton field: implications for soil biological quality. Soil Tillage Res 167:30–38

Liang Y, Tang J, Luo Y, Kaczmarek MB, Li X, Daroch M (2019) Thermosynechococcus as a thermophilic photosynthetic microbial cell factory for CO_2 utilisation. Bioresour Technol 278: 255–265

Lin B, Li X (2011) The effect of carbon tax on per capita CO_2 emissions. Energy Policy 39:5137–5146

Lin L, Xu Y, Wang Z, Diao C, Dong W, Xie SP (2018) Changes in extreme rainfall over India and China attributed to regional aerosol-cloud interaction during the late 20th century rapid industrialization. Geophys Res Lett 45(15):7857–7865

Long BM, Rae BD, Rolland V, Förster B, Price GD (2016) Cyanobacterial CO_2-concentrating mechanism components: Function and prospects for plant metabolic engineering. Curr Opin Plant Biol 31:1–8

Lu Y, Xu J (2015) Phytohormones in microalgae: a new opportunity for microalgal biotechnology? Trends Plant Sci 20(5):273–282

Luo D, Hu Z, Choi DG, Thomas VM, Realff MJ, Chance RR (2010) Life cycle energy and greenhouse gas emissions for an ethanol production process based on blue-green algae. Environ Sci Technol 44(22):8670–8677

Mahanty T, Bhattacharjee S, Goswami M, Bhattacharyya P, Das B, Ghosh A, Tribedi P (2017) Biofertilizers: a potential approach for sustainable agriculture development. Environ Sci Pollut Res 24:3315–3335

Majumdar K (2015) Bio-fertilizer use in Indian agriculture. Indian J Res 4(6):377–381

MalamIssa O, Trichet J, Defarge C, Coute A, Valentine C (1999) Morphology and microstructure of microbiotic soil crusts on a tiger bush sequence (Niger, Sahel). Catena 37:175–196

MalamIssa O, Bissonnais YL, Defarge C, Trichet J (2001) Role of a microbial cover on structural stability of a sandy soil in Sahelian part of western Niger. Geoderma 101:15–30

Manjunath M, Prasanna R, Nain L, Dureja P, Singh R, Kumar A, Jaggi S, Kaushik BD (2010) Biocontrol potential of cyanobacterial metabolites against damping off disease caused by *Pythium aphanidermatum* in solanaceous vegetables. Arch Phytopathol Plant Protect 43(7):666–677

Manjunath M, Kanchan A, Ranjan K, Venkatachalam S, Prasanna R, Ramakrishnan B, Hossain F, Nain L, Shivay YS, Rai AB, Singh B (2016) Beneficial cyanobacteria and eubacteria synergistically enhance bioavailability of soil nutrients and yield of okra. Heliyon 2(2):e00066

Marazzi F, Bellucci M, Rossi S, Fornaroli R, Ficara E, Mezzanotte V (2019) Outdoor pilot trial integrating a sidestream microalgae process for the treatment of centrate under non optimal climate conditions. Algal Res 39:101430

Markou G, Chatzipavlidis I, Georgakakis D (2012) Cultivation of *Arthrospira (Spirulina) platensis* in olive-oil mill wastewater treated with sodium hypochlorite. Bioresour Technol 112:234–241

Markou G, Iconomou D, Muylaert K (2016) Applying raw poultry litter leachate for the cultivation of *Arthrospira platensis* and *Chlorella vulgaris*. Algal Res 13:79–84. https://doi.org/10.1016/j.algal.2015.11.018

Mazhar S, Cohen JD, Hasnain S (2013) Auxin producing non-heterocystous cyanobacteria and their impact on the growth and endogenous auxin homeostasis of wheat. J Basic Microbiol 53:996–1003

Mazor G, Kidron GJ, Vanshak A, Abeliovich A (1996) The role of cyanobacterial exopolysaccharides in structuring desert microbial crusts. FEMS Microbiol Ecol 21:121–130

Mazur H, Konop A, Synak R (2001) Indole-3-acetic acid in the culture medium of two axenic green microalgae. J Appl Phycol 13(1):35–42

McKnight DM, Morel FM (1980) Copper complexation by siderophores from filamentous blue-green algae. Limnol Oceanogr 25(1):62–71

Mezynska M, Brzoska MM (2018) Environmental exposure to cadmium—A risk for health of the general population in industrialized countries and preventive strategies. Environ Sci Pollut Res 25(4):3211–3232

Miranda J, Krishnakumar G, Gonsalves R (2012) Lead sorption by living biomass of *Chroococcus multicoloratus* and *Oscillatoria trichoides*: Kinetics and equilibrium studies. Ann Microbiol 63(2):591–605

Mishra A, Kavita K, Jha B (2011) Characterization of extracellular polymeric substances produced by micro-algae *Dunaliella salina*. Carbohydr Polym 83(2):852–857

Mishra U, Choudhry KK, Pabbi S, Dhar DW, Singh PK (2005) Influence of blue green algae and *Azolla* inoculation on specific soil enzymes under paddy cultivation. Asian J Microbiol Biotechnol Environ Sci 7(1):9–12

Mogor AF, Ordog V, Lima GPP, Molnar Z, Mogor G (2018) Biostimulant properties of cyanobacterial hydrolysate related to polyamines. J Appl Phycol 30:453–460

Morweiser M, Kruse O, Hankamer B, Posten C (2010) Developments and perspectives of photobioreactors for biofuel production. Appl Microbiol Biotechnol 87(4):1291–1301

Mubarak M, Shaija A, Suchithra TV (2019) Flocculation: An effective way to harvest microalgae for biodiesel production. J Environ Chem Eng 7:103221

Mukherjee C, Chowdhury R, Sutradhar T, Begam M, Ghosh SM, Basak SK, Ray K (2016) Parboiled rice effluent: A wastewater niche for microalgae and cyanobacteria with growth coupled to comprehensive remediation and phosphorus biofertilization. Algal Res 19:225–236

Mulbry W, Kondrad S, Pizarro C, Kebede-Westhead E (2008) Treatment of dairy manure effluent using freshwater algae: Algal productivity and recovery of manure nutrients using pilot-scale algal turf scrubbers. Bioresour Technol 99:8137–8142

Munoz-Rojas M, Roman JR, Roncero-Ramos B, Erickson TE, Merritt DJ, Aguila Carricondo P, Canton Y (2018) Cyanobacteria inoculation enhances carbon sequestration in soil substrates used in dryland restoration. Sci Total Environ 636:1149–1154

Nain L, Rana A, Joshi M, Jadhav SD, Kumar D, Shivay YS, Paul S, Prasanna R (2010) Evaluation of synergistic effects of bacterial and cyanobacterial strains as biofertilizers for wheat. Plant Soil 331(1):217–230

Nielsen AZ, Mellor SB, Vavitsas K, Wlodarczyk AJ, Gnanasekaran T, Perestrello Ramos H de Jesus M, King BC, Bakowski K, Jensen PE (2016) Extending the biosynthetic repertoires of cyanobacteria and chloroplasts. Plant J 87:87–102

Nisha R, Kaushik A, Kaushik CP (2007) Effect of indigenous cyanobacterial application on structural stability and productivity of an organically poor semi-arid soil. Geoderma 138(1–2):49–56

Nisha R, Kiran B, Kaushik A, Kaushik CP (2018) Bioremediation of salt-affected soils using cyanobacteria in terms of physical structure, nutrient status and microbial activity. Int J Environ Sci Technol 15:571–580

Odjadjare EC, Mutanda T, Olaniran AO (2017) Potential biotechnological application of microalgae: a critical review. Crit Rev Biotechnol 37:37–52

Oliver JWK, Atsumi S (2014) Metabolic design for cyanobacterial chemical synthesis. Photosynth Res 120:249–261

Osman MEH, El-Sheekh MM, El-Naggar AH, Gheda SF (2010) Effect of two species of cyanobacteria as biofertilizers on some metabolic activities, growth, and yield of pea plant. Biol Fertil Soils 46(8):861–875

Ozkan A, Kinney K, Katz L, Berberoglu H (2012) Reduction of water and energy requirement of algae cultivation using an algae biofilm photobioreactor. Bioresour Technol 114:542–548

Pabbi S (2008) Cyanobacterial biofertilizers (Review). J Eco-Friendly Agric 3(2):95–111

Pabbi S (2015) Blue-green algae: a potential biofertilizer for rice. In: Sahoo DB, Seckbach J (eds) The Algae world. Springer India Pvt. Ltd., New Delhi, pp 449–466

Pacheco MM, Hoeltz M, Moraes MSA, Schneider RCS (2015) Microalgae: Cultivation techniques and wastewater phycoremediation. J Environ Sci Heal Part A 50:585–601

Pandey KD, Kashyap AK, Gupta RK (1992) Nitrogen fixation by cyanobacteria associated with moss communities in Schirmacher Oasis, Antarctica. Isr J Bot 41:187–198

Pandey KD, Shukla PN, Giri DD, Kashyap AK (2005) Cyanobacteria in alkaline soil and the effect of cyanobacteria inoculation with pyrite amendments on their reclamation. Biol Fertil Soils 41(6):451–457

Pandey VD (2017) Cyanobacteria mediated heavy metal remediation. In: Singh J, Seneviratne G (eds) Agro-environmental sustainability. Springer, Cham. https://doi.org/10.1007/978-3-319-49727-3_6

Patel A, Matsakas L, Rova U, Christakopoulos P (2019) A perspective on biotechnological applications of thermophilic microalgae and cyanobacteria. Bioresour Technol 278:424–434

Pavasant P, Apiratikul R, Sungkhum V, Suthiparinyanont P, Wattanachira S, Marhaba TF (2006) Biosorption of Cu^{2+}, Cd^{2+}, Pb^{2+}, and Zn^{2+} using dried marine green macroalga *Caulerpa lentillifera*. Bioresour Technol 97:2321–2329

Pawar SB (2016) Effectiveness mapping of open raceway pond and tubular photobioreactors for sustainable production of microalgae biofuel. Renew Sust Energ Rev 62:640–653

Pereira I, Ortega R, Barrientos L, Moya M, Reyes G, Kramm V (2009) Development of a biofertilizer based on filamentous nitrogen-fixing cyanobacteria for rice crops in Chile. J Appl Phycol 21(1):135–144

Prasanna R, Nain L, Tripathi R, Gupta V, Chaudhary V, Middha S, Joshi M, Ancha R, Kaushik BD (2008) Evaluation of fungicidal activity of extracellular filtrates of cyanobacteria – possible role of hydrolytic enzymes. J Basic Microbiol 48(3):186–194

Prasanna R, Joshi M, Rana A, Nain L (2010) Modulation of IAA production in cyanobacteria by tryptophan and light. Pol J Microbiol 59(2):99–105

Prasanna R, Chaudhary V, Gupta V, Babu S, Kumar A, Shivay YS, Nain L (2013) Cyanobacteria mediated plant growth promotion and bioprotection against *Fusarium* wilt in tomato. Eur J Plant Pathol 136:337–353

Prasanna R, Bidyarani N, Babu S, Hossain F, Shivay YS, Nain L, Moral MT (2015) Cyanobacterial inoculation elicits plant defense response and enhanced Zn mobilization in maize hybrids. Cogent Food Agric 1(1):998507

Prasanna R, Kanchan A, Ramakrishnan B, Ranjan K, Venkatachalam S, Hossain F, Shivay YS, Krishnan P, Nain L (2016a) Cyanobacteria-based bioinoculants influence growth and yields by modulating the microbial communities favourably in the rhizospheres of maize hybrids. Eur J Soil Biol 75:15–23

Prasanna R, Kanchan A, Kaur S, Ramakrishnan B, Ranjan K, Singh MC, Hasan M, Saxena AK, Shivay YS (2016b) Chrysanthemum growth gains from beneficial microbial interactions and fertility improvements in soil under protected cultivation. Hortic Plant J 2(4):229–239

Priya H, Prasanna R, Ramakrishnan B, Bidyarani N, Babu S, Thapa S, Renuka N (2015) Influence of cyanobacterial inoculation on the culturable microbiome and growth of rice. Microbiol Res 171:78–89

Pruvost J, Pottier L, Legrand J (2006) Numerical investigation of hydrodynamic and mixing conditions in a torus photobioreactor. Chem Eng Sci 61(14):4476–4489

Qin S, Lin H, Jiang P (2012) Advances in genetic engineering of marine algae. Biotechnol Adv 30(6):1602–1613

Ragush CM, Poltarowicz JM, Lywood J, Gagnon GA, Truelstrup Hansen L, Jamieson RC (2017) Environmental and operational factors affecting carbon removal in model arctic waste stabilization ponds. Ecol Eng 98:91–97

Rana A, Joshi M, Prasanna R, Shivay YS, Nain L (2012) Biofortification of wheat through inoculation of plant growth-promoting rhizobacteria and cyanobacteria. Eur J Soil Biol 50: 118–126

Rana A, Kabi SR, Verma S, Adak A, Pal M, Shivay YS, Prasanna R, Nain L (2015) Prospecting plant growth-promoting bacteria and cyanobacteria as options for enrichment of macro and micronutrients in grains in rice - wheat cropping sequence. Cogent Food Agric 1:1037379

Renuka N, Prasanna R, Sood A, Ahluwalia AS, Bansal R, Babu S, Singh R, Shivay YS, Nain L (2016) Exploring the efficacy of wastewater-grown microalgal biomass as a biofertilizer for wheat. Environ Sci Pollut Res 23(7):6608–6620

Renuka N, Prasanna R, Sood A, Bansal R, Bidyarani N, Singh R, Shivay YS, Nain L, Ahluwalia AS (2017) Wastewater grown microalgal biomass as inoculants for improving micronutrient availability in wheat. Rhizosphere 1:150–159

Renuka N, Guldhe A, Prasanna R, Singh P, Bux F (2018) Microalgae as multi-functional options in modern agriculture: current trends, prospects and challenges. Biotechnol Adv 36(4):1255–1273

Rodgers GA, Bergman B, Henriksson E, Urdis M (1979) Utilisation of blue green algae as biofertilizers. Plant Soil 52:99–107

Rodriguez A, Stella A, Storni M, Zulpa G, Zaccaro M (2006) Effects of cyanobacterial extracellular products and gibberellic acid on salinity tolerance in *Oryza sativa* L. Saline Syst 2(1):7

Rogers SL, Burns RG (1994) Changes in aggregate stability, nutrient status, indigenous microbial populations and seedling emergence following inoculation of soil with *Nostoc muscorum*. Biol Fertil Soils 18:209–215

Romanenko EA, Kosakovskaya IV, Romanenko PA (2015) Phytohormones of microalgae: biological role and involvement in the regulation of physiological processes. Pt I. auxins, abscisic acid, ethylene. Int J Algae 17(3):275–289

Ronga D, Biazzi E, Parati K, Carminati D, Carminati E, Tava A (2019) Microalgal biostimulants and biofertilisers in crop productions. Agronomy 9:192

Roselet F, Vandamme D, Roselet M, Muylaert K, Abreu PC (2015) Screening of commercial natural and synthetic cationic polymers for flocculation of freshwater and marine microalgae and effects of molecular weight and charge density. Algal Res 10:183–188

Rossi S, Díez-Montero R, Rueda E, Castillo Cascino F, Parati K, García J, Ficara E (2020) Free ammonia inhibition in microalgae and cyanobacteria grown in wastewaters: Photo-respirometric evaluation and modelling. Bioresour Technol 305:123046

Saavedra R, Muñoz R, Taboada ME, Vega M, Bolado S (2018) Comparative uptake study of arsenic, boron, copper, manganese and zinc from water by different green microalgae. Bioresour Technol 263:49–57

Schnurr PJ, Allen DG (2015) Factors affecting algae biofilm growth and lipid production: a review. Renew Sust Energ Rev 52:418–429

Schreiber C, Schiedung H, Harrison L, Briese C, Ackermann B, Kant J, Nedbal L (2018) Evaluating potential of green alga *Chlorella vulgaris* to accumulate phosphorus and to fertilize nutrient-poor soil substrates for crop plants. J Appl Phycol 30(5):2827–2836

Sergeeva E, Liaimer A, Bergman B (2002) Evidence for production of the phytohormone indole-3-acetic acid by cyanobacteria. Planta 215(2):229–238

Singh JS, Pandey VC, Singh DP (2011) Efficient soil microorganisms: a new dimension for sustainable agriculture and environmental development. Agric Ecosyst Environ 140(3–4):339–353

Singh JS (2014) Cyanobacteria: a vital bio-agent in eco-restoration of degraded lands and sustainable agriculture. Climate Change Environ Sustain 2:133–137

Singh JS, Kumar A, Rai AN, Singh DP (2016a) Cyanobacteria: a precious bio-resource in agriculture, ecosystem and environmental sustainability. Front Microbiol 7:529

Singh JS, Seneviratne G (2017) Agro-environmental sustainability. Manag Environ Pollut 2:1–257

Singh NK, Dhar DW (2007) Nitrogen and phosphorous scavenging potential in microalgae. Indian J Biotechnol 6:52–56

Singh NK, Dhar DW (2010) Cyanobacterial reclamation of salt-affected soil. In: Lichtfouse E (ed) Genetic engineering, biofertilisation, soil quality and organic farming, Sustainable Agriculture Reviews, vol 4. Springer, Dordrecht

Singh RN (1961) Role of blue-green algae in nitrogen economy of Indian agriculture. Indian Council of Agricultural Research, New Delhi, p 175

Singh S, Datta P (2007) Outdoor evaluation of herbicide resistant strains of *Anabaena variabilis* as biofertilizer for rice plants. Plant Soil 296(1–2):95–102

Singh SK, Sundaram S, Sinha S, Rahman MA, Kapur S (2016b) Recent advances in CO_2 uptake and fixation mechanism of cyanobacteria and microalgae. Crit Rev Environ Sci Technol 46:1297–1323. https://doi.org/10.1080/10643389.2016.1217911

Singh VP, Trehan T (1973) Effect of extracellular products of *Aulosira fertilissima* on the growth of rice seedlings. Plant Soil 38.457–464

Solovchenko A, Verschoor AM, Jablonowski ND, Nedbal L (2016) Phosphorus from wastewater to crops: an alternative path involving microalgae. Biotechnol Adv 34(5):550–564

Spadaro D, Gullino ML (2005) Improving the efficacy of biocontrol agents against soil borne pathogens. Crop Prot 24(7):601–613

Stirk WA, Ordog V, Van Staden J, Jager K (2002) Cytokinin- and auxin-like activity in cyanophyta and microalgae. J Appl Phycol 14(3):215–221

Su CM, Hsueh HT, Tseng CM, Ray DT, Shen YH, Chu H (2017) Effects of nutrient availability on the biomass production and CO_2 fixation in a flat plate photobioreactor. Aerosol Air Qual Res 17:1887–1897

Subramanian G, Sundaram SS (1986) Induced ammonia release by the nitrogen-fixing cyanobacterium *Anabaena*. FEMS Microbiol Lett 37:151–154

Sui Q, Cao X, Lu S, Zhao W, Qiu Z, Yu G (2015) Occurrence, sources and fate of pharmaceuticals and personal care products in the groundwater: a review. Emerg Contam 1:14–24

Suleiman AK, Lourenço KS, Clark C, Luz RL, Silva GH, Vet LE, Kuramae EE (2020) From toilet to agriculture: Fertilization with microalgal biomass from wastewater impacts the soil and rhizosphere active microbiomes, greenhouse gas emissions and plant growth. Resour Conserv Recycl 161:104924

Sutherland DL, Turnbull MH, Craggs RJ (2014) Increased pond depth improves algal productivity and nutrient removal in wastewater treatment high rate algal ponds. Water Res 53:271–281

Swain SS, Paidesetty SK, Padhy RN (2017) Antibacterial, antifungal and antimycobacterial compounds from cyanobacteria. Biomed Pharmacother 90:760–776

Swapnil P, Singh M, Singh S, Sharma NK, Rai AK (2015) Recombinant glycine betaine improves metabolic activities, ionic balance and salt tolerance in diazotrophic freshwater cyanobacteria. Algal Res 11:194–203

Swarnalakshmi K, Prasanna R, Kumar A, Pattnaik S, Chakravarty K, Shivay YS, Singh R, Saxena AK (2013) Evaluating the influence of novel cyanobacterial biofilmed biofertilizers on soil fertility and plant nutrition in wheat. Eur J Soil Biol 55:107–116

Tang CC, Tian Y, Liang H, Zuo W, Wang ZW, Zhang J, He ZW (2018) Enhanced nitrogen and phosphorus removal from domestic wastewater via algae-assisted sequencing batch biofilm reactor. Bioresour Technol 250:185–190

Tang D, Shi S, Li D, Hu C, Liu Y (2007) Physiological and biochemical responses of *Scytonema javanicum* (cyanobacterium) to salt stress. J Arid Environ 71(3):312–320

Tapia JFD, Lee JY, Ooi REH, Foo DCY, Tan RR (2018) A review of optimization and decision-making models for the planning of CO_2 capture, utilization and storage (CCUS) systems. Sustain Prod Consum 13:1–15

Thitakamol B, Veawab A, Aroonwilas A (2012) Environmental impacts of absorption-based CO_2 capture unit for post-combustion treatment of flue gas from coal-fired power plant. Int J Greenh Gas Control 1:318–342

Torzillo G, Zittelli GC (2015) Tubular photobioreactors. In: Algal biorefineries. Springer, Cham, pp 187–212

Trebuch L, Oyserman B, Janssen M, Wijffels R, Vet L, Fernandes T (2020) Impact of hydraulic retention time on community assembly and function of photogranules for wastewater treatment. Water Res 173:115506

Tredici MR, Chini Zittelli G, Rodolfi L (2009) Photobioreactors. Encyclopedia of industrial biotechnology: bioprocess, bioseparation, and cell technology. pp 1–15

Troschl C, Meixner K, Fritz I, Leitner K, Romero AP, Kovalcik A, Sedlacek P, Drosg B (2018) Pilot-scale production of poly-β-hydroxybutyrate with the cyanobacterium *Synechocytis* sp. CCALA192 in a non-sterile tubular photobioreactor. Algal Res 34:116–125

Tripathi RD, Dwivedi S, Shukla MK, Mishra S, Srivastava S, Singh R, Rai UN, Gupta DK (2008) Role of blue-green algae biofertilizer in ameliorating the nitrogen demand and fly-ash stress to the growth and yield of rice (*Oryza sativa* L.) plants. Chemosphere 70(10):19191929

Ugwu CU, Aoyagi H, Uchiyama H (2008) Photobioreactors for mass cultivation of algae. Bioresour Technol 99(10):4021–4028

Venkataraman GS (1972) Algal biofertilizers and rice cultivation. Today and Tomorrow's Printers and Publishers, New Delhi, p 75

Vo HN, Ngo HH, Guo W, Nguyen TM, Liu Y, Liu Y, Nguyen DD, Chang SW (2019) A critical review on designs and applications of microalgae-based photobioreactors for pollutants treatment. Sci Total Environ 651:1549–1568

Vijayaraghavan R, Ellappan V, Dharmar P, Lakshmanan U (2018) Preferential adsorption of uranium by functional groups of the marine unicellular cyanobacterium *Synechococcus elongatus* BDU130911. 3 Biotech 8(3):170

Weiss TL, Roth R, Goodson C, Vitha S, Black I, Azadi P, Rusch J, Holzenburg A, Devarenne TP, Goodenough U (2012) Colony organization in the green alga *Botryococcus braunii* (Race B) is specified by a complex extracellular matrix. Eukaryot Cell 11(12):1424–1414

Wolf AM, Baker DE, Pionke HB, Kunichi HM (1985) Soil test for estimating labile, soluble and algal available phosphorus in agricultural soils. J Environ Qual 14:341–348

Woods RP, Legere E, Moll B, Unamunzaga C, Mantecon E (2010) U.S. Patent No. 7,682,821. Washington, DC: U.S. Patent and Trademark Office

Woese CR (1987) Bacterial evolution. Microbiol Rev 51:221–271

Wu L, Qian L, Deng Z, Zhou X, Li B, Lan S, Yang L, Zhang Z (2020) Temperature modulating sand-consolidating cyanobacterial biomass, nutrients removal and bacterial community dynamics in municipal wastewater. Bioresour Technol 301:122758

WWAP (United Nations World Water Assessment Programme) (2017) The United Nations World Water Development Report 2017: Wastewater, The Untapped Resource. UNESCO, Paris

Xiao R, Zheng Y (2016) Overview of microalgal extracellular polymeric substances (EPS) and their applications. Biotechnol Adv 34(7):1225–1244

Xiong JQ, Kurade MB, Jeon BH (2018) Can microalgae remove pharmaceutical contaminants from Water? Trends Biotechnol 36:30–44

Xiong W, Morgan JA, Ungerer J, Wang B, Maness PC, Yu J (2015) The plasticity of cyanobacterial metabolism supports direct CO_2 conversion to ethylene. Nat Plants 1:1–6

Yandigeri MS, Pabbi S (2005) Response of diazotrophic cyanobacteria to alternative sources of phosphorus. Indian J Microbiol 45(2):132–134

Yandigeri MS, Yadav AK, Meena KK, Pabbi S (2010) Effect of mineral phosphate on growth and nitrogen fixation of diazotrophic cyanobacteria *Anabaena variabilis* and *Westiellopsis prolifica*. Antonie Van Leeuwenhoek 97(3):297–306

Yandigeri MS, Kashyap S, Pabbi S (2011) Studies on mineral phosphate solubilization by cyanobacteria *Westiellopsis* and *Anabaena*. Microbiology 80(4):558–565

Young P, Taylor M, Fallowfield HJ (2017) Mini-review: high rate algal ponds, flexible systems for sustainable wastewater treatment. World J Microbiol Biotechnol 33:1–13

Yu JT, Bouwer EJ, Coelhan M (2006) Occurrence and biodegradability studies of selected pharmaceuticals and personal care products in sewage effluent. Agric. Water Manage 86:72–80

Zhang HJ, Hu CM, Jia XY, Xu Y, Wu CJ, Chen L, Wang F (2012) Characteristics of γ-hexachlorocyclohexane biodegradation by a nitrogen-fi xing cyanobacterium, Anabaena azotica. J Appl Phycol 24:221–225

Zhang H, Jiang X, Xiao W, Lu L (2014) Proteomic strategy for the analysis of the polychlorobiphenyl degrading cyanobacterium Anabaena PD-1 exposed to Aroclor 1254. PLoS One 9(3):e91162. https://doi.org/10.1371/journal.pone.0091162

Zhang Q, Yu Z, Zhu L, Ye T, Zuo J, Li X, Xiao B, Jin S (2018) Vertical-algal-biofilm enhanced raceway pond for cost-effective wastewater treatment and value-added products production. Water Res 139:144–157

Zittelli GC, Biondi N, Rodolfi L, Tredici MR (2013) Photobioreactors for mass production of microalgae. In: Handbook of microalgal culture: applied phycology and biotechnology, vol 2, 2nd edn. Wiley Online Library, pp 225–266. https://doi.org/10.1002/9781118567166.ch13

Antioxidant, Anti-aging and Anti-neurodegenerative Biomolecules from Cyanobacteria

15

Mukesh Ghanshyam Chaubey, Stuti Nareshkumar Patel, Ravi R. Sonani, Niraj Kumar Singh, Rajesh Prasad Rastogi, and Datta Madamwar

Abstract

Cyanobacteria are considered the oldest photoautotrophs and possess the capacity to survive in extreme habitats like deserts, hot springs, deep oceans and arctic regions. They produce a wide range of unique biomolecules like phycobiliproteins (PBPs), mycosporin like amino acids (MAAs), and scytonemin playing an important role in light-harvesting, photosynthesis and other physiological activities. Besides having significant roles in cyanobacteria, these biomolecules are recognised to have a variety of biomedical applications; for instance, in the therapy of diseases, in medical diagnosis, in the formulation of nutraceuticals and cosmetics. Interestingly, many of these biomolecules have

Mukesh Ghanshyam Chaubey and Stuti Nareshkumar Patel contributed equally to this work.

M. G. Chaubey · N. K. Singh
Shri A. N. Patel P. G. Institute of Science and Research, Anand, Gujarat, India

S. N. Patel
Post Graduate Department of Biosciences, Sardar Patel University, Bakrol, Anand, Gujarat, India

R. R. Sonani
Malopolska Centre of Biotechnology, Jagiellonian University, Krakow, Poland

R. P. Rastogi
Ministry of Environment, Forest and Climate Change, New Delhi, Delhi, India

D. Madamwar (✉)
P. D. Patel Institute of Applied Sciences, Charotar University of Science and Technology, Changa, Anand, Gujarat, India
e-mail: dattamadamwar.as@charusat.ac.in

R. P. Rastogi (ed.), *Ecophysiology and Biochemistry of Cyanobacteria*,
https://doi.org/10.1007/978-981-16-4873-1_15

been noticed to possess antioxidant potential. As oxidative stress is a well-established cause of aging and neurodegenerative diseases, these cyanobacterial molecules have been tested for their anti-aging and neuroprotective potential by many researchers (including authors of this chapter) in the last decade. The present chapter discusses fundamental attributes, responsible for antioxidant, anti-aging and neuroprotective potential of various cyanobacterial biomolecules.

Keywords

Cyanobacterial biomolecules · Phycobiliproteins · Mycosporin like amino acids · Scytonemin · Antioxidant · Anti-aging · Neurodegenerative diseases

15.1 Introduction

Cyanobacteria are one of the oldest organisms on the earth and are found in a wide range of environmental conditions. Cyanobacteria do not possess a nucleus and organelles surrounded by a membrane and are categorised under the eubacteria domain (Seckbach 2007). During evolution, they were exposed to many stress-inducing natural extremities like anaerobiosis, exposure to high radiation, extreme heat and cold. Therefore, they have developed biochemical and physiological capacities to mitigate these stress conditions. They can survive in abiotic stresses such as pH, temperature, radiation, salinity and have been reported to colonise in diverse niches having extreme environmental conditions (Seckbach et al. 2007; Seckbach and Oren 2007). For instance, cyanobacterial genera *Microcoleus*, *Lyngbya*, *Oscillatoria*, *Phormidium*, *Lyngbya* and *Nostoc* were reported as *psychrophiles* found in lakes of the dry valley and in the soil of Antartica (Mandal and Rath 2014). Some cyanobacterial genera like *Chroococcidiopsis* are found even below rocks and in cracked rocks of Antartica (Vincent 2000, 2004; Caiola and Billi 2007). The thermophilic cyanobacteria like *Thermosynechococcus*, *Phormidium*, *Mastiglocadus laminosus* and *Oscillatoria* are extensively reported in various hotsprings (Ward et al. 2000, 2012). Various filamentous and unicellular cyanobacteria like *Microcoleus chthonoplastes*, *Halospirulina tapeticola*, *Aphanothece halophytica*, *Phormidium sp.*, *Oscillatoria sp.* and *Synechococcus sp.* were reported to colonise hypersaline habitats (Javor 1989; Garcia-Pichel and Belnap 1996; Nübel et al. 2000; Golubic et al. 2010; Oren 2015). Several rare species of cyanobacteria, *Oscillatoria limnothrix*, *Spirulina sp.*, *Aphanocapsa sp.*, and various *Chroococcus sp.* are found in a highly acidic environments (Steinberg et al. 1998; Gimmler 2001; Freeman et al. 2020). On the other hand, several cyanobacteria, *Spirulina plantensis*, *Anabaenopsis sp.*, *Synechococcus sp.*, *Gloecapsa sp.*, *Gloethece linaris*, *Microcystis aeruginosa* and *Plectonema nostocorum* are noticed to survive in highly alkaline conditions (Tindall and Grant 1986; Boussiba 2000; Jorjani̇ et al. 2020). Besides this, several cyanobacteria were also noticed to grow in the presence of extreme radiation and highlight (Castenholz and Garcia-Pichel 2012; Cassier-Chauvat et al. 2017). The existence of cyanobacteria in such a wide range of habitats indicates their capacity to mitigate 'reactive oxygen species (ROS)' mediated cellular damages, as the ROS are considered as a chief effector of any stress-induced deformities (Rastogi et al., 2020).

Furthermore, ROS is well-known by-product of photosynthesis, the oxygenic autotrophic cell possesses a higher level of ROS as compared to heterotrophic cells (Freeman et al. 2020; Ritter et al. 2020; Moore et al. 2020). Thus, cyanobacteria also possess advanced and extensive means to alleviate ROS damage. Cyanobacteria developed direct and indirect approaches to eliminate ROS. They possess a wide range of enzymatic and non-enzymatic biomolecules with a capacity to detoxify the ROS. These biomolecules contain a wealth of conjugated pi-double bonds, which permits them to accept the electrons from ROS and detoxify them (Sonani et al. 2017b). It has been also reported during the last few decades that numerous cyanobacterial compounds (of different sizes, compositions and solubility) exhibited their antioxidants activity in vitro and in vivo. The diverse range of cyanobacterial species is reported to produce antioxidant biomolecules such as polyunsaturated fatty acid (da Costa et al. 2020), phycobiliproteins (PBPs) (Galetović and Dufossé 2020), carotenoids (Lopes et al. 2020), mycosporin like amino acids (MAAs) (Rastogi and Incharoensakdi 2014a), scytonemins (Singh et al. 2010; Rastogi et al. 2015) and phlorotannins (Singh et al. 2017). In this chapter, we describe three antioxidant biomolecules PBPs, MAA and scytonemin with their occurrence, antioxidant activity, and anti-aging, neuroprotective and other biomedical applications.

15.2 Cyanobacteria: A Natural Source of Antioxidants

Exposure to wide ranges of environmental conditions over years leads to extensive genetic and epigenetic adaptations in cyanobacteria. For instance, the *Synechocystis* sp. PCC 6803 has been studied to possess an enriched level of stress response genes, whose expressions are elevated under stress conditions (Table 15.1) (Mironov et al. 2019). To counteract stress conditions, cyanobacteria have developed the stress-defence system, which consists of antioxidants and associated molecules. Antioxidants play a key role in the protection against oxidative stress by diverse modes of action (He and Häder 2002; Latifi et al. 2009). Cyanobacterial antioxidants can be classified into two categories, enzymatic and non-enzymatic (Latifi et al. 2009; Banerjee et al. 2013) (Table 15.2). The enzymatic antioxidants are substrate-specific, require co-factor(s) to perform and are recyclable; whereas, non-enzymatic antioxidants are not substrate-specific and recyclable (Haida and Hakiman 2019). The biomolecule categorised as enzymatic antioxidants is superoxide dismutase (SOD), catalase, peroxidases, peroxiredoxins, and DNA-binding proteins (DPS) (Rezayian et al. 2019). The non-enzymatic antioxidants are *chlorophyll* a, ascorbic acid, α-tocopherol, reduced glutathione, polyunsaturated fatty acids (PUFAs), vitamins, PBPs, MAAs and scytonemins (Sonani et al. 2017b; Mironov et al. 2019). They act either alone or in combination with other antioxidants to combat

Table 15.1 The list of genes expressed (transcriptome analysis) under different abiotic stress conditions in *Synechocystis* sp. PCC 6803 (Mironov et al. 2019)

Sr. no.	Stress	Optimum condition to generate stress	Exposure time (min)	Gene expressed
1	Heat stress	42–45 °C	15–20	*hik34, sigB, groES, groEL1, groEL2, hspA, dnaK2, dnaJ, htpG, clpB1, htrA, ctpA, sll1621, smtA, slr1674, ocpA, sbtA, sodB, sll0528, hypA1, sll3044, nblB1, sll0939, slr0967, slr1686, slr1603, sll1853, sll0846, sll1884, frpC, cbiA, sll0441, sll1892, slr0670, sll0982, slr1127*
2	Light stress + Heat stress	20–300 μM quanta/ m^2s	30	*groES, groEL1, groEL2, hspA, dnaK2, htpG, clpB1, ctpA, sll1621, smtA, slr1674, ocpA, sbtA, sodB, sll0528, hypA1, sll3044, nblB1, slr1686, sll0846, sll1884*
3	UV-B stress + Heat stress	Not available	30	*hik34, sigB, groES, groEL1, hspA, dnaK2, dnaJ, htpG, clpB1, htrA, ctpA, sll1621, slr1674, ocpA, sodB, sll0528, hypA1, sll3044, nblB1, sll0939, slr0967, slr1686, slr1603, sll0846*
4	Salt stress + Heat stress	0.5 M NaCl	15–20	*hik34, sigB, groES, groEL1, groEL2, hspA, dnaK2, dnaJ, htpG, clpB1, htrA, sll1621, smtA, slr1674, sodB, sll0528, hypA1, sll3044, nblB1, sll0939, slr0967, slr1686, slr1603, sll1853, sll0846, sll1884*
5	Osmo stress + Heat stress	0.5 M sorbitol	15–20	*hik34, sigB, groES, groEL1, groEL2, hspA, dnaK2, dnaJ, htpG, clpB1, htrA, ocpA, sodB, sll0528, hypA1, sll3044, nblB1, sll0939, slr0967, slr1603, sll0846, sll1884,*
6	pH stress + Heat stress	Low pH ~4.0	30	*hik34, sigB, hspA, dnaK2, dnaJ, htrA, ctpA, slr1674, sodB, sll0528, hypA1, sll3044, nblB1, sll0939, slr0967, sll0846*
7	Cold stress	22 °C	30	*hik31, rre5, crhR, rbpA1, tig, rpsL, rimO, typA, sbtA, syc2, ndhD2, hliA, hliB, lilA, nblB, nusG, rpoA, sigD*[1], *ssl3304*[1], *desA, desB, gpx2, sll1483, sll1057, sll1863, sll1862, sll1853, slr0551, slr0959, slr1686, slr1687*
8	Light stress + cold stress	20–300 μM photon/ m^2s	30	*rre5, crhR, rpsL, rimO, typA, sbtA, syc2, ndh2, hliA, hliB, lilA, nblB, nusG, rpoA, sigD*[1], *ssl3044*[1], *desA, desB, gpx2, sll1483, slr0551, slr0959, slr1686, slr1687*
9	UV-B stress + cold stress	Exposure for 30 min	30	*syc2, ndhD2, hliA, hliB, lilA, nblB*

(continued)

Table 15.1 (continued)

Sr. no.	Stress	Optimum condition to generate stress	Exposure time (min)	Gene expressed
10	Salt stress + cold stress	0.5 M NaCl	15–20	*crhr, rimO, hliA, hliB, lilA, nblB, nusG, sigD*[1], *ssl3044*[1], *sll1483, sll0157, sll1863, sll1862, sll1853, slr0959, slr1686, sl1687*
11	Osmotic stress + cold stress	0.5 M sorbitol	15–20	*crhR, tig, rpsL, rimO, typA, hliA, hliB, lilA, nblB, nusG, rpoA, ssl3304*[1], *sll1483, sll1057, sll1863, sll1862, slr0551*
12	pH stress + cold stress	Low pH 4.0	30	*hliB, lilA, nblB, sigD*[1], *ssl3304*[1], *sll1483*

against various stresses (Latifi et al. 2009; Tan et al. 2018). These antioxidant molecules have been demonstrated for their applicability in the pharmaceutical and food industry for the past few decades. Specifically, several research groups showed that the antioxidant assets of these biomolecules can be potentially used in the formulation of anti-aging, neuroprotective, anti-carcinogenic and many more therapeutic compositions.

15.3 Cyanobacterial Biomolecules

15.3.1 Mycosporine Like Amino Acids (MAA)

Exposure to ultraviolet radiation causes cellular damages in cyanobacteria. Cyanobacteria is evolved to produce UV-screening micro-molecules, known as mycosporine-like amino acids (MAAs), as their chemical structure resembles mycosporine. The MAAs are colourless, small (<400 Da) and water-soluble molecules (Rastogi et al. 2017). The MAAs has a high molar extinction coefficient and absorbs maximally in the UV light region (310–362 nm) of the solar spectrum. Structurally, it is similar to cyclohexenone or cyclohexenimine ring containing chromophore conjugated with nitrogen substituent of an amino acid or imino alcohol (Sinha and Häder 2008; Lawrence et al. 2018). The difference in absorption maxima of MAAs is observed (Table 15.3) owing to minor differences in their chemical structure as shown in Fig. 15.1.

The derivatives of the first part of the shikimate pathway and pentose pathway are involved in the biosynthesis of MAAs (Singh et al. 2008). There are four genes found to be involved in the biosynthesis of MAAs in cyanobacteria. The product of these genes is 3-dehydroquinate synthase (DHQS), O-methyltransferase (O-MT), adenosine triphosphate (ATP)-grasp enzyme and non-ribosomal peptide synthase

Table 15.2 Details of enzymatic and non-enzymatic molecules in defence system of cyanobacteria (Moussa et al. 2019; Rezayian et al. 2019; Latifi et al. 2009)

Sr. No.	Antioxidants	Catalytic centre	Substrate	Mode of action	Cyanobacteria
1	Superoxide dismutase (SOD)	Fe, Mn, Ni, Zn, Cu	O^{2-}	SOD is a metalloprotein and acts against oxidative stress by catalysing the dismutation of superoxide radicals into oxygen and hydrogen peroxide	*Anabena cylindrical, Synechococcus* PCC 7942, *Anabena* PCC 7120, *Nostoc commune*
2	Catalase	Heme (monofunctional, Bifunctional)	H_2O_2 and ROOH (in vitro)	Catalase is a tetrameric compound consisting of heme and mainly transform H_2O_2 into H_2O and O_2	*Anacystis nidulans, Synechocystis* PCC 6803, *Synechococcus* PCC 7942
3	Peroxiredoxin (Prx)	Single or double cystein(s)- (SH)	H_2O_2, peroxy-nitrite, ROOH	Peroxiredoxins belong to the ubiquitin family and it is commonly involved in the catalytic activity in which active cysteins are oxidized into sulfenic acid by the substrate of peroxide.	*Synechococcus* PCC 7942
4	Glutathione Peroxidise (GPX)	Cysteine (plants) Selenocysteine (mammals)	H_2O_2, ROOH	GPx decomposes H_2O_2 to H_2O	*Synechocystis* PCC 6803
5	DNA-binding protein (Dps)	Fe, Heme	H_2O_2	Dps consume H_2O_2 and act as like ferritin where Fe (II) oxidation is achieved by H_2O_2 and oxidized form Fe (III) is then mineralized and stored as Fe (III) (Insoluble).	*Thermosynechococcus elongates, Anabena* PCC 7120, *Synechococcus* PCC 7942
6	Caretenoids		$^1O_2$	Carotenoids due to their triplet energy level closer to the $^1O_2$, it act as a strong physical quencher of $^1O_2$	*Cyanobiumgracile, Nodosilinea (Leptolyngbya) antartica, Cuspidothrixissatschenkoi Leptolyngbya-like sp., Alkalinema aff., Pantanalense*
7	Scytonemin		UV-A/B	Scytonemin act as photoprotective against UV-A/B generated oxidative stress (free radicals)	*Nostoc commune* *Nostoc punctiforme*
8	MAA				*Nostoc commune*

			APPH radical and ABTS radical	MAA scavenging activities based on the skeletal structure and also act as a photoprotective	
9	PBPs		APPH, DPPH, reduction of Fe^{3+}	Scavenge radicals form APPH, DPPH, free radicals, reduction of Fe^{3+}, Partial inhibition of NADPH oxidase, Inhibit $ONOO^{-}$ mediated DNA degradation, Quench peroxy radicals	*Spirulina plantesis*, *Halomicronema sp.*, *Spirulina sp.*, *Arthospira maxima*, *Aphanizomenon flos-aquae*

Table 15.3 Chemical structure and absorption maximum of different MAAs from cyanobacteria

MAAs	Chemical structure	Absorption maxima (nm)
Mycosporine-2-Glycine (M2G)	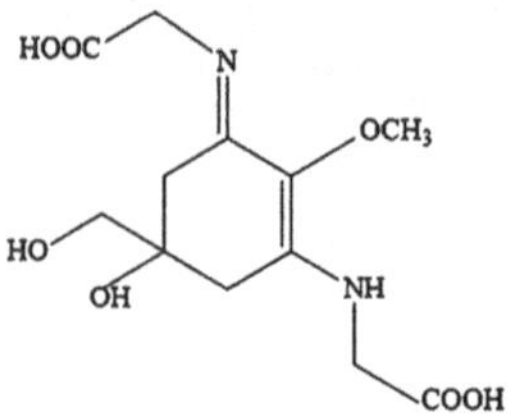	334 nm
Mycosporine–glycine (MG)	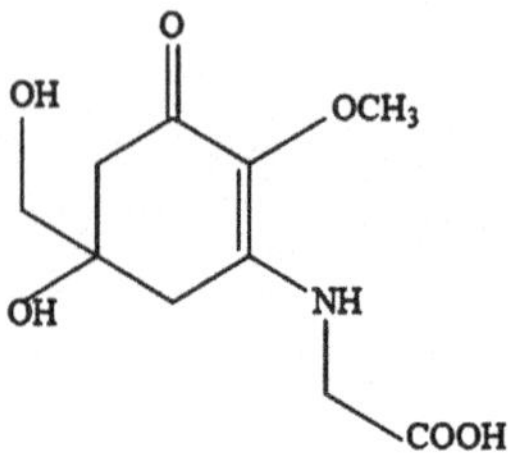	310 nm
Mycosporine–methylamin–threonine	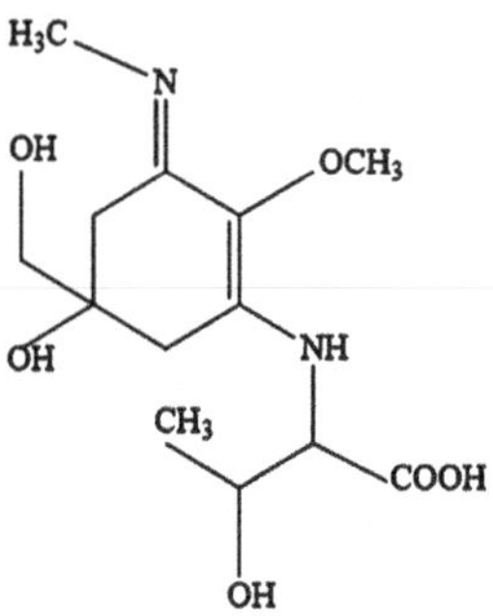	(327 nm)
Mycosporine–taurine	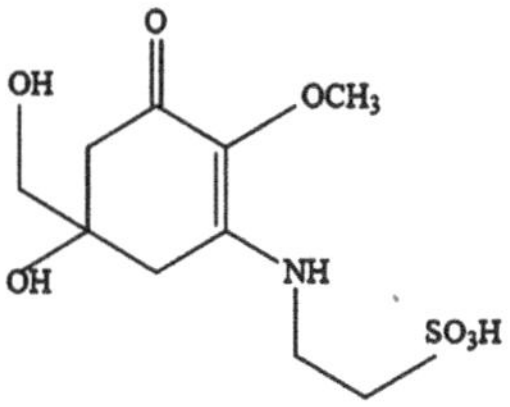	(309 nm)
Palythine	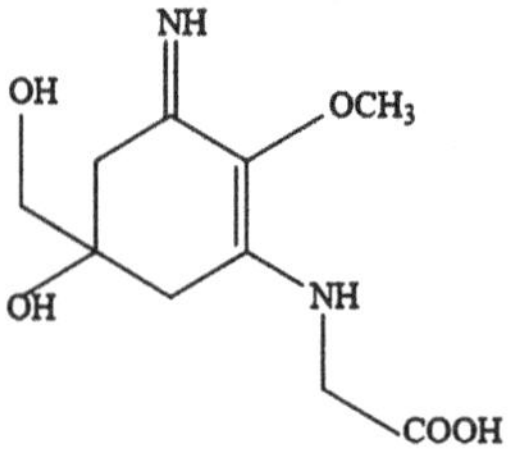	(320 nm)
Palythine–serin–sulphate		(320 nm)

(continued)

Table 15.3 (continued)

MAAs	Chemical structure	Absorption maxima (nm)
	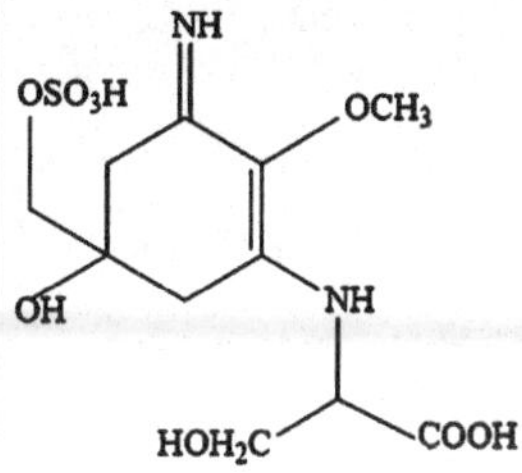	
Palythinol	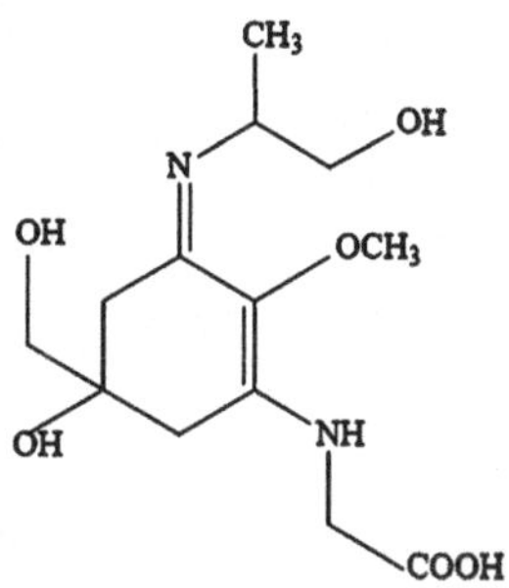	(332 nm)
Porphyra-334	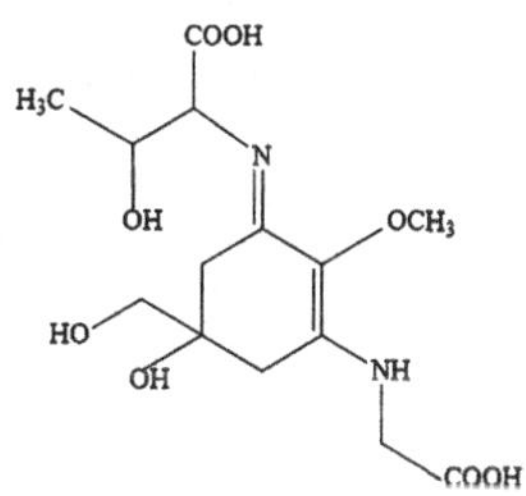	(334 nm)
Shinorine	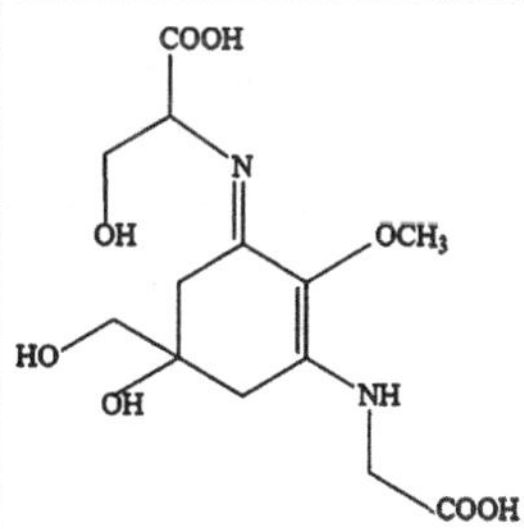	(334 nm)

(continued)

Table 15.3 (continued)

MAAs	Chemical structure	Absorption maxima (nm)
Usujirene	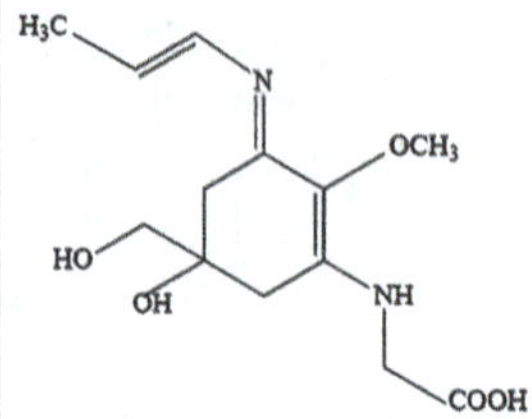	(357 nm)

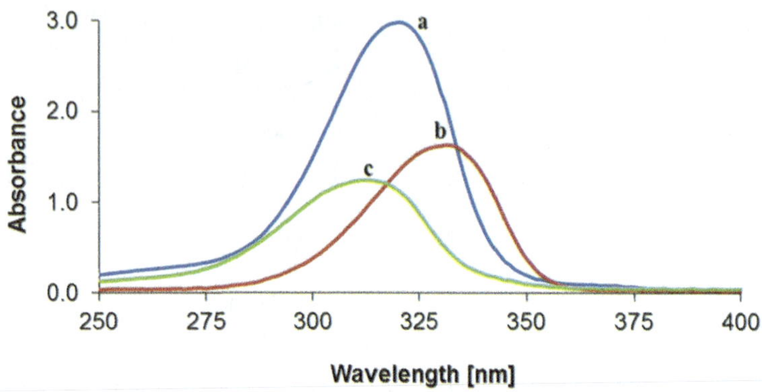

Fig. 15.1 The UV/Visible absorption maxima of different MAAs such as (**a**) palythine (320 nm), (**b**) asterina-330 (320 nm) and (**c**) M-312 (312 ± 1 nm). (Adapted from Rastogi and Incharoensakdi 2014b)

(NRPS)-like protein (also known as D-Ala-D-Ala ligase) (Kageyama and Waditee-Sirisattha 2019). The DHQS and O-MT are involved in the synthesis of the precursor of MAAs, 4-deoxygadusol (4-DG). The ATP-grasp enzyme forms the imine linkage and produces the mycosporine-glycine (MG). Moreover, the NRPS-like protein is involved in the synthesis of di-substituted MAAs (Balskus and Walsh 2010) as shown in Fig. 15.2.

There are 20 different types of MAAs reported. Type(s) of the MAA present in particular cyanobacteria may vary according to their niche (Lawrence et al. 2018). The MAAs biosynthesis is influenced by light and the osmotic environment (Oren 1997). The MAAs biosynthesis was reported and characterised in many cyanobacteria like *Calothrix*, *Chlorogloeopsis*, *Gloeocapsa*, *Synechococcus*, *Nostoc*, *Cyanothece, Scytonema, Rivularia* and *Anabaena* (Rastogi et al. 2012; Noreña-Caro and Benton 2018). MAAs act as photoprotective molecules. They absorb UV light without producing reactive oxygen species (ROS). The photoprotection efficacy of MAAs in cyanobacteria depends on the type of MAA and the location of MAA in the cell (Ehling-Schulz and Scherer 1999). It has been reported that UV

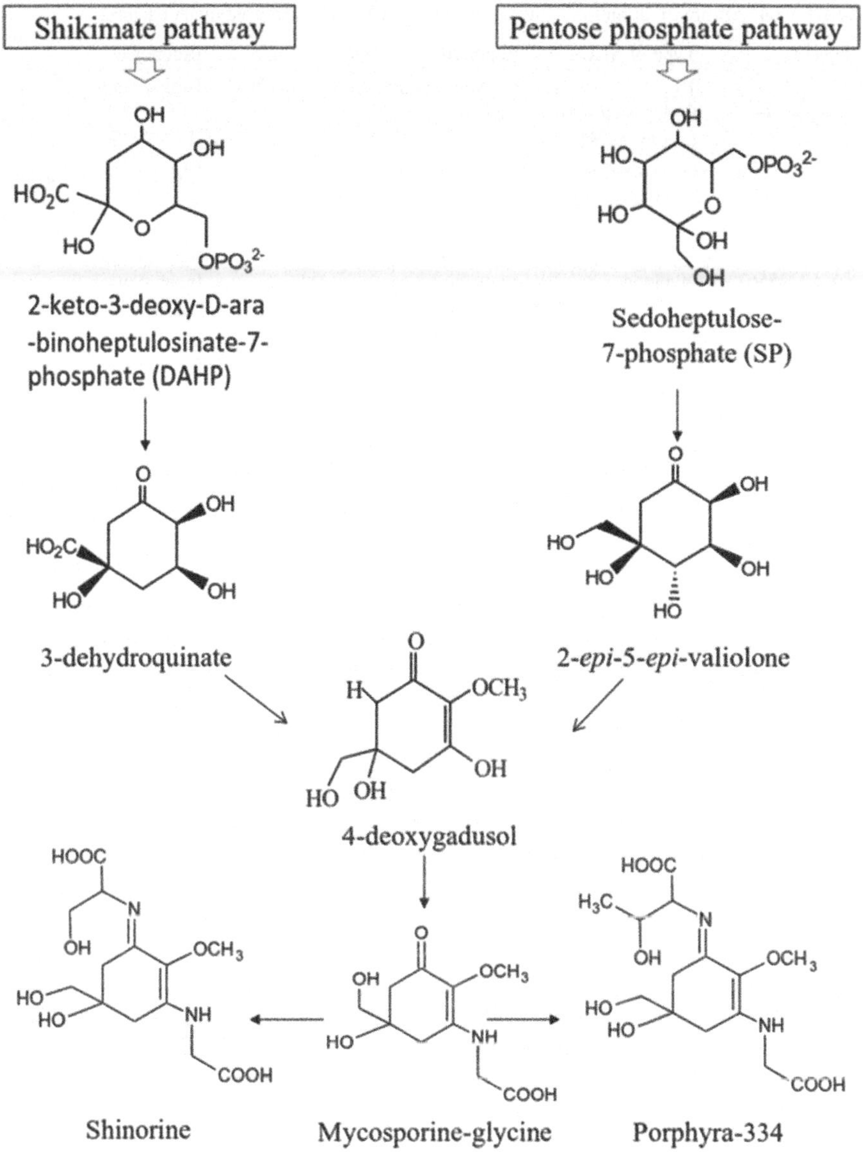

Fig. 15.2 Proposed biosynthesis pathway of MAAs. (Adapted from Rastogi and Incharoensakdi 2014b)

light induces ROS generation which in turn triggers elevated MAAs biosynthesis (Wada et al. 2013).

Along with photoprotection, MAAs also showed antioxidant activity and high stability, which make them useful for application in the cosmetic and pharmaceutical industries. The MAAs from different cyanobacteria was proved to show significant

Table 15.4 Summary of antioxidant activities of MAAs from different cyanobacteria

Species	Type of MAA	Reported Activities	References
Nostoc sphaericum	13-O-β-galactosyl-porphyra-334	• Antioxidant activity (in vitro analysis).	Ishihara et al. (2017)
Recombinant strain *E.coli*	Mycosporine-2-glycine (M2G)	• Antioxidant activity (in vitro analysis using human melanoma cell line).	Cheewinthamrongrod et al. (2016)
Arthrospira sp. CU2556	Mycosporine–glycine	• Antioxidant activity (*in vitro* analysis). • ROS scavenging activity (in vivo analysis).	Rastogi and Incharoensakdi (2014b)
Gloeocapsa sp. CU-2556	Shinorine and M-307	• Antioxidant activity (in vitro analysis).	Rastogi and Incharoensakdi (2014c)
Nostoc sp. R76DM	Palythine, asterina, porphyra, palythene	• Antioxidant activity (in vitro analysis). • ROS scavenging activity (in vivo analysis).	Rastogi et al. (2016)

free radical scavenging potential and reducing ability (Table 15.4). The shinorine and M-307, purified from *Gloeocapsa* sp. CU-2556 showed antioxidant activity, analysed by in vitro DPPH scavenging assay (Rastogi and Incharoensakdi 2014b). The *Arthrospira* sp. CU2556 was reported to produce mycosporine-glycine (M-Gly). The M-Gly shows good stability under various physiological stresses and is reported to possess the DPPH scavenging activity (Rastogi and Incharoensakdi 2014a). The production of M-Gly in cyanobacteria was noticed to increase upon UV-light exposure. Furthermore, the increase in M-Gly level is inversely related to the intracellular ROS level, indicating the ROS scavenging potential of M-Gly (Rastogi and Incharoensakdi 2014c). The MAAs namely M2G was analysed for radical scavenging activity using in vitro and in vivo methods. The in vivo analysis on human melanoma A375 showed the protective effect of M2G against oxidative stress generated by hydrogen peroxide (Cheewinthamrongrod et al. 2016). The four different MAAs (Palythine, Asterina, Porphyra and Palythene) from *Nostoc* sp. R76DM showed the potential antioxidant and radical scavenging activity (Rastogi and Incharoensakdi 2014b). The novel MAA, 13-O- β-galactosyl-porphyra-334 (β-Gal-P334), found in *Nostoc sphaericum* shows the UV protective effect in human keratinocytes (Ishihara et al. 2017). Despite being a potential antioxidant, the anti-aging and neuroprotective potential of MAAs is rarely investigated and explored.

15.3.2 Scytonemin

Scytonemin is an alkaloid pigment molecule, protecting cyanobacteria from harmful UV radiation. It is yellow to brown in colour and water-insoluble. Scytonemin is located on the outer sheath of cyanobacteria cell wall and can absorb the light in the UV range, from 100 to 564 nm (Rastogi et al. 2015; Mandal et al. 2020) (Fig. 15.3). In cyanobacteria, various forms of scytonemin have been reported, like reduced form, oxidized form and methoxylated form (tetramethoxy scytonemin) based on surrounding redox conditions (Table 15.5).

It has been proposed that scytonemin may have been synthesised from metabolites of aromatic amino acid biosynthesis pathways like tryptophan and tyrosine derivatives (Sinha and Häder 2008; Rastogi et al. 2015). The biosynthesis of scytonemin is influenced by various abiotic factors like UV radiation, temperature, salinity, oxidative stress, desiccation and nitrogen sources (Fleming and Castenholz 2007, 2008; Rastogi et al. 2015). It is known that the biosynthesis of scytonemin is activated by UV-A/B light exposure in *Nostoc plunctiforme* (Wada et al. 2013). Genes involved in the scytonemin biosynthesis are recognised in some cyanobacteria. The cluster of 18 genes (ORFs: NpR127-NpR1259) are known to regulate the biosynthesis of scytonemin. Furthermore, eight genes (out of 18 reported genes) are also known to involve in the biosynthesis of tryptophan and tyrosine, whereas the activity of other genes showed insignificant similarity with functionally characterised proteins (Rastogi et al. 2015) as shown in Fig. 15.4.

The absorption maximum of scytonemin is 370 nm in vivo, whereas it shifts to a longer wavelength of 384 nm in organic solvent (Garcia-Pichel and Castenholz 1991). Furthermore, it is of note that the molecular extinction coefficient of

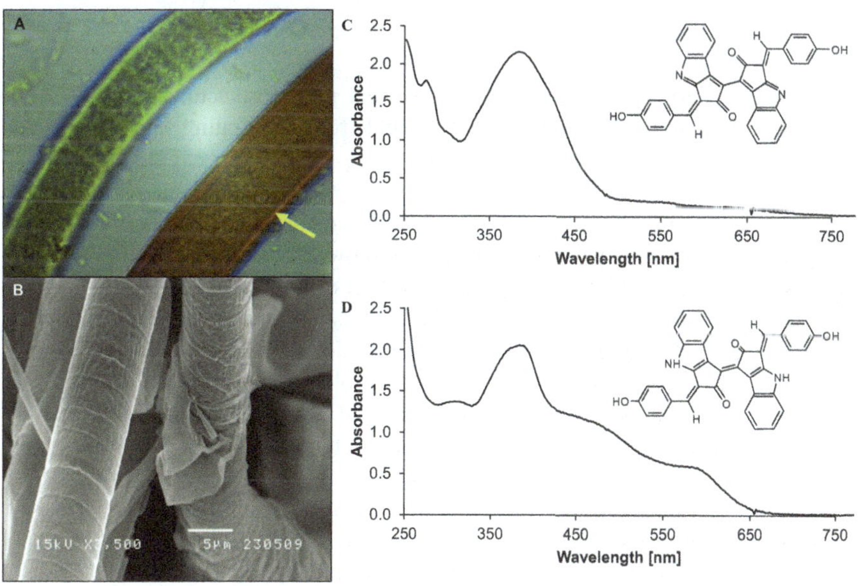

Fig. 15.3 *Lyngbya* sp. as observed under light (**a**) and electron (**b**) microscope, showing orange-brown coloured scytonemin compound (shown by arrow) in the extracellular sheath. (**c**, **d**): UV-Visible spectrum of scytonemin and reduced scytonemin, respectively. (Adapted from Rastogi and Incharoensakdi 2014b)

Table 15.5 Chemical structure of Scytonemin from cyanobacteria

Reduced Scytonemin	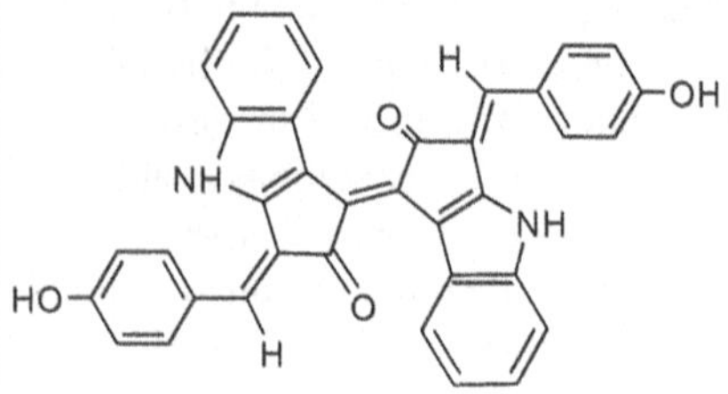
Oxidized Scytonemin	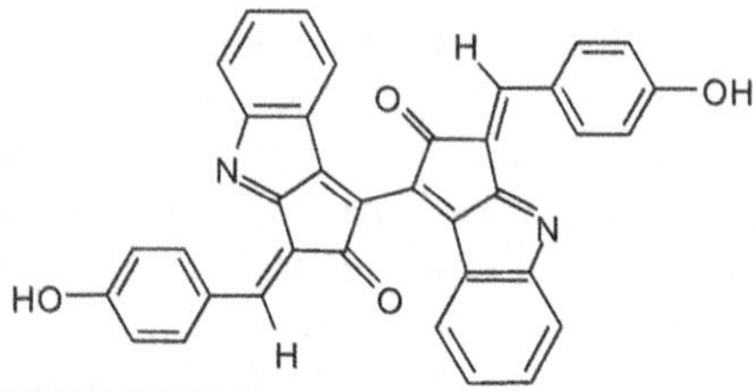
Tetramethoxyscytonemin	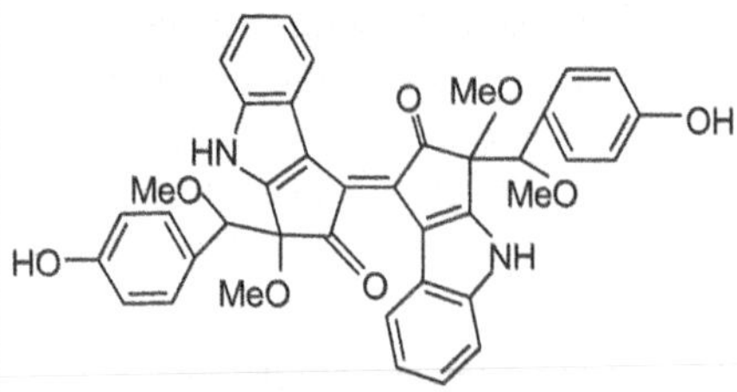

scytonemin is large, which makes them efficient photo-protector against UV-A/B light. Scytonemin blocks ~90% of UV-A light from penetrating into cells (Wada et al. 2013). Matsui et al. (2012) analysed the radical scavenging potential of scytonemin via electron spin resonance spectroscopy and obtained a 36 μM IC_{50} value of scytonemin against standard ABTS radical. Additionally, scytonemin extracted from various cyanobacterial species showed good free radical scavenging activity (Table 15.6) (Matsui et al. 2012; Rastogi et al. 2014). Another biomedical potential of scytonemin such as anti-inflammatory and anti-proliferative activities are also reported owing to their capacity to selectively inhibit various kinases (Mishra et al. 2015). Scytonemin prevents the proliferation of the endothelial cells, fibroblasts and tumour cells (Abd El-Hack et al. 2019). Even apoptosis can be induced in osteosarcoma cells by scytonemin (Mishra et al. 2015). Thus scytonemin is considered as a potential anti-cancerous compound in the therapy of myeloma. Scytonemin is reported to act as an inhibitor of polo-like protein kinase1 (PLK1). PLK1 is a serine/threonine kinase protein that plays a significant role in the G2-M transition phase and spindle formation in mitotic cell division (Pezuk et al. 2013). It is also reported that the cancer cell cycle is regulated by PLK1 by increasing the rate of proliferation in bladder cancer cells which can be modulated by using scytonemin as an inhibitor (Zhang et al. 2013). It can prevent the growth of osteosarcoma cells and cause apoptosis (Mishra et al. 2015). Scytonemin is also used as a cosmeceutical agent in skin moisturizers and sun-screen creams (Morone et al. 2019). Thus, the

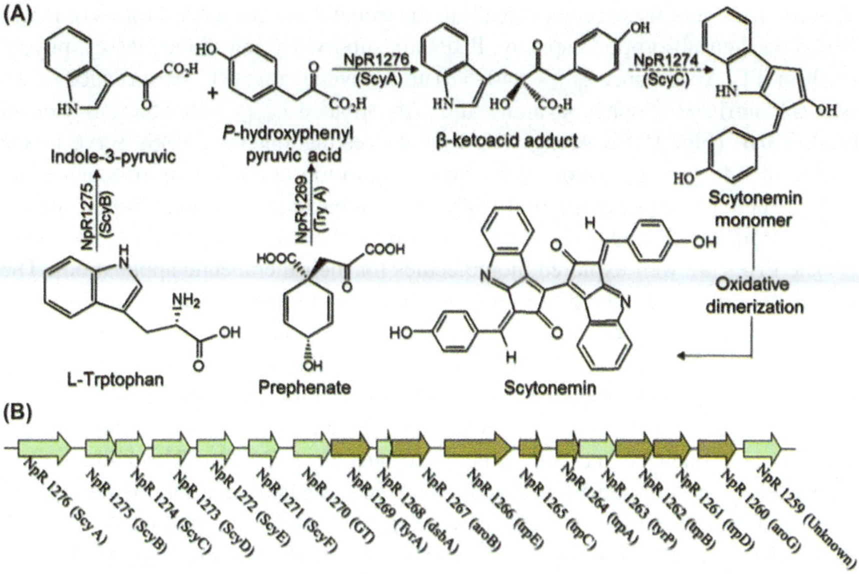

Fig. 15.4 (**a**) Proposed biosynthesis pathway of Scytonemin. (**b**) Schematic representation of genes involved in scytonemin biosynthesis. (Adapted from Rastogi et al. 2017)

Table 15.6 The antioxidant activities of scytonemin from different cyanobacteria

Cyanobacteria	Type of scytonemin	Reported activities	References
Nostoc commune	Scytonemin	Antioxidant activity (analysed by in vitro method)	Matsui et al. (2012)
Lyngbya sp. CU2555	Scytonemin, reduced scytonemin	Antioxidant activity (analysed by in vitro method)	Rastogi and Incharoensakdi (2014b)
Scytonema sp. R77DM	Scytonemin, reduced scytonemin	Antioxidant activity (analysed by in vitro method)	Rastogi et al. (2014)

biomedical applications of scytonemin studied by various research groups clearly indicate syctonemin as the multifunctional molecule. The reported functions of scytonemin include radical scavenging, protection of cells against UV radiation, anticancer and UV-induced ROS generation. However, the anti-aging and neuroprotective potential of MAAs is yet to be investigated.

15.3.3 Phycobiliproteins (PBPs)

PBPs are accessory light-harvesting proteins that enable cyanobacteria to maintain their light-harvesting capacity in extreme habitats. Structurally, they are made up of two subunits, named as α subunit and β subunit. These subunits are covalently attached to chromophores called phycobilins (Singh et al. 2015). The oligomerization of these subunits leads to the trimeric and hexameric forms of PBPs. Based on their absorption capacity, PBPs are subdivided into three classes: phycoerythrin (PE, Absorbance λ_{max} – 540–570 nm), phycocyanin (PC, Absorbance λ_{max} – 610–620 nm) and allophycocyanin (APC, Absorbance λ_{max} – 650–655 nm) (Sonani et al. 2016) (Fig. 15.5). Along with light-harvesting function, PBSs serve as the source of nitrogen in cyanobacteria under starvation conditions (Grossman et al. 1993). The type and proportion of PBPs in cyanobacteria vary according to species and environmental conditions, like temperature and availability of light (Adir 2005).

The PBPs are well-explored biomolecules for their therapeutic applications. The PBPs showed a wide range of biomedically useful properties such as antioxidant, anti-aging, neuroprotective, anti-inflammatory, hepatoprotective, and anticancer activities (Sonani et al. 2015). The PBPs from different cyanobacteria species have been widely reported for their in vitro and in vivo antioxidant properties. The free radical neutralizing capacity and antioxidant potential of PBPs depend on their amino-acid sequences and 3-D structure (Patel et al. 2018b). The amino acids, responsible for antioxidant potential were investigated and the percentage of residues for contributing the radical scavenging activity was studied (Patel et al. 2018a, b). The PC from *Synechococcus* sp. R42DM, *Geitlerinema* sp. TRV57, *Lyngbya* sp. A09DM and *Leptolyngbya* sp. N62DM was reported to show in vivo and in vitro antioxidant activities (Sonani et al. 2014; Singh et al. 2016; Sonani et al. 2017a; Renugadevi et al. 2018). The thermostable PBPs purified from hot spring cyanobacteria namely *Cyanosarcina* sp. SK40, *Phormidium* sp. PD40–1, *Scytonema* sp. TP40 and *Leptolyngbya* sp. KC45 have been reported to display antioxidant activities (Pumas et al. 2011). It is widely known that stress and resistance against stress require various homeostatic habituations such as oxidants, neurotransmitters, hormones and other mediators. Imbalance in these homeostatic habituations or reduced endogenous expression of stress-induced hormones causes damaged to biomolecule due to increased production of oxidants. Hence, elevated levels of oxidants and decreased endogenous antioxidants influenced the normal metabolism and generate an accumulation of harmful oxidants in mitochondria which leads to aging (Liu and Mori 1999). Thus, oxidative damage leads to the accelerated rate of aging and causes age-associated disease (a neurodegenerative disease). Therefore, antioxidant and anti-aging effect of PE from different cyanobacteria like *Lyngbya* sp. A09DM and *Halomicronema* sp. R31DM was analysed using *C. elegans* as a model organism (Sonani et al. 2017c; Patel et al. 2018b). The supplementation of PE showed an increase in the lifespan and health span of *C. elegans*. The in silico and in vivo study reveals the neuroprotective effect of PBPs (Chaubey et al. 2019, 2020). Chaubey et al. (2019) reported that PE from *Phormidium* sp. showed strong binding against BACE1 which is the rate-limiting step enzyme in the progression of AD. Furthermore, the study of PE on the Aβ generation showed decreased Aβ formation in PE-treated *C. elegans* as compared to untreated *C. elegans* (Chaubey et al. 2019). Similarly, Chaubey et al. (2020) explored the effect of APC on the anti-aging and Aβ formation. This study described the new findings on that the APC-mediated longevity in *C. elegans* is daf-16 dependent and skn-1 independent (Chaubey et al. 2020). The docking study of PC and PE αβ-monomer with β-secretase enzyme (enzyme responsible for Alzheimer's disease) pointed out that

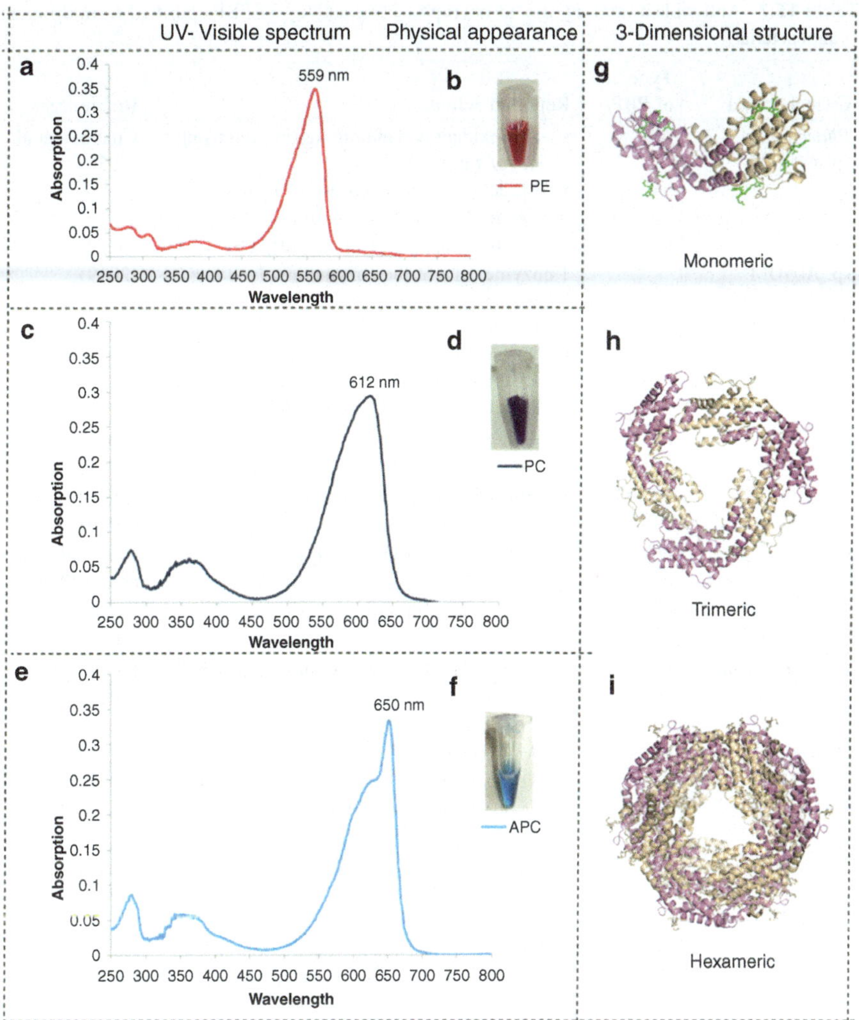

Fig. 15.5 Spectral properties and three-dimensional structures of phycobiliproteins. (**a**, **b**) UV–visible absorption spectra and appearance of phycoerythrin aqueous solution. (**c**, **d**) UV–visible absorption spectra and appearance of phycocyanin aqueous solution. (**e**, **f**) UV–visible absorption spectra and appearance of allophycocyanin in aqueous solution. (**g–i**) Cartoon representation of the three-dimensional structure of phycoerythrin monomer ($\alpha\beta$), trimer ($\alpha_3\beta_3$) and hexamer ($\alpha_6\beta_6$), respectively

the PC and PE can be promising drug-candidate for Alzheimer's disease (Singh et al. 2014; Chaubey et al. 2019). The neuroprotective effect of PC against neuronal cell death was analysed in gebrils (Pentón-Rol et al. 2011). The APC from *Phormidium* sp. A09DM showed the neuroprotective effect in the Alzheimer's disease model of *C. elegans* (Chaubey et al. 2020). The antioxidant activity, anti-aging activity and neuroprotective potential of PBPs from various cyanobacteria, reported so far are summarised in Table 15.7. The therapeutic potential along with the fluorescence

Table 15.7 Antioxidant, anti-aging and neuroprotective activity of PBPs purified from different cyanobacteria

Cyanobacteria	Type of PBPs	Reported activities	References
Phormidium sp. A09DM	APC	• Antioxidant and anti-aging (in vivo using *C. elegans* model) • Moderate Aβ-induced paralysis in the transgenic *C. elegans* CL4176	Chaubey et al. (2020)
Lyngbya sp. A09DM	PE	• In silico docking of PE with BACE 1 enzyme (one of the enzyme involved for Alzheimer's disease) • In vitro interaction by surface plasma resonance and isothermal titration • Histopathological staining of Aβ aggregate in Alzheimer model *C. elegans* CL4176	Chaubey et al. (2019)
Spirulina platensis	PC	• Antioxidant (in vitro) • Biocompatible (Wistar Rat model)	Namasivayam et al. (2019)
Nostoc sp. R76DM	PC	• Antioxidant (in vitro and in vivo using *C. elegans* model)	Sonani et al. (2019)
Geitlerinema sp. TRV57	PC	• Antioxidant (in vitro analysis)	Renugadevi et al. (2018)
Geitlerinema sp. H8DM	PC	• Antioxidant (in vitro and in silico)	Patel et al. (2018a)
Halomicronema sp. R31DM	PE	• Antioxidant (in vitro and in vivo using *C. elegans* model) • Anti-aging activity (*C. elegans*)	Patel et al. (2018b)
Phormidium sp. A09DM	PE	• Antioxidant and anti-aging (in vivo using *C. elegans* model, *D. melanogaster* model) • Delay in paralysis of CL4176 (*C. elegans* Alzheimer's model)	Sonani et al. (2017c)
Synechococcus sp. R42DM	PC	• Antioxidant (in vitro and in vivo using *C. elegans* model) • Anti-aging activity (*C. elegans*)	Sonani et al. (2017a)
Arthrospira platensis	PC	• Improves lifespan and locomotion in *D. melanogaster* Parkinson's disease model DJ-1$\beta^{\Delta 93}$	Kumar et al. (2017)
Leptolyngbya sp. N62DM	PC	• Antioxidant and anti-aging (in vivo using *C. elegans* model) • Suppress neuronal toxicity (*C. elegans* AM141 Huntington disease model)	Singh et al. (2016)
Commercial	PC	• Antioxidant and neuroprotective in 3D astrocyte model	Min et al. (2015)
Lyngbya sp. A09DM	PC, PE, APC	• Antioxidant (in vitro and in vivo using *C. elegans* model) • Anti-aging activity (*C. elegans*)	Sonani et al. (2014)
Leptolyngbya sp. N62DM	PC	• Molecular docking of PC αβ-dimer with the enzyme β-secretase • Significantly delayed in paralysis in Alzheimer model *C. elegans* CL4176	Singh et al. (2014)
	PC		

(continued)

Table 15.7 (continued)

Cyanobacteria	Type of PBPs	Reported activities	References
Oscillatoria tenuis		• Antioxidant and anti-proliferative activity against human cancer cell lines	Thangam et al. (2013)
Arthrospira platensis	PC	• Neuroprotective (gerbils model used)	Pentón-Rol et al. (2011)

nature make PBPs suitable for nutraceutical industries, biomedical applications and pharmaceutical industries (Sonani et al. 2015; Sonani et al. 2016; Singh et al. 2016; Namasivayam et al. 2019).

15.4 Concluding Remarks

Cyanobacteria are a rich source of natural antioxidants. The antioxidant biomolecules from cyanobacteria showed potential in the field of cosmetics, pharmaceuticals, food and other therapeutics. Different cyanobacterial species that occurred with the antioxidant characteristics raised the interest in various research groups to find the compounds with high bioactivity. In recent years the cyanobacterial biomolecules scytonemin, MAAs and PBPs have been investigated and found promising ingredients to be used for antioxidant, anti-aging and neuroprotective formulations. However, many questions like, "How do exactly these biomolecules interrupt with ROS associated disease?"; "How do they delay aging?"; 'What can be the possible economic ways for large-scale production of these molecules?" are yet to be answered in this field.

Acknowledgments Datta Madamwar is grateful to the Department of Biotechnology, New Delhi (BT/PR15686/AAQ/3/811/2016) for financial support. Ravi R. Sonani acknowledges the NAWA-Ulam fellowship grant (grant PPN/ULM/2019/1/00175) funded by Polish National Agency for Academic Exchange (NAWA).

References

Abd El-Hack ME, Abdelnour S, Alagawany M, Abdo M, Sakr MA, Khafaga AF, Mahgoub SA, Elnesr SS, Gebriel MG (2019) Microalgae in modern cancer therapy: current knowledge. Biomed Pharmacother 111:42–50

Adir N (2005) Elucidation of the molecular structures of components of the phycobilisome: reconstructing a giant. Photosynthesis Res 85:15–32

Balskus EP, Walsh CT (2010) The genetic and molecular basis for sunscreen biosynthesis in cyanobacteria. Science 329:1653–1656

Banerjee M, Raghavan PS, Ballal A, Rajaram H, Apte SK (2013) Oxidative stress management in the filamentous, heterocystous, diazotrophic cyanobacterium, Anabaena PCC7120. Photosynth Res 118:59–70

Boussiba S (2000) Carotenogenesis in the green alga *Haematococcus pluvialis*: cellular physiology and stress response. Physiol Plantarum 108:111–117

Caiola MG, Billi D (2007) Chroococcidiopsis from desert to mars. In: Algae and cyanobacteria in extreme environments. Springer, pp 553–568

Cassier-Chauvat C, Dive V, Chauvat F (2017) Cyanobacteria: photosynthetic factories combining biodiversity, radiation resistance, and genetics to facilitate drug discovery. Appl Microbiol Biotechnol 101:1359–1364

Castenholz RW, Garcia-Pichel F (2012) Cyanobacterial responses to UV radiation. In: Ecology of cyanobacteria II. Springer, pp 481–499

Chaubey MG, Patel SN, Rastogi RP, Srivastava PL, Singh AK, Madamwar D, Singh NK (2019) Therapeutic potential of cyanobacterial pigment protein phycoerythrin: in silico and in vitro study of BACE1 interaction and in vivo Aβ reduction. Int J Biol Macromol 134:368–378

Chaubey MG, Patel SN, Rastogi RP, Madamwar D, Singh NK (2020) Cyanobacterial pigment protein allophycocyanin exhibits longevity and reduces Aβ-mediated paralysis in *C. elegans*: complicity of FOXO and NRF2 ortholog DAF-16 and SKN-1. 3 Biotech 10:1–11

Cheewinthamrongrod V, Kageyama H, Palaga T, Takabe T, Waditee-Sirisattha R (2016) DNA damage protecting and free radical scavenging properties of mycosporine-2-glycine from the Dead Sea cyanobacterium in A375 human melanoma cell lines. J Photochem Photobiol B Biol 164:289–295

da Costa E, Amaro HM, Melo T, Guedes AC, Domingues MR (2020) Screening for polar lipids, antioxidant, and anti-inflammatory activities of *Gloeothece* sp. lipid extracts pursuing new phytochemicals from cyanobacteria. J Appl Phycol 32:3015–3030

Ehling-Schulz M, Scherer S (1999) UV protection in cyanobacteria. Eur J Phycol 34:329–338

Fleming ED, Castenholz RW (2007) Effects of periodic desiccation on the synthesis of the UV-screening compound, scytonemin, in cyanobacteria. Environ Microbiol 9:1448–1455

Fleming ED, Castenholz RW (2008) Effects of nitrogen source on the synthesis of the UV-screening compound, scytonemin, in the cyanobacterium Nostoc punctiforme PCC 73102. FEMS Microbiol Ecol 63:301–308

Freeman EC, Creed IF, Jones B, Bergström A-K (2020) Global changes may be promoting a rise in select cyanobacteria in nutrient-poor northern lakes. Glob Chang Biol 26:4966–4987

Galetović A, Dufossé L (2020) Phycobiliproteins as food additives. In: Pigments from microalgae handbook. Springer, pp 559–573

Garcia-Pichel F, Belnap J (1996) Microenviroments and Microscale productivity of Cyanobacterial desert crusts 1. J Phycol 32:774–782

Garcia-Pichel F, Castenholz RW (1991) Characterization and biological implications of scytonemin, a cyanobacterial sheath pigment. J Phycol 27:395–409

Gimmler H (2001) Acidophilic and acidotolerant algae. In: Algal adaptation to environmental stresses. Springer, pp 259–290

Golubic S, Abed RM, Palińska K, Pauillac S, Chinain M, Laurent D (2010) Marine toxic cyanobacteria: diversity, environmental responses and hazards. Toxicon 56:836–841

Grossman AR, Schaefer MR, Chiang GG, Collier JL (1993) The phycobilisome, a light-harvesting complex responsive to environmental conditions. Microbiol Mol Biol Rev 57:725–749

Haida Z, Hakiman M (2019) A comprehensive review on the determination of enzymatic assay and nonenzymatic antioxidant activities. Food Sci Nutr 7:1555–1563

He Y-Y, Häder D-P (2002) Reactive oxygen species and UV-B: effect on cyanobacteria. Photochem Photobiol Sci 1:729–736

Ishihara K, Watanabe R, Uchida H, Suzuki T, Yamashita M, Takenaka H, Nazifi E, Matsugo S, Yamaba M, Sakamoto T (2017) Novel glycosylated mycosporine-like amino acid, 13-O-(-β-galactosyl)-porphyra-334, from the edible cyanobacterium Nostoc sphaericum-protective activity on human keratinocytes from UV light. J Photochem Photobiol B Biol 172:102–108

Javor B (1989) Cyanobacteria. In: Hypersaline environments. Springer, pp 134–146

Jorjani S, Etesami E, Shokravi S (2020) Ecophysiological study on soil cyanobacterium in combination of pH and salinity conditions at limited irradiance. JENAS J Environ Nat Stud 2: 34–45

Kageyama H, Waditee-Sirisattha R (2019) Antioxidative, anti-inflammatory, and anti-aging properties of mycosporine-like amino acids: molecular and cellular mechanisms in the protection of skin-aging. Mar Drugs 17:222

Kumar A, Christian PK, Panchal K, Guruprasad BR, Tiwari AK (2017) Supplementation of Spirulina (Arthrospira platensis) improves lifespan and locomotor activity in paraquat-sensitive DJ-1βΔ93 flies, a Parkinson's disease model in Drosophila melanogaster. J Diet Suppl 14:573–588

Latifi A, Ruiz M, Zhang C-C (2009) Oxidative stress in cyanobacteria. FEMS Microbiol Rev 33: 258–278

Lawrence KP, Long PF, Young AR (2018) Mycosporine-like amino acids for skin photoprotection. Curr Med Chem 25:5512–5527

Liu J, Mori A (1999) Stress, aging, and brain oxidative damage. Neurochem Res 24:1479–1497

Lopes G, Clarinha D, Vasconcelos V (2020) Carotenoids from cyanobacteria: a biotechnological approach for the topical treatment of psoriasis. Microorganisms 8:302

Mandal S, Rath J (2014) Extremophilic cyanobacteria for novel drug development. Springer

Mandal MK, Chanu NK, Chaurasia N (2020) Chapter 5: Cyanobacterial pigments and their fluorescence characteristics: applications in research and industry. In: Singh PK, Kumar A, Singh VK, Shrivastava AK (eds) . Academic Press, Advances in cyanobacterial biology, pp 55–72

Matsui K, Nazifi E, Hirai Y, Wada N, Matsugo S, Sakamoto T (2012) The cyanobacterial UV-absorbing pigment scytonemin displays radical-scavenging activity. J Gen Appl Microbiol 58:137–144

Min SK, Park JS, Luo L, Kwon YS, Lee HC, Shim HJ, Kim I-D, Lee J-K, Shin HS (2015) Assessment of C-phycocyanin effect on astrocytes-mediated neuroprotection against oxidative brain injury using 2D and 3D astrocyte tissue model. Sci Rep 5:14418

Mironov KS, Sinetova MA, Shumskaya M, Los DA (2019) Universal molecular triggers of stress responses in cyanobacterium synechocystis. Life 9:67

Mishra A, Tandon R, Kesarwani S, Singh R, Tiwari GL (2015) Emerging applications of cyanobacterial ultraviolet protecting compound scytonemin. J Appl Phycol 27:1045–1051

Moore KA, Altus S, Tay JW, Meehl JB, Johnson EB, Bortz DM, Cameron JC (2020) Mechanical regulation of photosynthesis in cyanobacteria. Nat Microbiol 5:757–767

Morone J, Alfeus A, Vasconcelos V, Martins R (2019) Revealing the potential of cyanobacteria in cosmetics and cosmeceuticals—a new bioactive approach. Algal Res 41:101541

Moussa Z, Judeh ZM, Ahmed SA (2019) Non-enzymatic exogenous and endogenous antioxidants. In: Organic chemistry. IntechOpen

Namasivayam SKR, Shivaramakrishnan K, Bharani RSA (2019) Potential antioxidative protein-pigment complex Spirulina platensis mediated food grade phycocyanin C -Extraction, purification, antioxidative activity and biocompatibility. Indian J Biochem Biophys 56(3):230–239

Noreña-Caro D, Benton MG (2018) Cyanobacteria as photoautotrophic biofactories of high-value chemicals. J CO2 Util 28:335–366

Nübel U, Garcia-Pichel F, Clavero E, Muyzer G (2000) Matching molecular diversity and ecophysiology of benthic cyanobacteria and diatoms in communities along a salinity gradient. Environ Microbiol 2:217–226

Oren A (1997) Mycosporine-like amino acids as osmotic solutes in a community of halophilic cyanobacteria. Geomicrobiol J 14:231–240

Oren A (2015) Cyanobacteria in hypersaline environments: biodiversity and physiological properties. Biodivers Conserv 24:781–798

Patel HM, Rastogi RP, Trivedi U, Madamwar D (2018a) Structural characterization and antioxidant potential of phycocyanin from the cyanobacterium Geitlerinema sp. H8DM. Algal Res 32:372–383

Patel SN, Sonani RR, Jakharia K, Bhastana B, Patel HM, Chaubey MG, Singh NK, Madamwar D (2018b) Antioxidant activity and associated structural attributes of Halomicronema phycoerythrin. Int J Biol Macromol 111:359–369

Pentón-Rol G, Marín-Prida J, Pardo-Andreu G, Martínez-Sánchez G, Acosta-Medina EF, Valdivia-Acosta A, Lagumersindez-Denis N, Rodríguez-Jiménez E, Llópiz-Arzuaga A, López-Saura PA, Guillén-Nieto G, Pentón-Arias E (2011) C-Phycocyanin is neuroprotective against global cerebral ischemia/reperfusion injury in gerbils. Brain Res Bull 86:42–52

Pezuk JA, Brassesco MS, Morales AG, de Oliveira JC, de Paula Queiroz RG, Machado HR, Carlotti CG, Neder L, Scrideli CA, Tone LG (2013) Polo-like kinase 1 inhibition causes decreased proliferation by cell cycle arrest, leading to cell death in glioblastoma. Cancer Gene Ther 20: 499–506

Pumas C, Vacharapiyasophon P, Peerapornpisal Y, Leelapornpisid P, Boonchum W, Ishii M, Khanongnuch C (2011) Thermostablility of phycobiliproteins and antioxidant activity from four thermotolerant cyanobacteria. Phycol Res 59:166–174

Rastogi RP, Incharoensakdi A (2014a) UV radiation-induced biosynthesis, stability and antioxidant activity of mycosporine-like amino acids (MAAs) in a unicellular cyanobacterium Gloeocapsa sp. CU2556. J Photochem Photobiol B Biol 130:287–292

Rastogi RP, Incharoensakdi A (2014b) Characterization of UV-screening compounds, mycosporine-like amino acids, and scytonemin in the cyanobacterium *Lyngbya* sp. CU2555. FEMS Microbiol Ecol 87:244–256

Rastogi RP, Incharoensakdi A (2014c) Analysis of UV-absorbing photoprotectant mycosporine-like amino acid (MAA) in the cyanobacterium Arthrospira sp. CU2556. Photochem Photobiol Sci 13:1016–1024

Rastogi RP, Kumari S, Richa HT, Sinha RP (2012) Molecular characterization of hot spring cyanobacteria and evaluation of their photoprotective compounds. Can J Microbiol 58:719–727

Rastogi RP, Sonani RR, Madamwar D (2014) The high-energy radiation protectant extracellular sheath pigment scytonemin and its reduced counterpart in the cyanobacterium *Scytonema* sp. R77DM. Bioresour Technol 171:396–400

Rastogi RP, Sonani RR, Madamwar D (2015) Cyanobacterial sunscreen Scytonemin: role in photoprotection and biomedical Research. Appl Biochem Biotechnol 176:1551–1563

Rastogi RP, Sonani RR, Madamwar D, Incharoensakdi A (2016) Characterization and antioxidant functions of mycosporine-like amino acids in the cyanobacterium *Nostoc* sp. R76DM. Algal Res 16:110–118

Rastogi RP, Sonani RR, Madamwar D (2017) UV photoprotectants from algae-synthesis and bio-functionalities. In: Algal Green chemistry. Elsevier, pp 17–38

Rastogi RP, Madamwar D, Nakamoto H, Incharoensakdi A (2020) Resilience and self-regulation processes of microalgae under UV radiation stress. J Photochem Photobiol C: Photochem Rev 43:100322

Renugadevi K, Valli Nachiyar C, Sowmiya P, Sunkar S (2018) Antioxidant activity of phycocyanin pigment extracted from marine filamentous cyanobacteria Geitlerinema sp TRV57. Biocatal Agric Biotechnol 16:237–242

Rezayian M, Niknam V, Ebrahimzadeh H (2019) Oxidative damage and antioxidative system in algae. Toxicol Rep 6:1309–1313

Ritter SPA, Lewis AC, Vincent SL, Lo LL, Cunha APA, Chamot D, Ensminger I, Espie GS, Owttrim GW (2020, 1864) Evidence for convergent sensing of multiple abiotic stresses in cyanobacteria. Biochim Biophys Acta Gen Subj:129462

Seckbach J (2007) Algae and cyanobacteria in extreme environments. Springer Science & Business Media

Seckbach J, Oren A (2007) Oxygenic photosynthetic microorganisms in extreme environments. In: Algae and cyanobacteria in extreme environments. Springer, pp 3–25

Seckbach J, Chapman DJ, Garbary D, Oren A, Reisser W (2007) Algae and Cyanobacteria under environmental extremes. In: Algae and cyanobacteria in extreme environments. Springer, pp 781–786

Singh SP, Kumari S, Rastogi RP, Singh KL, Sinha RP (2008) Mycosporine-like amino acids (MAAs): chemical structure, biosynthesis and significance as UV-absorbing/screening compounds. Indian J Exp Biol 46:7–17
Singh SP, Kumari S, Rastogi RP, Singh KL, Sinha RP (2010) Photoprotective and biotechnological potentials of cyanobacterial sheath pigment, scytonemin. Afr J Biotechnol 9:580–588
Singh NK, Hasan SS, Kumar J, Raj I, Pathan AA, Parmar A, Shakil S, Gourinath S, Madamwar D (2014) Crystal structure and interaction of phycocyanin with β-secretase: A putative therapy for Alzheimer's disease. CNS Neurol Disord Drug Targets 13:691–698
Singh NK, Sonani RR, Rastogi RP, Madamwar D (2015) The phycobilisomes: an early requisite for efficient photosynthesis in cyanobacteria. EXCLI J 14:268–289
Singh NK, Sonani RR, Awasthi A, Prasad B, Patel AR, Kumar J, Madamwar D (2016) Phycocyanin moderates aging and proteotoxicity in *Caenorhabditis elegans*. J Appl Phycol 28:2407–2417
Singh R, Parihar P, Singh M, Bajguz A, Kumar J, Singh S, Singh VP, Prasad SM (2017) Uncovering potential applications of cyanobacteria and algal metabolites in biology, agriculture and medicine: current status and future prospects. Front Microbiol 8:515
Sinha RP, Häder D-P (2008) UV-protectants in cyanobacteria. Plant Sci 174:278–289
Sonani RR, Singh NK, Kumar J, Thakar D, Madamwar D (2014) Concurrent purification and antioxidant activity of phycobiliproteins from Lyngbya sp. A09DM: an antioxidant and anti-aging potential of phycoerythrin in *Caenorhabditis elegans*. Process Biochem 49:1757–1766
Sonani RR, Rastogi RP, Madamwar D (2015) Antioxidant potential of phycobiliproteins: role in anti-aging research. Biochem Anal Biochem 4(172):2161–1009
Sonani RR, Rastogi RP, Patel R, Madamwar D (2016) Recent advances in production, purification and applications of phycobiliproteins. World J Biol Chem 7:100–109
Sonani RR, Patel S, Bhastana B, Jakharia K, Chaubey MG, Singh NK, Madamwar D (2017a) Purification and antioxidant activity of phycocyanin from *Synechococcus* sp. R42DM isolated from industrially polluted site. Bioresour Technol 245:325–331
Sonani RR, Rastogi RP, Madamwar D (2017b) Chapter 5 - Natural antioxidants from algae: a therapeutic perspective. In: Rastogi RP, Madamwar D, Pandey A (eds) Algal Green chemistry. Elsevier, Amsterdam, pp 91–120
Sonani RR, Rastogi RP, Singh NK, Thadani J, Patel PJ, Kumar J, Tiwari AK, Devkar RV, Madamwar D (2017c) Phycoerythrin averts intracellular ROS generation and physiological functional decline in eukaryotes under oxidative stress. Protoplasma 254:849–862
Sonani RR, Rastogi RP, Patel SN, Chaubey MG, Singh NK, Gupta GD, Kumar V, Madamwar D (2019) Phylogenetic and crystallographic analysis of *Nostoc* phycocyanin having blue-shifted spectral properties. Sci Rep 9(1):1–10
Steinberg CEW, Schäfer H, Beisker W (1998) Do acid-tolerant cyanobacteria exist? Acta Hydrochim Hydrobiol 26:13–19
Tan BL, Norhaizan ME, Liew W-P-P, Sulaiman Rahman H (2018) antioxidant and oxidative stress: a mutual interplay in age-related diseases. Front Pharmacol 9:1162
Thangam R, Suresh V, Asenath Princy W, Rajkumar M, Senthilkumar N, Gunasekaran P, Rengasamy R, Anbazhagan C, Kaveri K, Kannan S (2013) C-Phycocyanin from Oscillatoria tenuis exhibited an antioxidant and in vitro antiproliferative activity through induction of apoptosis and G0/G1 cell cycle arrest. Food Chem 140:262–272
Tindall BJ, Grant WD (1986) The anoxygenic phototrophic bacteria. In: Society for Applied Bacteriology Symposium Series. 1986
Vincent WF (2000) Cyanobacterial dominance in the polar regions. In: The ecology of cyanobacteria. Springer, pp 321–340
Vincent WF (2004) Microbial ecosystems of Antarctica. Cambridge University Press

Wada N, Sakamoto T, Matsugo S (2013) Multiple roles of photosynthetic and sunscreen pigments in cyanobacteria focusing on the oxidative stress. Meta 3:463–483

Ward DM, Castenholz RW, Whitton BA, Potts M (2000) The ecology of Cyanobacteria. Springer, Dordrecht, Netherlands

Ward DM, Castenholz RW, Miller SR (2012) Cyanobacteria in geothermal habitats. In: Ecology of cyanobacteria II. Springer, pp 39–63

Zhang Z, Zhang G, Kong C (2013) High expression of polo-like kinase 1 is associated with the metastasis and recurrence in urothelial carcinoma of bladder. In: Urologic oncology: seminars and original investigations. Elsevier, pp 1222–1230

Mukesh Ghanshyam Chaubey was born on 20th May 1991 at Varanasi, India. He was pursuing a Ph.D. in Biotechnology from the Department of Biotechnology, Shri A. N. Patel P. G. Institute of Science and Research, Anand, Gujarat, India. Recently, he had submitted his Ph.D. thesis on 23rd October 2020. Unfortunately, Mr. Mukesh passed away on 23rd November 2020 due to Covid-19. We have lost one of our sincere and young colleagues. He will always remain and alive within our hearts. We pray that his soul rests in peace and may God give enough strength to the bereaved family to bear the irreparable loss. We don't have any words to express our grief. This publication is sincerely dedicated to our young colleague who sadly left us too early.

Engineering Challenges of Carbon Dioxide Capture and Sequestration by Cyanobacteria

16

Alexander Dimitrov Kroumov, Maya Margaritova Zaharieva,
Fabiano Bisinella Scheufele, Vessela Balabanova,
and Hristo Najdenski

Abstract

Cyanobacterial strains (also known as blue-green microalgae) have been applied to sequester CO_2 because of their high efficient bioconversion into biomass. The system is extremely complex and needs understanding and knowledge about many subsystems such as synthesis and further extraction of biomolecules—proteins, carbohydrates, lipids, and high-value products. Carbon dioxide sequestration by using cyanobacteria requires special engineering specifications such as the design of photobioreactors (PBRs), cultivation techniques under different working conditions, etc. The strain tolerance to the high CO_2 concentrations, which are available in waste gases (e.g., flue gas up to 20% and biogas up to 45%) is considered especially important. All other key control parameters of the system are light intensity, temperature, pH, and inoculum size. Maximization of CO_2 sequestration and maximum productivity of biomass and valuable metabolites are not easy tasks. Many advanced approaches and innovative constructions of PBRs are recently designed based on computational fluid dynamics software. This very

A. D. Kroumov (✉)
Department of Biotechnology, The Stephan Angeloff Institute of Microbiology, Bulgaria Academy of Sciences, Sofia, Bulgaria

M. M. Zaharieva · H. Najdenski
Department of Infectious Microbiology, The Stephan Angeloff Institute of Microbiology, Bulgarian Academy of Sciences, Sofia, Bulgaria

F. B. Scheufele
Universidade Tecnológica Federal do Paraná—UTFPR, Engenharia de Bioprocessos e Biotecnologia—COEBB, Toledo, Paraná, Brazil
e-mail: fabianob@utfpr.edu.br

V. Balabanova
Department of Pharmacognosy, Faculty of Pharmacy, Medical University of Sofia, Sofia, Bulgaria
e-mail: vbalabanova@pharmfac.mu-sofia.bg

R. P. Rastogi (ed.), *Ecophysiology and Biochemistry of Cyanobacteria*,
https://doi.org/10.1007/978-981-16-4873-1_16

powerful tool opens new opportunities and perspectives to robustly study hydrodynamics, algal behavior, and kinetics under dynamic changes of fluid movement. On the other hand, the microalgae kinetics of cyanobacteria is fundamental for the success of the overall process. Hence, the present book chapter discusses the complex approach of modeling and analysis of the system by starting with medium optimization and going through many steps up to scaling up the process and PBRs in order to help society to reach a better and greener world.

Keywords

Cyanobacteria · CO_2 sequestration · Photobioreactors design · Cultivation techniques · Computational fluid dynamics · Kinetics · Medium design · Optimization

16.1 Introduction

Carbon dioxide makes up 68% of the estimated total greenhouse gas emissions (Harrington and Foster 2010). Under the Kyoto Protocol, 37 industrialized countries and the European Community (the European Union-15, made up of 15 states at the time of the Kyoto negotiations) commit themselves to binding targets for GHG emissions (United Nations 2011).

Potential pathways to help to minimize the GHG emissions require an overview of existing technologies and sequestration options. The option of sequestering CO_2 by means of a solar energized photobioreactor system is the aim of this book chapter. PBR design and innovative methods to achieve optimal light penetration and distribution are analyzed. Hence, the maximization of CO_2 sequestration and maximum productivity of biomass and valuable metabolites is a milestone for microalgal technologies.

16.2 Options for Capture and Sequestration

Advanced options such as mineral carbonation, copolymerization of CO_2, electrochemical conversion of CO_2, photocatalytic and photothermal catalytic conversion of CO_2, catalytic conversion of CO_2, bioconversion of CO_2 from waste industrial CO_2, etc. could be found in the literature (Zhu 2019). Many recent reviews analyzed in detail chemical and physical methods for CO_2 capturing and storage (Li et al. 2013; Dutcher et al. 2015; Gardarsdottir et al. 2019; Guandalini et al. 2019; Lai et al. 2019; Song et al. 2019a; Ayittey et al. 2020). The concept behind most disposal methods is to offset the immediate effect on the levels of carbon dioxide in the atmosphere by relocation, i.e., by injection into either geologic or oceanic sinks. In estimating two scenarios, a "High Carbon World Case" and a "Low Carbon World Case," the conclusion was reached that under the best scenario, carbon sequestration is essential to prevent irreparable damage to the environment. An estimated

80 million tons a year in 2020 compared to nearly 900 million tons a year by 2050 would have to be sequestered under the "Low Carbon World Case" as energy efficiency and fuel switching options are pressed to maximum capabilities (Beecy and Kuuskraa 2001).

The best way to minimize damages to the environment lies on:

- improving the efficiency of energy production,
- reducing the carbon content of fuels (Zhu 2019),
- sequestration of CO_2 from the waste gases such as flue gases, biogas, waste gases from fermentation, etc. (Klinthong et al. 2015).

16.3 Economic Considerations

The fundamental issue of any technology is its competitiveness in the market. We may provide various analyses of experts about applying it for CO_2 emissions (Lee et al. 2019; Adedoyin and Zakari 2020; Wei et al. 2020). For carbon sequestration and transportation to be accomplished economically, carbon capture needs to result in a relatively pure stream of gas. Although power plants emit more than one-third of the carbon dioxide worldwide, the cost of capture is significant because the concentration levels are low, typically 13–15% in coal-based power plants (Herzog 2001).

16.4 Cost of CO_2 Sequestration

The cost of biological sequestration is a daily issue during the epoch of climate changes and global warming concerns (Eloka-Eboka et al. 2019; Song et al. 2019b; Yu et al. 2020).

A source suggests that to be economic, the cost of transportation and injection of carbon dioxide should be around the US $5–$15 per ton of CO_2 avoided (Herzog 1999). CO_2 capture via photosynthesis and under microalgae biorefinery concept to directly fix carbon into microalgae has attracted the attention of researchers worldwide (Singh and Dhar 2019; Wu and Chang 2019).

16.5 An Integrated Concept of CO_2 Sequestration

A detailed study of CO_2 capture and sequestration can be found elsewhere (Razzak et al. 2013). Bio-sequestration is not only physical absorption and dissociation of CO_2 into the liquid phase. Macro- and micro-elements of the nutrient medium and pH play important role in CO_2 capture and stay as the first step in process development strategy. Combination of CO_2 sequestration with sources of phosphorus (P) and nitrogen (N) from wastewaters is a visible cost-effective option (Molazadeh et al. 2019). On the other hand, medium design may offer CO_2 storage in the culture medium by using sodium bicarbonate as a depot (Kroumov et al. 2015, 2016, 2017; Scheufele et al. 2019; Zhu et al. 2020).

Researchers worldwide have focus attention on many different directions of CO_2 capture and storage. Physicochemical methods are highly developed and summarized in comprehensive and excellent reviews focusing on our concerns about climate changes because of global warming (An et al. 2019; Borhani and Wang 2019; Garcia and Berghout 2019; Li et al. 2019; Lian et al. 2019; Sifat and Haseli 2019; Adamu et al. 2020; Shah et al. 2020).

A very important approach in scientific development is multidisciplinary hybridization. Combining knowledge from different fields in order to setup a new concept and technology is always very beneficial (Song et al. 2019b).

In CO_2 capture and utilization by cyanobacteria/microalgae as a biological option, the milestone for developing technology optimal for the market is considered the PBR design. It means, the PBR should be studied by using modern logical and mathematical methods and strategies.

16.6 The PBR as a System

The PBR system (Kroumov et al. 2016) makes use of the natural process known as photosynthesis to convert light, heat, and carbon dioxide into useful products, such as carbohydrates, hydrogen, and oxygen.

$$6CO_2(aq) + 6H_2O + \text{light} + \text{heat} \rightarrow C_6H_{12}O_6(aq) + 6O_2(g)$$

The above equation illustrates a photosynthetic reaction. The type of product generated, in this case, glucose depends highly on the biological agent used in the photobioreactor.

16.7 S_f/V ratio Milestone for Evaluation of Natural Light Illumination

Light is the only source of energy for the growth of photoautotrophic cyanobacteria/microalgae. Therefore, one of the major concerns in microalgae mass production is to achieve a sufficient natural supply of light in the culture. It means that the direct light flux penetrates from the wall to the center in the case of column PBR and to the opposite wall in the radial direction. When the biomass concentration increases in the center of PBR a dark zone may occur in some critical values. Hence, the PBR productivity decreases according to Buge–Lambert–Bear law. In this case, the key understanding of S_f/V (S_f— is PBR's surface area, V is the PBR's volume) ratio is necessary and more importantly, when linked with microalgae physiology it focuses on the question of how it influences the PBR performance. There are many studies on the S_f/V ratio as a key parameter, but our group was the first that developed and presented such a model (Kroumov et al. 2013). Because of its fundamental importance, we are going to discuss here this phenomenon in detail. The forthcoming analysis will prove why many new techniques for internal light illumination of PBRs with biophotonics are studied and developed (Carvalho et al. 2011; Dye et al. 2011).

Hence, an optimal design of a PBR includes light unlimited growth kinetics, specific to each photosynthetic microorganism, which must be related to maximum possible penetration of the light into the liquid volume of the PBR (Molina Grima et al. 1999; Rubio et al. 2003). This is possible only if the culture medium is very transparent to radiant energy within the wavelength range from 400 to 700 nm, used for photosynthesis (Heldt and Heldt 2005a, b).

16.8 Light Criterion

The light criterion assumes that gas–liquid flow inside the PBR is uniform and the mixing conditions are optimal; light covers the whole PBR surface area S_f.

One may write the simple form of surface to volume ratio:

$$S_f/V = \left(\pi.D_{pbr}.H\right)/\left(H.\pi.\left({D_{pbr}}^2/4\right)\right) = 4/D_{pbr} = 4/0.01 = 400\ \mathrm{m}^{-1}$$

where S_f is PBR surface area, m^2, V is the PBR's liquid volume, m^3; S_f/V stands for specific surface area, m^{-1}; H stands for the height of PBR, m; D_{pbr} is the tubular PBR's diameter, m.

The ratio $S_f/V = 4/D_{pbr}$ showed that light availability depends only on tubular PBR's diameter (D_{pbr}). PBR designs resulting in a theoretical maximum ratio value of 400 m^2/m^3 for $D_{pbr} = 0.01$ m were state-of-the-art in the year 2008 (Kunjapur and Eldridge 2010). It must be noted, the value $S_f/V = 400$ m^{-1} can be considered as a controlled border with which any realistic Lab (Pilot plant) PBR design can be compared.

Note: In all calculations, V stands for the liquid volume in the solar receiver.

16.9 S_f/V Ratio Linked with Algal Physiology

The relationship between specific growth rate (SGR) and S_f/V ratio is fundamental. Our group first considered and described in a model this link in order to quantitatively evaluate the tubular PBR potential of productivity (Kroumov et al. 2013). In the literature, there was no published analysis considering the population level of microalgae growth. The details about the model assumption can be found in the paper. The model was built as follows:

Kinetic model,

$$\mu = \mu_{\max} \frac{S}{(K_s + S)} \tag{16.1}$$

$$\frac{dX(t)}{dt} = \mu * X(t) \tag{16.2}$$

where μ stands for specific growth rate, h^{-1}; $\mu_{\max}$ stands for maximum specific growth rate, h^{-1}; S stands for substrate concentration, kg m^{-3}; X is the biomass concentration, kg m^{-3}; and K_s is the half-saturation constant, kg m^{-3}.

The key point in model creation is the presentation of the S (under light limiting conditions) as a function of I_0 and S_f/V ratio as follows:

$$S = I_0 * \left(\frac{S_f}{V}\right) * \varepsilon * X \tag{16.3}$$

where:

S_f is the illuminated surface area of the reactor, m^2; V is PBR's liquid volume, m^3; I_0 is the intensity of the incident light on the tubular PBR surface, μmol m^{-2} s^{-1}; and ε is the extinction coefficient of the culture.

Substituting Eq. (16.3) into kinetic model resulted in:

$$\mu = \mu_{\max} \frac{\left(I_0 * \left(\frac{S_f}{V}\right) * \varepsilon * X\right)}{\left(K_s + \left(I_0 * \left(\frac{S_f}{V}\right) * \varepsilon * X\right)\right)} \tag{16.4}$$

Because average light intensity (I_{av}) is a more representative parameter of light availability in Tubular PBRs than I_0, we used it further as follows:

$$I_{av} = \frac{I_0 * E_0}{X^n} * \left(1 - e^{-E_1 X^n}\right) \tag{16.5}$$

where E_0—parameter in the I_{av} (Eq. 16.5) light model [g^n/L^n]; E_1—parameter in the light model [g^n/L^n]; n—is a constant with evaluated value [−]; I_{av} stands for average light intensity, μmol m^{-2} s^{-1}.

Lambert–Beer's law, extensively used in photometry, is based on three assumptions: (1) the direction of the incident radiation does not change as it crosses through the culture; (2) the radiation is monochromatic; and (3) scattering effect due to solid particles is negligible compared to absorption. The Lambert–Beer's law adjustment of light attenuation is not appropriate for high biomass concentrations due to the existence of different scattering and selective absorption effects (Acién Fernández et al. 2001). Equation (16.5) was successfully used for the online evaluation of algal growth dynamics (Li 2002). In this thesis, the semiempirical approach is used when derived Eq. (16.5), as well as assumptions made and benefits when applied to this model were explained in detail. The E_0 and E_1 coefficients take into account the above mention effects. The simplicity of the model is very useful for PBRs design, optimization, and online control.

Model using the I_{av} parameter:

$$\mu = \mu_{\max}\frac{\left(I_{\text{av}} * \left(\frac{S_{\text{f}}}{V}\right) * \varepsilon * X\right)}{\left(K_{\text{s}} + \left(I_{\text{av}} * \left(\frac{S_{\text{f}}}{V}\right) * \varepsilon * X\right)\right)} \tag{16.6}$$

This specific growth rate (SGR) model can be used for quantitative evaluation of the SGR versus S_f/V relationship when I_{av} does not exceed $I_{av(critical)}$ where the inhibition by light takes place.

Response surface analysis of this model was performed where I_{av} and D_{pbr} ($S_f/V = 4/D_{pbr}$) are independent variables and SGR is a function of them. Details could be found (Kroumov et al. 2013) wherever linear growth can be achieved under given conditions and values of kinetic parameters.

The equation of I_{av} as a function of $X(t)$ is:

$$I_{\text{av}} = \frac{I_0 * E_0 * \left(1 - e^{-E_1 X(t)^{n_1}}\right)}{X(t)^{n_1}} \tag{16.7}$$

where $X(t)$—is the biomass concentration [kg/m^3] in time; n_1—is a constant [−].

If we describe algae growth inhibition by light in the Aiba's form, we have:

$$SGR_{\text{Iav}} = \frac{I_{\text{av}}}{Ks_{\text{Iav}} + I_{\text{av}} + \frac{I_{\text{av}}^2}{K_{\text{Iavmax}}}} \tag{16.8}$$

$$\text{SGR}_{\text{Iav}} = \frac{\left(I_0 * E_0\left(1 - e^{-E_1 X(t)^{n_1}}\right)\right)}{\left(X(t)^{n_1}\left(Ks_{\text{Iav}} + \frac{I_0 E_0\left(1-e^{-E_1 X(t)^{n_1}}\right)}{X(t)^{n_1}} + \frac{I_0^2 E_0^2\left(1-e^{-E_1 X(t)^{n_1}}\right)^2}{\left(X(t)^{n_1}\right)^2 K_{\text{Iavmax}}}\right)\right)} \tag{16.9}$$

Then, SGR as a function of S_f/V and I_{av} is written as follows:

$$\text{SGR}_{\text{total}} = \mu_{\max} \cdot \text{SGR}_{(S_{\text{f}}/V)} * \text{SGR}(_{\text{Iav}}) \tag{16.10}$$

where $\text{SGR}_{(Sf/V)}$—is specific growth rate as a function of S_f/V ratio, h^{-1}; $\text{SGR}(_{\text{Iav}})$—is specific growth rate as a function of average light intensity where inhibition by light takes place, h^{-1}; $\text{SGR}_{\text{total}}$—is the overall specific growth rate, h^{-1}.

After substitution, we obtain:

$$\text{Substrate} = \frac{I_{\text{av}} * S_{\text{f}}}{V \;\; * \varepsilon * \;\; X(t)} \tag{16.11}$$

$$\text{SGR}_{(S_{\text{f}}/V)} = \frac{I_{\text{av}} * S_{\text{f}}}{V * \varepsilon * X(t) * \left(Ks_{S_{\text{f}}V} + \frac{I_{\text{av}} * \;\; S_{\text{f}}}{V * \;\; \varepsilon * \;\; X(t)}\right)} \tag{16.12}$$

Hence, the $\text{SGR}_{\text{total}}$ can be written as follows:

$$\mathrm{SGR}_{(S_f/V)} = \frac{I_{av} * S_f}{V * \varepsilon * X(t) * \left(Ks_{S_f V} + \frac{I_{av} * \quad S_f}{V * \quad \varepsilon * \quad X(t)} \right)} \tag{16.13}$$

where

$$A_{\mathrm{SGR_{total}}} = \mu_{max} * I_0^2 * E_0^2 * S_f * \left(1 - e^{-E_1 * X(t)^{n_1}} \right)^2$$

$$B_{\mathrm{SGR_{total}}} = \left(X(t)^{n_1} \right)^2 V * \varepsilon$$

$$C_{\mathrm{SGR_{total}}} = K_{S_{S_f V}} + \frac{I_0 * E_0 * \left(1 - e^{-E_1 * X(t)^{n_1}} \right) S_f}{X(t)^{n_1} * V * \varepsilon * X(t)}$$

$$D_{\mathrm{SGR_{total}}} = K_{S_{IaV}} + \frac{I_0 * E_0 * \left(1 - e^{-E_1 * X(t)^{n_1}} \right)}{X(t)^{n_1}} + \frac{I_0^2 * E_0^2 * \left(1 - e^{-E_1 * X(t)^{n_1}} \right)^2}{\left(X(t)^{n_1} \right)^2 * K_{\mathrm{Iav\ max}}}$$

Furthermore, the biomass balance is written as follows:

$$\frac{dX(t)}{dt} = SGR_{\mathrm{total}} * X(t) \tag{16.14}$$

Hence:

$$\frac{dX(t)}{dt} = \frac{A_{\mathrm{SGR_{total}}}}{B_{\mathrm{SGR_{total}}} * C_{\mathrm{SGR_{total}}} * D_{\mathrm{SGR_{total}}}} \tag{16.15}$$

Mass balance presented as Eq. (16.15) completed the model. Solving the equation (16.15) for different initial conditions (X_0—inoculum) and given process time ($t = 14$ days), we may evaluate different industrial tubular PBRs' performances by substituting their real PBR diameters (S_f/V ratios, respectively). We evaluate the industrial PBRs from the mini-review of (Pulz 2001). Hence, the S_f/V ratios of industrial PBRs are in the range $S_f/V = 6.7\ \mathrm{m}^{-1}$ raceway ponds to $S_f/V = 86.7\ \mathrm{m}^{-1}$ and $D_{pbr} = 0.046$ m for tubular PBRs. Maximum D_{pbr} reported in the literature is about $D_{pbr} = 0.40$ m, but achieved biomass concentration is very low and cannot be increased by any manipulations of the regime parameters (conditions assume only natural source of light!). Table 16.1 represents the Simulation results with the kinetic model.

It must be understood, that the example gave a clear understanding of how the S_f/V criterion influenced the PBR performance. The simulation results can be performed for any given SGR and particular strain culturing in each plant. By changing the kinetics parameters the researcher may obtain different values of PBR productivity for a 14-day period or calculate the stationary phase conditions. Nevertheless, the trend of predictions cannot change which is fundamental.

The results from the above analysis can be interpreted as follows:

Table 16.1 Simulation results with the kinetic model where SGR is a function of S_f/V ratio (D_{pbr}) and light illumination (light limitation and inhibition) is considered

Conditions—$X_0 = 0.1$ [kg m^{-3}], $SGR_{max} = 0.11$ [h^{-1}], $I_0 = 3270.0$ [μmol m^{-2} s^{-1}], time = 14 [day]		
Tubular PBR diameter [m]	Biomass concentration [kg m^{-3}]	S_f/V ratio [m^{-1}]
$d = 0.40$	$X = 2.48$	$S_f/V = 10$
$d = 0.20$	$X = 3.5$	$S_f/V = 20$
$d = 0.14$	$X = 4.15$	$S_f/V = 28.6$
$d = 0.08$	$X = 5.38$	$S_f/V = 50$
$d = 0.0467$	$X = 6.85$	$S_f/V = 85.7$
$d = 0.03$	$X = 8.3$	$S_f/V = 133$
$d = 0.01$	$X = 13.12$	$S_f/V = 400$

Assumptions: No mixing and no mass transfer limitations; no limitation by nutrients; no inhibition by O_2 concentration; no limitation and inhibition by CO_2 concentration (with permission from the Journal)

Note: The kinetics constants in the calculations are taken from experiments performed with *Chlorella vulgaris* species (unpublished results). The process time duration responded to the stationary phase of the growth curve. The simulation results from the table are in agreement with the results published elsewhere (Molina et al. 2000; Pulz 2001)

- for low light intensities tubular PBRs with $D_{pbr} > 0.40$ m are not effective,
- for high light intensities $I_0 = 3270.0$ [μmol m^{-2} s^{-1}] (hot summers; tropical regions, deserts). Tubular PBRs with diameters up to $D_{pbr} = 0.20$ m can be successfully used,
- Tubular PBRs with diameters 0.05–0.15 m have great theoretical potential in tropical regions (US deserts, Brazil, North Africa, etc.)
- algae strains resistant to shear stress and having high SGR may successfully be used in tubular PBRs where biomass concentration up to 6 kg m^{-3} can be achieved easily.

It must be noted that 20 kg m^{-3} biomass concentration, during photoautotrophic growth and direct natural light reported in the literature is possible to be achieved ONLY in very thin 0.01 m (or less) tubular PBRs which is unrealistic for industrial applications.

Additionally, this analysis shows that PBRs with diameters up to 0.20 m have great potential in terms of light availability in very hot tropical regions and for fast-growing shear stress-resistant cyanobacteria/microalgae species. With tubular PBRs, we may safely expect to achieve a biomass concentration in the range 2–4 kg m^{-3} for liquid velocities up to 0.8 m s^{-1}. This is in agreement with reported optimal liquid velocities 0.20 m s^{-1} to 0.50 m s^{-1} and biomass concentrations published elsewhere (Molina et al. 2000).

This requires the following measures to be undertaken:

- any engineering solution which is going to minimize the dark liquid volume in the scheme must be utilized (Hinterholz et al. 2017).

– increasing the radial mixing in the tubular PBR may be very beneficial for Light/Dark (L/D) cycles' improvement "Flash Light Effects" (FLE) (where liquid velocities are in the range between 0.20 and 0.50 m s^{-1} or above depending on PBR design).

It must be noticed that the kinetics model for the description of SGR was used for the evaluation of a novel Fibonacci-type PBR (Diaz et al. 2021). The authors applied our model for analysis of their novel PBR performance with 1250 L reactor volume under semi-continuous mode of cultivation. The used strain was microalgae *Dunaliella salina* (code 007). The Fibonacci-type PBR has been scaled up at a site in the Atacama Desert where the average maximum solar radiation was supposed to be 1752 μE m^{-2} s^{-1}. The authors claimed that biomass concentration achieved in this study was three times higher than that achieved with the same strain in a raceway pond. Moreover, their results showed that the novel Fibonacci-type PBR allows 1.6 times more capturing than solar radiation on the horizontal surface and can be successfully used for large-scale conditions.

As a conclusion from this work we may say that the novel model will be extremely useful (already successfully applied (Diaz et al. 2021) in obtaining appropriate growth kinetics in a novel tubular PBR or solar receiver tube, which in turn will enable to predict constraints of biomass productivity by simulations. Furthermore, the model can be applied for optimization and scale-up of the PBR depending on the selected species and their light-dependent behavior. Moreover, having a catalog of kinetics models of cyanobacteria/microalgae where SGR is like a function of light intensity (Kroumov et al. 2016), one may calculate many scenarios by substituting them in our model. On the other hand, the conditions of ultra-high-density culture (UHDC) must be considered as well, because of their importance for industrial applications (Hu et al. 1998). The optimal light regime in PBRs is studied in detail by authors (Pruvost et al. 2016a, b, c). Analyzing the results of the J. Provost's group, the researchers may find new challenging opportunities and alternatives where definitive complex analysis is required for the light-transfer phenomena which take place in PBRs (Kandilian et al. 2016; Pruvost et al. 2016a, b, c, 2017).

A comparative study between PBRs and perspectives by using microalgae for CO_2 capture can be found elsewhere (Sivasangari et al. 2019; Yadav et al. 2020).

16.10 Flashing Light Effect (FLE) as Another Key Parameter for Optimal Biomass Production

In the above paragraph, the influence of FLE was mentioned as an important key for increasing PBR productivity (Iluz and Abu-Ghosh 2016; Kroumov et al. 2016; Khadim et al. 2018). FLE occurs through repeatedly cycling cells from the dimly lit interior of the PBR to the higher illumination of the exterior. The FLE highly increases the conversion of light to biomass. The simplest way to design effective

PBR is to use static mixers/baffles (Degen et al. 2001). Ugwu et al. (2002) apply baffles to create FLE. Using baffles can increase productivity in two ways:

1. By increasing the residence time of gas bubbles in the reactor, but in this case, CO_2 is supplied on demand and oxygen does not accumulate, thus negating the effects of the baffles.
2. By affecting the mixing, i.e., nutrients are mixed better.

Another observation from this work was that microalgae production rates depend on the length of the light path. If the length of the light path changes from 30 cm to 15 cm then the productivity rate increases up to 2.5 times higher.

Hence, extensive research has been done to improve this technique and many new findings emphasized the availability and strong benefits by applying (FLE) in PBRs (Hu and Sato 2017; Scheufele et al. 2019; Ali Kubar et al. 2020; Cheng et al. 2020; Cui et al. 2020; Guo et al. 2020). A very interesting study shows different approaches to creating FLE in PBRs (Luzi et al. 2019).

Following these thoughts, a new technique as computational fluid dynamics (CFD) is widely used recently as a mathematical tool to improve understanding for FLE and PBR design, optimization, and functioning (Moberg et al. 2012; Gómez-Pérez et al. 2015; Soman and Shastri 2015; Pires et al. 2017; Hinterholz et al. 2019). By using such methods, the FLE can be calculated and can be controlled in any internal space of the studied PBRs (Hinterholz et al. 2019). Model simulations with CFD have minimized efforts and expenses for PBR design. Moreover, any attempts in this field are greatly appreciated and necessary because the simulations can be checked only with a few experiments in order to discriminate the competing hypothesis (Sato et al. 2010; Hinterholz et al. 2019). An example of using CFD simulations when studying the application of static mixers in order to ensure FLE in PBRs was described by the authors (Cheng et al. 2016). The authors claimed that in such a system the fluid mixing and average velocity along the light direction had been increased by nearly 1000 times and turbulent kinetic energy increased by 1.3 times in PBRs with novel static mixers. Hence, the Light/Dark (L/D) cycle frequency was increased by 85.21% and 95.77% at a fluid inlet velocity of 0.3 and 0.4 m s^{-1}, respectively. The model simulations results from this work once again proved unconditionally that the CFD method has a great potential for optimization of construction of the novel, tubular PBRs.

Our group experimentally proved the potential and the power of CFD simulation applied to flat plate PBR design (Hinterholz et al. 2019).

16.11 CO_2 Fixation by Cyanobacteria/Microalgae

The design of a system that combines solar energy collection and a fiber optic light delivery system is an innovative approach in PBR illumination. In order to promote uniform growth of the organisms, the distribution of photosynthetic photon flux light in the wavelength range of 400–700 nm needs to be delivered to the bioreactor

(Kremer et al. 2000). Beer's law suggests that a particulate-laden flue gas would result in a large loss of photon flux due to scattering. Clearly, the efficient distribution of light throughout the PBR will affect the CO_2 uptake rates. This needs to be further investigated in order to achieve optimal growth rates for these microorganisms (Benemann 1997). Kremer et al. (2000 pointed out that despite the 50 years of development of closed PBR systems, commercial viability has not yet been achieved. At present, open pond systems produce around 100 tons of biomass annually at a cost of around $US10.000/Mt. Although the central role of technology limitations has not yet been properly addressed, a number of facilities capable of mitigating over a million tons of CO_2 would have a significant impact on the reduction of greenhouse gas emissions.

16.12 Studies on CO_2 Utilization in PBRs by Using Microalgae

Many different interrelated factors need to be considered in order to design an overall system. Usui and Ikenouchi (1997) suggest three types of solar energized photobioreactor system designs. All three systems are efficient in the sense that they collect the sunrays and transfer the light energy required for photosynthesis to the photobioreactor. The suggested systems incorporate three different types of light distribution systems: a flat plate type, an internally luminous stirrer type, and a fountain type. In various experimentation conditions, the flat plate type of reactor had been used due to its ease of scaling up. Usui and Ikenouchi (1997) used a concave mirror to collect the light and passed the IR radiation to the bioreactor by fiber optics. The authors realized an improvement of 55% in overall efficiency by developing the light transmission system. These included developing a more sophisticated protective cover for the system, selecting an appropriate fiber optic diameter, and optimizing the ratio of fiber core area to bundle size. The light was distributed throughout the reactor by means of illumination plates consisting of a lattice of strips between two frosted acrylic plates. The experimental system experienced illumination of 10 h a day with 14 h darkness and achieved a CO_2 fixation rate almost 10 times higher than that of a tree undergoing a natural photosynthetic process. The first step in designing such a system is to invent a similar small-scale system capable of fixing carbon dioxide, producing useful products, and remaining energetically efficient. Some of these issues will now be addressed in terms of a current literature survey.

The second generation of internally irradiated PBRs (Olivieri et al. 2014) and their designs incorporating parabolic optical fibers as internal light sources can be found elsewhere (An and Kim 2000; Suh and Lee 2001; Zijffers et al. 2008). The action of solar collectors includes harvesting light from the sun and concentrating it about 10,000 times (An and Kim 2000). Further, this light is transmitted into optical fibers and into the PBR system (Xue et al. 2013). The disadvantages of using optical fibers remain the economic considerations.

16.13 Cultivation of Cyanobacteria and Influence of Working Parameters on the Process

Bacteria and cyanobacteria/microalgae are able to sequestrate CO_2 via photosynthetic reactions. Each of these strains requires certain conditions to ensure optimal growth and synthesis rates. Catalogs showing features of these microorganisms and their products are very helpful for choosing engineering specifications and startup of development algal technology.

Experimental verification is one approach, but modeling as a method is more reliable and discriminates the competing hypotheses fast and robust. The complex relation between irradiation conditions, PBR design, internal light attenuation conditions, and resulting growth cannot be studied straightforward without a modeling approach. The work of Pruvost et al. (2015) demonstrates this clearly.

The authors introduced a model, which deals with light capture and its influence on process performances. The model can predict even theoretical limits of PBR productivity under given light conditions and PBR location as well. It was applied to simulate the physiology of the microalgae species *Chlamydomonas reinhardtii* in PBRs in the facilities in Nantes, France. The comparison against the cyanobacterium *Arthrospira platensis* showed that not only light transmission but also dark volumes were found to negatively affect biomass productivity. Their findings were proved with deep mathematical analysis.

Experimental methods to study cyanobacteria/microalgae physiology for CO_2 capturing from waste gases (especially flue gas) in different PBR scales and working conditions can be found in the literature.

Some examples:

An airlift PBR for flue gas purification (CO_2 absorption and utilization) by using cyanobacteria was investigated. *Spirulina platensis* culture was fed with CO_2 and NOx, simulating a flue gas. The preliminary test proved good cell productivity (86.8 mg $L^{-1}d^{-1}$) and CO_2 utilization (229 mg d^{-1}). Dosages of flue gas used in fed-batch mode achieved a high CO_2 capture (407 mg d^{-1}), 90.0% removal of NOx, and a biomass productivity of 188.7 mg L^{-1} d^{-1} (Kajiwara et al. 1997). Arata et al. (2013) found that *Synechococcus* achieved a maximum CO_2 uptake rate of 0.025 g L^{-1} h^{-1} or 0.6 g L^{-1} daily at a cell mass concentration of 0.286 g L^{-1}. If the scaling up procedure was correct, this would result in a bioreactor of size 4000 m^3 with an average utilization rate of 1 ton CO_2 h^{-1} from emission sources.

One problem with cyanobacteria/microalgae is their tolerance to high CO_2 content in the gas stream and requires hard work of screening and strain isolation. Murakami et al. (1997), using *Synechocystis aquatilis* in a 5 L bioreactor and optimized conditions, reached a maximum CO_2 fixation rate of 1.5 g CO_2 /L/day. The not only potential of microalgae to utilize CO_2 is important but also the synthesis of BAC with the high value which meets the requirements of the integral biorefinery concept (Gonçalves et al. 2019a, b; Schuelter et al. 2019).

16.14 Internal Illumination of PBRs as a Key for Optimization of Biomass Production

The mathematical analysis of the S_f/V ratio clearly showed that to obtain high-density culture or ultrahigh density culture conditions the engineering approach should consider an option of internal light supply among other working conditions. Hence, internal illumination of PBRs for example, by using light-emitting diode (LED) has been studied very extensively (Yeh and Chung 2009; Jacobi et al. 2012; Heining and Buchholz 2015; Hu and Sato 2017).

A study (Amaral et al. 2020a, b) showed integration of two configurations of PBRs—such as tubular and bubble columns in a single system, combining the benefits of each configuration. The evaluation was performed in a comparative test between the performances of the two PBRs. Both of them were illuminated with blue light-emitting diodes (LED). The measurable parameters were the specific growth rate (μ_{max}) and cell productivity (P.X) of the microalgae *Chlorella minutissima*. The results showed that μ_{max} and P.X in an integral PBR were at least 42.85 and 58.06% higher than in a bubble column PBR, respectively. The first was 36% more efficient in terms of electricity consumption than the second one. The study showed that the combination of known PBR types might result in unexpected benefits.

The same group presented another study (Amaral et al. 2020a, b) where the functioning of newly developed internally illuminated and integrated PBR was investigated by applying the method of Taguchi. The best values of operational parameters such as volumetric biomass productivity and volumetric lipid productivity of *Chlorella minutissima* microalgae cultivated under the autotrophic mode of cultivation were found. For the given working conditions of illumination (blue, white, and red); photoperiod, etc. under the Taguchi method, the authors succeeded to increase biomass concentration, volumetric biomass productivity, and volumetric lipid productivity in *Chlorella minutissimain* with 8.6%, 42%, and 143%, respectively.

The authors (Malapascua et al. 2019) present a very interesting experiment on microalgae kinetics. Tests were performed in an internally illuminated 10-L PBR to find out the link between photosynthesis activity and algae growth in *Chlorella vulgaris R-117* (CCALA 1107). The growth conditions included very high irradiance values by using LED submerged in the culture. In the first test when the strain was cultured under 2.500 μmol (photon) m^{-2} s^{-1} the following parameter values were obtained: doubling time of 3.5 days and biomass density of 3.5 g (DM) L^{-1} after the about 10-day period. Under 3.500 μmol (photon) m^{-2} s^{-1} light conditions, the culture reached values: doubling time of 1.7 days and biomass density of ~5.5 g L^{-1} before entering the stationary phase. The information from this study has crucial importance for the design of a large-scale PBR and performance optimization by taking into account the physiological response and kinetics of the particular strain.

In terms of the integral biorefinery concept, not only maximization of biomass yield and productivity matter. The internal high-value products have been the key to cost-effective technologies. A recent investigation of life cycle assessment (LCA)

(Onorato and Rösch 2020) provided valuable information by comparing three types of PBRs—the flat panel airlift, the green wall panel, and the unilayer horizontal tubular PBR. The light supply for these three systems was as follows: (1) the flat panel had a double-sided LED illumination with a location inside a building; (2) the green wall panel was equipped with one-side LED, located outside; (3) the unilayer horizontal tubular without any artificial lighting with a location outside. The strain producer of astaxanthin was *Haematococcus pluvialis*. The total final volume of the three systems was 93 m^3. The measurable parameters which the authors considered were 1 kg of *Haematococcus pluvialis* (80% dw) and 1 kg of astaxanthin. The LCA results showed that the system with the lowest environmental impact was the unilayer horizontal tubular.

16.15 Construction Types of PBRs and Large-Scale Application of the Photosynthetic Carbon Fixing Method

There are many configurations and types of PBRs, which were extensively studied during the years. Their optimization of construction and functioning is presented elsewhere (Acién Fernández et al. 1999; Suh and Lee 2001; Ugwu et al. 2008; Xu et al. 2009; Pfaffinger et al. 2019; Uddin et al. 2020).

One new evaluation of PBR configurations is demonstrated in a review (Sero et al. 2020). Light as the most influential factor in the PBR system mainly due to challenges associated with its distribution and control is discussed from the point of view of photonics achievements. The authors analyzed conventional and new unconventional PBR designs in deep detail. Special attention was given to the hybrid PBR systems (open ponds coupled with closed PBRs) where all their advantages and disadvantages were analyzed. In this work, light as a criterion was analyzed in deep detail along with its distribution into the PBRs. Pyramid PBR design gets our attention because of the historical nostalgia for Egyptian pyramids. Plants based on this type of PBR have been successfully used by culturing *Spirulina* (Płaczek et al. 2017). The biomass concentration achieved in this plant was four times higher than that in the open ponds system.

Improvement of photon capture is analyzed in detail (Sero et al. 2020). There are three types of filters that have been considered in this review as having cost-effective application in microalgae cultures: the colored/absorbing glass filters (inexpensive and stable type of filter), thin-film coatings (technically superior, enhance algal proliferation), and thermochromic solar control film materials (in PBRs they can be used for their thermoregulation properties). The authors conclude that the application of light filtration technologies on PBRs most probably will result in novel PBRs designs where the production of cyanobacteria/microalgae biomass can be increased tremendously both in quantitative and qualitative production of biologically active compounds (BAC).

16.16 Conclusion

The biological CO_2 capture techniques need complex evaluation and many novel approaches to improve units operations, which have to be included in the system development and design. Firstly, the isolation of industrial strains must meet criteria such as high tolerance to high concentrations of CO_2 contained in the waste industrial gases (e.g., flue gases, biogas, etc.) and strong environmental survival under contamination conditions. Strain engineering by using molecular and recombinant techniques must be considered too in order to achieve overproduction of targeted metabolites. Further, medium optimization as one of the major concerns should be designed by using frontiers of modern mathematical approaches and the elemental composition of the selected strain. Thus, the studies can be performed fast and robustly in lab and pilot plant scale where the application of scale up and scale down strategies for process development determines the success. PBRs design as a complex system cannot be studied straightforward hence application of system analysis theory is a powerful tool for PBRs construction and optimization of their functioning. Light as the most influential factor in the PBR system should be collected, transmitted, and delivered into the PBRs volumes in an optimal way. FLE based on hydrodynamics achievements and light-transmitting devices may tremendously increase the microalgae productivity and quality of biomass by supporting the production of BAC.

Finally, if the scale-up procedure solves successfully the economic restrictions of the market, then the photosynthetic solution as an option to capture CO_2, the utilization of CO_2 from waste industrial gases, and production of microalgae biomass and BAC can be competitive or even superior compared to the other technical options.

16.17 Future Prospects

The future of microalgae technologies lies in the full utilization of the integral biorefinery concept. This requires an extremely complex approach where all unit operation and their functioning must be optimized under the strong requirements of the market and environmental regulations. New concepts in systems biology, metabolic and chemical engineering, biophotonics, etc. should be rapidly utilized in cyanobacteria/microalgae systems. This will result in novel PBR designs and competitiveness on the market of high-value products.

Acknowledgments This work was performed under the grant КП-06-Н37/12 of the Bulgarian National Science Foundation, which the authors gratefully acknowledge. The authors are very grateful to Dr. Yana Ilieva for the professional reading and revising of the MS.

References

Acién Fernández FG, García Camacho F, Chisti Y (1999) Photobioreactors: light regime, mass transfer, and scaleup. In: Osinga R, Tramper J, Burgess JG et al (eds) . Elsevier, Progress in industrial microbiology, pp 231–247

Acién Fernández FG, Fernández Sevilla JM, Sánchez Pérez JA, Molina Grima E, Chisti Y (2001) Airlift-driven external-loop tubular photobioreactors for outdoor production of microalgae: assessment of design and performance. Chem Eng Sci 56(8):2721–2732

Adamu A, Russo-Abegão F, Boodhoo K (2020) Process intensification technologies for CO_2 capture and conversion – a review. BMC Chem Eng 2(1):2

Adedoyin FF, Zakari A (2020) Energy consumption, economic expansion, and CO_2 emission in the UK: the role of economic policy uncertainty. Sci Total Environ 738:140014

Ali Kubar A, Cheng J, Guo W, Kumar S, Song Y (2020) Development of a single helical baffle to increase CO_2 gas and microalgal solution mixing and *Chlorella PY-ZU1* biomass yield. Bioresour Technol 307:123253

Amaral MS, Loures CCA, Silva MB, Prata AMR (2020a) Adjustment of the operational parameters of an unconventional integrated and illuminated Internally Photobioreactor (ILI-PBR) for the batch autotrophic cultivation of the *Chlorella minutissima*, using the Taguchi method. Appl Biochem Biotechnol 191(1):245–257

Amaral MS, Loures CCA, Naves FL, Baeta BEL, Silva MB, Prata AMR (2020b) Evaluation of cell growth performance of microalgae *Chlorella minutissima* using an internal light integrated photobioreactor. J Environ Chem Eng 8(5):104200

An J-Y, Kim B-W (2000) Biological desulfurization in an optical-fiber photobioreactor using an automatic sunlight collection system. J Biotechnol 80(1):35–44

An J, Middleton RS, Li Y (2019) Environmental performance analysis of cement production with CO_2 capture and storage technology in a life-cycle perspective. Sustainability 11(9):2626

Arata S, Strazza C, Lodi A, Del Borghi A (2013) Spirulina platensis culture with flue gas feeding as a cyanobacteria-based carbon sequestration option. Chem Eng Technol 36(1):91–97

Ayittey FK, Obek CA, Saptoro A, Perumal K, Wong MK (2020) Process modifications for a hot potassium carbonate-based CO_2 capture system: a comparative study. Greenhouse Gases: Sci Technol 10(1):130–146

Beecy DJ, Kuuskraa VA (2001) Status of U.S. Geologic Carbon Sequestration Research and Technology. Environ Geosci 8(3):152–159

Benemann JR (1997) CO_2 mitigation with microalgae systems. Energy Convers Manag 38:S475–S479

Borhani TN, Wang M (2019) Role of solvents in CO_2 capture processes: the review of selection and design methods. Renew Sust Energ Rev 114:109299

Carvalho AP, Silva SO, Baptista JM, Malcata FX (2011) Light requirements in microalgal photobioreactors: an overview of biophotonic aspects. Appl Microbiol Biotechnol 89 (5):1275–1288

Cheng W, Huang J, Chen J (2016) Computational fluid dynamics simulation of mixing characteristics and light regime in tubular photobioreactors with novel static mixers. J Chem Technol Biotechnol 91(2):327–335

Cheng J, Lai X, Ye Q, Guo W, Zhou J (2020) Numerical simulation on optimizing flow field and flashing-light effect in jet-aerated tangential swirling-flow plate photobioreactor to improve microalgal growth. Chem Eng Sci 215:115371

Cui X, Yang J, Feng Y, Zhang W (2020) Simulation of a novel tubular microalgae photobioreactor with aerated tangent inner tubes: improvements in mixing performance and flashing-light effects. Archaea (Vancouver, BC) 2020:8815263

Degen J, Uebele A, Retze A, Schmid-Staiger U, Trösch W (2001) A novel airlift photobioreactor with baffles for improved light utilization through the flashing light effect. J Biotechnol 92 (2):89–94

Diaz JP, Inostroza C, Acien FG (2021) Scale-up of a Fibonacci-type photobioreactor for the production of *Dunaliella salina*. Appl Biochem Biotechnol 193(1):188–204
Dutcher B, Fan M, Russell AG (2015) Amine-based CO_2 capture technology development from the beginning of 2013: a review. ACS Appl Mater Interfaces 7(4):2137–2148
Dye D, Muhs J, Wood B, Sims R (2011) Design and performance of a solar photobioreactor utilizing spatial light dilution. J Solar Energy Eng 133(1)
Eloka-Eboka AC, Bwapwa JK, Maroa S (2019) Biomass for CO_2 sequestration. In: Hashmi S, Choudhury IA (eds) Encyclopedia of renewable and sustainable materials. Elsevier, Oxford, pp 277–290
Garcia M, Berghout N (2019) Toward a common method of cost-review for carbon capture technologies in the industrial sector: cement and iron and steel plants. Int J Greenhouse Gas Control 87:142–158
Gardarsdottir SO, De Lena E, Romano M, Roussanaly S, Voldsund M, Pérez-Calvo J-F, Berstad D, Fu C, Anantharaman R, Sutter D et al (2019) Comparison of technologies for CO_2 capture from cement production-Part 2: Cost analysis. Energies 12:542
Gómez-Pérez CA, Espinosa J, Montenegro Ruiz LC, van Boxtel AJB (2015) CFD simulation for reduced energy costs in tubular photobioreactors using wall turbulence promoters. Algal Res 12:1–9
Gonçalves VD, Fagundes-Klen MR, Goes Trigueros DE, Kroumov AD, Módenes AN (2019a) Statistical and optimization strategies to carotenoids production by *Tetradesmus acuminatus* (LC192133.1) cultivated in photobioreactors. Biochem Eng J 152:107351
Gonçalves VD, Fagundes-Klen MR, Trigueros DEG, Schuelter AR, Kroumov AD, Módenes AN (2019b) Combination of Light Emitting Diodes (LEDs) for photostimulation of carotenoids and chlorophylls synthesis in *Tetradesmus* sp. Algal Res 43:101649
Guandalini G, Romano MC, Ho M, Wiley D, Rubin ES, Abanades JC (2019) A sequential approach for the economic evaluation of new CO_2 capture technologies for power plants. Int J Greenhouse Gas Control 84:219–231
Guo W, Cheng J, Song Y, Kumar S, Ali KA, Wang Y, Li X, Yang W (2020) Improving flashing light frequency and CO_2 fixation rate with vortex movement of algal cells in raceway pond with conic baffles. Chem Eng Sci 216:115536
Harrington L, Foster R (2010). Australian residential building sector greenhouse gas emissions 1990–2010. Final Report, Energy Efficient Strategies. Australian Greenhouse Office
Heining M, Buchholz R (2015) Photobioreactors with internal illumination – a survey and comparison. Biotechnol J 10(8):1131–1137
Heldt H-W, Heldt F (2005a) The use of energy from sunlight by photosynthesis is the basis of life on earth. In: Heldt H-W, Heldt F (eds) Plant biochemistry, 3rd edn. Academic Press, Burlington, pp 45–66
Heldt H-W, Heldt F (2005b) 3 - Photosynthesis is an electron transport process. In: Heldt H-W, Heldt F (eds) Plant biochemistry, 3rd edn. Academic Press, Burlington, pp 67–114
Herzog H (1999) An introduction to CO_2 separation and capture technologies. Energy Laboratory Working Paper Massachusetts Institute of Technology, Cambridge
Herzog HJ (2001) What future for carbon capture and sequestration? Environ Sci Technol 35 (7):148a–153a
Hinterholz C, Schuelter A, Módenes AN, Trigueros DE, Borba C, Espinoza-Quiñones F, Kroumov A (2017) Microalgae Flat Plate-Photobioreactor (FP-PBR) system development: computational tools to improve experimental results. Acta Microbiol Bulg 33(3):119–124
Hinterholz CL, Trigueros DEG, Modenes AN, Borba CE, Scheufele FB, Schuelter AR, Kroumov AD (2019) Computational fluid dynamics applied for the improvement of a flat-plate photobioreactor towards high-density microalgae cultures. Biochem Eng J 151:107257
Hu J-Y, Sato T (2017) A photobioreactor for microalgae cultivation with internal illumination considering flashing light effect and optimized light-source arrangement. Energy Convers Manag 133:558–565

Hu Q, Kurano N, Kawachi M, Iwasaki I, Miyachi S (1998) Ultrahigh-cell-density culture of a marine green alga *Chlorococcum littorale* in a flat-plate photobioreactor. Appl Microbiol Biotechnol 49(6):655–662

Iluz D, Abu-Ghosh S (2016) A novel photobioreactor creating fluctuating light from solar energy for a higher light-to-biomass conversion efficiency. Energy Convers Manag 126:767–773

Jacobi A, Steinweg C, Sastre RR, Posten C (2012) Advanced photobioreactor LED illumination system: Scale-down approach to study microalgal growth kinetics. Eng Life Sci 12(6):621–630

Kajiwara S, Yamada H, Ohkuni N, Ohtaguchi K (1997) Design of the bioreactor for carbon dioxide fixation by *Synechococcus PCC7942*. Energy Convers Manag 38:S529–S532

Kandilian R, Pruvost J, Artu A, Lemasson C, Legrand J, Pilon L (2016) Comparison of experimentally and theoretically determined radiation characteristics of photosynthetic microorganisms. J Quant Spectrosc Radiat Transf 175:30–45

Khadim SR, Singh P, Singh AK, Tiwari A, Mohanta A, Asthana RK (2018) Mass cultivation of *Dunaliella salina* in a flat plate photobioreactor and its effective harvesting. Bioresour Technol 270:20–29

Klinthong W, Yang Y-H, Huang C-H, Tan C-S (2015) A review: microalgae and their applications in CO_2 capture and renewable energy. Aerosol Air Qual Res 15(2):712–742

Kremer G, Vis M, Prudich M, Bayless D, Inventors (2000) Practical photosynthetic carbon dioxide mitigation. USA

Kroumov AD, Gacheva G, Iliev I, Alexandrov S, Pilarski P, Petkov G (2013) Analysis of S_f/V ratio of photobioreactors linked with algal physiology. Genet Plant Physiol 3(1–2):55–64

Kroumov AD, Módenes AN, Trigueros DEG (2015) A complex theoretical approach for algal medium optimization for CO_2 fixation from flue gas. Acta Microbiol Bulg 31(1):61–70

Kroumov AD, Modenes AN, Trigueros DEG, Espinoza-Quinones FR, Borba CE, Scheufele FB, Hinterholz CL (2016) A systems approach for CO_2 fixation from flue gas by microalgae – theory review. Process Biochem 51(11):1817–1832

Kroumov AD, Scheufele FB, Trigueros DEG, Modenes AN, Zaharieva MM, Najdenski HM (2017) Chapter 11: Modeling and techno-economic analysis of algae for bio-energy and co-products. In: Rastogi (ed) Algal green chemistry: recent progress in biotechnology. Elsevier, pp 202–241

Kunjapur AM, Eldridge RB (2010) Photobioreactor design for commercial biofuel production from microalgae. Ind Eng Chem Res 49:3516–3526

Lai Q, Kong L, Gong W, Russell AG, Fan M (2019) Low-energy-consumption and environmentally friendly CO_2 capture via blending alcohols into amine solution. Appl Energy 254:113696

Lee S-Y, Lee I-B, Han J (2019) Design under uncertainty of carbon capture, utilization and storage infrastructure considering profit, environmental impact, and risk preference. Appl Energy 238:34–44

Li J. (2002) On-line state estimation of microalgal photobioreactors [Master of Science in Biosystems Engineering]. University of Hawaii

Li B, Duan Y, Luebke D, Morreale B (2013) Advances in CO_2 capture technology: a patent review. Appl Energy 102:1439–1447

Li A, Wang J, Bao B (2019) High-efficiency CO_2 capture and separation based on hydrate technology: a review. Greenhouse Gases: Sci Technol 9(2):175–193

Lian X, Xu L, Chen M, Wu CE, Li W, Huang B, Cui Y (2019) Carbon dioxide captured by metal organic frameworks and its subsequent resource utilization strategy: a review and prospect. J Nanosci Nanotechnol 19(6):3059–3078

Luzi G, McHardy C, Lindenberger C, Rauh C, Delgado A (2019) Comparison between different strategies for the realization of flashing-light effects – pneumatic mixing and flashing illumination. Algal Res 38:101404

Malapascua JR, Ranglova K, Masojídek J (2019) Photosynthesis and growth kinetics of *Chlorella vulgaris R-117* cultured in an internally LED-illuminated photobioreactor. Photosynthetica 57 (1):103–112

Moberg AK, Ellem GK, Jameson GJ, Herbertson JG (2012) Simulated cell trajectories in a stratified gas–liquid flow tubular photobioreactor. J Appl Phycol 24(3):357–363

Molazadeh M, Ahmadzadeh H, Pourianfar HR, Lyon S, Rampelotto PH (2019) The use of microalgae for coupling wastewater treatment with CO_2 bio-fixation. Front Bioeng Biotechnol 7:42–42

Molina Grima E, Fernández FGA, García Camacho F, Chisti Y (1999) Photobioreactors: light regime, mass transfer, and scale-up. J Biotechnol 70(1):231–247

Molina E, Acién Fernández FG, García Camacho F, Camacho Rubio F, Chisti Y (2000) Scale-up of tubular photobioreactors. J Appl Phycol 12(3):355–368

Murakami M, Ikenouchi M (1997) The biological CO_2 fixation and utilization project by rite (2) - Screening and breeding of microalgae with high capability in fixing CO_2. Energy Convers Manag 38:S493–S497

Olivieri G, Salatino P, Marzocchella A (2014) Advances in photobioreactors for intensive microalgal production: configurations, operating strategies and applications. J Chem Technol Biotechnol 89(2):178–195

Onorato C, Rösch C (2020) Comparative life cycle assessment of astaxanthin production with *Haematococcus pluvialis* in different photobioreactor technologies. Algal Res 50:102005

Pfaffinger CE, Severin TS, Apel AC, Göbel J, Sauter J, Weuster-Botz D (2019) Light-dependent growth kinetics enable scale-up of well-mixed phototrophic bioprocesses in different types of photobioreactors. J Biotechnol 297:41–48

Pires JCM, Alvim-Ferraz MCM, Martins FG (2017) Photobioreactor design for microalgae production through computational fluid dynamics: a review. Renew Sust Energ Rev 79:248–254

Płaczek M, Patyna A, Witczak S (2017) Technical evaluation of photobioreactors for microalgae cultivation. E3S Web Conf 19:02032

Pruvost J, Cornet JF, Le Borgne F, Goetz V, Legrand J (2015) Theoretical investigation of microalgae culture in the light changing conditions of solar photobioreactor production and comparison with cyanobacteria. Algal Res 10:87–99

Pruvost J, Cornet J-F, Pilon L (2016a) Large-scale production of algal biomass: photobioreactors. In: Bux F, Chisti Y (eds) Algae biotechnology: products and processes. Springer International Publishing, Cham, pp 41–66

Pruvost J, Le Borgne F, Artu A, Cornet J-F, Legrand J (2016b) Chapter 5: Industrial photobioreactors and scale-up concepts. In: Legrand J (ed) Advances in chemical engineering. Academic Press, pp 257–310

Pruvost J, Le Gouic B, Lepine O, Legrand J, Le Borgne F (2016c) Microalgae culture in building-integrated photobioreactors: Biomass production modeling and energetic analysis. Chem Eng J 284:850–861

Pruvost J, Le Borgne F, Artu A, Legrand J (2017) Development of a thin-film solar photobioreactor with high biomass volumetric productivity (AlgoFilm©) based on process intensification principles. Algal Res 21:120–137

Pulz O (2001) Photobioreactors: production systems for phototrophic microorganisms. Appl Microbiol Biotechnol 57(3):287–293

Razzak SA, Hossain MM, Lucky RA, Bassi AS, de Lasa H (2013) Integrated CO_2 capture, wastewater treatment and biofuel production by microalgae culturing: a review. Renew Sust Energ Rev 27:622–653

Rubio FC, Camacho FG, Sevilla JM, Chisti Y, Grima EM (2003) A mechanistic model of photosynthesis in microalgae. Biotechnol Bioeng 81(4):459–473

Sato T, Yamada D, Hirabayashi S (2010) Development of virtual photobioreactor for microalgae culture considering turbulent flow and flashing light effect. Energy Convers Manag 51 (6):1196–1201

Scheufele FB, Hinterholz CL, Zaharieva MM, Najdenski HM, Modenes AN, Trigueros DEG, Borba CE, Espinoza-Quinones FR, Kroumov AD (2019) Complex mathematical analysis of photobioreactor system. Eng Life Sci 19(12):844–859

Schuelter AR, Kroumov AD, Hinterholz CL, Fiorini A, Trigueros DEG, Vendruscolo EG, Zaharieva MM, Modenes AN (2019) Isolation and identification of new microalgae strains with antibacterial activity on food-borne pathogens. Engineering approach to optimize synthesis of desired metabolites. Biochem Eng J 144:28–39

Sero ET, Siziba N, Bunhu T, Shoko R, Jonathan E (2020) Biophotonics for improving algal photobioreactor performance: a review. Int J Energy Res 44(7):5071–5092

Shah S, Shah M, Shah A, Shah M (2020) Evolution in the membrane-based materials and comprehensive review on carbon capture and storage in industries. Emerg Mater 3(1):33–44

Sifat NS, Haseli Y (2019) A critical review of CO_2 capture technologies and prospects for clean power generation. Energies 12:4143

Singh J, Dhar DW (2019) Overview of carbon capture technology: microalgal biorefinery concept and state-of-the-art [mini review]. Front Mar Sci 6(29)

Sivasangari S, VelRajan T, Nandhini J (2019) A comparative study on the performance of conventional photobioreactors and ALGADISK in CO_2 sequestration – a review. Energy Sources, Part A: Recov Util Environ Effects:1–6

Soman A, Shastri Y (2015) Optimization of novel photobioreactor design using computational fluid dynamics. Appl Energy 140:246–255

Song C, Liu Q, Deng S, Li H, Kitamura Y (2019a) Cryogenic-based CO_2 capture technologies: state-of-the-art developments and current challenges. Renew Sust Energ Rev 101:265–278

Song C, Liu Q, Qi Y, Chen G, Song Y, Kansha Y, Kitamura Y (2019b) Absorption-microalgae hybrid CO_2 capture and biotransformation strategy: a review. Int J Greenhouse Gas Control 88:109–117

Suh IS, Lee SB (2001) Cultivation of a cyanobacterium in an internally radiating air-lift photobioreactor. J Appl Phycol 13(4):381–388

Uddin D, Gani O, Mahato A, Sakib I, Mony R (2020) Spirulina (*Spirulina platensis*) production in different photobioreactors on rooftop. Int J Business Soc Scientific Res 8(1):15–19

Ugwu C, Ogbonna J, Tanaka H (2002) Improvement of mass transfer characteristics and productivities of inclined tubular photobioreactors by installation of internal static mixers. Appl Microbiol Biotechnol 58(5):600–607

Ugwu CU, Aoyagi H, Uchiyama H (2008) Photobioreactors for mass cultivation of algae. Bioresour Technol 99(10):4021–4028

United Nations. (2011). Kyoto Protocol to the United Nations Framework Convention on Climate Change (UNFCCC)

Usui N, Ikenouchi M (1997) The biological CO_2 fixation and utilization project by RITE(1) – highly-effective photobioreactor system. Energy Convers Manag 38:S487–S492

Wei X, Manovic V, Hanak DP (2020) Techno-economic assessment of coal- or biomass-fired oxy-combustion power plants with supercritical carbon dioxide cycle. Energy Convers Manag 221:113143

Wu W, Chang J-S (2019) Integrated algal biorefineries from process systems engineering aspects: a review. Bioresour Technol 291:121939

Xu L, Weathers PJ, Xiong X-R, Liu C-Z (2009) Microalgal bioreactors: challenges and opportunities. Eng Life Sci 9(3):178–189

Xue S, Zhang Q, Wu X, Yan C, Cong W (2013) A novel photobioreactor structure using optical fibers as inner light source to fulfill flashing light effects of microalgae. Bioresour Technol 138:141–147

Yadav G, Dubey BK, Sen R (2020) A comparative life cycle assessment of microalgae production by CO_2 sequestration from flue gas in outdoor raceway ponds under batch and semi-continuous regime. J Clean Prod 258:120703

Yeh N, Chung J-P (2009) High-brightness LEDs—energy efficient lighting sources and their potential in indoor plant cultivation. Renew Sust Energ Rev 13(8):2175–2180

Yu Q, Wang H, Li X, Yin Y, Qin S, Ge B (2020) Enhanced biomass and CO_2 sequestration of *Chlorella vulgaris* using a new mixotrophic cultivation method. Process Biochem 90:168–176

Zhu Q (2019) Developments on CO_2-utilization technologies. Clean Energy 3(2):85–100

Zhu C, Zhai X, Xi Y, Wang J, Kong F, Zhao Y, Chi Z (2020) Efficient CO_2 capture from the air for high microalgal biomass production by a bicarbonate Pool. J CO2 Util 37:320–327

Zijffers J-WF, Janssen M, Tramper J, Wijffels RH (2008) Design process of an area-efficient photobioreactor. Mar Biotechnol (NY) 10(4):404–415

Engineering Cyanobacteria Cell Factories for Photosynthetic Production of Sucrose

17

Shanshan Zhang, Huili Sun, Jiahui Sun, Quan Luo, Guodong Luan, and Xuefeng Lu

Abstract

Biorefinery technology serves as an important alternative route to alleviate the energy and environmental crisis and to promote sustainable development. The sufficient supply of sugar feedstocks is the basis and prerequisite for the economic feasibility of modern biorefinery systems. Cyanobacterial photosynthetic

S. Zhang · H. Sun · J. Sun · Q. Luo
Key Laboratory of Biofuels, Qingdao Institute of Bioenergy and Bioprocess Technology, Chinese Academy of Sciences, Qingdao, China

Shandong Energy Institute, Qingdao, China

Qingdao New Energy Shandong Laboratory, Beijing, China

G. Luan (✉)
Key Laboratory of Biofuels, Qingdao Institute of Bioenergy and Bioprocess Technology, Chinese Academy of Sciences, Qingdao, China

Shandong Energy Institute, Qingdao, China

Qingdao New Energy Shandong Laboratory, Beijing, China

Dalian National Laboratory for Clean Energy, Dalian, China
e-mail: luangd@qibebt.ac.cn

X. Lu (✉)
Key Laboratory of Biofuels, Qingdao Institute of Bioenergy and Bioprocess Technology, Chinese Academy of Sciences, Qingdao, China

Shandong Energy Institute, Qingdao, China

Qingdao New Energy Shandong Laboratory, Beijing, China

Dalian National Laboratory for Clean Energy, Dalian, China

Laboratory for Marine Biology and Biotechnology, Qingdao National Laboratory for Marine Science and Technology, Qingdao, China
e-mail: lvxf@qibebt.ac.cn

R. P. Rastogi (ed.), *Ecophysiology and Biochemistry of Cyanobacteria*,
https://doi.org/10.1007/978-981-16-4873-1_17

production of sucrose provides a promising green route for sugar feedstock supply, channeling carbon dioxide and solar energy directly into sucrose in a single platform and a single step. Sucrose works as an essential intracellular osmoprotective substance for a large portion of cyanobacterial species to survive salts stress. Cyanobacterial sucrose synthesis pathways are sensitive to salts stress and are rapidly regulated on transcription and enzymatic levels by osmo- and salts signals. To construct efficient photosynthetic cell factories for sucrose production, types of metabolic engineering strategies have been developed and adopted. The introduction of sucrose transporters into cyanobacteria cells permitted the secretory synthesis of sucrose and significantly improved the production performances. Comprehensive modifications on the metabolism network rewired more carbon flow toward sucrose synthesis and removed the dependence on salts induction. In recent years, cocultivation strategy further optimized the continuity, robustness, and economic feasibility of the process for photosynthetic production of sucrose. This chapter reviews the development history of engineering cyanobacteria for sucrose production, highlights the recent progress on disclosing the sucrose synthesis mechanisms, constructing sucrose production cell factories and developing co-cultivation systems based on cyanobacterial sucrose production, and prospects the future directions of the technology.

Keywords

Cyanobacteria · Sucrose · Co-cultivation · Metabolic Engineering · Cell factories

17.1 Introduction

Petrochemical refinery technology provides the material basis for the development and prosperity of modern society. However, the sustainability of the mode is severally threatened by the accompanying irreversible resource shortage and environmental pollutions. The biorefinery technology system, converting renewable resources into biochemicals and biofuels through environment-friendly processes with microbial fermentation and conversion technologies, provides an alternative route for promoting sustainable development (Zhang et al. 2009). For traditional biorefinery technology utilizing cell factories derived from heterotrophic microbes, for example, *Escherichia coli, Bacillus subtilis*, and *Saccharomyces cerevisiae*, abundant and sustainable supply of sugar feedstocks are important prerequisites for the feasible industrial applications of the technology (Dien et al. 2003). At present, there are three main sources of carbohydrate feedstocks supply: starch food crops (such as corn), lignocellulosic biomass (such as straw), and sucrose plants (such as sugarcane), all of which are facing several potential restrictions and bottlenecks (Hays and Ducat 2015; Melis 2012). As for food crop sourced sugar production, the consumption of farmland and food is attracting social controversies. Lignocellulose could serve as a nonfood carbohydrate substrate for the biorefinery process; however, the necessary pretreatments dependent on complicated and expensive enzymatic hydrolysis and saccharification elevated the economic costs and decreased the environmental friendliness of the whole technology. Sugar plants could be a promising source for carbohydrate feedstock production, while the restrictions on climate and environments of the cultivation make it hard for promotion as a widely applied route (Rastogi et al. 2018). Therefore, developing novel sugar synthesis technology to achieve sustainable, economically competitive, and nonregional dependent sugar feedstocks production routes is of great significance for reducing the overall process costs and promoting the industrialization of biorefinery technologies.

Cyanobacteria are the only known prokaryotic microorganisms that can perform oxygenic photosynthesis. They are widely distributed in diverse ecosystems, including ocean, land, freshwater, and some extreme environments, playing an essential role in the global cycle of carbon, phosphate, and oxygen (Waterbury et al. 1979). Compared with higher plants and eukaryotic microalgae, cyanobacteria possess simpler structures, shorter life cycles, and more convenient genetic manipulation systems. Through modifications of the native metabolism network or the introduction of heterologous metabolic pathways, photosynthetic carbon flux and energy flux in cyanobacteria cells can be reallocated for enhancing the production of natural and non-natural metabolites (Lu 2010; Melis 2012). Sugars are important forms of carbohydrates in cyanobacteria (Hays and Ducat 2015). Cyanobacteria can naturally synthesize monosaccharaides such as glucose and fructose, disaccharides such as sucrose and trehalose, and many types of macromolecular polysaccharides such as glycogen. Engineering cyanobacteria for photosynthetic production of sugars could serve as a promising alternative route for feedstock supply for the biorefinery industry. Compared with traditional sugar production routes, photosynthetic production of sugars facilitated the direct conversion from solar energy and carbon dioxide into final carbohydrate forms, reducing the economic and environmental costs from the cultivation, collection, pretreatment, and refinery of the plant biomass.

Sucrose is an important sugar feedstock for microbial fermentation and a sweetener widely used in the food industry (Du et al. 2013; Ducat et al. 2012b; Lowe et al. 2017), and it is also one of the most representative sugar metabolites in cyanobacteria. A large portion of cyanobacteria species would synthesize and accumulate sucrose as an osmoprotective compound to resist environmental salts stress (Hagemann 2011). Metabolic engineering of cyanobacteria further facilitates secretion of the intracellular sucrose (Ducat et al. 2012b), which removed the potential sink limitation effects, elevated the overall production, and improved the application potentials of the sucrose photosynthetic production technology in industrial processes. Here in this chapter, we will summarize the physiological and metabolic mechanisms of cyanobacterial sucrose synthesis, the metabolic engineering strategies and tools to develop efficient photosynthetic cell factories for sugars production, and the technology extensions of the photosynthetic driven sugars production technology. The trends and directions of the related technologies in this area would also be prospected.

17.2 Physiological and Metabolic Background of Cyanobacterial Sucrose Synthesis

17.2.1 Physiological Significance of Cyanobacterial Sucrose Synthesis

Cyanobacteria are widely distributed in diverse habitats and have evolved effective physiological and metabolic strategies and mechanisms to acclimate fluctuating environments and nutrition (Waterbury et al. 1979). Salt stress caused by high concentration metal ions (Na^+, K^+, etc.) is a common and typical environmental stress for microbes. Salt stress usually challenges cyanobacterial cellular homeostasis via two effects, high ionic concentrations and associated external osmotic pressure, which both pose serious threats to the structural stability and function maintenance of intracellular proteins and membrane systems. In response to high salt stress, an important survival strategy adopted by cyanobacteria cells is to rapidly synthesize and accumulate osmoprotective compounds to balance the osmotic pressure difference between the intracellular and the extracellular environments. The osmoprotective compounds were also called the compatible solute, referring to a class of small molecular weight metabolites synthesized by microbial cells, that could improve intracellular water activity and maintain turgor pressure and cell volume (Brown and Simpson 1972). Among cyanobacteria, a close correlation was found between the salt tolerance capacities and types of the major compatible solute. Freshwater strains (such as *Anabaena* sp. PCC 7120 and *Synechococcus elongatus* PCC 7942, hereafter PCC 7942) with low tolerance to salt stress (meaning can only survive at a salt concentration not higher than 0.6 M NaCl) accumulate sucrose and/or trehalose as major compatible solutes. Moderately halotolerant strains (such as *Synechocystis* PCC 6803, hereafter PCC 6803, which can survive at a salt concentration not higher than 1.7 M NaCl) are characterized to synthesize glucosylglycerol as a main compatible solute, whereas halophilic strains (such as *Microcystis aeruginosa* and *Aphanothece halophytica*, which can tolerate high salts concentrations as high as 3.0 M NaCl) that can grow in saturated salt concentrations usually synthesize glycine betaine or glutamate betaine (Hagemann 2011; Mackay et al. 1984; Reed and Stewart 1985). The preference of different compatible solutes at different salinity levels might be correlated with the specific degree of protection for the stabilization of macromolecules such as enzymes or membranes. However, there is not always a strict correspondence between the types of compatible solutes and salt stress tolerances. For example, a marine cyanobacteria strain *Crocosphaera watsonii* solely accumulates trehalose (Pade et al. 2012), and most oceanic *Prochlorococcus* species mainly accumulate sucrose (Klähn et al. 2010). Sucrose is the most representative compatible solute for a majority of freshwater cyanobacterial strains and some marine cyanobacterial strains. More than 60 cyanobacterial species have been characterized to synthesize sucrose as a compatible solute responding to salt stress (Hagemann 2011). The salt-induced accumulation of sucrose was first reported in *S. elongatus* PCC6301 (Blumwald et al. 1983) and *Anabaena cariabilis (Erdmann* 1983*).* The phylogenetic analysis indicated that the synthesis of sucrose occurred in the early phase of cyanobacterial evolution as the initial mechanism for compatible solute accumulation in ancient cyanobacteria (Blank 2013). Sucrose remains to be the dominant compatible solute for cyanobacteria species in freshwater habitats as well as trehalose. In addition, sucrose also occurs in some moderately salt-tolerant strains such as PCC 6803 as a secondary or complementary compatible solute in addition to GG.

In addition to the role of a compatible solute against salt stress, sucrose is also synthesized and accumulated as a complementary carbon pool in cyanobacteria. To adapt to the circadian rhythm and environmental changes during day–night cycles, photosynthesis-sourced energy and organic carbon excess the requirements for normal growth and maintenance would be stored as the carbon sink, which would support cell growth and survival in dark conditions and starvation status. Natural carbon sink mechanisms include the synthesis of polysaccharide macromolecules (e.g., glycogen and polyhydroxybutyrate) and some small molecular weight solutes, such as glucosylglycerol and sucrose (Ball and Morell 2003; Damrow et al. 2016; Yasunori et al. 2005). Under normal conditions, glycogen metabolism works as the most important natural carbon sink mechanism, and the existence of sucrose and glucosylglycerol synthesis mechanisms provides a guarantee for the plasticity and flexibility of the cyanobacterial carbon sink network. When glycogen synthesis and accumulation are blocked in the marine strain *Synechococcus* sp. PCC 7002, sucrose synthesis in glycogen-deficient mutants was significantly increased to buffer the carbon and energy overflow (Guerra et al. 2013; Hendry et al. 2017). When the glycogen-deficient mutants of PCC 7002 are cultivated in dark conditions, increased intracellular sucrose would also work as important substrates for heterologous fermentation and respiratory to partially replace the functions of the natural carbon sink mechanisms.

17.2.2 Metabolic Mechanisms of Cyanobacterial Sucrose Synthesis

The metabolism pathway of sucrose synthesis in cyanobacteria has been clearly elucidated. Sucrose is synthesized from fructose-6-phosphate and UDP-glucose through the sequential reactions catalyzed by sucrose phosphate synthase (SPS, EC 2.4.1.14) and sucrose-phosphate phosphatase (SPP, EC3.1.3.24). First, the intermediate sucrose-6-phosphate would be synthesized by SPS with UDP-glucose and fructose-6-phosphate and a UDP molecule would be released. Sucrose-6-phosphate would subsequently be hydrolyzed into sucrose and inorganic phosphate (Pi) by sucrose-phosphate phosphatase (SPP). Two structurally different forms of SPS have been identified in cyanobacteria. The SPS from *Nostoc* (*Anabaena*) sp. PCC7119 and some other strains only require the glucosyltransferase domain for catalytic activity (Porchia and Salerno 1996; Salerno and Curatti 2003), while, as for the more common SPS form, the glucosyltransferase domain was found to be fused with an inactive phosphohydrolase domain, although the second reaction would be catalyzed by a separate SPP protein. Some of the fused type SPS (containing both the glucosyltransferase and phosphohydrolase domains) could work as bifunctional SPS-SPP enzyme, which can catalyze the two steps for synthesis of sucrose (Martínez-Nol et al. 2013). In addition to the SPS-SPP pathway, another pathway for sucrose biosynthesis was found in some filamentous cyanobacterial strains such as *Nostoc* sp. PCC 7120, PCC 7119, ATCC 29413, in which sucrose synthase (SuS, EC 2.4.1.13) catalyzes the reversible conversion from UDP-glucose and D-fructose to sucrose and UDP or vice versa. However, SuS is mainly responsible for sucrose hydrolysis instead of sucrose synthesis under physiological conditions (Curatti et al. 2000, 2006; Porchia et al. 1999).

17.2.3 Regulatory Mechanism of Sucrose Synthesis in Cyanobacteria Under Salt Stress

The metabolic pathway for sucrose synthesis has been clearly characterized, while the detailed regulatory mechanisms of salt-induced sucrose synthesis are yet to be elucidated. As mentioned earlier, environmental salt stress inhibited the physiological homeostasis of cyanobacterial cells in two ways, concentrated ionic stress and osmotic pressure stress. Increasing evidence indicate that concentrated ionic stress and osmotic pressure stress from environmental stress challenged cyanobacteria cells in different ways. Murata et al. reported that salt stress (high concentrations of NaCl) and osmotic stress (high concentrations of sorbitol) caused different influences and changes on the transcriptomics pattern in PCC 6803, indicating that the two stress signals might induce the different regulatory responses of genes transcriptions (Kanesaki et al. 2002). Additionally, it was also found that osmotic stress (high concentration of sorbitol) and salt stress (high concentration of NaCl), with the same osmotic potentials, brought different influences on the cytoplasmic volume of PCC 7942. Sorbitol stress decreased the cytoplasmic volume by 55%, whereas salt stress only caused 15% decrease in the initial volume. Besides the reduced cytoplasmic volume, salt stress and hyperosmotic stress also cause other different effects on cellular physiology and metabolism. Under hyperosmotic stress, the cytoplasm of PCC 7942 shrunk quickly, and the photosystem would be rapidly but reversibly inactivated. The activities of the photosystem will be restored when the cells were put back into the isotonic solution. The high concentrations of NaCl stress would also bring in rapid and reversible inhibition on the cyanobacterial photosystem through hyperosmotic stress, and then the influx of Na^+ ions through K^+/Na^+ channels would irreversibly damage and inactivate PSI and PSII (Allakhverdiev et al. 2000a, b). The abovementioned results indicated that the hyperosmotic effects and concentrated ionic effects functioned quite differently on cyanobacterial cells during the salt stress process, and the detailed roles of the two factors specifically in inducing sucrose synthesis remain to be revealed.

Currently, it has not been clearly elucidated how salt stress signals are sensed by cyanobacteria to trigger the expression regulation of multiple genes. At the transcriptional level, salt stress would induce the elevation in the *sps* expression level within minutes. Recently, negative regulation by response regulator 39 (Rre39; Slr1588) on *sps* expression has been identified in PCC 6803, and the expression of *spsA* was enhanced in the Rre39 deletion mutant (Song et al. 2017). Two typical types of signal transduction systems are identified to be involved in perception, transduction, and the response of environmental changes in microbial cells, that is, one-component transduction system (serine/threonine kinases, STK) and two-component transduction system (histidine kinases and the cognate response regulators, His-Rre). His-Rre typed two-component transduction systems have been reported to participate in responding to salt stress, osmotic stress, metal ion stress, the changes of temperature, and light intensity in cyanobacteria cells (Ginerlamia et al. 2012; Liu et al. 2015; Narikawa et al. 2011). There have been four kinds of His-Rre two-component transduction systems (Hik33-Rre31, Hik34-Rre1, Hik-Hik41-Rre17, and Hik10-Rre3) and one-component transduction system (SpkG) identified to be possibly related with perceiving and transducing signals of salt stress and hyperosmotic stress in PCC 6803 (Liang et al. 2011). And a two-component response regulator OrrA (Alr3768) in *Nostoc* sp. PCC 7120 was confirmed to control sucrose synthesis (Ehira et al. 2014). However, little is known about the clear pathways for perceiving and transducing salt stress signals for the majority of sugar-producing cyanobacterial strains.

In 2020, Liang et al. reported that SPS, the rate-limiting enzyme for sucrose synthesis, would be expressed and maintained at a basic and constant level in PCC 7942 even under non-salt stress conditions. When the cell was treated with concentrated salts, the intracellular ion concentrations would increase rapidly, and the SPS enzyme would be activated, initializing the synthesis and accumulation of sucrose to maintain the osmotic balance inside and outside the cell. When the environmental salinity was reduced, the intracellular ion concentrations would also decrease, which convert the SPS enzymes back to the low activity state and decrease sucrose synthesis rates. Interestingly, activities of invertase, the enzyme degrading sucrose into glucose and fructose, shows an opposite responsive mode comparing with SPS, that is, high concentrations of ions inhibit invertase activities, while the lowered ion concentrations would elevate the activities. Overall, the dynamic ion concentrations in cyanobacteria cells regulate the key enzymes of sucrose synthesis and degradation in an opposite way, thus realizing the dynamic response of cyanobacteria cells to the change in environmental salinity (Liang et al. 2020).

Theoretically, the transduction of salt stress signals in cyanobacterial cells will initiate the regulation of sucrose synthesis pathways. Previous studies have shown that this regulation may proceed on multiple levels such as transcriptional, translational, enzyme activity, and so on. In PCC 6803, the amount of *sps* (coding for sucrose-phosphate synthase) transcripts increased rapidly in PCC 6803 cells after the salt shock and reached a maximum after 0.5 h (Desplats et al. 2005). Whereas in *Synechococcus* sp. PCC 7002, the transcript level of *sps* and *spp* were also increased significantly after 24 h of salt treatment (Cumino et al. 2010). At the translation level, Western blot assays revealed that the expression level of SPS in *Nostoc* sp. PCC 7120 was increased significantly after salt treatment for 6 h and was reversed to control values when cells were returned to basal-medium growth conditions. And correspondingly, SPS activity would be increased threefold after 80 mM NaCl stress for 6 h. The high SPS activity in salt-stressed cells decreased to control values when cells were transferred to the basic culture medium (Salerno et al. 2004). However, the stimulation of SPS activity by NaCl was just confirmed in PCC 7942 and the SPS activities in PCC 6803 were not regulated by salts, suggesting that the biochemical activation mechanism of sucrose synthesis might not be a conservative one (Kirsch et al. 2019).

17.2.4 Sucrose Synthesis in Cyanobacteria Under Salt-Free Conditions

As mentioned earlier, sucrose phosphate synthase (SPS) is a key enzyme in the sucrose synthesis process of cyanobacteria and plays a key role in carbon metabolism. The ion effect of salt stress also plays an important part in inducing sucrose synthesis. The increased ion concentration directly activates SPS and inhibits sucrose-degrading enzyme INV, which leads to the rapid accumulation of sucrose (Liang et al. 2020). Thus, many cyanobacteria need high salt environment to activate the sucrose biosynthetic pathway. However, salt stress will cause a certain burden on the growth and metabolism of algae cells. Moreover, salt stress is also faced with the disadvantage of increasing cost. It is particularly important to find a method for producing sucrose without salt stress, which is of great significance for the large-scale production of sucrose. In 2020, Pakrasi et al. achieved salt-free sucrose production for the first time in cyanobacteria by expressing the PCC 6803 *sps* and *spp* genes in the mutant of UTEX 2973 (Lin et al. 2020). At present, the idea of "salt-stress" independent sucrose synthesis through synthetic biology and metabolic engineering transformation has been realized. Obviously, it could be expected that the engineering salt stress independent sucrose-synthesizing strains could further expand the use of cyanobacteria in coculture systems.

17.2.5 Metabolic Mechanisms of Cyanobacterial Sucrose Degradation

At present, three sucrose degradation pathways have been identified in cyanobacteria (Kolman et al. 2015), and the corresponding three enzymes are sucrose synthase (SuS), invertase (Inv), and amylosucrase, respectively. SuS catalyze both synthesis and degradation of sucrose but is mainly active in the direction of sucrose hydrolysis instead of the synthesis reactions in vivo (Porchia et al. 1999). Sus was mainly identified in heterocyst-forming cyanobacteria and played an important role in N_2-fixation (Kolman et al. 2015). The most common sucrose degradation pathway in cyanobacteria is catalyzed by the enzyme Inv, which catalyzes the irreversible breakdown of sucrose into monosaccharides glucose and fructose. Phylogenetic studies indicated that invertases were part of the original sucrose metabolism and have been transferred from cyanobacteria to plants (Vargas and Salerno 2010). The invertases can be classified as acid invertases (Ac-Inv) and alkaline/neutral invertases (A/N-Inv) based on the pH optimum for enzyme activity. The optimum pH for Ac-Inv is 4.5 while A/N-Inv has a broad pH optimum from 6.5 to 8. The Ac-Inv belongs to the family of β-fructofuranosidases (EC 3.2.1.26) which can not only cleave sucrose but also other β-fructose-containing oligosaccharides such as raffinose and stachyose (Sturm 1999). They exist mainly in heterotrophic bacteria, yeasts, and plants and have been widely studied due to its great significance for the food industry (Nadeem et al. 2015). Unlike Ac-Inv, A/N-Inv was only found in cyanobacteria and plants (Vargas et al. 2003), degrading sucrose by cleaving the α, β-1,2-glycosidic linkage of sucrose (Vargas and Salerno 2010). In vitro enzymatic assays revealed that the activity of PCC 6803 Inv would be increasingly inhibited when elevated concentrations of NaCl were supplemented into the reaction system, indicating a potential regulatory mechanism to enhance sucrose accumulation in salt-stresses cells (Kirsch et al. 2018). A significant amount of sucrose was found in the cells of the Δ*inv* mutant under salt-free conditions, which indicates that active sucrose turnover and explains the low steady-state content of sucrose in wild-type cells. When NaCl is added to the culture medium, the accumulation of sucrose is induced immediately in the wild-type strain as well as in the mutant. However, it is a transient accumulation with a maximum abundance achieved after 6–12 h salt shock and the intracellular sucrose concentration would drop again almost to the initial level with 24 h past. The Δ*inv* mutant showed constantly enhanced sucrose content even after long-term salt acclimation (Kirsch et al. 2018), indicating that the decrease of sucrose accumulation in salt-stressed wild-type cells was owing to the action of sucrose degradation via Inv (Kirsch et al. 2019). In the Inv, amylosucrase (AMS) also splits sucrose into free fructose and glucose and then linked glucose to oligosaccharides or glycogen via glycosidic bond, which has recently been identified in *Synechococcus sp.* PCC 7002 (Perez-Cenci and Salerno 2014). In PCC 7002, the *amsA* gene encoding amylosucrase was grouped into the same transcriptional unit with several genes participating in sucrose synthesis and metabolism, including *spsA*, *sppA*, and *frkA* (encoding fructokinase catalyzing the phosphorylation of fructose). And it has been reported that the expression of these genes would be increased in PCC 7002 cells facing increased NaCl concentrations, confirming the important role of sucrose metabolism in PCC 7002 cell adaptation to salts stress (Perez-Cenci and Salerno 2014).

17.2.6 Strategies for Cyanobacteria in Response to Decreased Salinity in the Environment

At present, lots of efforts have been put in exploring the synthesis mechanisms of compatible solutes in cyanobacterial cells under salt stress, but little is known about the cellular response when environmental salinity decreases. In fact, reduced salinity in extracellular space may be also a big challenge for cyanobacteria. The rapidly decreased osmotic potential of the cytosol when salinity stress was removed would generate a severe burden on the cytoplasmic membrane and the cell wall. To protect the cells from bursting, cyanobacteria cells have to remove the intracellularly enriched compatible solutes to balance the intracellular and the extracellular osmotic pressure. A strategy widely used by cyanobacteria is to open mechanosensitive channels and release compatible solutes from the cells (Levina et al. 2014). Another more advantageous and economical strategy is to enzymatically degrade and metabolize compatible solutes, which allow the large amounts of organic carbon or nitrogen incorporated into compatible solutes to be reused for cellular metabolism.

17.3 Metabolic Engineering Strategies for Cyanobacteria Based Photosynthetic Production of Sucrose

In natural cyanobacterial strains, sucrose is mainly synthesized as a compatible solute to resist osmotic pressure imbalance caused by high salt stress. When the sucrose abundances accumulated in the cell are sufficient to rebalance the osmotic pressure between intracellular and extracellular environments, the concentrations of sucrose will not further increase, and its synthesis and degradation will be in a state of dynamic equilibrium. This natural sucrose metabolism regulation mode fundamentally limits the sucrose production capacities of cyanobacteria. As shown in Fig. 17.1, through systematic metabolic engineering manipulations, breaking the natural regulation mode of sucrose synthesis under salt stress is an important choice to strengthen the ability of sucrose synthesis and to enhance the application potentials of cyanobacterial photosynthetic cell factories for sucrose production.

17.3.1 Introduction of Sucrose Transporter

The key step to remove the bottleneck of sucrose synthesis in cyanobacteria is to achieve extracellular secretion of sucrose, and the most breakthrough progress in the development of cyanobacterial photosynthetic cell factories for sucrose production is obtained through the introduction of sucrose transporters. In 2012, Ducat et al. introduced the *cscB* gene from *E. coli* into PCC 7942 and realized the extracellular secretion of sucrose for the first time (Ducat et al. 2012a). CscB protein is a kind of sucrose permease that can transport protons/sucrose in the same direction. In *E. coli* cells, CscB protein promotes the absorption of sucrose and protons from the acidic environment, while the cultivation environment of cyanobacteria is usually alkaline, which is conducive to the simultaneously pumping out of sucrose and protons synthesized under salt stress. Ducat et al. used IPTG-induced P*tac* promoter to control the expression of *cscB* gene. Under 150 mM NaCl stress and 1 mM IPTG inducing condition, the sucrose synthesis efficiency of the engineered PCC 7942 strain reached 28 mg/L/h, and the yield reached 2.7 g/L after 168 h of salt stress cultivation (Ducat et al. 2012a).

In 2016, Song et al. engineered a fast-growing *Synechococcus* strain, UTEX 2973, which can grow rapidly under high temperature and high light, for photosynthetic production of sucrose. It was found that the introduction of *cscB* gene can also promote the secretion of sucrose in UTEX 2973 (Song et al. 2016). UTEX 2973 is a recently characterized cyanobacterial strain with great significant industrial

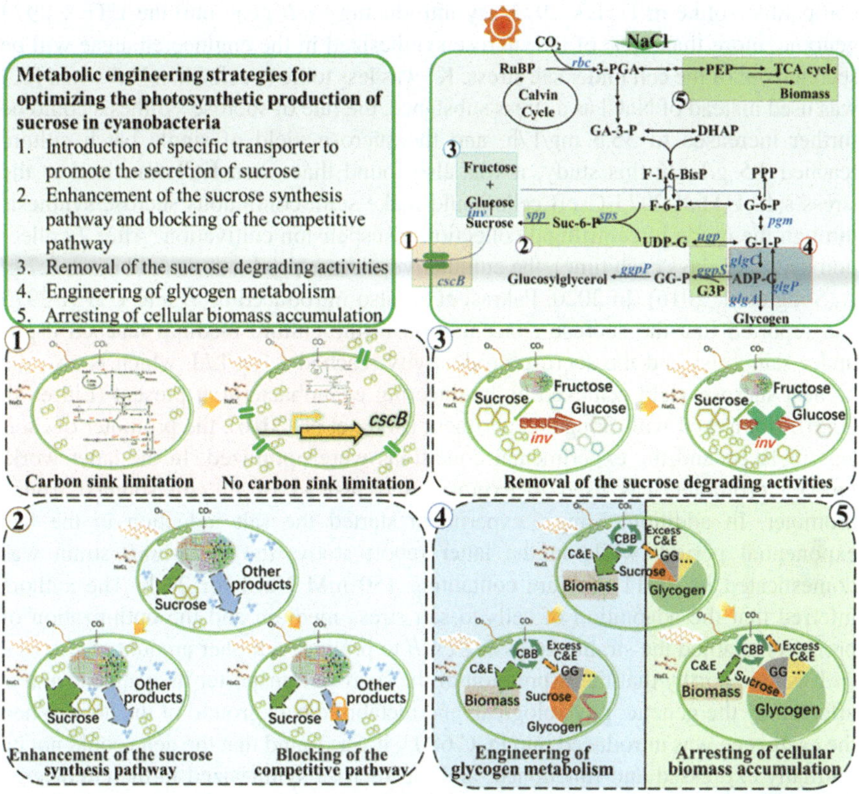

RuBP, Ribulose-1,5-Biphosphate; 3-PGA, 3-Phosphoglyceric acid; PEP, phosphoenolpyruvate; TCA cycle, tricarboxylic acid cycle; DHAP, dihydroxyacetone phosphate; GA-3-P, 3-phosphoglyceraldehyde; F-1,6-BisP, Fructose-1,6-biphosphate; F-6-P, Fructose-6-phosphate, G-6-P, glucose-6-phosphate; G-1-P, glucose-1-phosphate; ADP-G, ADP-Glucose; UDP-G, UDP-Glucose; Suc-6-P, sucrose-6-phosphate; GG-P, glucosylglycerol-phosphate.

Fig. 17.1 Metabolic engineering strategies to optimize sucrose synthesis in cyanobacteria. (1) Introduction of sucrose transporter to weaken carbon sink limitation; (2) Pathway engineering to strengthen sucrose synthesis and block the competitive products synthesis; (3) Removal of the sucrose degrading activities by knocking out the hydrolase gene *invA*; (4) Disturbance of glycogen metabolism; (5) Redirecting more photosynthetic carbon flux to metabolites production by blocking cell biomass synthesis pathway

properties. The genomic similarity between the strain and PCC 7942 is 99.8% (the difference only includes 55 SNPs, a 188.6 kb fragment position reversal, and a fragment deletion). However, the strain shows a significantly improved growth rate and adaptabilities to high temperatures and strong illuminations. Under high light (500 μmol photons m^{-2} s^{-1}) and high temperature (41 °C) cultivation conditions, the doubling time is only 2.1 h, far exceeding the previously reported various model strains (Yu et al. 2015). It was also found that the biomass of glycogen could reach about 50% of dry weight under the condition of 250 μmol photons m^{-2} s^{-1} and 38 °C. Under salt stress, sucrose was also synthesized and accumulated as the main compatible solute in UTEX 2973. By introducing *cscB* gene into the UTEX 2973 genome, more than 95% of the sucrose synthesized in the engineered algae will be secreted out of the cell under salt stress. K^+ was less toxic to cells than Na^+ when KCl was used instead of NaCl as a stress substance, the rate of sucrose synthesis could be further increased to 35.5 mg/L/h, and the sucrose yield of single batch culture reached 3.5 g/L. In this study, it was also found that when KCl was used as the stress salt, UTEX 2973-CscB cells could make semi-continuous sucrose synthesis through the mode of centrifugal collection-resuspension cultivation. After 7 collection cycles (3 days each time), the cumulative sucrose production could reach 8.7 g/L (Song et al. 2016). In 2020, Pakrasi et al. also introduced *cscB* into UTEX 2973 and reported that the sucrose concentration in the culture medium reached 8 g/L under salt stress and the sucrose productivity reached 1.9 g/L/d, which is also the highest sucrose yield achieved by engineering cyanobacteria at present (Lin et al. 2020). Compared with Song's experiment (Song et al. 2016), the promoter of *cscB* was replaced and the experimental conditions were optimized. In the latter work, *lacUV5* promoter was used to express *cscB* instead of *E. coli* sourced trp/lac promoter. In addition, Song's experiment started the salt induction in the late exponential period, while in the latter report study, the 2973-*cscB* strain was domesticated in BG11 medium containing 150 mM NaCl for 24 h. The authors inferred that the adaptation of cells to salt stress medium and the optimization of promoter enabled the strain expressing *cscB* to produce a higher amount of sucrose.

It is noteworthy that the expression of the sucrose transporter in cyanobacteria is affected by the genetic, physiological, and metabolic background of the host. When the *cscB* gene was introduced into PCC 6803, it was found that the gene could not be normally expressed and functioned, and the sucrose synthesized in the engineered strain could not be efficiently secreted out of the cell (Du et al. 2013). This indicates that for the development of photosynthetic cell factories for sucrose production in the future, the adaptability between the secreted protein and the chassis algae strain is also a problem to be taken into consideration. Improving the expression and activities of sucrose secreted proteins in specific cyanobacterial chassis strains through resource mining and enzyme engineering will become an important strategy to enhance sucrose secretion and strengthen sucrose synthesis.

17.3.2 Enhancing Sucrose Synthesis Pathway and Weakening the Degradation Pathways

In addition, to facilitate sucrose secretion, the modification of node genes involved in sucrose metabolism in cyanobacteria is also an important strategy to improve the efficiency of sucrose synthesis, and enhancing the sucrose synthesis pathway and inhibiting the sucrose degradation pathway are two main approaches. Du et al. found that enhancing the expression of key genes in the sucrose synthesis pathway was of great significance to improve sucrose production of PCC 6803 (Du et al. 2013). The co-expression of three genes, sucrose synthase *sps*, sucrose phosphate synthase *spp*, and UDP-glucose pyrophosphatase *ugp*, can increase the sucrose yield of the engineered algae strain by twofold, while blocking the key gene *ggpS* of glycerol glucoside synthesis, the competitive pathway of sucrose synthesis, can increase the sucrose yield of the algae strain by 1.5-fold. When the two strategies are combined, the sucrose yield of the obtained engineering algae strain would be increased by fourfold compared with the PCC 6803 wild-type strain (Du et al. 2013). Qiao et al. further reported that in PCC7942 strain carrying the *cscB* gene, overexpression of the sucrose synthase gene *sps* could increase the sucrose yield of the engineered strain by 74%. In addition, in the strain overexpressing both *cscB* and *glgC*, the overexpression of *sps* can increase the sucrose synthesis capacities of the recombinant cells by threefold (from 590 mg/L to more than 1760 mg/L) (Qiao et al. 2018). For cyanobacterial algal strains under salt stress, when there is sufficient carbon source supply, the catalytic activity of the sucrose synthesis pathway becomes the rate-limiting step of sucrose synthesis, and enhancing the expression of key proteins in the synthesis pathway can effectively improve the production efficiency and actual yield of sucrose.

As mentioned earlier, sucrose, as a natural compatible solute in cyanobacterial cells, has an endogenous balance mechanism for its own metabolism. The presence of sucrose hydrolase invertase can rapidly degrade the excessive accumulation of sucrose into glucose and fructose and then undergo phosphorylation and enter the central metabolism. In 2012, Ducat et al. group demonstrated for the first time that knocking out the hydrolase gene *invA* in the PCC 7942 algae strain into which the *cscB* gene was introduced could further increase sucrose production by 15%. When the knockout of *invA* is combined with the knockout of the glycogen synthesis node gene *glgC*, the final sucrose yield is increased by 25% (Ducat et al. 2012a). Other studies also found that knocking out the hydrolase gene in PCC 6803 engineering algae strain (Du et al. 2013) with an enhanced sucrose synthesis pathway can further increase sucrose yield by 40% (Kirsch et al. 2018).

17.3.3 Disturbance of Glycogen Metabolism

A common feature of photosynthetic microalgae is the presence of natural carbon sink mechanisms within the cells to store the carbon source and energy fixed by the Calvin cycle, which are beyond the requirements of normal cell growth and maintenance. The most important and representative carbon sink mechanism in cyanobacteria is glycogen metabolism. The existence of glycogen metabolism is of great significance for cyanobacteria to resist environmental stress and adapt to the dynamic changes of nutrients. However, for the photosynthetic production of chemicals, glycogen synthesis is generally regarded as a major competitive route (Zhou et al. 2016). Xu et al. reported that by knocking out two glycogen synthase genes (*glgAI* and *glgAII*), glycogen synthesis and accumulation in PCC 7002 cells can be completely blocked. And when mutant strains lacking glycogen synthesis capacity encounter salt stress, the contents of sucrose and glycerol glucoside accumulated in the cells would be increased, the content of sucrose being increased by about threefold (Xu et al. 2013). In PCC 7942 recombinant strain carrying *cscB* gene, Ducat et al. knocked out *glgC* gene encoding ADP glucose pyrophosphorylase, which was the rate-limiting enzyme of glycogen synthesis, and increased the sucrose production of engineering strain by 5%–10% (Ducat et al. 2012a). However, it should be noticed that the knockout of the *glgC* gene would cause significant impacts on the cellular physiological robustness. Under salt stress, the doubling time of the mutant cells was extended from 12 h to 43.5 h, and even after domestication, it would still reach 20 h (Ducat et al. 2012a). Considering the comprehensive benefits of the whole process of carbon fixation and sugar production by photosynthetic production, whether it is reasonable to block glycogen synthesis by knocking out *glgC* gene remains to be evaluated.

Qiao et al. proposed a different role of glycogen metabolism in salt-stress induced sucrose synthesis and reported that in a two-stage mode of sucrose production (cyanobacteria cells would be cultivated to late logarithm phase and then treated with salts stress for sucrose production), glycogen synthesis and glycogen contents were positively related to final sucrose titers (Qiao et al. 2018). The different discoveries comparing with the previous reports might be caused by the unique cultivation mode. In Ducat's experiment, NaCl stress was performed since the beginning of the culture, meaning that sucrose synthesis and cell biomass accumulation in the engineered strain is synchronized and coupled. In this mode, blocking glycogen accumulation will force cells to distribute "spilled" carbon flow and energy flow to sucrose synthesis, an "alternative" carbon sink pathway, so that sucrose synthesis could be increased. While in Qiao's work, cyanobacteria cells are cultivated to the end of the logarithm phase and then subjected to salt stress. At this time, the biomass accumulation of engineered strain has basically stopped, and the carbon source already stored in glycogen can be used as a "reserve carbon pool," supplemental carbon flow other than the Calvin cycle is provided for sucrose synthesis, thus improving sucrose synthesis. The comparison between the two groups also shows that in the design and development of the cyanobacterial photosynthetic cell factories, it is of great significance to improve the adaptation degree between the natural carbon sink mechanism of cells and the artificial carbon sink pathway and to reasonably optimize the carbon flow distribution according to different environmental, physiological and metabolic conditions.

17.3.4 Biomass Accumulation Arresting Strategy

Similar to the glycogen metabolism disturbance strategy, arresting the accumulation of cellular biomass so as to maximize the photosynthetic carbon flow to target metabolites is a recently proposed strategy for engineering cyanobacterial photosynthetic cell factories. Ducat et al. realized the effective inhibition of biomass accumulation in engineering algal cells through overexpression of an important responsive regulatory factor RpaB (Regulator of Phycobilisome-Associated B) in PCC 7942, which means that photosynthetic carbon flow leading to cell biomass synthesis pathway is blocked. Under these conditions, the photosynthetic growth of the mutant cells will be severely inhibited, and the photosynthetic efficiency will be greatly reduced. When the induced overexpression of sucrose synthase *sps* is carried out simultaneously in *rpaB* overexpressing algae strains, feedback inhibition of photosynthesis in engineering algae strains can be eliminated due to the new "outlet" of photosynthetic carbon flow, and sucrose synthesis efficiency is increased by twofold (Abramson et al. 2018). Recently, researchers from the Delft University of Technology applied a similar strategy to the development of cyanobacterial cell factories for ethanol photosynthetic production and achieved good results, proving that optimizing carbon flow control and distribution by limiting biomass accumulation has a good application prospect in the development of cyanobacterial photosynthetic cell factories (Kiyan et al. 2018).

17.3.5 Reform Photosynthetic Electron Flux of Cyanobacteria

As photoautotrophs, cyanobacteria evolved to have diverse regulatory mechanisms to cope with rapid changes in environmental conditions, so as to protect photosystems from physiological impairments, among which alternative electron transfer pathways are one of the most essential and representative mechanisms. Cyanobacteria uses C-type flavodiferritin protein (FDP) as a powerful photoprotective electronic compound, which permits redundant photosynthetic electrons from PSI downstream to O_2 as part of the strategy to adapt to different environmental conditions (Kati et al. 2019). Although it is important to adjust the distribution of electron flux between photosystems, the solar energy obtained by cells will be distributed to many cell functions other than the synthesis of the desired products, thus reducing the yield of target end-products in engineered cyanobacterial cells.

Kati et al. found that under different light conditions, the deletion and inactivation of Flv1/3 (an alternative acceptor flavodiiron protein) in the engineered PCC 6803 strain, affected the production of sucrose, glycogen, and related photosynthetic gas flux (Kati et al. 2019). In strains lacking Flv1/3, the excited electrons generated by photosynthetic water division can be rewired to increase the relative metabolic flux toward the target product, such as sucrose. Compared with the control strain, the accumulation of total sucrose increased about threefold under low light (50 μmol photons m^{-2} s^{-1}). However, the sucrose production of this strain decreased under high light (200 μmol photons m^{-2} s^{-1}), which means that the adaptability of the cells without Flv1/3 may be impaired when the cultivation broth getting close to the phase of high culture density. In the same work, additional genetic modifications were introduced to enhance the sucrose pathway. The deletion of *ggpS* (essential for the formation of glucosylglycerol) is carried out simultaneously in *sps* and *cscB* overexpressing strains to enhance the overall sucrose productivity. Interestingly, the deletion of Flv1/3 resulted in a decrease in sucrose production, which indicated that sucrose was not the best final product of deficient strains under that experimental condition.

Although the elimination of the native Flv1/3 reaction can improve photosynthetic production of the required products in cyanobacteria under certain conditions, careful selection of engineered targeting pathways consistent with intracellular redox balance must be carried out.

17.3.6 Prospect of Metabolic Engineering Strategies for Sugar Production by Cyanobacteria

For the development of microbial cell factories, the control and optimization of metabolic flow are extremely important and effective strategies. Enhancing the distribution of carbon flow and energy flow to target metabolites and improving the atom economy of the whole synthesis route are the keys to improve the economic competitiveness of biosynthetic and biorefinery technology systems. However, from the perspective of metabolic flow, the photosynthetic carbon flow allocated to sucrose synthesis can reach about 60% of the total photosynthetic carbon fixation and the carbon flow ratio used for cell biomass synthesis only accounts for about 40% in the present cell factory with high synthesis level (UTEX 2973 overexpressing *cscB*). In addition, glycogen synthesis, the main natural carbon sink mechanism in cyanobacteria and generally recognized as the carbon pool for mobilization in cyanobacteria metabolic engineering, generally accounts for 30–50% of dry cell weight and 5–20% of total photosynthetic carbon sequestration (Song et al. 2016). From this point of view, aiming to optimize the technology of cyanobacterial photosynthetic sucrose production, the key to further improve the rate of sucrose synthesis and to enhance the application potential of the technology lies in "opening sources" rather than "cutting expenses." Through the optimization of carbon source distribution mode, it is difficult to achieve further significant improvement in the sucrose synthesis performances, and it might impair the robustness and fitness of normal cell growth and metabolism. Excessive emphasis on the increase of carbon flow proportion of the target product at the expense of the reduction of growth speed and photosynthetic activity often results in the reduction of the efficiency of the whole photosynthetic production process. The breakthrough in the synthesis level of the cyanobacterial photosynthetic production of sucrose technology should rely on the improvement of the photosynthetic production efficiency of cyanobacteria chassis cells and engineered strains. Fundamental breakthroughs on understanding and engineering the mechanism bottleneck limiting the efficiency and activity of natural photosynthetic carbon fixation system would be a prerequisite to achieving a substantial increase in sucrose yield through the overall strengthening of the fixed carbon flow in photosynthesis.

17.4 Synthetic Light-Driven Consortia Based on Cyanobacterial Photosynthetic Sucrose Production

17.4.1 The Proof of Concept of Synthetic Light-Driven Consortia System

As mentioned earlier, the introduction of sucrose secretory proteins promoted the successful development of efficient cyanobacterial photosynthetic cell factories for sucrose production, which realized the direct synthesis and secretion of sucrose from solar energy and carbon dioxide in the environment based on a single photosynthetic platform. However, the effective secretion of sucrose also increases the risk of biocontamination of environment-sourced heterotrophic microorganisms. Besides the adoption of strict selective cultivation strategies to inhibit the invasion and proliferation of the contaminating microorganisms, simulating the symbiotic system of autotrophs and heterotrophs such as lichens under natural conditions provides an alternative solution by coculturing the sucrose-secreting cyanobacteria and heterotrophic microorganisms that can use sucrose as carbon source. The artificially constructed light-driven consortia facilitated in situ utilization of the photosynthetic synthesized sucrose and the further expansion and elongation of the photosynthetic metabolism network.

In 2012, Ducat et al. introduced the *E. coli* sourced *cscB* gene into PCC 7942 and realized the secretion of sucrose under salt stress. For the first time, they explored the possibility of survival and growth of *S. cerevisiae* utilizing the cyanobacteria synthesized sucrose. It is confirmed that the BG11 medium utilized for PCC 7942 cultivation could also support the growth of *S. cerevisiae* cells after supplementation of nitrogen sources and 2% sucrose. Furthermore, it was found that in the culture system of sucrose-producing cyanobacteria (*cscB*-expressing PCC7942), the sucrose synthesized and secreted by cyanobacterial cells could maintain the survival of *S. cereviсiae* and division for at least twice (Ducat et al. 2012a). These results preliminarily confirmed the feasibility of maintaining the synthetic light-driven consortia based on cyanobacterial photosynthetic carbon sequestration and sugar production system.

17.4.2 Development and Application of Synthetic Light-Driven Consortia Synthesis System

Ducat et al. further explored the feasibility of coculturing the sucrose-synthesizing cyanobacterial cell factories with three classical heterotrophic industrial microorganisms, *E. coli*, *B. subtilis*, and *S. cerevisiae*, respectively (Hays et al. 2017). The results show that the constructed light-driven consortia can be maintained stably for long terms ranging from weeks to months and show strong robustness to fluctuations of light intensities and light rhythms. In the constructed consortia, the survival and growth of heterotrophic microorganisms are completely dependent on the sucrose secreted by cyanobacteria. It is noteworthy that the

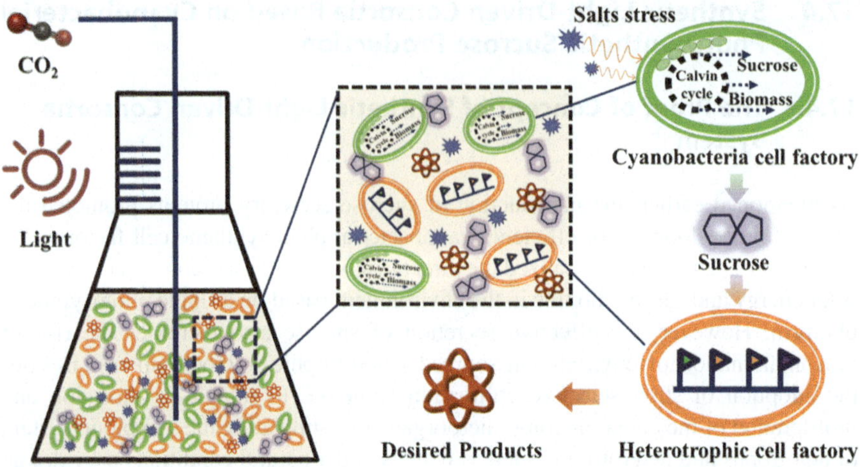

Fig. 17.2 Schematic representation of the artificial light-driven consortium based on cyanobacterial photosynthetic production of sucrose. Co-cultivation of sucrose synthesizing cyanobacterial cell factories and heterotrophic microbial cell factories can facilitate in situ utilization of the secreted sucrose and the synthesis of desired products with CO_2 and light as the primary inputs

metabolism of the heterotrophs can reversely promote the photosynthetic carbon fixation and growth of sugar-producing cyanobacteria cells and therefore further enhance the stability and robustness of the consortia. As shown in Fig. 17.2, when the wild-type strain was replaced by heterotrophic microbial cell factories, the synthetic photosynthetic consortia system can also support the biosynthesis of high value-added chemicals (alpha-amylase synthesized by *B. subtilis* and polyhydroxybutyrate synthesized by *E. coli*) (Hays et al. 2017).

In 2017, the research team further optimized the synthetic light-driven consortia by using sodium alginate as the matrix to coat sugar-producing cyanobacterial cells. The alginate encapsulation has little effect on the survival and sugar-producing activities of the cyanobacterial cells, while can greatly limit cell proliferation, so that the portion of photosynthetic carbon flow oriented to sucrose synthesis could be maximized in the encapsulated cells comparing with the planktonic cells, the sucrose production efficiency of sodium alginate coated cells was increased by two- to threefold (Weiss et al. 2017). When the alginate encapsulated cyanobacterial sucrose-producing cells were cocultivated with *Halomonas boliviensis*, a natural PHB-synthesizing microorganism, the updated light-driven consortia showed significantly improved resistance to environmental disturbance and biological pollution. Continual accumulation of total biomass (mainly *H. boliviensis*) and target product (PHB) could be maintained during long-term cultivations of up to 5 months, during which no exogenous antibiotics and other selective strategies were needed. In addition, comparing with planktonic cyanobacterial cells, there is a significant difference in the sedimentation coefficient between sodium alginate–coated cells and *H. boliviensis* cells, which means that the biomass of *H. boliviensis* cells containing PHB can be selectively collected from coculture system. The researchers also confirmed that the sodium alginate coating can also be used for coculture of *E. coli* derived cell factories that can synthesize PHB, indicating that this strategy could be universally adopted (Weiss et al. 2017).

It is noteworthy that the synthetic light-driven consortia synthesis system derived from the existing sucrose synthesizing cyanobacterial cell factories requires that the heterotrophic microbial cell factories must have the ability to utilize sucrose. Heterotrophic cells deficient in sucrose absorbing and metabolizing capacities could not be used for the construction of the light-driven consortia before getting a tailored genetic modification of the sugar utilizing a metabolism network. Comparing with other heterotrophic microorganisms as mentioned earlier, *Pseudomonas aeruginosa*, generally recognized as a promising biotechnological chassis with significant tolerances to abiotic stress, has the best adaptability to the medium used by sugar-producing cyanobacteria, requiring only small amounts of nutritional supplement, which is very suitable for cocultivation [43]. However, natural *P. putida* strains could not use sucrose as carbon source. To solve this problem, Lowe et al. introduced *E. coli*-sourced *cscA-cscB* module into the *P. putida* to engineer the sucrose utilization capacities. In the new consortia consisting of the engineered *P. putida* and sucrose-synthesizing PCC 7942, *P. putida*-secreted CscA enzyme would hydrolyze extracellular sucrose into glucose and fructose, while the CscB could promote the absorption of sugars, and finally facilitate synthesis and accumulation of PHB in *P. putida* cells (Löwe et al. 2017).

Recently, Ducat et al. expanded the application of the optimized synthetic light-driven consortia system to the area of bioremediation and proofed the concept of the coupled bioremediation and biosynthesis. As introduced earlier, *P. putida* was engineered to consume sucrose by the introduction of the *cscA-cscB* system. Besides, the engineered *P. putida* could also be endowed with the capacities to degrade 2,4-DNT by introducing the DNT gene cluster required for 2,4-DNT biotransformation from *Burkholderia* sp. R34. Meanwhile, the researchers confirmed that the growth and sucrose synthesis capacity of alginate-encapsulated *cscB*-expressing PCC 7942 were almost unaffected by the 2,4-DNT toxicities. Through the cocultivation of alginate-encapsulated *cscB*-expressing PCC 7942 and engineered *Pseudomonas putida* (*P. putida*) that can utilize sucrose and degrade the environmental pollutant 2,4-dinitrotoluene (2,4-DNT), a synthetic light-driven consortia synthesis system for bioremediation was constructed. And the synthetic light-driven consortia could convert the industrial pollutant 2,4-DNT over an extended time range with CO_2 and light as the primary inputs. Furthermore, the researchers further explored the possibility of both removing 2,4-DNT and simultaneously producing a valuable product polyhydroxyalkanoate (PHA), by culturing the synthetic consortia in a low nitrogen medium although the specific volumetric productivities were only ca. 5 mg PHA/L/d (Fedeson et al. 2020).

In some cases, the heterotrophic microbial cell factories have the ability to utilize high concentrations of sucrose, which also needs to be improved because the sucrose yield of cyanobacteria in the cocultivation system was not sufficient. Li et al. constructed a light-driven consortium including the fast-growing cyanobacterium

S. elongatus UTEX 2973 and *E. coli* to produce fine chemical 3-hydroxy-propionic acid (3-HP) using CO_2 and light as the main input. The sucrose-secreting fast-growing cyanobacterium *S. elongatus* UTEX 2973 strain with a productivity of 612.0 mg/L in 6 days was first obtained by introducing the *cscB* gene into the *S. elongatus* UTEX 2973. Then, the malonyl-CoA-dependent 3-HP biosynthetic pathway was introduced into *E. coli* to realize the synthesis of 3-HP. However, it was reported that *E. coli* can only grow under a minimal sucrose concentration of 1.2 g/L, which is much higher than the productivity of *S. elongatus* UTEX *cscB*$^+$ 2973. Therefore, essential genes for efficient sucrose metabolism consisting of *cscB* (ECW_m2594), *cscK* (ECW_m2595), and *cscA* (ECW_m2596) were introduced into *E. coli* to improve the efficiency of sucrose utilization. Eventually, a synthetic light-driven consortia synthesis system consisting of the fast-growing *S. elongatus* UTEX *cscB*$^+$ 2973 and engineered *E. coli* was successfully constructed, which can convert sucrose to 3-HP in one step under photoautotrophic growth conditions (Zhang et al. 2020).

17.4.3 Engineering and Understanding the Mutual Interaction Mechanisms in the Synthetic Light-Driven Consortia

Most of the artificially constructed light-driven consortia are one-way supported, meaning that energy and materials supporting the consortia were completely based on cyanobacterial photosynthesis. Recently, Smith et al. designed a two-way cocultivation system, using a widely used sucrose-producing cyanobacteria strain (PCC 7942 overexpressing *cscB*) and a diazotrophic microorganism, *Azotobacter vinelandii*, to construct consortia based on a new mutualism relationship (Smith and Francis 2016). In this cocultivation system, salts stress would force the *cscB*-expressing PCC 7942 strain to synthesize and secret sucrose, providing organic carbon source to diazotrophic microorganism, *A. vinelandii*, while *A. vinelandii* would grow with sucrose as the sole carbon source and provide organic nitrogen sources for PCC 7942 cells. Compared with the previously reported light-driven consortia working in a unidirectional feeding mode, this bidirectional mutual beneficial interaction between the autotrophs and heterotrophs in the light-driven consortia could be maintained with basic nutrients and can synthesize valuable metabolites without the addition of any organic nitrogen and carbon sources (Smith and Francis 2016).

The cocultivation of cyanobacterial photosynthetic sucrose-synthesizing cell factories with heterotrophic microbial cells synthesizing biochemicals provides an artificial consortia solution realizing the full chain for high-value utilization of carbon dioxide (Li et al. 2017). Comparing with the single-platform synthesis mode (assembling the whole synthetic pathways for the final product in single cyanobacteria strain), the artificial constructed light-driven consortia distribute and buffers the metabolic and physiological burden by organically integrating different microbial components with diversified metabolic and physiological characteristics. This strategy would be conducive to the realization of a stable and sustainable biosynthesis process. Theoretically, a certain portion of the carbon flow fixed by cyanobacterial photosynthesis will be redirected to the biomass accumulation of heterotrophic cells, resulting in a decrease of the cell numbers performing solar energy utilization and carbon fixation. However, such a loss can be partially compensated by the "back-feeding" effects of heterotrophic microbial metabolism on cyanobacteria growth and metabolism in the cocultivation system. Previously, it has been reported that inoculating heterotrophic microorganisms from the respective natural environment into the microalgae culture broth can effectively promote the photosynthetic growth and metabolic activity of microalgae cells. For the development of synthetic light-driven consortia, it was also found that cocultivation with heterotrophic microorganisms (including *E. coli*, *B. subtilis*, and *S. cerevisiae*) had significant promoting effects on the growth and stress resistance of cyanobacterial cells. Although detailed mechanisms are yet to be disclosed, potential mutual interactions have been found. Li et al. reported that active photosynthesis of cyanobacteria in high-density culture would lead to the accumulation of reactive oxygen species in the culture medium and further to the inhibition of the cyanobacteria growth. In a cocultivation system, *Rhodotorula glutinis* cells could effectively eliminate the ROS, reduce the physiological impairments of the cyanobacteria cells, and facilitate better growths (Li et al. 2017). Similarly, Li et al. found that the cell growth of a sucrose-synthesizing strain of *S. elongatus* UTEX 2973 would be enhanced when cultivated in a light-driven consortium with a 3-hydroxy-propionic acid-producing *E. coli* strain. A possible mechanism could be that the engineered *E. coli* cells might quickly quench the reactive oxygen species (ROS) in the cocultivation system and relieve the oxygenic stress to the cyanobacteria, thus leading to improved photosynthesis activities (Zhang et al. 2020). In the future, it is necessary to further explore the design principles and effective regulation schemes of artificial consortia, aiming to achieve metabolic complementarity and mutual benefit between photoautotrophic and heterotrophic microorganisms (Luan and Lu 2018).

17.5 Summary and Prospect

Globally, cyanobacteria provide about 20% of the total organic carbon in the biosphere through efficient photosynthesis and are extremely important primary productivity (Flombaum et al. 2013; Rousseaux and Gregg 2014). Synthetic biology and metabolic engineering technology have made the process to be more intensive and controllable through artificial modifications and regulations at the levels of protein, pathway, and modules. As a result, solar energy and carbon dioxide could be directly converted into energetic organic molecules in photoautotrophic cell factories. On this basis, the development of photosynthetic biomanufacturing technology is to achieve the efficient, high-throughput, and directional transformation of energy and matter in tailored cyanobacteria and microalgae cell populations in large scales and longtime courses. Cyanobacterial photosynthetic carbon sequestration and sucrose synthesis technology is not only a representative photosynthetic biological manufacturing technology but also a promising route for the supply of sugar feedstock for traditional biorefinery technology. Compared with the traditional route of "plant-planting > biomass-collection > raw material-pretreatment > sugar extraction," the route of "microalgae cultivation–sugar–production" provided a simpler procedure and the final product is clearer. The cultivation of cyanobacteria for sugar production could be performed on deserted and marginal lands, like saline soil and tidal flat, reaching the effects of "non-competition for foods with people, non-competition for land with grain." Previously, it is proposed that when sucrose could be synthesized at the rate of 36 mg/L/h by cyanobacteria cell factories under the large-scale cultivation system, the total productivity of 55 tons of sucrose per year could be achieved, which would far exceed the actual productivity of sugarcane planting in the same area (Ducat et al. 2012a). However, to promote the industrial application of cyanobacterial photosynthetic sucrose production technology, there are still many problems to be solved.

First, the detailed regulatory mechanisms of salt stress-responsive sucrose synthesis in cyanobacteria are yet to be disclosed. Although salt stress independent sucrose production has been achieved with engineered UTEX 2973 strain carrying PCC 6803 sourced *sps* and *spp* gene, the yield is still lower than that obtained under salt stress. At present, to achieve high performance of photosynthetic sucrose production in cyanobacteria, salt stress is still a more general and popular model to be adopted, which increases the technical complexities for light-driven carbon sequestration and sugar production. To reduce the dependence on salt stress and maintain an efficient sucrose production process, systematic physiological and biochemical assays should be performed to reveal the induction and activation mechanism for the sucrose biosynthesis pathway and the essential enzymes.

Second, the stress tolerance and other industrial properties of cyanobacterial chassis cells and recombinant strains for sugar production still need to be improved. Compared with the stable cultivation conditions in the laboratory, stressful environmental factors, including high temperature, strong illuminations, extreme pH, and so on may be encountered in the large-scale cultivation of cyanobacteria cells in the industrial process. In particular, the extracellular sucrose accumulation would further increase the risk of biological contamination. Thus, various selective strategies have been adopted to restrict biological pollutions, which would make the robustness of cyanobacteria strains more necessary for feasibilities of the photosynthetic sugar production technology. Physiological robustness of several important model cyanobacteria strains, PCC 6803, PCC 7942, *Nostoc* sp. PCC 7120, *Synechococcus* sp. PCC 7002, generally utilized for the construction of photosynthetic cell factories cannot meet the requirements from large-scale cultivation in outdoor conditions. Thus, the screening and development of next-generation chassis cells with stronger adaptabilities to environmental stress and industrial conditions or comprehensive metabolic engineering manipulations to enhance the robustness of the typical chassis cells will be a prerequisite to develop cyanobacterial cell factories for sugar production in the future (Luan and Lu 2018).

Third, strategies and instruments facilitating efficient separation and harvest of sucrose from the cyanobacterial cultivation broth are yet to be developed. As mentioned earlier, although the strategies of developing synthetic light-driven consortia can achieve the effects of in situ utilization of secreted sucrose and partially solve the problem of potential biocontamination caused by the accumulation of sugars. However, considering the requirements to scale up the technology of cyanobacterial photosynthetic carbon sequestration and production of sugars in the future in the industrial process, the development of convenient and cost-effective systems that enable efficient sucrose recovery in large volumes and long terms would be an urgent issue. The development of strong and specific adsorbent resin and permeable membrane could be expected to provide promising solutions.

In the future, through the systematic adoption of the strategies and tools of synthetic biology, system biology, and process engineering technology, the detailed genetic and metabolic mechanisms of sugar production in cyanobacteria cells would be disclosed, and the bottlenecks holding control over the photosynthetic conversion efficiency from intracellular materials and energy to sucrose would be removed. Combining the innovative process and equipment, significantly updated technologies and industries of photosynthetic carbon sequestration and sugar production based on advanced cyanobacterial cell factories could be expected.

References

Abramson BW, Josh L, Lin YT, Emily J, Ducat DC (2018) Redirecting carbon to bioproduction via a growth arrest switch in a sucrose-secreting cyanobacterium. Algal Res 33:248–255

Allakhverdiev SI, Sakamoto A, Nishiyama Y, Inaba M, Murata N (2000a) Ionic and osmotic effects of NaCl-induced inactivation of photosystems I and II in Synechococcus sp. Plant Physiol 123 (3):1047–1056

Allakhverdiev SI, Sakamoto A, Nishiyama Y, Murata N (2000b) Inactivation of Photosystems I and II in response to osmotic stress in synechococcus. Contribution of water channels. Plant Physiol 122(4):1201–1208

Ball SG, Morell MK (2003) From bacterial glycogen to starch: understanding the biogenesis of the plant starch granule. Annu Rev Plant Biol 54(1):207–233

Blank CE (2013) Phylogenetic distribution of compatible solute synthesis genes support a freshwater origin for cyanobacteria. J Phycol 49(5):880–895

Blumwald E, Mehlhorn RJ, Packer L (1983) Studies of osmoregulation in salt adaptation of cyanobacteria with ESR spin-probe techniques. Proc Natl Acad Sci U S A 80(9):2599–2602

Brown AD, Simpson JR (1972) Water relations of sugar-tolerant yeasts: the role of intracellular polyols. Microbiology 72(3):589–591

Cumino AC, Perezcenci M, Giarrocco LE, Salerno GL (2010) The proteins involved in sucrose synthesis in the marine cyanobacterium Synechococcus sp. PCC 7002 are encoded by two genes transcribed from a gene cluster. FEBS Lett 584(22):4655–4660

Curatti L, Porchia AC, Herreraestrella L, Salerno GL (2000) A prokaryotic sucrose synthase gene (susA) isolated from a filamentous nitrogen-fixing cyanobacterium encodes a protein similar to those of plants. Planta 211(5):729–735

Curatti L, Giarrocco LE, Salerno GL (2006) Sucrose synthase and RuBisCo expression is similarly regulated by the nitrogen source in the nitrogen-fixing cyanobacterium Anabaena sp. Planta 223 (5):891–900

Damrow R, Maldener I, Zilliges Y (2016) The multiple functions of common microbial carbon polymers, glycogen and PHB, during stress responses in the non-diazotrophic cyanobacterium Synechocystis sp. PCC 6803. Front Microbiol 7:966. https://doi.org/10.3389/fmicb.2016.00966

Desplats P, Folco EJ, Salerno GL (2005) Sucrose may play an additional role to that of an osmolyte in Synechocystis sp. PCC 6803 salt-shocked cells. Plant Physiol Biochem 43(2):133–138

Dien BS, Cotta MA, Jeffries TW (2003) Bacteria engineered for fuel ethanol production: current status. Appl Microbiol Biotechnol 63(3):258–266

Du W, Liang F, Duan Y, Tan X, Lu X (2013) Exploring the photosynthetic production capacity of sucrose by cyanobacteria. Metab Eng 19:17–25

Ducat DC, Avelar-Rivas JA, Way JC, Silver PA (2012a) Rerouting carbon flux to enhance photosynthetic productivity. Appl Environ Microbiol 78(8):2660–2668. https://doi.org/10.1128/AEM.07901-11

Ducat DC, Avelarrivas JA, Way JC, Silver PA (2012b) Rerouting carbon flux to enhance photosynthetic productivity. Appl Environ Microbiol 78(8):2660–2668

Ehira S, Kimura S, Miyazaki S, Ohmori M (2014) Sucrose synthesis in the nitrogen-fixing cyanobacterium Anabaena sp. strain PCC 7120 is controlled by the two-component response regulator OrrA. Appl Environ Microbiol 80(18):5672–5679

Erdmann N (1983) Organic osmoregulatory solutes in blue-green algae. Z Pflanzenphysiol 110 (2):147–155

Fedeson DT, Saake P, Calero P, Nikel PI, Ducat DC (2020) Biotransformation of 2,4-dinitrotoluene in a phototrophic co-culture of engineered *Synechococcus elongatus* and *Pseudomonas putida*. Microb Biotechnol 13(4):997–1011. https://doi.org/10.1111/1751-7915.13544

Flombaum P, Gallegos JL, Gordillo RA, Rincón J, Zabala LL, Jiao N, Karl DM, Li WK, Lomas MW, Veneziano D, Vera CS, Vrugt JA, Martiny AC (2013) Present and future global distributions of the marine cyanobacteria prochlorococcus and synechococcus. Proc Natl Acad Sci U S A 110(24):9824–9829. https://doi.org/10.1073/pnas.1307701110

Ginerlamia J, Lopezmaury L, Reyes JC, Florencio FJ (2012) The CopRS two-component system is responsible for resistance to copper in the cyanobacterium synechocystis sp. PCC 6803. Plant Physiol 159(4):1806–1818

Guerra LT, Xu Y, Bennette N, Mcneely K, Bryant DA, Dismukes GC (2013) Natural osmolytes are much less effective substrates than glycogen for catabolic energy production in the marine cyanobacterium Synechococcus sp. strain PCC 7002. J Biotechnol 166(3):65–75

Hagemann M (2011) Molecular biology of cyanobacterial salt acclimation. FEMS Microbiol Rev 35(1):87–123

Hays SG, Ducat DC (2015) Engineering cyanobacteria as photosynthetic feedstock factories. Photosynth Res 123(3):285–295

Hays SG, Yan LLW, Silver PA (2017) Synthetic photosynthetic consortia define interactions leading to robustness and photoproduction. J Biol Eng 11:4

Hendry JI, Prasannan CB, Ma F, Mollers KB, Jaiswal D, Digmurti MG, Allen DK, Frigaard N, Dasgupta S, Wangikar PP (2017) Rerouting of carbon flux in a glycogen mutant of cyanobacteria assessed via isotopically non-stationary 13C metabolic flux analysis†. Biotechnol Bioeng 114(10):2298–2308

Kanesaki Y, Suzuki I, Allakhverdiev SI, Mikami K, Murata N (2002) Salt stress and hyperosmotic stress regulate the expression of different sets of genes in Synechocystis sp. PCC 6803. Biochem Biophys Res Commun 290(1):339–348

Kati T, Pekka P, Csaba N, Duncan F, Nicolas P (2019) Redirecting photosynthetic electron flux in the cyanobacterium Synechocystis sp. PCC 6803 by the deletion of flavodiiron protein Flv3. Microb Cell Factories 18(1):189–189

Kirsch F, Luo Q, Lu X, Hagemann M (2018) Inactivation of invertase enhances sucrose production in the cyanobacterium Synechocystis sp. PCC 6803. Microbiology 164(10):1220–1228

Kirsch F, Klähn S, Hagemann M (2019) Salt-regulated accumulation of the compatible solutes sucrose and glucosylglycerol in cyanobacteria and its biotechnological potential. Front Microbiol 10:2139

Kiyan S, Josefine A, Emil L, Michael J, Lun Y, Hudson EP (2018) Targeted repression of essential genes to arrest growth and increase carbon partitioning and biofuel titers in cyanobacteria. ACS Synth Biol 7(7):1669–1675

Klähn S, Steglich C, Hess WR, Hagemann M (2010) Glucosylglycerate: a secondary compatible solute common to marine cyanobacteria from nitrogen-poor environments. Environ Microbiol 12(1):83–94

Kolman MA, Nishi CN, Perez-Cenci M, Salerno GL (2015) Sucrose in cyanobacteria: from a salt-response molecule to play a key role in nitrogen fixation. Life (Basel) 5(1):102–126. https://doi.org/10.3390/life5010102

Levina N, TãTemeyer S, Stokes NR, Louis P, Jones MA, Booth IR (2014) Protection of Escherichia coli cells against extreme turgor by activation of MscS and MscL mechanosensitive channels: identification of genes required for MscS activity. EMBO J 18(7):1730–1737

Li T, Li CT, Butler K, Hays SG, Guarnieri MT, Oyler GA, Betenbaugh MJ (2017) Mimicking lichens: incorporation of yeast strains together with sucrose-secreting cyanobacteria improves survival, growth, ROS removal, and lipid production in a stable mutualistic co-culture production platform. Biotechnol Biofuels 10:55. https://doi.org/10.1186/s13068-017-0736-x

Liang C, Zhang X, Chi X, Guan X, Li Y, Qin S, Shao H (2011) Serine/threonine protein kinase SpkG is a candidate for high salt resistance in the unicellular cyanobacterium synechocystis sp PCC 6803. PLoS One 6(5):e18718

Liang Y, Zhang M, Wang M, Zhang W, Qiao C, Luo Q, Lu X (2020) Freshwater cyanobacterium *Synechococcus elongatus* PCC 7942 adapts to an environment with salt stress via ion-induced enzymatic balance of compatible solutes. Appl Environ Microbiol 86(7):e02904–e02919. https://doi.org/10.1128/aem.02904-19

Lin PC, Zhang F, Pakrasi HB (2020) Enhanced production of sucrose in the fast-growing cyanobacterium *Synechococcus elongatus* UTEX 2973. Sci Rep 10(1):390. https://doi.org/10.1038/s41598-019-57319-5

Liu ZX, Li HC, Wei YP, Chu WY, Chong YL, Long XH, Liu ZP, Qin S, Shao H (2015) Signal transduction pathways in Synechocystis sp. PCC 6803 and biotechnological implications under abiotic stress. Crit Rev Biotechnol 35(2):269–280

Löwe H, Hobmeier K, Moos M, Kremling A, Pflüger-Grau K (2017) Photoautotrophic production of polyhydroxyalkanoates in a synthetic mixed culture of Synechococcus elongatus cscB and Pseudomonas putida cscAB. Biotechnol Biofuels 10:190. https://doi.org/10.1186/s13068-017-0875-0

Lowe H, Hobmeier K, Moos M, Kremling A, Pflugergrau K (2017) Photoautotrophic production of polyhydroxyalkanoates in a synthetic mixed culture of Synechococcus elongatus cscB and Pseudomonas putida cscAB. Biotechnol Biofuels 10(1):190

Lu X (2010) A perspective: photosynthetic production of fatty acid-based biofuels in genetically engineered cyanobacteria. Biotechnol Adv 28(6):742–746

Luan GD, Lu XF (2018) Tailoring cyanobacterial cell factory for improved industrial properties. Biotechnol Adv 36(2):430–442. https://doi.org/10.1016/j.biotechadv.2018.01.005

Mackay MA, Norton RS, Borowitzka LJ (1984) Organic osmoregulatory solutes in cyanobacteria. Microbiology 130(9):2177–2191

Martínez-Nol GMA, Cumino AC, Kolman MDLA, Salerno GL (2013) First evidence of sucrose biosynthesis by single cyanobacterial bimodular proteins. FEBS Lett 587(11)

Melis A (2012) Photosynthesis-to-fuels: from sunlight to hydrogen, isoprene, and botryococcene production. Energy Environ Sci 5(2):5531–5539

Nadeem H, Rashid MH, Siddique MH, Azeem F, Muzammil S, Javed MR, Ali MA, Rasul I, Riaz M (2015) Microbial invertases: a review on kinetics, thermodynamics, physiochemical properties. Process Biochem 50(8):1202–1210

Narikawa R, Suzuki F, Yoshihara S, Higashi S, Watanabe M, Ikeuchi M (2011) Novel photosensory two-component system (PixA–NixB–NixC) involved in the regulation of positive and negative phototaxis of cyanobacterium synechocystis sp. PCC 6803. Plant Cell Physiol 52 (12):2214–2224

Pade N, Compaoré J, Klhn S, Stal LJ, Hagemann M (2012) The marine cyanobacterium Crocosphaera watsonii WH8501 synthesizes the compatible solute trehalose by a laterally acquired OtsAB fusion protein. Environ Microbiol 14(5):1261–1271

Perez-Cenci M, Salerno GL (2014) Functional characterization of *Synechococcus amylosucrase* and fructokinase encoding genes discovers two novel actors on the stage of cyanobacterial sucrose metabolism. Plant Sci 224:95–102. https://doi.org/10.1016/j.plantsci.2014.04.003

Porchia AC, Salerno GL (1996) Sucrose biosynthesis in a prokaryotic organism: presence of two sucrose-phosphate synthases in Anabaena with remarkable differences compared with the plant enzymes. Proc Natl Acad Sci U S A 93(24):13600–13604

Porchia AC, Curatti L, Salerno GL (1999) Sucrose metabolism in cyanobacteria: sucrose synthase from Anabaena sp. strain PCC 7119 is remarkably different from the plant enzymes with respect to substrate affinity and amino-terminal sequence. Planta 210(1):34–40

Qiao C, Duan Y, Zhang M, Hagemann M, Luo Q, Lu X (2018) Effects of reduced and enhanced glycogen pools on salt-induced sucrose production in a sucrose-secreting strain of *Synechococcus elongatus* PCC 7942. Appl Environ Microbiol 84(2). https://doi.org/10.1128/AEM.02023-17

Rastogi RP, Pandey A, Larroche C, Madamwar D (2018) Algal green energy – R&D and technological perspectives for biodiesel production. Renew Sustain Energy Rev 82:2946–2969

Reed RH, Stewart WDP (1985) Osmotic adjustment and organic solute accumulation in unicellular cyanobacteria from freshwater and marine habitats. Mar Biol 88(1):1–9

Rousseaux CS, Gregg WW (2014) Interannual variation in phytoplankton primary production at a global scale. Remote Sens 6(1):1–19. https://doi.org/10.3390/rs6010001

Salerno GL, Curatti L (2003) Origin of sucrose metabolism in higher plants: when, how and why? Trends Plant Sci 8(2):0–69

Salerno GL, Porchia AC, Vargas WA, Abdian PL (2004) Fructose-containing oligosaccharides: novel compatible solutes in Anabaena cells exposed to salt stress. Plant Sci 167(5):1003–1008

Smith MJ, Francis MB (2016) A designed *A. vinelandii-S. elongatus* coculture for chemical photoproduction from air, water, phosphate, and trace metals. ACS Synth Biol 5(9):955–961. https://doi.org/10.1021/acssynbio.6b00107

Song K, Tan X, Liang Y, Lu X (2016) The potential of Synechococcus elongatus UTEX 2973 for sugar feedstock production. Appl Microbiol Biotechnol 100(18):7865–7875. https://doi.org/10.1007/s00253-016-7510-z

Song K, Martin H, Tan X, Lu X (2017) The response regulator Slr1588 regulates spsA but is not crucial for salt acclimation of Synechocystis sp. PCC 6803. Front Microbiol 8:1176

Sturm A (1999) Invertases. Primary structures, functions, and roles in plant development and sucrose partitioning. Plant Physiol 121(1):1–8

Vargas WA, Salerno GL (2010) The Cinderella story of sucrose hydrolysis: alkaline/neutral invertases, from cyanobacteria to unforeseen roles in plant cytosol and organelles. Plant Sci 178(1):0–8

Vargas WA, Cumino AC, Salerno GL (2003) Cyanobacterial alkaline/neutral invertases. Origin of sucrose hydrolysis in the plant cytosol? Planta 216(6):951–960

Waterbury JB, Watson SW, Guillard RRL, Brand LE (1979) Widespread occurrence of a unicellular, marine, planktonic, cyanobacterium. Nature 277(5694):293–294

Weiss TL, Young EJ, Ducat DC (2017) A synthetic, light-driven consortium of cyanobacteria and heterotrophic bacteria enables stable polyhydroxybutyrate production. Metab Eng 44:236–245. https://doi.org/10.1016/j.ymben.2017.10.009

Xu Y, Tiago Guerra L, Li Z, Ludwig M, Charles Dismukes G, Bryant DA (2013) Altered carbohydrate metabolism in glycogen synthase mutants of Synechococcus sp. strain PCC 7002: cell factories for soluble sugars. Metab Eng 16(Complete):56–67

Yasunori N, Jun-ichiro T, Aya S, Yumiko I, Eiji S, Satoko N, Shoko F, Mikio T, Hideaki M, Hisato I (2005) Some cyanobacteria synthesize semi-amylopectin type α-polyglucans instead of glycogen. Plant Cell Physiol 3:3

Yu J, Liberton M, Cliften PF, Head RD, Jacobs JM, Smith RD, Koppenaal DW, Brand JJ, Pakrasi HB (2015) *Synechococcus elongatus* UTEX 2973, a fast growing cyanobacterial chassis for biosynthesis using light and CO(2). Sci Rep 5:8132 doi:https://doi.org/10.1038/srep08132
Zhang Y, Zhu Y, Zhu Y, Li Y (2009) The importance of engineering physiological functionality into microbes. Trends Biotechnol 27(12):664–672
Zhang L, Chen L, Diao J, Song X, Shi M, Zhang W (2020) Construction and analysis of an artificial consortium based on the fast-growing cyanobacterium *Synechococcus elongatus* UTEX 2973 to produce the platform chemical 3-hydroxypropionic acid from CO(2). Biotechnol Biofuels 13:82. https://doi.org/10.1186/s13068-020-01720-0
Zhou J, Zhu T, Cai Z, Li Y (2016) From cyanochemicals to cyanofactories: a review and perspective. Microb Cell Factories 15(1):2

Optimal Biomass Production by Cyanobacteria, Mathematical Evaluation, and Improvements in the Light of Biorefinery Concept

18

Alexander Dimitrov Kroumov, Fabiano Bisinella Scheufele, Maya Margaritova Zaharieva, Dimitrina Zheleva-Dimitrova, and Hristo Najdenski

Abstract

Cyanobacteria like other microalgal species are considered a key element in the new bioenergy concepts. Nevertheless, despite their enormous potential, this is still not enough to compete with natural fossil fuels for the production of microalgae as biofuels, with technical–economic competitiveness to other commercial technologies. Financial problems connected with the steps of the upstream and downstream processes must be overcome. Recent techno-economic analyses and life cycle assessments of microalgae-based production systems have suggested that the only most possible way for scaling up the cyanobacteria biomass technology passes through complete and optimal utilization of the cell components in an integrated biorefinery setup. This chapter provides a comprehensive analysis of the present CO_2 biofixation approaches and technologies using cyanobacteria under the strategy of biorefinery with cells treatment. Herein are discussed various cultivation techniques to maximize desirable products of

A. D. Kroumov (✉)
Department of Biotechnology—Laboratory of Bioconversion and Biosynthesis of Microbial Metabolites, The Stephan Angeloff Institute of Microbiology—Bulgarian Academy of Sciences, Sofia, Bulgaria

F. B. Scheufele
Graduation Program of Biotechnology and Bioprocess Engineering, Federal University of Technology—Paraná—UTFPR, Toledo, Paraná, Brazil
e-mail: fabianob@utfpr.edu.br

M. M. Zaharieva · H. Najdenski
Department of Infectious Microbiology, The Stephan Angeloff Institute of Microbiology—Bulgarian Academy of Sciences, Sofia, Bulgaria

D. Zheleva-Dimitrova
Department of Pharmacognosy, Faculty of Pharmacy, Medical University of Sofia, Sofia, Bulgaria
e-mail: dzheleva@pharmfac.mu-sofia.bg

R. P. Rastogi (ed.), *Ecophysiology and Biochemistry of Cyanobacteria*,
https://doi.org/10.1007/978-981-16-4873-1_18

cyanobacteria biorefinery. Advanced methods for metabolites isolation are analyzed to ensure the stable quantity and quality of cyanobacteria-based metabolites. Optimal biomass production in advanced closed photobioreactors raises many scientific and engineering problems, which occur when the scale increased. Scale-up is the last and most difficult technological step where all hypotheses are checked. Biorefinery deals with this problem in each stage of setup. The complexity and sustainability of this approach are the foremost concerns of specialists analyzing trends in the environmental, technological, and economic dynamic changes.

Keywords

Cyanobacteria · Biorefinery concept · Photobioreactors · Optimization · Scale-up · High-value metabolites

Notations (For Ethanol Model)

Enz	stands for the concentration of enzyme (U m^{-3})
Et	is the concentration of ethanol (kg m^{-3})
Glu	is the concentration of glucose (kg m^{-3})
k_i	stands for rate constant (kg U^{-1} h^{-1})
K_j	are the inhibition and saturation constants (kg m^{-3})
K_m	stands for the Michaelis-Menten constant (kg m^{-3})
S_0	represents the initial concentration of starch (kg m^{-3})
S_{total}	is the concentration of total starch (kg m^{-3})
Sus, Res, Total	stands for the concentrations of susceptible, resistant, and total starch, respectively
q_p	is the rate of specific ethanol production (h^{-1})
R_{st}	is the rate of starch utilization (kg m^{-3} h^{-1})
R_{Enz}	stands for the rate of enzyme synthesis (U m^{-3} h^{-1})
$R_{(Glu,formation)}$	is the rate of glucose formation (kg m^{-3} h^{-1})
$R_{(Glu,utilization)}$	represents the rate of glucose utilization (kg m^{-3} h^{-1})
X	is the concentration of biomass (kg m^{-3})
$Y_{(p/s)}$	stands for the yield coefficient of the product (kg kg^{-1})
$Y_{(x/s)}$	is the yield coefficient of cell growth (kg/kg)
β	is the rate of enzyme degradation (h^{-1})
μ	is the rate of specific cell growth (h^{-1})
t	is the time (h)

18.1 Introduction

The milestone of culturing cyanobacteria/microalgae for industrial application is to grow them under high concentrations (real or simulated) of flue gas as well on pure gaseous CO_2 and soluble carbonate (bicarbonate). The results are increased carbon biofixation and high biomass productivity (Aslam et al. 2017; Kuo et al. 2017).

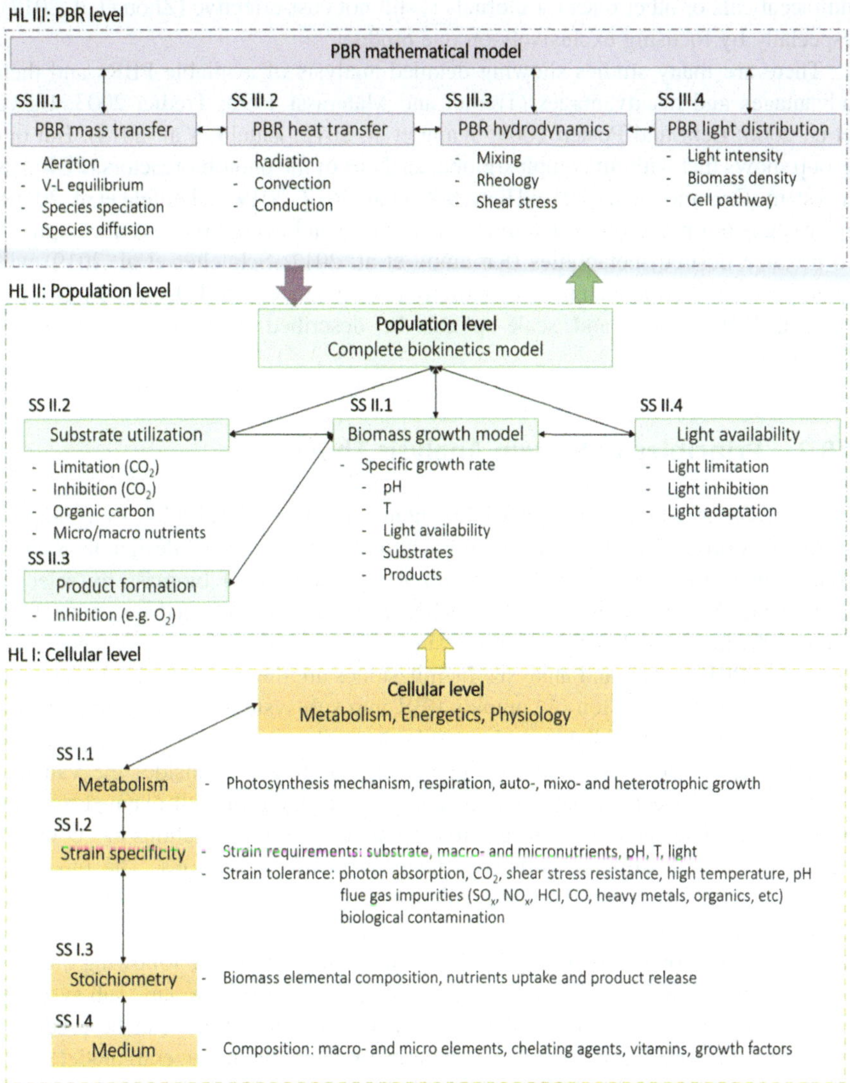

Fig. 18.1 Scheme of complex PBR model based on System Analysis Theory (HL—hierarchical levels and SS—subsystems). Adapted from Kroumov et al. (2016)

Published articles give more details on the subject (Cheah et al. 2015; Kroumov et al. 2015; Thomas et al. 2016; Vuppaladadiyam et al. 2018). The microalgal cell components can result in multiple valuable products as shown in Fig. 18.1.

The supplied carbon for the cyanobacteria eventually will be transformed into lipids, proteins, sugars, and pigments through different metabolic transformations determined by the environmental conditions. However, the production of microalgal and cyanobacterial cheap sources for food/feed products such as fatty acids for nutraceuticals or other ones for biofuels is still not cost-effective (Zhou et al. 2017), especially by focusing exclusively on one product.

There are many studies showing detailed analysis of available PBRs and their advantages and disadvantages (Tredici and Materassi 1992; Tredici 2003; Ugwu et al. 2008; Lehr and Posten 2009; Wang et al. 2012; Zittelli et al. 2013). But our group shows that without complex/global analysis of the photobioreactors (PBRs) as a system (Kroumov et al. 2016; Hinterholz et al. 2017, 2019; Scheufele et al. 2019), microalgae optimal growth (Kroumov et al. 2015) and overall process development to produce desired metabolites (Kroumov et al. 2017; Schuelter et al. 2019) will never reach a feasible technical–economic status in any closed PBR design. Briefly, optimal PBR design and scale-up can be described as published elsewhere (Kroumov et al. 2016).

18.2 Principles of System Analysis Theory

Any researcher in biotechnology and bioengineering faces the problem of a complex study of bioprocesses. Hence, in algology, photobioreactor (PBR) design plays a key role in minimizing the overall costs for the production of biomass by selected microalgae. A powerful tool to solve such a problem is the so-called system analysis theory (Kaffarov et al. 1979, 1985). Using principles of analogy, this theory was proved in PBR design and analysis. Its milestones are the principles of decomposition which offer a division of complex PBR into sub-systems and hierarchic levels for further robust and reliable investigation. It means that any part is analyzed and modeled separately. The next step in the algorithm is to consider the valuable relationships between the sub-systems and to study them in well-planned relevant experiments. The penultimate step is to draw up differential equations in a complex model which validation is obligatory in experimental conditions. The final step requires optimization of PBR and prediction of its behavior for different scenarios of the working environment.

The purification of flue gas in PBRs could be analyzed in the following order:

First, the PBR functioning is simplified and divided into key sub-systems. Second, the subsystems are investigated, and their interactions can be predicted and verified by the planning of active experiments. Finally, the overall model of the PBR can be developed including all reliable knowledge from the sub-systems, such as microalgal kinetics, light irradiation effects (photo-limitation, photo-inhibition, etc.). It is important to evaluate the links between microalgal physiology with gas-liquid mass-transfer processes. Computational fluid dynamics (CFD) tremendously would help to study such interactions in working PBRs (Perner-Nochta and Posten 2007; Bitog et al. 2011; Bari et al. 2015). It must be noted that the description of the processes which occurred on the population level is the reliable fundament for modeling the processes of photosynthesis of cyanobacteria cultivated in PBRs. In this context, the authors have developed phenomenological model of the column PBR. Extensive research has been realized for modeling different construction of PBRs (Pruvost et al. 2008, 2016a, b; Slegers et al. 2011).

Understanding and description of CO_2 fixation from flue gas by microalgae with high-density culture (HDC) or ultra-high-density culture (HDC) in a tubular PBR are obligatory.

The definitions of HDC and UHDC lie below and above the value of 10 gdw L^{-1} (Hu et al. 1998; Alagesan et al. 2013); HDC safely can be considered in the range between 5 and 10 gdw L^{-1}. Of course, the terms HCD and/or UHDC are relative in any state of the art. Evidently, these values continuously change and differ from species to species depending on the growth conditions (i.e., photoautotrophic, heterotrophic, or mixotrophic growth) and PBR construction (Kroumov et al. 2016).

18.3 Overview of UHDC Cultivation Techniques

The fast development of new culturing techniques and monitoring of microalgae physiological behavior in closed PBR together with the application of innovative illumination technologies based on biophotonics allow the continuous increase of biomass production. In one report, biomass concentration achieved under photoautotrophic conditions reach 84 g L^{-1} (Hu et al. 1998). This value can be considered as the highest one, but it must be pointed out that such concentration can be achieved in very thin-film PBR. When the substrate is CO_2 from flue gas, the biomass concentration (X) reaches 5 up to 10 g L^{-1}. Comparatively, the cultivation of microalgae under heterotrophic conditions yields biomass concentration from 100 to 150 g L^{-1} (Bumbak et al. 2011). It is well known from previous studies that under mixotrophic modes, biomass yields higher values for similar (Ogbonna and Tanaka 2000; Heredia-Arroyo et al. 2011; Mohamed et al. 2014) and different trophic conditions (Sansawa and Endo 2004; Bumbak et al. 2011). By providing adequate control of flashing light effects (FOR), PBR systems increase continuously their effectiveness and competitiveness (Luzi et al. 2019; Straka and Rittmann 2019; Cui et al. 2020; Guo et al. 2020). All the studies on FLE as a methodology to maximize biomass production and to improve biochemical composition of the cyanobacterial/microalgae cells have to be considered as well, with the potential of the PBR system to provide conditions for HDC. The HDC and UHDC achievements are discussed in detail (Hu et al. 1998).

In terms of PBR system modeling, such information is also valuable for the search of process optimum (Kroumov et al. 2016) and would help to reach maximum productivity of PBRs (Kroumov et al. 2016). Additionally, one example of how to use empirical correlations to achieve robust results is reported by Greenwald et al. (2012).

18.4 Applications of the Principles of System Analysis Theory to PBR Design and Scale-Up

The PBR system relatively could be divided into biological, chemical, and physical sub-systems (Olivieri et al. 2015a, b). The scheme which fully presents our understanding of the PBR system is shown in Fig. 18.1.

Analysis of the scheme showed that physical and chemical processes, cell metabolism, flue gas composition, and light penetration and trajectory in PBR must be considered and their relationships must be modeled and applied for fast process development and scaling-up.

The crucial parameter of closed PBR is light availability. The surface-to-volume ratio and optimal light–dark (L/D) cycles in PBRs are the key factors to achieve HDC (Kroumov et al. 2013). The bioengineering solutions should take into account the interaction between the flue gas composition and cyanobacteria/microalgal physiology. Hence, the modeling of the overall PBR system allows to plan active experiments and, on such base, to achieve optimal process development. Scientific efforts of three leading groups (managed by J. Merchuk, C. Posten, and J. Pruvost) were guided by a systematic approach applicable to kinetics, hydrodynamics, mass-, and light-transfer phenomena by analyzing the reactor system dividing it into sub-systems as we described earlier (see Fig. 18.1). For several decades, these teams gave their valuable contribution not only in bioreactor design but also in photobioreactors optimization and scaling-up (Merchuk et al. 2007; Posten 2009; Pruvost et al. 2016a, b). Hence, the potential of bioreactors/photobioreactors engineering is extremely well developed being a robust base for further innovative technology application.

18.5 Modeling, Optimization, and Scale-Up of PBRs

System analysis theory allows the fast and robust transfer of knowledge from lab to pilot and industrial scale. The complex PBR model is a base for searching optimal parameters values and shows the link between the light intensity and algal physiology in dynamics. We demonstrated it in a process of CO_2 utilization from flue gas (Kroumov et al. 2016; Hinterholz et al. 2019). Using a complex PBR model is a great possibility to have fast success in process development and PBR design.

The complex phenomena involved between the multiphase system in a PBR can be described through adequate modeling of the:

(i) Conservation principles of the species (i.e., CO_2, O_2, SO_2, NO_2, etc.)
(ii) Thermodynamics (e.g., vapor–liquid equilibrium—VLE).
(iii) Mass transfer mechanisms and between the phases (gas to liquid and liquid to cell).
(iv) Stoichiometry.

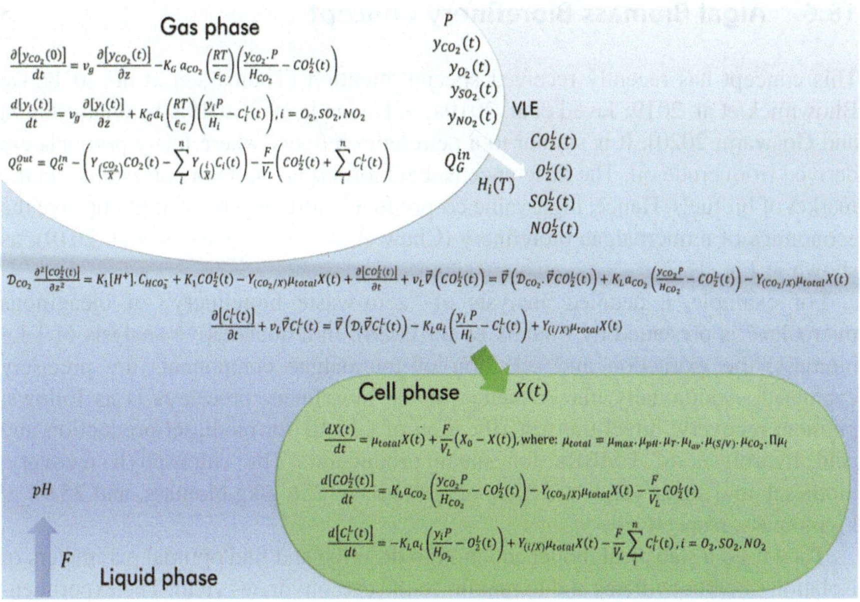

Fig. 18.2 Example of a complex PBR model. Adapted from Kroumov et al. (2016)

(v) Cell growth and product generation kinetics (which is dependent to several factors, such as, metabolism, physiology, energetics, nutrient availability, limitation, and inhibition effects).

(vi) Fluid dynamics.

(vii) Among other phenomena, that may apply for each PBR setup.

An example of such a complex engineering problem is represented in Fig. 18.2, which summarizes some of the abovementioned phenomena. More detail can be found elsewhere (Kroumov et al. 2016, 2017; Hinterholz et al. 2019; Scheufele et al. 2019).

New findings in the area have to take into account the milestone study of Merchuk and Wu (2003). They clearly demonstrated that the physiology of microorganisms and their intimate biochemical features during the light–dark cycles including the microalgae growth are related to the bioreactor geometry and cell behavior. This means that photobioreactor design and scale-up depend critically on the physiology and growth kinetics of microorganisms which must be modeled precisely (Olivieri et al. 2015a, b; Pires et al. 2017; Ali et al. 2019; del Rio-Chanona et al. 2019; Gao et al. 2018). Consequently, microalgae kinetics is the coordinating task of any PBR optimization procedure.

18.6 Algal Biomass Biorefinery Concept

This concept has recently received special attention (Thomassen et al. 2018; De Bhowmick et al. 2019; Javed et al. 2019a, b; Koyande et al. 2019a, b; Bhattacharya and Goswami 2020). It is similar to a petroleum refinery where many products are derived from crude oil. The difference is that biofuels are still not competitive in the market of biofuels. Hence, high-value co-products must be generated to improve the economics of a microalgae biorefinery (Chew et al. 2017; Chandra et al. 2019), as shown in Fig. 18.3.

For example, a detailed analysis of "zero-waste biorefinery" of oleaginous microalgae is presented by Mandik et al. (2020). In a quantitative analysis of 1 kg biomass, the extraction and selection of microalgae components are precisely evaluated. Additionally, the sequence of the biorefinery processes is as follows: pigment recovery, direct transesterification of CEMB for biodiesel production and acid hydrolysis of LMBRs for sugar production. The chlorophyll recovery, biodiesel, and sugar yields are 27.8 mg/g biomass, 256 g/kg biomass, and 25.98 g/kg biomass, respectively.

They used a statistical model equation to describe and find optimal parameters of isolation of carbohydrates and to maximize the carbohydrates yield. The experimental sugar yield of 44.8 g/kg–LMBRs is achieved what is close to the model-predicted value. Hence, the innovative and precise bio-refinery process for oleaginous microalgae biomass as shown by the authors may greatly increase the competitiveness of the microalgae-based industries. More specifically, another team (Wang et al. 2018) presented in detail a study where six eustigmatophycean microalgae had been evaluated for biodiesel production. Strains, standards, and biodiesel properties were

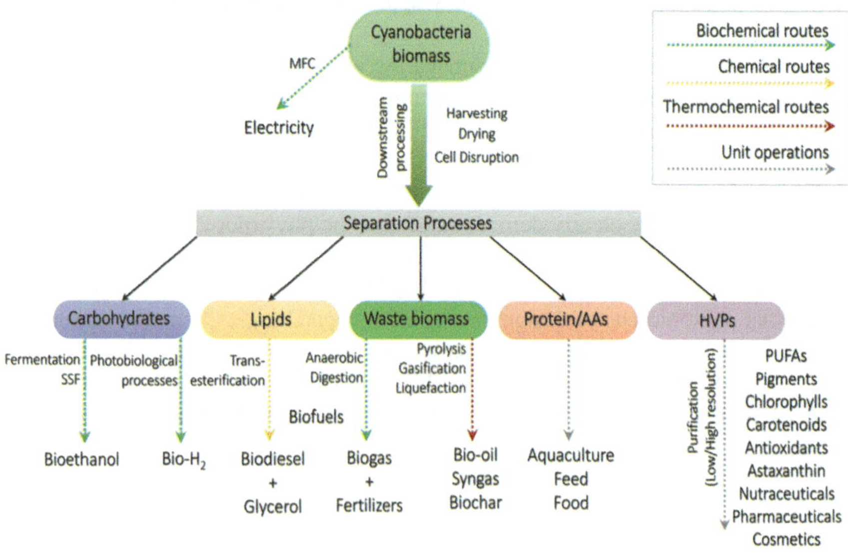

Fig. 18.3 Cyanobacterial/microalgae biomass potential products and/or application

analyzed and systematized. The conclusion was that the potential of six eustigmatophytes was good enough for their use as biorefinery feedstock to coproduce palmitoleic acid and biodiesel under different working conditions.

In the same recursion, the investigation of the most well-known *Chlamydomonas sp.* as feedstock for the production of methyl ester and ε-polylysine was presented as an alternative of a biorefinery concept (Sivaramakrishnan et al. 2019). A systematic analysis of commercially used microalgae strains, culturing conditions, photobioreactors, downstream processing, and the applications of final products of biorefinery can be found elsewhere (Javed et al. 2019a, b). From this article, it could be concluded that the major cost of the microalgae process is the nutrient supply which necessitates the use of wastewater nutrients for microalgae cultivation.

Overall, summarizing the important value-added products from microalgae should be emphasized their applicability in the food, nutraceutical, cosmetic, and pharmaceutical industries. On the other hand, microalgal biomass is considered to be the source of third-generation biofuels (Lee and Lavoie 2013).

18.7 Downstream Bioprocessing of Microalgae Biomass

Downstream processing of microalgae cells utilizes various unit operations to extract valuable bioactive compounds (BAC). The biorefinery of microalgae is a promising waste-free technology (Mandik et al. 2020) that is very well developed in chemical engineering and biotechnology as well. This approach excellently fits to alleviate global warming concerns where CO_2 emission must be minimized to preserve the environment which recently is an enormous danger (Juan et al. 2011).

In a downstream procedure, avoiding the loss of fragments and valuable metabolites is critical. Hence, downstream processing of algae to get valuable BAC is the current challenge. The extraction techniques (Salinas-Salazar et al. 2019) were analyzed elsewhere (Kroumov et al. 2017). Other processes have also been utilized (Jacob-Lopes et al. 2015). Hence, the economic impact of such techniques on a big scale could not be neglected (Chew et al. 2017). In the bio-refinery, the microalgae biomass must be completely utilized including heat and water flows.

18.7.1 Carbohydrate Fraction

Cyanobacteria can reach carbohydrate contents (starch or glycogen) between 25 and 50+ % (d.w.) strongly depending on working conditions (Gouveia et al. 2008; Becker 2013; Cerón-García et al. 2013). Microalgae polysaccharides are important in many aspects of medicine (Matsui et al. 2003; Tannin-Spitz et al. 2005; Schepetkin and Quinn 2006; Park et al. 2011; Kim et al. 2012). Therefore, sulfated polysaccharides' activities can be utilized as drugs (Yen et al. 2013).

The polysaccharides extracted from microalgae have a wide variety of applications in the pharmacy and food industries (Arad and Levy-Ontman 2010).

The extracellular ones found in any microalgae including cyanobacteria are beneficial because the step of cell disruption is avoided. Despite the obvious advantages of microalgal polysaccharides, they cannot compete with the cheaper sources such as xanthan gum, agar, guar gum, and carrageenan (Bleakley and Hayes 2017).

18.7.2 Protein Fraction

The content of proteins in microalgae biomass is in the range between 40% and 70% (Gouveia et al. 2008; Christaki et al. 2011; Cai et al. 2013), which varies depending on strains and working conditions. Conventional sources of proteins include meat, dairy, eggs, soybean, and so on. The potential of microalgae biomass is a promising raw material for the production of these valuable and cheap sources, but the challenges to find niches in the market remain (Udayan et al. 2017). This study evidences that soy, whey, and fish proteins can be replaced by microalgae ones (Kose et al. 2017). The economic and environmental effectiveness of cultivation techniques such as autotrophic and heterotrophic ones offer definitive benefits for the production of cyanobacterial/microalgal protein sources for any market (Smetana et al. 2017). In this regard especially important is the application of extraction methods to isolate the microalgal proteins of interest. More details about the proteins extraction methods and working conditions can be found elsewhere (Bjornsson et al. 2012; Vanthoor-Koopmans et al. 2013; Chew et al. 2017; Koyande et al. 2019a, b). It must be noted that microalgae protein can be cheaper than pea protein and soybean protein among others (Smetana et al. 2017).

A life cycle assessment (LCA) clearly shows the difference between microalgae food products and pork, chicken, and beef, respectively (De Vries and De Boer 2010).

18.7.3 Pigments

Cyanobacteria have the potential to synthesize chlorophylls, carotenoids, phycocyanin, and astaxanthin (Begum et al. 2016; Sonani et al. 2016). The pigments from microalgae are valuable sources for pharmaceuticals and cosmetics (Singh et al. 2010; Yen et al. 2013; Wichuk et al. 2014; Rastogi et al. 2015; Sonani et al. 2015; Begum et al. 2016).

The importance of pigments resulted in their approval by administrations such as Food and Drug Administration (FDA) and European Food Safety Authority (EFSA) for the replacement of synthetic colorants (Rahman et al. 2017). Hence, the pigment production from cyanobacteria shows many benefits in the production of the pigments from fruits and vegetables (García-López et al. 2020). As we discussed earlier in detail, the scheme of microalgae cultivation system, its optimal functioning depends on PBRs configuration where operating conditions provide a base of maximum specific growth and production rates of cyanobacteria (Yen et al. 2014).

C-phycocyanin is emblematic for cyanobacteria and red microalgae (Pan-utai and Iamtham 2019). Phycocyanin is a natural dye and is a valuable product for the food and cosmetic industry because of its beautiful blue color and stability. Besides this application, phycocyanin is valuable with its functional activities as a drug (Vernès et al. 2015). Astaxanthin from *Haematococcus* is recognized by the Food and Drug Administration (FDA) as Generally Regarded as Safe (GRAS) for use as an ingredient in several food categories (Jacob-Lopes et al. 2019). β-carotene obtained from *Dunaliella* is recognized by the FDA as GRAS and it is used for the nutrition of animals and humans (Sui and Vlaeminck 2020).

The chemical and biochemical synthesis is well studied and documented in the scientific literature. Today, the largest industrial producer of pigments is BASF®. The company produces various pigments. Hence, the future of industrial production of pigments most probably will focus on their photoautotrophic synthesis (Cardoso et al. 2017).

The overall pigments industrial process and its techno-economic analysis are well documented (Acién Fernández et al. 2012; Vernès et al. 2015; Ouada and Ammar 2017; Borowitzka 2018; Depra et al. 2018; AstaReal 2019; Hu 2019; Jacob-Lopes et al. 2019). The astaxanthin production is discussed by Schultz (2016). It is worthy to mention some companies which industrially produce pigments, namely, Astaxanthin and C-phycocyanin—Cyanotech (location USA) and β-carotene Betatene (BASF) (location Australia). A huge list of companies in the business is available on the internet platforms. Hence, pigments from microalgae are steadily winning niches in the industry. In the future decades, to expand the market for pigments, R&D efforts should solve the following problems: competitive price, which is a result of production cost reduction.

18.8 Utilization of Biomass to Biofuels

The state of the art of microalgae biomass conversion is shown in Fig. 18.3:

- Biochemical processes.
- Thermochemical processes.
- Transesterification.
- Microalgal fuel cell (MFC).

The selection of the reliable process is determined from the available budget and the characteristics of crude biomass.

18.8.1 Biochemical Process

Biochemical processes have a goal to transfer microalgae feedstock into biofuels. The conversion steps pass through fermentation, anaerobic digestion, and photobiological hydrogen generation. The fermentation of carbohydrates from microalgae to alcohol is via different techniques by using yeast (Brennan and Owende 2010; Miranda et al. 2012; Chen et al. 2013).

The carbohydrates obtained from cyanobacterial biomass and further their utilization for biofuels is discussed elsewhere (Chochois et al. 2009). Chemical and enzymatic methods of polysaccharides treatment are discussed in Simas-Rodrigues et al. (2015). An interesting approach is to produce bioethanol via hydrolysis and fermentation by directly using microalgae itself (Ueno et al. 1998; Markou et al. 2012). The intracellular ethanol synthesis is possible when using *Chlamydomonas reinhardtii* (Hirano et al. 1997). Advantages and disadvantages of this process compared with the conventional one are analyzed in Chen et al. (2013).

Recombinant strains proved to be robust in the fermentation of starch. For instance, our studies of modeling such a process are indicative (Kroumov et al. 2006). Hence, any cultivation cyanobacteria/microalgae process resulting in high carbohydrate contents in the cells may be set up in the bioethanol production chain.

It is important to study a cultivation technique such as simultaneous saccharification and fermentation of starch to ethanol (SSF-SE). This approach is essential in biotechnology and is widely introduced in ethanol production schemes. The SSF-SE technique will be greatly beneficial for the cyanobacterial biorefinery concept.

We were the first to develop a model of the SSF-SE process by a recombinant strain of *Saccharomyces cerevisiae YPB-G*. Here we are going to highlight the principles of modeling SSF-SE based on our personal experience (Kroumov et al. 2006). The methodology of model development for the description of this complex process is divided into two hierarchical levels being shortly discussed below.

18.8.1.1 First Hierarchic Level

The model equations describing phenomena of starch hydrolysis by glucoamylase, in recombinant strain, formed the knowledge about the first hierarchic level. This stage is fundamental for the overall kinetic model.

The second phase of starch degradation is slow and determines the rate of hydrolysis. Hence, it is convenient to divide the starch structure into susceptible and resistant fractions. A model can be represented as developed by authors (Polakovic and Bryjak 2004). The model is suitable to describe glucoamylase activities of the recombinant strain *S. cerevisiae YPB-G* (Kroumov et al. 2006) and can be written as follows:

The degradation rate of susceptible starch (R_{sus}) (phase one):

$$R_{\mathrm{sus}} = \frac{k_{\mathrm{sus}}\mathrm{Enz}(t)S_{\mathrm{sus}}(t)}{K_{\mathrm{m}}\left(1+\frac{\mathrm{Glu}(t)}{K_{\mathrm{Glu}}}\right)+\frac{S_{\mathrm{sus}}(t)^2}{K_{\mathrm{starch}}}+S_{\mathrm{sus}}(t)+S_{\mathrm{res}}(t)} \tag{18.1}$$

The degradation rate of resistible starch (R_{res}) fraction (phase two):

$$R_{\mathrm{res}} = \frac{k_{\mathrm{res}}\mathrm{Enz}(t)S_{\mathrm{res}}(t)}{K_{\mathrm{m}}\left(1+\frac{\mathrm{Glu}(t)}{K_{\mathrm{Glu}}}\right)+\frac{S_{\mathrm{res}}(t)^{2}}{K_{\mathrm{starch}}}+S_{\mathrm{sus}}(t)+S_{\mathrm{res}}(t)})\tag{18.2}$$

Equations representing the dynamic changes of susceptible (S_{sus}) and resistant (S_{res}) starch fractions:

$$\frac{dS_{\mathrm{sus}}(t)}{dt} = -R_{\mathrm{sus}}\tag{18.3}$$

$$\frac{dS_{\mathrm{res}}(t)}{dt} = -R_{\mathrm{res}}\tag{18.4}$$

Equation representing the dynamic changes of total starch (S_{total}) degradation:

$$\frac{dS_{\mathrm{total}}(t)}{dt} = \frac{dS_{\mathrm{sus}}(t)}{dt} + \frac{dS_{\mathrm{res}}(t)}{dt}\tag{18.5}$$

Dynamic of glucose production during susceptible and resistant starch degradation:

$$\frac{d\mathrm{Glu}(t)}{dt} = 1.111(R_{\mathrm{sus}} + R_{\mathrm{res}})\tag{18.6}$$

The model assumptions (Polakovic and Bryjak 2004) for the description of starch saccharification by one additive enzyme activity are as follows:

1. The glucoamylase and alpha-amylase activity to hydrolyze starch by *S. cerevisiae YPB-G* is presented as a sum of activities of both enzymes.
2. The starch structure conditionally is divided into two fractions with different degradation rates.
3. The susceptible hydrolysis rate is a function of the concentration of two starch fractions and is inhibited by both the action of glucose and the concentrations of susceptible starch.
4. The resistant hydrolysis rate is a function of the concentration of two starch fractions and is inhibited by both the action of glucose and the concentrations of the resistant starch fraction.
5. Mass transfer limitations and conformation changes of the enzyme structure are out of consideration.

18.8.1.2 Second Hierarchic Level

At this level, microbial kinetics represents the behavior of all populations. The recombinant strain activity is represented by simple models from the kinetic models database (Dourado et al. 1987; Birol et al. 1998; Nishiwaki and Dunn 1999). The specific growth rate (SGR) of recombinant strain is a function of glucose and initial and total starch concentrations. This best fitted with the experimental kinetics data. Specific ethanol production rate (SEPR) depends on the glucose, initial starch, and ethanol concentrations. It is obvious that this rate can be analyzed as a nonlinear function of SGR.

The second level is formalized as follows:

SGR (μ) model:

$$\mu = \frac{\mu_{\max}\mathrm{Glu}(t)\left(\frac{S_{\mathrm{total}}(t)}{S_0}\right)}{K_s + \mathrm{Glu}(t)} \tag{18.7}$$

SEPR (q_p) model:

$$q_p = \frac{q_{p,\max}\,\mathrm{Glu}(t)Et(t)\left(1 - \frac{Et(t)}{Et_{\max}}\right)}{(K_{s1} + \mathrm{Glu}(t))\left(K_{ps1} + Et(t) + \frac{Et(t)^2}{K_{pi}}\right)} \tag{18.8}$$

Biomass (X) mass balance:

$$\frac{dX(t)}{dt} = \mu X(t) \tag{18.9}$$

Product (E) mass balance:

$$\frac{dEt(t)}{dt} = q_p X(t) \tag{18.10}$$

Glucose (Glu) mass balance:

$$\frac{d\mathrm{Glu}(t)}{dt} = R_{\mathrm{Glu,formation}} - R_{\mathrm{Glu,utilization}} \tag{18.11}$$

where

$$R_{\mathrm{Glu,formation}} = 1.111(R_{\mathrm{sus}} + R_{\mathrm{res}}) \tag{18.12}$$

$$R_{\mathrm{Glu,utilization}} = \frac{1}{Y_{x/s}}\frac{dX(t)}{dt} + \frac{1}{Y_{p/s}}\frac{dEt(t)}{dt} \tag{18.13}$$

Note: The coefficient used in Eqs. (18.6) and (18.12) is theoretical yield $\left(Y_{\frac{\mathrm{Glu}}{S}} = 1.111\right)$, representing glucose production from 1 gram of starch (Huang et al. 2005).

Enzyme (Enz) balance:

$$\frac{d\mathrm{Enz}(t)}{dt} = R_{\mathrm{Enz}} - (\mu + \beta)\mathrm{Enz}(t) \tag{18.14}$$

$$R_{\mathrm{Enz}} = \frac{(\mu_{\max} + \beta)\mathrm{Enz}_{\max} S_{\mathrm{total}}(t)}{K_{\mathrm{Enz}} + S_{\mathrm{total}}(t)} \quad (18.15)$$

R_{Enz} synthesis rate shows enzyme induction multiplied by S_{total} (t). The catabolite repression by glucose is neglected because of the low glucose concentration during the process dynamics. The balance of key enzyme synthesis is as follows:

$$\frac{d\mathrm{Enz}(t)}{dt} = R_{\mathrm{Enz}} - \mu\mathrm{Enz}(t) - \beta\mathrm{Enz}(t) \quad (18.16)$$

This model of SSF-SE was used to describe and analyze microbial growth, amylolytic enzyme synthesis, glucose synthesis and utilization, and ethanol overproduction of the recombinant strain *S. cerevisiae* YPB-G. The response surface analysis (RSA) helps to discriminate the kinetic hypotheses of the rate models. A hybrid genetic algorithm was applied for the search of values of model parameters (Kroumov et al. 2006). The proposed is validated on a set of experimental data and showed excellent flexibility for different operational conditions of the SSF-SE process.

Based on the principle of analogy, it can be used to successfully describe the physiological behavior of other genetically modified strains. Moreover, the developed SSF-SE model can be used to control the process of cyanobacterial/microalgal starch utilization and ethanol production on an industrial scale through a biochemical route aiming at the integral microalgae biomass utilization.

Recently, anaerobic digestion of microalgal biomass into biogas is favored in large-scale facilities. The composition of biogas obtained from microalgae can be found elsewhere (Acién Fernández et al. 2012), wherein a circular approach may be applied for the simultaneous microalgae cultivation/biogas upgrading.

18.8.2 Thermochemical Conversion

The thermochemical conversion of microalgae biomass can be seen in Fig. 18.3. The pyrolysis process can be realized in large-scale facilities. During thc gasification of the organics from the cyanobacteria, syngas is produced. It is a precursor for the synthesis of biofuels or can be used alone in turbines and engines. Details about the production of syngas from microalgae biomass by utilization of a high-temperature tubular furnace were done by Raheem et al. (2015). Liquefaction converts wet microalgae biomass to biofuel. The operational conditions are published by authors (Goyal et al. 2008). The combustion of microalgae biomass and production of electricity in hybrid plans operations is analyzed elsewhere (Kadam 2002).

18.8.3 Transesterification

Transesterification of microalgae lipids to fatty acid methyl esters (FAME) is a promising process. Several chemical reactions proceed to produce FAME and glycerol. The process can be carried out in the presence or absence of a catalyst. We will describe the transesterification of lipids to biodiesel (Wenzel et al. 2006) which fits well to cyanobacterial lipids.

The reversible transesterification reactions were modeled and are presented below (Freedman et al. 1986):

$$\mathrm{TG} + \mathrm{ROH} \overset{k_1/k_{-1}}{\leftrightarrow} \mathrm{DG} + \mathrm{BD} \tag{18.17}$$

$$\mathrm{DG} + \mathrm{ROH} \overset{k_2/k_{-2}}{\leftrightarrow} \mathrm{DG} + \mathrm{BD} \tag{18.18}$$

$$\mathrm{TG} + \mathrm{ROH} \overset{k_3/k_{-3}}{\leftrightarrow} \mathrm{GL} + \mathrm{BD} \tag{18.19}$$

The overall reaction is:

$$\mathrm{TG} + 3\mathrm{ROH} \rightarrow \mathrm{GL} + 3\mathrm{BD} \tag{18.20}$$

where

TG—stands for the triglycerides; DG—are the diglycerides; BD—is the biodiesel; GL—is the glycerol.

Note: The same applies for mono-alcohol ethyl esters and methyl esters depending on the used alcohol (ROH) in the reaction.

The assumption of the mechanism is as follow:

1. Reversible first-order kinetics for each reaction.
2. Higher order chemical reactions are not considered for simplicity.
3. Mass transfer and transport phenomena limitations are out of consideration.
4. Kinetics constants are a function of temperature and are described by the Arrhenius model.

The equilibrium or forward transesterification reactions depending on operational conditions, catalyst (if any is involved), and molar ratio between Alcohol–Soybean oil (A:SO). Hence, the balance equations describing the kinetics of chemical system are as follows:

$$\frac{dC_{\mathrm{TG}}}{dt} = -k_1 C_{\mathrm{TG}}(t) C_{\mathrm{ROH}}(t) + k_{-1}(t) C_{\mathrm{DG}}(t) C_{\mathrm{BD}}(t) \tag{18.21}$$

$$\frac{dC_{DG}}{dt} = k_1 C_{TG}(t) C_{ROH}(t) - k_{-1}(t) C_{DG}(t) C_{BD}(t) - k_2(t) C_{DG}(t) C_{ROH}(t) + k_{-2}(t) C_{MG}(t) C_{BD}(t) \quad (18.22)$$

$$\frac{dC_{MG}}{dt} = k_2 C_{DG}(t) C_{ROH}(t) - k_{-2}(t) C_{MG}(t) C_{BD}(t) - k_3(t) C_{MG}(t) C_{ROH}(t) + k_{-3}(t) C_{GL}(t) C_{BD}(t) \quad (18.23)$$

$$\frac{dC_{GL}}{dt} = k_3(t) C_{MG}(t) C_{ROH}(t) - k_{-3}(t) C_{GL}(t) C_{BD}(t) \quad (18.24)$$

$$\frac{dC_{BD}}{dt} = -3\left(\frac{dC_{TG}}{dt}\right) - 2\left(\frac{dC_{DG}}{dt}\right) - \left(\frac{dC_{MG}}{dt}\right) \quad (18.25)$$

$$\frac{dC_{ROH}}{dt} = -\frac{dC_{BD}}{dt} \quad (18.26)$$

The system of ordinary differential equations (ODEs) (Eqs. 18.21–18.26) is rearranged by representing each component concentration as a mass ratio X_i (kg component/kg ester) as proposed elsewhere (Darnoko and Cheryan 2000).

$$\frac{dX_{TG}}{dt} = -k_1 X_{TG}(t) C_{ROH}(t) + \frac{MW_{TG}}{MW_{DG}} k_{-1}(t) X_{DG}(t) C_{BD}(t) \quad (18.27)$$

$$\frac{dX_{DG}}{dt} = \left(\frac{MW_{DG}}{MW_{TG}} k_1 X_{TG}(t) - k_2(t) X_{DG}(t)\right) C_{ROH}(t) + \left(\frac{MW_{DG}}{MW_{MG}} k_{-2}(t) X_{MG}(t) - k_{-1}(t) X_{DG}(t)\right) C_{BD}(t) \quad (18.28)$$

$$\frac{dX_{MG}}{dt} = \left(\frac{MW_{MG}}{MW_{DG}} k_2 X_{DG}(t) - k_3(t) X_{MG}(t)\right) C_{ROH}(t) + \left(\frac{MW_{MG}}{MW_{GL}} k_{-3}(t) X_{GL}(t) - k_{-2}(t) X_{MG}(t)\right) C_{BD}(t) \quad (18.29)$$

$$\frac{dX_{GL}}{dt} = \frac{MW_{GL}}{MW_{MG}} k_3(t) X_{MG}(t) C_{ROH}(t) - k_{-3}(t) X_{GL}(t) C_{BD}(t) \quad (18.30)$$

$$X_{BD}(t) = 1 - X_{TG}(t) - X_{DG}(t) - X_{MG}(t) \quad (18.31)$$

where,

$$C_{ROH}(t) = \frac{n_{ROH} - 3n_{TG}(1 - X_{TG}(t) - X_{DG}(t) - X_{MG}(t))}{(MW_{TG} n_{TG} + MW_{ROH} n_{ROH})/1000} \quad (18.32)$$

$$C_{\mathrm{BD}}(t) = \frac{3n_{\mathrm{TG}}(1 - X_{\mathrm{TG}}(t) - X_{\mathrm{DG}}(t) - X_{\mathrm{MG}}(t))}{(\mathrm{MW}_{\mathrm{TG}}n_{\mathrm{TG}} + \mathrm{MW}_{\mathrm{ROH}}n_{\mathrm{ROH}})/1000} \tag{18.33}$$

In which, n_{TG}—is the number of SBO mols; n_{ROH}—represents the number of alcohol mols; and MW_i—stands for the molecular weight of "i" component. The rate constants in the forward (k_i) and reverse reactions (k_{-i}) are temperature dependent as follows:

$$k_i = k_{i,0}e^{-\frac{E_{a,i}}{\mathrm{RT}}} \tag{18.34}$$

$$k_{-i} = k_{-i,0}e^{-\frac{E_{a,-i}}{\mathrm{RT}}} \tag{18.35}$$

In Eqs. (18.34) and (18.35), $E_{a,\ i}$—stands for the activation energy for the *ith* forward reaction and $E_{a,\ -\ i}$ the activation energy for reverse reactions, R—is the universal gas constant and T—is the absolute temperature (K), and $k_{i,\ 0}$ and $k_{-i,\ 0}$—are the initial rate constants.

The model (see Eqs. 18.27–18.33) was used to describe/fit experimental data of transesterification processes under a set of different working conditions.

A new mathematical model of transesterification of soybean oil to biodiesel has been developed. The RSA methodology helped to maximize the soybean oil conversion rate. The kinetic model was verified on various sets of experimental data. As a result, an excellent agreement between the model simulations and experimental data was achieved. The developed new kinetics model is promising and can be successfully used for experimental design, optimization of biodiesel production based on cyanobacteria lipids. Especially beneficial this model will be applied for educational aims.

Further, an enzymatic transesterification process was modeled by our group (Kroumov et al. 2007). It can be noticed that catalog of models of any biotechnological process is crucial for the scale-up and industrial setup of the technology.

Other beneficial characteristics of algal lipids are linked with medical applications because of their anti-inflammatory and anticarcinogenic effect on humans (Jaswir and Hammed 2011).

18.8.4 Microalgal Cultivation and Microbial Fuel Cell (MFC) Systems

Studies on microalgae and MFC predict promising integration into microalgae-MFC (mMFC) system. The mMFC system presents unique features by converting solar energy into electricity through the photosynthetic reactions of microalgae. Other applications are the production of bioelectricity, CO_2 fixation from air and waste industrial gases, and water purification (Strik et al. 2008; Cui et al. 2014; Lee et al. 2015; Shukla and Kumar 2018; Kakarla and Min 2019).

Recently, examples for innovative approaches concerning MFC application in power generation, wastewater treatment, bioelectricity production from kitchen wastewaters, and enhanced treatment of landfill leachate by hybrid MFC system were discussed in detail elsewhere (Khandelwal et al. 2018; Yang et al. 2018; Mohamed et al. 2020; Elmaadawy, et al. 2020).

It is remarkable that the mMFC accumulates all the positive and negative characteristics of both microalgae and MFC processes. Hence, a cost-effective mMFC requires further research and scientific efforts to study both systems. The current state of the art demonstrated an improvement in this area. The developments on mFMC systems will open new frontiers for light conversion to electricity avoiding the liberation of CO_2. Further R&D will discover the full potential of the mMFC technologies.

18.9 Techno-Economic Analysis and Life Cycle Analysis (LCA)

Biorefinery approach for high value-added products from algal biomass is promising for all possible applications in medicine, food, and drugs (Faried et al. 2017; Barsanti and Gualtieri 2018). The net energy ratio (NER) is the ratio between the energy involved to obtain a product from microalgae and the total energy stored in the final product. The life cycle analysis (LCA) applied to Nannochloropsis sp. with NER evaluation for different scenarios is published (Jorquera et al. 2010). The obtained NER values were as follows: 8.34, 4.5, and 0.2 for open/raceway ponds, flat plate photobioreactors, and tubular ones. The authors (Tredici et al. 2015) present LCA for the system where *Tetraselmis suecica* is cultivated in flat photobioreactors with different engineering specifications. The study compared NER of photovoltaic panels 1.73–0.82 with no support by external energy input. The comparative analysis of these results with the NER of 3.71, 4.11, and 7.57 for soybean, corn, and cassava showed the need for further improvement of microalgal technologies.

It must be noticed that many studies on the NER values of biofuels from algae biomass differ because the systems differ from the fossil fuel NER values. The study of authors conducted on four scenarios is indicative (Chowdhury and Franchetti 2017). This LCA is performed on energy generation from algae biomass by utilizing different substrates and methods for the transformation of the energetic components to products. The results from the study showed that NER values were as follows: 0.35, 0.48, 0.50, and 0.68. The overall conclusion from this study is that the biorefinery approach is imperative when aiming at cost-effective and feasible process; otherwise, sole production of biofuel is very costly.

LCA and cost-effective microalgae biorefinery are a winning scenario for the industrial scale. Hoffman et al. (2017) studied and compared biodiesel products between Algal Turf Scrubber (ATS) and Open Raceway Ponds (ORP). Analysis of these data demonstrated that the cost from ATS is $8.34 per gallon of biodiesel and the cost from ORP is $6.27 per gallon of biodiesel, respectively. Hence, these prices are not economically competitive. A complex study of authors (Dasan et al. 2019) for given systems showed that the involved capital cost in tubular and bubble column PBRs accounts for nearly 47.5–86.2% of the total cost.

In open ponds, the value of total costs for operations and maintenance was 45.73%. Further, the authors analyzed the production of bioethanol as a by-product and concluded that the bioethanol plant does not give expected economic benefits. In contrast, biorefinery studies by Lam et al. (2017) predicted that the maximum total income from microalgae biomass is approximately €31 per kg of dry weight compared to the production cost of €6–7 per kg of dry weight. The condition to achieve such values requires minimization of the cost for downstream processing. Therefore, the research activities pass through technical economical criteria where simple and cost-effective downstream processing techniques have to be developed and applied. Hence, complex multilayer investigations on microalgae biorefinery are still necessary prior to realization in the market.

18.10 Challenges and Future Prospects

Overall, the development of cost-effective scenarios for downstreaming involving minimum possible unit operations is obligatory to achieve a technical–economic competitive microalgae biorefinery. Hence, the multiple and cascade extraction of desirable fractions of the whole microalgae, namely, proteins followed by lipids, carbohydrates, and, eventually, other high-value products. For example, the application of mild liquid-based extraction results in negligible damages to other fractions. From these results, the recovery of the maximum number of products from microalgae is a function of the success of the extraction technology. The products from algae already are obtained in industrial-scale plants. Then they are used for the production of BAC as well as in the aquaculture industry and animal feed industry among other applications (Smetana et al. 2017). It must be taken into the consideration the following facts: investments for startup a new facility depend on the advisability and risk assessment because microalgae products are not highly competitive in the particular market. As a principle that investors look for income under the requirements of the market. It means that they search for a minimum risk on the market in long-term competition for any new investment (Caporgno and Mathys 2018).

The biomass composition depends on microalgae species and the working conditions (Brennan and Owende 2010; Chacón-Lee and González-Mariño 2010). In the biomass production processes, downstream operations are commonly costly; therefore, they significantly impact the cost of the final product. It must be noted that the scale of the plant and cultivation medium regulates the costs, as well (Fasaei et al. 2018).

Currently, the most robust engineering specification for algae culturing is large-scale open pond/lagoon (Smetana et al. 2017). Their advantages are simplicity for realization and operation which results in cheaper production on huge scales. Because the ponds are open, the contamination is a serious problem, and the algal physiology is not constant because of variation in the operational conditions during the day (Xu et al. 2009). The reliable production of cyanobacteria/microalgae biomass definitely passed through the development of cost-effective closed large-scale PBRs offering innovative systems of light penetration and distribution inside throughout the reactor. Flashing light effects are the key in this field to tremendously increased CO_2 utilization and as a result maximization of biomass productivity (Kroumov et al. 2016; Hinterholz et al. 2019).

Detailed theoretical analysis of closed PBRs reveals the potential of the modern mathematical approaches based on system analysis theory done by Kroumov et al. (2016). In this review-research article, the PBR was considered as a sophisticated system including many sub-systems and different hierarchic levels. Special attention was given to light irradiance as a PBR sub-system. The current state of the art has overcome this issue because of the application of optical fibers to the optimal distribution of internal illumination of the PBRs. (Glemser et al. 2016; Sun et al. 2016).

Biorefinery concept for successful microalgae technology requires modern and robust approaches for optimization of extraction techniques of BAC at the given engineering specification. Therefore, if the cyanobacteria/microalgae biomass is going to be utilized completely, this will result in high profits for humans in the long term. Environmental protection needs a lot of the application of microalgae to reduce the greenhouse gas CO_2 released by coal-fired plants as well as by all fermentation plants, especially by plants for the production of ethanol from molasses and waste wood.

18.11 Conclusions

The optimal biomass production process can be considered as a complex system where many subsystems are involved with their sophisticated interactions. Nutrients supply, PBRs design, and downstream operations are crucial sub-systems in which optimal parameters are base to reach zero-waste biorefinery concept. Scale-up is the last and most difficult task to be solved before the technology reaches the competitive free market.

The future of cyanobacteria/microalgae production mostly dependents on the development of innovative closed large-scale PBRs with novel light penetration trajectories inside throughout the reactor. Detailed theoretical analysis of closed PBRs, based on the modern mathematical approaches and techniques, reveals their potential under the system analysis theory done by (Kroumov et al. 2016).

Biorefinery concept for successful zero-waste microalgae technology requires modern and robust approaches for further optimization of extraction procedures of high value-added bio-components and nutraceuticals at the given engineering specification. Hence, the strategy will open new opportunities for industrial production of many BAC needed for human society.

Acknowledgments The authors gratefully acknowledge the support of the Bulgarian National Science Foundation under the grant for fundamental scientific studies № КП-06-Н37/12.

References

Acién Fernández FG, González-López CV, Fernández Sevilla JM, Molina Grima E (2012) Conversion of CO_2 into biomass by microalgae: how realistic a contribution may it be to significant CO_2 removal? Appl Microbiol Biotechnol 96(3):577–586

Alagesan S, Gaudana SB, Krishnakumar S, Wangikar PP (2013) Model based optimization of high cell density cultivation of nitrogen-fixing cyanobacteria. Bioresour Technol 148:228–233

Ali H, Solsvik J, Wagner JL, Zhang D, Hellgardt K, Park CW (2019) CFD and kinetic-based modeling to optimize the sparger design of a large-scale photobioreactor for scaling up of biofuel production. Biotechnol Bioeng 116(9):2200–2211

Arad S, Levy-Ontman O (2010) Red microalgal cell-wall polysaccharides: biotechnological aspects. Curr Opin Biotechnol 21(3):358–364

Aslam A, Thomas-Hall SR, Mughal TA, Schenk PM (2017) Selection and adaptation of microalgae to growth in 100% unfiltered coal-fired flue gas. Bioresour Technol 233:271–283

AstaReal (2019) Our company. http://astarealusa.com/about-us/; (accessed 2019 August 28)

Bari GS, Suess TN, Anderson GA, Gent SP (2015) Hydrodynamic and heat transfer effects of varying sparger spacing within a column photobioreactor using computational fluid dynamics. J Fuel Cell Sci Technol 12(1):011004

Barsanti L, Gualtieri P (2018) Is exploitation of microalgae economically and energetically sustainable. Algal Res-Biomass Biofuels Bioprod 31:107–115

Becker EW (2013) Microalgae for human and animal nutrition. In: Richmond A, Hu Q (eds) Handbook of microalgal culture. John Wiley & Sons, Ltd, pp 461–503

Begum H, Yusoff FM, Banerjee S, Khatoon H, Shariff M (2016) Availability and utilization of pigments from microalgae. Crit Rev Food Sci Nutr 56(13):2209–2222

Bhattacharya M, Goswami S (2020) Microalgae: a green multi-product biorefinery for future industrial prospects. Biocatal Agric Biotechnol 25:101580

Birol G, Doruker P, Kirdar B, Önsan Zİ, Ülgen K (1998) Mathematical description of ethanol fermentation by immobilised *Saccharomyces cerevisiae*. Process Biochem 33(7):763–771

Bitog JP, Lee IB, Lee CG, Kim KS, Hwang HS, Hong SW, Seo IH, Kwon KS, Mostafa E (2011) Application of computational fluid dynamics for modeling and designing photobioreactors for microalgae production: a review. Comput Electron Agric 76(2):131–147

Bjornsson WJ, MacDougall KM, Melanson JE, O'Leary SJB, McGinn PJ (2012) Pilot-scale supercritical carbon dioxide extractions for the recovery of triacylglycerols from microalgae: a practical tool for algal biofuels research. J Appl Phycol 24(3):547–555

Bleakley S, Hayes M (2017) Algal proteins: extraction, application, and challenges concerning production. Foods (Basel, Switzerland) 6(5):33

Borowitzka M (2018) Commercial-scale production of microalgae for bioproducts. In: La Barre S, Bates SS (eds) Blue Biotechnology: production and use of marine molecules. Wiley-VCH Verlag GmbH & Co. KGaA, pp 33–65

Brennan L, Owende P (2010) Biofuels from microalgae: a review of technologies for production, processing, and extractions of biofuels and co-products. Renew Sust Energ Rev 14(2):557–577

Bumbak F, Cook S, Zachleder V, Hauser S, Kovar K (2011) Best practices in heterotrophic high-cell-density microalgal processes: achievements, potential and possible limitations. Appl Microbiol Biotechnol 91(1):31–46

Cai T, Park SY, Li Y (2013) Nutrient recovery from wastewater streams by microalgae: status and prospects. Renew Sust Energ Rev 19:360–369

Caporgno MP, Mathys A (2018) Trends in microalgae incorporation into innovative food products with potential health benefits. Front Nutr 5:58

Cardoso LAC, Karp SG, Vendruscolo F, Kanno KYF, Zoz LIC, Carvalho JC (2017) Chapter 8, Biotechnological production of carotenoids and their applications in food and pharmaceutical products. In: Cvetkovic D, Nikolic G (eds) Carotenoids. InTechOpen, p 125–141

Cerón-García MC, Macías-Sánchez MD, Sánchez-Mirón A, García-Camacho F, Molina-Grima E (2013) A process for biodiesel production involving the heterotrophic fermentation of *Chlorella protothecoides* with glycerol as the carbon source. Appl Energy 103:341–349

Chacón-Lee TL, González-Mariño GE (2010) Microalgae for "Healthy" foods – possibilities and challenges. Compr Rev Food Sci Food Saf 9:655–675

Chandra R, Iqbal HMN, Vishal G, Lee H-S, Nagra S (2019) Algal biorefinery: a sustainable approach to valorize algal-based biomass towards multiple product recovery. Bioresour Technol 278:346–359

Cheah WY, Show PL, Chang JS, Ling TC, Juan JC (2015) Biosequestration of atmospheric CO_2 and flue gas-containing CO_2 by microalgae. Bioresour Technol 184:190–201

Chen C-Y, Zhao X-Q, Yen H-W, Ho S-H, Cheng C-L, Lee D-J, Bai F-W, Chang J-S (2013) Microalgae-based carbohydrates for biofuel production. Biochem Eng J 78:1–10

Chew KW, Yap JY, Show PL, Suan NH, Juan JC, Ling TC, Lee D-J, Chang J-S (2017) Microalgae biorefinery: high value products perspectives. Bioresour Technol 229:53–62

Chochois V, Dauvillee D, Beyly A, Tolleter D, Cuine S, Timpano H, Ball S, Cournac L, Peltier G (2009) Hydrogen production in *Chlamydomonas*: photosystem II-dependent and -independent pathways differ in their requirement for starch metabolism. Plant Physiol 151(2):631–640

Chowdhury R, Franchetti M (2017) Life cycle energy demand from algal biofuel generated from nutrients present in the dairy waste. Sustain Prod Consump 9:22–27

Christaki E, Florou-Paneri P, Bonos E (2011) Microalgae: a novel ingredient in nutrition. Int J Food Sci Nutr 62(8):794–799

Cui Y, Rashid N, Hu N, Rehman MSU, Han J-I (2014) Electricity generation and microalgae cultivation in microbial fuel cell using microalgae-enriched anode and bio-cathode. Energy Convers Manag 79:674–680

Cui X, Yang J, Feng Y, Zhang W (2020) Simulation of a novel tubular microalgae photobioreactor with aerated tangent inner tubes: improvements in mixing performance and flashing-light effects. Archaea (Vancouver, BC) 2020:8815263

Darnoko D, Cheryan M (2000) Kinetics of palm oil transesterification in a batch reactor. J Am Oil Chem Soc 77(12):1263–1267

Dasan YK, Lam MK, Yusup S, Lim JW, Lee KT (2019) Life cycle evaluation of microalgae biofuels production: effect of cultivation system on energy, carbon emission and cost balance analysis. Sci Total Environ 688:112–128

De Bhowmick G, Sarmah AK, Sen R (2019) Zero-waste algal biorefinery for bioenergy and biochar: a green leap towards achieving energy and environmental sustainability. Sci Total Environ 650:2467–2482

De Vries M, De Boer IJM (2010) Comparing environmental impacts for livestock products: a review of life cycle assessments. Livest Sci 128(1):1–11

del Rio-Chanona EA, Wagner JL, Ali H, Fiorelli F, Zhang D, Hellgardt K (2019) Deep learning-based surrogate modeling and optimization for microalgal biofuel production and photobioreactor design. AICHE J 65(3):915–923

Depra MC, dos Santos AM, Severo IA, Santos AB, Zepka LQ, Jacob-Lopes E (2018) Microalgal biorefineries for bioenergy production: can we move from concept to industrial reality? Bioenergy Res 11(4):727–747

Dourado A, Calvet JL, Sevely Y, Goma G (1987) Modeling and static optimization of the ethanol production in a cascade reactor. II Static optimization. Biotechnol Bioeng 29(2):195–203

Elmaadawy K, Hu J, Guo S, Hou H, Xu J, Wang D, Liang T, Yang J, Liang S, Xiao K, Liu B (2020) Enhanced treatment of landfill leachate with cathodic algal biofilm and oxygen-consuming unit in a hybrid microbial fuel cell system. Bioresource Technol:123420

Faried M, Samer M, Abdelsalam E, Yousef RS, Attia YA, Ali AS (2017) Biodiesel production from microalgae: processes, technologies and recent advancements. Renew Sust Energ Rev 79:893–913

Fasaei F, Bitter JH, Slegers PM, van Boxtel AJB (2018) Techno-economic evaluation of microalgae harvesting and dewatering systems. Algal Res 31:347–362

Freedman B, Butterfield RO, Pryde EH (1986) Transesterification kinetics of soybean oil 1. J Am Oil Chem Soc 63(10):1375–1380

Gao X, Kong B, Vigil RD (2018) Simulation of algal photobioreactors: recent developments and challenges. Biotechnol Lett 40:1311–1327

García-López DA, Olguína EJ, González-Portela RE, Sánchez-Galván G, De Philippis R, Lovitt RW, Llewellyn CA, Fuentes-Grünewald C, Parra Saldívar R (2020) A novel two-phase bioprocess for the production of *Arthrospira* (*Spirulina*) *maxima* LJGR1 at pilot plant scale during different seasons and for phycocyanin induction under controlled conditions. Bioresour Technol 298:122548

Glemser M, Heining M, Schmidt J, Becker A, Garbe D, Buchholz R, Brück T (2016) Application of light-emitting diodes (LEDs) in cultivation of phototrophic microalgae: current state and perspectives. Appl Microbiol Biotechnol 100(3):1077–1088

Gouveia L, Batista AP, Sousa I, Raymundo A, Bandarra NM (2008) Microalgae in novel food product. In: Papadopoulos KN (ed) Food chemistry research developments. Nova Science Publishers, Inc., New York, pp 1–37

Goyal HB, Seal D, Saxena RC (2008) Bio-fuels from thermochemical conversion of renewable resources: a review. Renew Sust Energ Rev 12(2):504–517

Greenwald E, Gordon JM, Zarmi Y (2012) Physics of ultra-high bioproductivity in algal photobioreactors. Appl Phys Lett 100(14):143703

Guo W, Cheng J, Song Y, Kumar S, Ali KA, Wang Y, Li X, Yang W (2020) Improving flashing light frequency and CO_2 fixation rate with vortex movement of algal cells in raceway pond with conic baffles. Chem Eng Sci 216:115536

Heredia-Arroyo T, Wei W, Ruan R, Hu B (2011) Mixotrophic cultivation of *Chlorella vulgaris* and its potential application for the oil accumulation from non-sugar materials. Biomass Bioenergy 35(5):2245–2253

Hinterholz CL, Schuelter AR, Módenes AN, Trigueros DEG, Borba CE, Espinoza-Quiñones FR, Kroumov AD (2017) Microalgae flat plate-photobioreactor (FP-PBR) system development: computational tools to improve experimental results. Acta Microbiol Bulg 33(3):119–124

Hinterholz CL, Trigueros DEG, Modenes AN, Borba CE, Scheufele FB, Schuelter AR, Kroumov AD (2019) Computational fluid dynamics applied for the improvement of a flat-plate photobioreactor towards high-density microalgae cultures. Biochem Eng J 151:107257

Hirano A, Ueda R, Hirayama S, Ogushi Y (1997) CO_2 fixation and ethanol production with microalgal photosynthesis and intracellular anaerobic fermentation. Energy 22(2):137–142

Hoffman J, Pate RC, Drennen T, Quinn JC (2017) Techno-economic assessment of open microalgae production systems. Algal Res 23:51–57

Hu IC (2019) Chapter 14: Production of potential coproducts from microalgae. In: Pandey A, Chang J-S, Soccol CR et al (eds) Biofuels from algae, 2nd edn. Elsevier, pp 345–358

Hu Q, Kurano N, Kawachi M, Iwasaki I, Miyachi S (1998) Ultrahigh-cell-density culture of a marine green alga *Chlorococcum littorale* in a flat-plate photobioreactor. Appl Microbiol Biotechnol 49(6):655–662

Huang LP, Jin B, Lant P, Zhou J (2005) Simultaneous saccharification and fermentation of potato starch wastewater to lactic acid by *Rhizopus oryzae* and *Rhizopus arrhizus*. Biochem Eng J 23 (3):265–276

Jacob-Lopes E, Mérida LGR, Queiroz MI, Zepka LQ (2015) Chapter 5, Microalgal biorefineries. In: Jacob-Lopes E, Zepka LQ (eds) Biomass production and uses. IntechOpen, p 81–106

Jacob-Lopes E, Maroneze MM, Depra MC, Sartori RB, Dias RR, Zepka LQ (2019) Bioactive food compounds from microalgae: an innovative framework on industrial biorefineries. Curr Opin Food Sci 25:1–7

Jaswir I, Hammed A (2011) Anti-inflammatory compounds of macro algae origin: a review. J Med Plants Res 5:7146–7154

Javed F, Aslam M, Rashid N, Shamair Z, Khan AL, Yasin M, Fazal T, Hafeez A, Rehman F, Rehman MSU et al (2019a) Microalgae-based biofuels, resource recovery and wastewater treatment: a pathway towards sustainable biorefinery. Fuel 255:115826

Javed F, Aslam M, Rashid N, Shamair Z, Khan AL, Yasin M, Fazal T, Hafeez A, Rehman F, Rehman MSU, Khan Z (2019b) Microalgae-based biofuels, resource recovery and wastewater treatment: a pathway towards sustainable biorefinery. Fuel 255:115826

Jorquera O, Kiperstok A, Sales EA, Embirucu M, Ghirardi ML (2010) Comparative energy life-cycle analyses of microalgal biomass production in open ponds and photobioreactors. Bioresour Technol 101(4):1406–1413

Juan JC, Kartika DA, Wu TY, Hin T-YY (2011) Biodiesel production from jatropha oil by catalytic and non-catalytic approaches: an overview. Bioresour Technol 102(2):452–460

Kadam KL (2002) Environmental implications of power generation via coal-microalgae cofiring. Energy 27(10):905–922

Kaffarov VV, Vinarov AJ, Gordeev LS (1979) Modeling biochemical reactors. Lesnaya Promishlenost, Forrest Industry, Moscow

Kaffarov VV, Vinarov AJ, Gordeev LS (1985) Modeling and system analysis of biochemical industrial production. Lesnaya Promishlenost, Forrest Industry, Moscow

Kakarla R, Min B (2019) Sustainable electricity generation and ammonium removal by microbial fuel cell with a microalgae assisted cathode at various environmental conditions. Bioresour Technol 284:161–167

Khandelwal A, Vijay A, Dixit A, Chhabra M (2018) Microbial fuel cell powered by lipid extracted algae: a promising system for algal lipids and power generation. Bioresour Technol 247:520–527

Kim M, Yim JH, Kim SY, Kim HS, Lee WG, Kim SJ, Kang PS, Lee CK (2012) In vitro inhibition of influenza A virus infection by marine microalga-derived sulfated polysaccharide p-KG03. Antivir Res 93(2):253–259

Kose A, Ozen MO, Elibol M, Oncel SS (2017) Investigation of in vitro digestibility of dietary microalga *Chlorella vulgaris* and cyanobacterium *Spirulina platensis* as a nutritional supplement. 3 Biotech 7(3):170–170

Koyande AK, Chew KW, Lim J-W, Lee SY, Lam MK, Show P-L (2019a) Optimization of protein extraction from *Chlorella vulgaris* via novel sugaring-out assisted liquid biphasic electric flotation system. Eng Life Sci 19(12):968–977

Koyande AK, Show P-L, Guo R, Tang B, Ogino C, Chang J-S (2019b) Bio-processing of algal bio-refinery: a review on current advances and future perspectives. Bioengineered 10 (1):574–592

Kroumov AD, MГidenes AN, Tait MCDA (2006) Development of new unstructured model for simultaneous saccharification and fermentation of starch to ethanol by recombinant strain. Biochem Eng J 28(3):243–255

Kroumov AD, Modenes AN, Wenzel BM (2007) Development of the enzymatic kinetics model of vegetable oils transesterification for biodiesel production. Acta Scientiarum Technol 29(1)

Kroumov AD, Gacheva G, Iliev I, Alexandrov S, Pilarski P, Petkov G (2013) Analysis of S_f/V ratio of photobioreactors linked with algal physiology. Genet Plant Physiol 3(1–2):55–64

Kroumov AD, Módenes AN, Trigueros DEG (2015) A complex theoretical approach for algal medium optimization for CO_2 fixation from flue gas. Acta Microbiol Bulg 31(1):61–70

Kroumov AD, Modenes AN, Trigueros DEG, Espinoza-Quinones FR, Borba CE, Scheufele FB, Hinterholz CL (2016) A systems approach for CO_2 fixation from flue gas by microalgae: theory review. Process Biochem 51(11):1817–1832

Kroumov AD, Scheufele FB, Trigueros DEG, Modenes AN, Zaharieva MM, Najdenski HM (2017) Chapter 11: Modeling and techno-economic analysis of algae for bio-energy and co-products. In: Rastogi RP, Madamwar D, Pandey A (eds) Algal green chemistry: recent progress in biotechnology. Elsevier, pp 202–241

Kuo CM, Lin TH, Yang YC, Zhang WX, Lai JT, Wu HT, Chang JS, Lin CS (2017) Ability of an alkali-tolerant mutant strain of the microalga *Chlorella* sp. AT1 to capture carbon dioxide for increasing carbon dioxide utilization efficiency. Bioresour Technol 244(Pt 1):243–251

Lee RA, Lavoie J-M (2013) From first- to third-generation biofuels: challenges of producing a commodity from a biomass of increasing complexity. Anim Front 3(2):6–11

Lee D-J, Chang J-S, Lai J-Y (2015) Microalgae—microbial fuel cell: a mini review. Bioresour Technol 198:891–895
Lehr F, Posten C (2009) Closed photo-bioreactors as tools for biofuel production. Curr Opin Biotechnol 20(3):280–285
Luzi G, McHardy C, Lindenberger C, Rauh C, Delgado A (2019) Comparison between different strategies for the realization of flashing-light effects вЂ—Pneumatic mixing and flashing illumination. Algal Res 38:101404
Mandik YI, Cheirsilp B, Srinuanpan S, Maneechote W, Boonsawang P, Prasertsan P, Sirisansaneeyakul S (2020) Zero-waste biorefinery of oleaginous microalgae as promising sources of biofuels and biochemicals through direct transesterification and acid hydrolysis. Process Biochem 95:214–222
Markou G, Angelidaki I, Georgakakis D (2012) Microalgal carbohydrates: an overview of the factors influencing carbohydrates production, and of main bioconversion technologies for production of biofuels. Appl Microbiol Biotechnol 96(3):631–645
Matsui MS, Muizzuddin N, Arad S, Marenus K (2003) Sulfated polysaccharides from red microalgae have antiinflammatory properties in vitro and in vivo. Appl Biochem Biotechnol 104(1):13–22
Merchuk J, Wu X (2003) Modeling of photobioreactors: application to bubble column simulation. J Appl Phycol 15:163–169
Merchuk JC, Garcia-Camacho F, Molina-Grima E (2007) Photobioreactor design and fluid dynamics. Chem Biochem Eng Q 21(4):345–355
Miranda JR, Passarinho PC, Gouveia L (2012) Bioethanol production from *Scenedesmus obliquus* sugars: the influence of photobioreactors and culture conditions on biomass production. Appl Microbiol Biotechnol 96(2):555–564
Mohamed MS, Tan JS, Kadkhodaei S, Mohamad R, Mokhtar MN, Ariff AB (2014) Kinetics and modeling of microalga *Tetraselmis* sp. FTC 209 growth with respect to its adaptation toward different trophic conditions. Biochem Eng J 88:30–41
Mohamed SN, Hiraman PA, Muthukumar K, Jayabalan T (2020) Bioelectricity production from kitchen wastewater using microbial fuel cell with photosynthetic algal cathode. Bioresour Technol 295:122226
Nishiwaki A, Dunn IJ (1999) Analysis of the performance of a two-stage fermentor with cell recycle for continuous ethanol production using different kinetic models. Biochem Eng J 4(1):37–44
Ogbonna JC, Tanaka H (2000) Production of pure photosynthetic cell biomass for environmental biosensors. Mater Sci Eng C 12(1):9–15
Olivieri G, Gargiulo L, Lettieri P, Mazzei L, Salatino P, Marzocchella A (2015a) Photobioreactors for microalgal cultures: a Lagrangian model coupling hydrodynamics and kinetics. Biotechnol Prog 31(5):1259–1272
Olivieri G, Gargiulo L, Lettieri P, Mazzei L, Salatino P, Marzocchella A (2015b) Photobioreactors for microalgal cultures: AL agrangian model coupling hydrodynamics and kinetics. Biotechnol Prog 31(5):1259–1272
Ouada HB, Ammar J, inventors. (2017). U.S. Patent Application No. 15/505,935
Pan-utai W, Iamtham S (2019) Extraction, purification and antioxidant activity of phycobiliprotein from *Arthrospira platensis*. Process Biochem 82:189–198
Park JK, Kim ZH, Lee CG, Synytsya A, Jo HS, Kim SO, Park JW, Park YI (2011) Characterization and immunostimulating activity of a water-soluble polysaccharide isolated from *Haematococcus lacustris*. Biotechnol Bioprocess Eng 16(6):1090–1098
Perner-Nochta I, Posten C (2007) Simulations of light intensity variation in photobioreactors. J Biotechnol 131(3):276–285
Pires JC, Alvim-Ferraz MC, Martins FG (2017) Photobioreactor design for microalgae production through computational fluid dynamics: a review. Renew Sust Energ Rev 79:248–254
Polakovic M, Bryjak J (2004) Modelling of potato starch saccharification by an *Aspergillus niger* glucoamylase. Biochem Eng J 18(1):57–63

Posten C (2009) Design principles of photobioreactors for cultivation of microalgae. Eng Life Sci 9 (3):165–177

Pruvost J, Cornet JF, Legrand J (2008) Hydrodynamics influence on light conversion in photobioreactors: an energetically consistent analysis. Chem Eng Sci 63(14):3679–3694

Pruvost J, Le Gouic B, Lepine O, Legrand J, Le Borgne F (2016a) Microalgae culture in building-integrated photobioreactors: biomass production modelling and energetic analysis. Chem Eng J 284:850–861

Pruvost J, Le Borgne F, Artu A, Cornet JF, Legrand J (2016b) Industrial photobioreactors and scale-up concepts. Elsevier. Adv Chem Eng 48:257–310, Photobioreaction Engineering. https://doi.org/10.1016/bs.ache.2015.11.002ff.ffhal-02539887

Raheem A, WAKG WA, Taufiq Yap YH, Danquah MK, Harun R (2015) Optimization of the microalgae *Chlorella vulgaris* for syngas production using central composite design. RSC Adv 5(88):71805–71815

Rahman DY, Sarian FD, van Wijk A, Martinez-Garcia M, van der Maarel MJEC (2017) Thermostable phycocyanin from the red microalga *Cyanidioschyzon merolae*, a new natural blue food colorant. J Appl Phycol 29(3):1233–1239

Rastogi RP, Sonani RR, Madamwar D (2015) Cyanobacterial sunscreen scytonemin: role in photoprotection and biomedical research. Appl Biochem Biotechnol 176:1551–1563

Salinas-Salazar C, Saul Garcia-Perez J, Chandra R, Castillo-Zacarias C, Iqbal HMN, Parra-Saldívar R (2019) Methods for extraction of valuable products from microalgae biomass. In: Alam M, Wang Z (eds) Microalgae biotechnology for development of biofuel and wastewater treatment. Springer, Singapore, pp 245–263

Sansawa H, Endo H (2004) Production of intracellular phytochemicals in *Chlorella* under heterotrophic conditions. J Biosci Bioeng 98(6):437–444

Schepetkin IA, Quinn MT (2006) Botanical polysaccharides: macrophage immunomodulation and therapeutic potential. Int Immunopharmacol 6(3):317–333

Scheufele FB, Hinterholz CL, Zaharieva MM, Najdenski HM, Modenes AN, Trigueros DEG, Borba CE, Espinoza-Quinones FR, Kroumov AD (2019) Complex mathematical analysis of photobioreactor system. Eng Life Sci 19(12):844–859

Schuelter AR, Kroumov AD, Hinterholz CL, Fiorini A, Trigueros DEG, Vendruscolo EG, Zaharieva MM, Modenes AN (2019) Isolation and identification of new microalgae strains with antibacterial activity on food-borne pathogens. Engineering approach to optimize synthesis of desired metabolites. Biochem Eng J 141:28–39

Schultz H. 2016. AstaReal stands by assertion that its process yields highest quality astaxanthin. https://www.nutraingredients-usa.com/article/2016/11/17/astareal-stands-by-assertion-that-its-process-yields-highest-quality-astaxanthin; (accessed 2019 August 28)

Shukla M, Kumar S (2018) Algal growth in photosynthetic algal microbial fuel cell and its subsequent utilization for biofuels. Renew Sust Energ Rev 82:402–414

Simas-Rodrigues C, Villela HD, Martins AP, Marques LG, Colepicolo P, Tonon AP (2015) Microalgae for economic applications: advantages and perspectives for bioethanol. J Exp Bot 66(14):4097–4108

Singh SP, Kumari S, Rastogi RP, Sinha RP (2010) Photoprotective and biotechnological potentials of cyanobacterial sheath pigment, scytonemin. Afr J Biotechnol 9(5):580–588

Sivaramakrishnan R, Suresh S, Incharoensakdi A (2019) *Chlamydomonas* sp. as dynamic biorefinery feedstock for the production of methyl ester and ε-polylysine. Bioresour Technol 272: 281–287

Slegers PM, Wijffels RH, van Straten G, van Boxtel AJB (2011) Design scenarios for flat panel photobioreactors. Appl Energy 88(10):3342–3353

Smetana S, Sandmann M, Rohn S, Pleissner D, Heinz V (2017) Autotrophic and heterotrophic microalgae and cyanobacteria cultivation for food and feed: life cycle assessment. Bioresour Technol 245(Pt A):162–170

Sonani RR, Rastogi RP, Madamwar D (2015) Antioxidant potential of phycobiliproteins: role in anti-aging research. Biochem Anal Biochem 4:2

Sonani RR, Rastogi RP, Patel R, Madamwar D (2016) Recent advances in production, purification and applications of phycobiliproteins. World J Biol Chem 7(1):100–109

Straka L, Rittmann BE (2019) Growth kinetics and mathematical modeling of *Synechocystis* sp. PCC 6803 under flashing light. Biotechnol Bioeng 116(2):469–474

Strik DP, Terlouw H, Hamelers HV, Buisman CJ (2008) Renewable sustainable biocatalyzed electricity production in a photosynthetic algal microbial fuel cell (PAMFC). Appl Microbiol Biotechnol 81(4):659–668

Sui Y, Vlaeminck SE (2020) *Dunaliella* microalgae for nutritional protein: an undervalued asset. Trends Biotechnol 38(1): 10–12

Sun Y, Huang Y, Liao Q, Fu Q, Zhu X (2016) Enhancement of microalgae production by embedding hollow light guides to a flat-plate photobioreactor. Bioresour Technol 207:31–38

't Lam GP, Vermuë MH, Eppink MHM, Wijffels RH, van den Berg C (2017) Multi-product microalgae biorefineries: from concept towards reality. Trends Biotechnol 36(2):216–227

Tannin-Spitz T, Bergman M, van- Moppes D, Grossman S, Arad S (2005) Antioxidant activity of the polysaccharide of the red microalga *Porphyridium* sp. J Appl Phycol 17(3):215–222

Thomas DM, Mechery J, Paulose SV (2016) Carbon dioxide capture strategies from flue gas using microalgae: a review. Environ Sci Pollut Res Int 23(17):16926–16940

Thomassen G, Van Dael M, Van Passel S (2018) The potential of microalgae biorefineries in Belgium and India: an environmental techno-economic assessment. Bioresour Technol 267:271–280

Tredici MR (2003) Mass production of microalgae: photobioreactors. In: Richmond A (ed) Handbook of microalgal culture: biotechnology and applied phycology. Wiley Online Library, pp 178–214

Tredici MR, Materassi R (1992) From open ponds to vertical alveolar panels: the Italian experience in the development of reactors for the mass cultivation of phototrophic microorganisms. J Appl Phycol 4(3):221–231

Tredici MR, Bassi N, Prussi M, Biondi N, Rodolfi L, Chini Zittelli G, Sampietro G (2015) Energy balance of algal biomass production in a 1-ha "Green Wall Panel" plant: how to produce algal biomass in a closed reactor achieving a high Net Energy Ratio. Appl Energy 154:1103–1111

Udayan A, Arumugam M, Pandey A (2017) Chapter 4: Nutraceuticals from algae and cyanobacteria. In: Rastogi RP, Madamwar D, Pandey A (eds) Algal green chemistry. Elsevier, Amsterdam, pp 65–89

Ueno Y, Kurano N, Miyachi S (1998) Ethanol production by dark fermentation in the marine green alga, *Chlorococcum littorale*. J Ferment Bioeng 86(1):38–43

Ugwu CU, Aoyagi H, Uchiyama H (2008) Photobioreactors for mass cultivation of algae. Bioresour Technol 99(10):4021–4028

Vanthoor-Koopmans M, Wijffels RH, Barbosa MJ, Eppink MH (2013) Biorefinery of microalgae for food and fuel. Bioresour Technol 135:142–149

Vernès L, Granvillain P, Chemat F, Vian M (2015) Phycocyanin from *Arthrospira platensis*. Production, extraction and analysis. Curr Biotechnol 4(4):481–491

Vuppaladadiyam AK, Yao JG, Florin N, George A, Wang X, Labeeuw L, Jiang Y, Davis RW, Abbas A, Ralph P et al (2018) Impact of flue gas compounds on microalgae and mechanisms for carbon assimilation and utilization. ChemSusChem 11(2):334–355

Wang B, Lan CQ, Horsman M (2012) Closed photobioreactors for production of microalgal biomasses. Biotechnol Adv 30(4):904–912

Wang F, Gao B, Huang L, Su M, Dai C, Zhang C (2018) Evaluation of oleaginous eustigmatophycean microalgae as potential biorefinery feedstock for the production of palmitoleic acid and biodiesel. Bioresour Technol 270:30–37

Wenzel B, Tait M, Módenes A, Kroumov A (2006) Modelling chemical kinetics of soybean oil transesterification process for biodiesel production: an analysis of molar ratio between alcohol and soybean oil temperature changes on the process conversion rate. Bioautomation 5:13–22

Wichuk K, Brynjólfsson S, Fu W (2014) Biotechnological production of value-added carotenoids from microalgae: emerging technology and prospects. Bioengineered 5(3):204–208

Xu L, Weathers PJ, Xiong X-R, Liu C-Z (2009) Microalgal bioreactors: Challenges and opportunities. Eng Life Sci 9(3):178–189

Yang Z, Pei H, Hou Q, Jiang L, Zhang L, Nie C (2018) Algal biofilm-assisted microbial fuel cell to enhance domestic wastewater treatment: nutrient, organics removal and bioenergy production. Chem Eng J 332:277–285

Yen H-W, Hu IC, Chen C-Y, Ho S-H, Lee D-J, Chang J-S (2013) Microalgae-based biorefinery: from biofuels to natural products. Bioresour Technol 135:166–174

Yen H-W, Hu IC, Chen C-Y, Chang J-S, Pandey A, Lee D-J, Chisti Y, Soccol CR (2014) Chapter 2: Design of photobioreactors for algal cultivation. In: Pandey A, Chang J-S, Soccol CR et al (eds) Biofuels from Algae. Elsevier, Amsterdam, pp 23–45

Zhou W, Wang J, Chen P, Ji C, Kang Q, Lu B, Li K, Liu J, Ruan R (2017) Bio-mitigation of carbon dioxide using microalgal systems: advances and perspectives. Renew Sust Energ Rev 76:1163–1175

Zittelli GC, Biondi N, Rodolfi L, Tredici MR (2013) Chapter 13, Photobioreactors for mass production of microalgae. In: Richmond A, Hu Q (eds) Handbook of microalgal culture: applied phycology and biotechnology, 2nd edn. Wiley-Blackwell, pp 225–266

Cyanobacteria as Renewable Sources of Bioenergy (Biohydrogen, Bioethanol, and Bio-Oil Production)

19

Ramachandran Sivaramakrishnan and Aran Incharoensakdi

Abstract

Cyanobacteria, formerly known as blue-green algae, are a diverse group of photosynthetic bacteria that are currently regarded as an important source of biofuels. This diversity of cyanobacteria makes them a prominent organism for a variety of applications. Several decades of cyanobacterial research have proven that cyanobacteria could produce various biomolecules including those for bioenergy with wide industrial application. Currently, the issues concerning the depletion of fossil fuels necessitate the need to find an alternative solution for global energy demand. Utilizing cyanobacteria for biofuel production has various benefits such as sequestration of CO_2, no requirement of arable land, and no competition with the food crop. Cyanobacteria have a high growth rate and embed valuable components like carbohydrates and lipids that are used for biofuel production. Cyanobacterial carbohydrates can be utilized to produce bioethanol or biohydrogen production, and lipids are considered for bio-oil production. Based on these advantages of cyanobacteria, this chapter highlights and exploits the recent trends in cyanobacteria biofuel production, such as biohydrogen, bioethanol, and bio-oil. The discussion and suggestions acquired from the recent studies are explored in this chapter which could provide insights

R. Sivaramakrishnan
Laboratory of Cyanobacterial Biotechnology, Department of Biochemistry, Faculty of Science, Chulalongkorn University, Bangkok, Thailand

A. Incharoensakdi (✉)
Laboratory of Cyanobacterial Biotechnology, Department of Biochemistry, Faculty of Science, Chulalongkorn University, Bangkok, Thailand

Academy of Science, Royal Society of Thailand, Bangkok, Thailand
e-mail: aran.i@chula.ac.th

R. P. Rastogi (ed.), *Ecophysiology and Biochemistry of Cyanobacteria*,
https://doi.org/10.1007/978-981-16-4873-1_19

into the utilization of cyanobacteria as renewable sources for future bioenergy demand.

Keywords

Biohydrogen · Bioethanol · Bio-oil · Cyanobacteria · Feedstocks · Fermentation · Pyrolysis · Zeolites

19.1 Introduction

It is known that 85% of the world's energy requirement is derived from fossil fuels. Continuous consumption will lead to future fuel depletion (Quintana et al. 2011). The carbon dioxide emission in the atmosphere is another setback caused by fossil fuels leading to climatic change and global warming issues (Rajneesh et al. 2017). In addition to carbon dioxide, other toxic components such as NO, SO_2, NO_2, polycyclic aromatic hydrocarbons, and black carbon are also produced by fossil fuels (Perera 2018). The greenhouse gas emission can affect human life directly and indirectly. Hence, renewable sources such as biomass, geothermal power, wind, and sunlight are being considered (Hosseini et al. 2018). Lignocellulose biomass is widely considered for biofuels despite various technological bottlenecks for large-scale commercial biofuel production (Ko et al. 2020). The photosynthetic microorganisms are attractively considered for the alternative feedstocks. Photosynthetic microorganisms utilize carbon dioxide and solar energy to synthesize various valuable products through photosynthesis (Li and Liao 2013; Robertson et al. 2011; Sivaramakrishnan and Incharoensakdi 2020). Hence, photosynthetic microorganisms are widely considered for alternative fuels which could be a promising alternative for fossil fuels (Demirbas 2008). Algal biomass is considered as the possible raw material for commercial biofuel production which could be sustainable, eco-friendly, and cost-effective. In addition, algal biomass can produce various bioproducts that add the value of biofuels and generate revenue other than that of the biofuel application (Kumar et al. 2020a, b). Alkane production from cyanobacteria could be another biofuel produced by the cyanobacteria in which the enzyme aldehyde deformylating oxygenase is responsible for the alkane production. Enhancing aldehyde deformylating oxygenase by various biotechnological strategies can give rise to increased alkane production which could increase the possibility of scale-up (Basri et al. 2020). Biofuels such as solid, gas, and liquids can be produced from photosynthetic microorganisms (Demirbas 2008). Photosynthetic biomass is the attractive choice for biorefinery applications that can produce multiple products from single feedstocks. Microalgae and cyanobacteria are the reservoirs of carbohydrates and lipids that can be further converted into biofuels such as bioethanol or biohydrogen and biodiesel, respectively. In addition to the biofuels, spent biomass can be utilized for biogas, bio-oil, and bio-char production.

Cyanobacteria are photosynthetic microorganisms and are classified as gram-negative. The origin of cyanobacteria on the Earth is around 3.5 billion years ago

(Mazard et al. 2016; Lindblad et al. 2012). Cyanobacteria are diverse in nature, unicellular, filamentous, or colony-forming organisms. The habitat of cyanobacteria is from geographically different environments including hot springs. Cyanobacteria perform CO_2 sequestration from the atmosphere and convert it into various valuable products that are used for biofuel production as well as being considered for the bioremediation processes (Bavandi et al. 2019; Patel et al. 2019). Genetic engineering and heterotrophic cultivation strategies are used to improve the valuable products available in cyanobacteria. Products that can be produced by cyanobacteria are hydrocarbons, fatty metabolites, carbohydrates, hydrogen, methane, proteins, diols, alcohols, terpenes, carboxylic acids, toxins, isoprenes, pigments, antioxidants, and vitamins (Oliver et al. 2016; Rastogi et al. 2017, 2018). The metabolic products produced by the cyanobacteria have different applications in the biofuel industries, pharmaceutical, cosmetics, and food industries (Rastogi and Sinha 2009; Rastogi et al. 2017). Hence, different strategies like stress induction, genetic engineering, and heterotrophic cultivation are being considered for efficient cyanobacteria products synthesis (de Farias Silva and Bertucco 2016).

19.2 Stages of Biofuel Development

Biofuels production is categorized into different generations depending on the type of feedstocks. The raw materials such as sugars or starch, food crops, and oil can serve as the feedstock for the biofuels which are considered as the first-generation biofuels. The major drawback of the first-generation biofuels is the requirement of huge fertile land, pesticides, water, and more importantly the competition with the food production (Rajneesh et al. 2017; Rosgaard et al. 2012). Hence, biofuels production from non-edible resources is considered second-generation biofuels. The nonedible second-generation feedstocks are agricultural materials or wastes, lignocellulosic, and cellulosic materials. Although second-generation feedstock does not compete with the food, the major drawbacks are the requirement of large arable area and the low biofuel yields. The mass production of feedstock in both first- and second-generation feedstocks is limited and cannot favor sustainable biofuel production (Quintana et al. 2011). Hence, researchers shifted the focus toward prokaryotic cyanobacteria and eukaryotic microalgae that are being considered as the third-generation biofuels (Demirbas and Fatih Demirbas 2011). Cyanobacteria have the efficiency to produce biofuels sustainably which could be considered for the future feedstock for biofuel production (Anahas and Muralitharan 2018). Cyanobacteria and microalgae are considered as the potential candidates for bioethanol or biohydrogen and biodiesel production due to the presence of carbohydrates and lipids as primary storage molecules (Khetkorn et al. 2017; Rastogi et al. 2018). Both cyanobacteria and microalgae have a high production rate when compared to other plant crops, and the requirement of land is much less, hence cyanobacteria and microalgae are considered as potential candidates for biofuels production. The combination of metabolic or genetic engineering with third-generation feedstock is considered as the fourth-generation biofuels (Rajneesh et al. 2017; Sarsekeyeva et al.

2015). It is worth noting that the photosynthetic rate of cyanobacteria is very high when compared to plants and microalgae. Cyanobacterial photosynthetic rate is 10% which is higher than those of the plants (1%) and microalgae (5%) (Sharma et al. 2011; Parmar et al. 2011).

19.3 Cyanobacterial Biohydrogen Production

Hydrogen is an important alternative fuel that has attracted great attention due to its clean nature. During combustion hydrogen releases only water as a by-product and hydrogen generates huge energy during combustion. However, current technologies involved in the hydrogen production from methane steam reforming or electrolysis using electricity have been criticized regarding sustainability (Ghosh et al. 2018). Microorganisms especially cyanobacteria and microalgae are considered as the sustainable resources for hydrogen production (Khetkorn et al. 2017; Monir et al. 2018). Comparing with physicochemical methods, the production of hydrogen from cyanobacteria and microalgae has been considered a promising route (Kumar et al. 2020a, b). By utilizing light energy, water molecules are split into electrons, protons, and oxygen, where the obtained protons are further converted into hydrogen (Manish and Banerjee 2008). Nitrogenase and hydrogenase are considered hydrogen-producing enzymes, which are the important enzymes involved in certain eukaryotic and prokaryotic metabolic regulations (Show et al. 2018).

However, the major setback in the hydrogen production from the microorganism is the low hydrogen yield. Various researches such as physiological modifications and process modifications have been carried out to improve the hydrogen production yield. In recent days, molecular approaches like cyanobacterial metabolic engineering showed potential toward improvement in photosynthesis efficiency and hydrogen production with oxygen tolerance. Different groups of microalgae and cyanobacteria have the ability to produce biohydrogen through photolysis, photocatabolism, and dark fermentation. The cyanobacterial biohydrogen production routes are presented in Fig. 19.1.

19.3.1 Utilization of Light for the Biohydrogen Production

19.3.1.1 Photolysis

Photolysis involves the utilization of light energy by microalgae for the splitting of water molecules into hydrogen and oxygen. The photosynthetic splitting of water molecules and conversion of the proton into hydrogen gas is represented by the equations below.

$$2H_2O \rightarrow 4H^+ + 4e^- + O_2$$

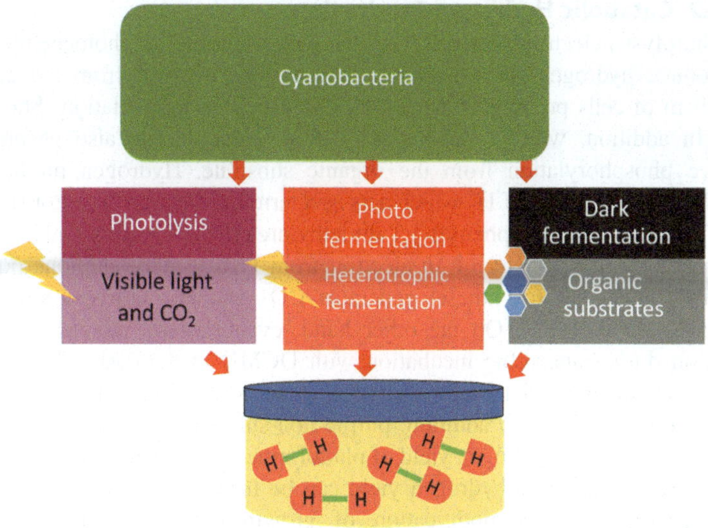

Fig. 19.1 Cyanobacterial biohydrogen production routes

$$4H^+ + 4e^- \rightarrow 2\,H_2$$

Cyanobacteria and microalgae are being considered photoautotrophs and their important energy metabolism is the production of hydrogen and oxygen. The conversion of proton into hydrogen gas is catalyzed by the hydrogenase, and its purity is about 98% (Hankamer et al. 2007).

The light-dependent photolysis is distinguished as direct and indirect photolysis. Direct photolysis is the transfer of light-driven electrons from the oxidation of water molecules into hydrogenase [Fe] which further reduces proton to H_2 (Florin et al. 2001). Protons [H^+] generated along with the electrons during photochemical water oxidation in the chloroplast of microalgae result in the simultaneous production of O_2 and H_2 gases. The protons [H^+] released from the photochemical water oxidation can also be utilized for the ATP synthesis by ATP synthase (Greenbaum et al. 1983). In some cases, cyanobacterial photosynthetic H_2 can be generated indirectly when the nutrient composition lacks sulfur. During sulfur deprivation, cyanobacteria utilize other internal substrates from catabolic reactions to produce hydrogen. Markov et al. (1997) reported that the *Anabaena variabilis* (1 g dry cell weight) can produce 12.5 mL H_2/h through direct photolysis. In the indirect photolysis method, changing the pH of *Gleocapsa alpicola* growth medium from 6.8 to 8.3 increased the H_2 production significantly (Troshina et al. 2002). A similar study reported that the alteration of temperature from 30 to 40 °C could double the volume of H_2 production. If scale-up of the production is considered, indirect photolysis could be the promising method for sustainable H_2 production. Altering the light capture capacity of the microorganism by genetic manipulations also increased the H_2 production (Turner et al. 2008).

19.3.1.2 Catabolic Hydrogen Production

In biophotolysis, electrons derived from the water splitting by photochemical reaction produce hydrogen. In cyanobacteria, electrons evolved from the catabolic metabolism of cells produce hydrogen by heterotrophic fermentation (Show et al. 2011). In addition, with photophosphorylation, electrons are also generated by oxidative phosphorylation from the organic substrate. Hydrogen production is driven by the integration of H^+ which acts as a terminal electron acceptor (Bennoun 2001). The 3-(3,4-dichlorophenyl)-1,1-dimethylurea (DCMU) acts as an inhibitor of PSII and blocks the transfer of an electron from PSII to plastoquinone pool, and hence incomplete photosynthesis generates CO_2 and H_2 with the ratio of 1:2 (Bamberger et al. 1982). On the other hand, cyanobacteria increased hydrogen production during anaerobic incubation with DCMU in the dark. This condition induced the cyanobacterial [Fe] – dehydrogenase which favors hydrogen production (Florin et al. 2001). In addition, purple non-sulfur bacteria (PNS) can significantly enhance the biohydrogen yield of photofermentation. The association of PNS and photofermentative biohydrogen yield can be further enhanced by altering the growth conditions. The modification of growth conditions involved in the biohydrogen production such as *C/N* ratio (substrate), light intensity, bacterial seed volume, and reactor design can improve the hydrogen production yields (Basak and Das 2007).

19.3.2 Dark Fermentative Biohydrogen Production

Photofermentation and biophotolysis require sunlight to generate energy. Apart from the sunlight, some anaerobic microorganisms, microalgae, and cyanobacteria produce hydrogen from the organic substrates under dark conditions through dark fermentation. Dark fermentation is an attractive route to produce hydrogen in the photobioreactors. The cost of hydrogen production through photobioreactors is comparatively lower than other methods; moreover, complex organic wastes can be used as feed (Marone et al. 2017). On the other hand, during anaerobic dark fermentation, some heterotrophic microbes occurring in the microalgal biomass can produce hydrogen anaerobically (Lakaniemi et al. 2011). Transformation of complex organic substrates into hydrogen by dark fermentation involves several steps including hydrolysis, acidogenesis, and acetogenesis. The obtained hydrolysate is further processed to produce hydrogen gas by acidogenesis. In addition to the hydrogen gas, other metabolic intermediates such as carbon dioxide and fatty acids are formed during dark fermentation. The fatty acids produced are further converted into acetate and hydrogen by the acetogenesis process. Later, the acetate produced is converted into carbon dioxide and methane by the methanogenesis process. Methanogenesis is performed by methanogens through the decarboxylation process. These methanogens utilize the little amount of hydrogen gas which acts as an electron donor for carbon dioxide reduction (Show et al. 2018). During dark fermentation, metabolic intermediates produced from the acidogenesis and acetogenesis process drive hydrogen production. On the other hand, hydrogen produced in the dark fermentation can be utilized by other metabolic processes such as methanogenesis or other hydrogen utilizing activities. Hence, for maximum hydrogen production, it is necessary to suppress the activities that utilize hydrogen. A study reported that the biohydrogen produced by *Scenedesmus* sp. is utilized by the anaerobic sludge culture. Lakaniemi et al. (2011) studied the utilization of initial hydrogen produced from *C. vulgaris* and *D. tertiolecta* by anaerobic colonies. Those anaerobic colonies considerably depleted the hydrogen produced from the microalgae. However, controlling hydrogen utilizing microbes by pH adjustment, chemical, or heat treatment improves the hydrogen production yield (Kim et al. 2014).

In general, dark fermentation produces hydrogen gas by two specific mechanisms such as catabolic transformation of formic acid and NADH reoxidation by hydrogenase (Show et al. 2011). The most important NADH-mediated reaction occurs in an anaerobic glycolytic pathway that produces NADH from NAD^+, and the reaction is represented by the following equation.

$$C_6H_{12}O_6 + 2\,NAD^+ \rightarrow 2\,CH_3\,CO\,COOH + 2\,NADH + 2H^+$$

The electrons generated in the pyruvate pathway by NADH-ferredoxin oxidoreductase and pyruvate-ferredoxin are utilized to reduce protons to hydrogen gas by hydrogenases. All these enzymes are influenced by environmental factors or NADH or acetyl-CoA. On the other hand, oxidation and reduction by these enzymes are also stabilized by other pathways such as lactate, butanol, and ethanol synthesis pathways through NADH transformation which in turn result in the reduction of the hydrogen production yield.

19.4 Cyanobacterial Bioethanol Production

The primary biochemical compositions of cyanobacteria are carbohydrates, proteins, and lipids. However, cyanobacterial lipids are widely utilized for biodiesel production. In addition, cyanobacterial carbohydrates could serve as an attractive source for bioethanol production (Zhu et al. 2014). The carbohydrate content could be increased by various strategies such as heterotrophic cultivation or stress induction or genetic manipulations (Wang et al. 2013a, b).

In general, cyanobacterial bioethanol production can be achieved in two stages, that is, the production of carbohydrates by utilizing sunlight and the conversion of carbohydrates into ethanol. There is a strong correlation between these two stages and their operation is directly affecting the production costs. Hence researchers have been focusing on single-step bioethanol production in cyanobacteria so that the production cost can be reduced considerably, which is amenable for commercial bioethanol production. The major steps involved in bioethanol production are hydrolysis and fermentation. Before the hydrolysis, pretreatment is the necessary step in which cell structure is disrupted and further subjected to hydrolysis. The hydrolysate is then fermented by bacteria or yeast to produce ethanol. The major

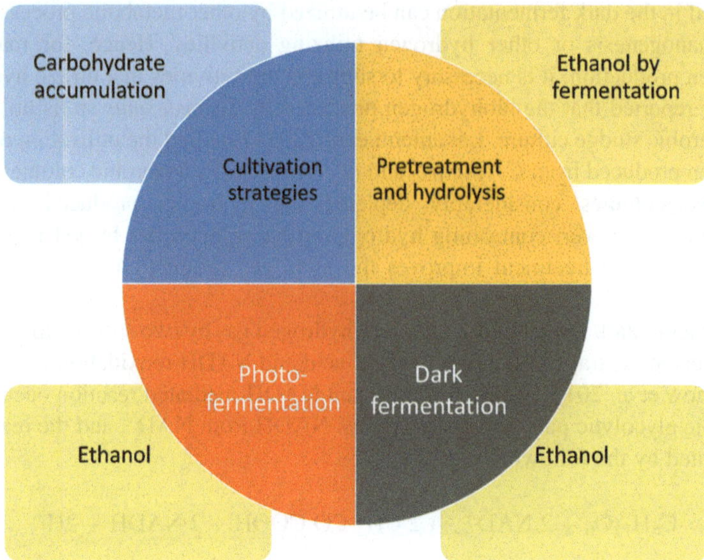

Fig. 19.2 Strategies involved in the cyanobacterial ethanol production routes

drawback of bioethanol production is due to its multistep process, which requires high energy, enzymes for hydrolysis, and fermentative microorganisms. The strategies involved in the cyanobacterial ethanol production routes are presented in Fig. 19.2.

19.4.1 Cyanobacterial Carbohydrate Accumulation

Cyanobacteria suitable for bioethanol production should have the capacity to accumulate high content of carbohydrates. The carbohydrate content of the cyanobacteria is dependent on the nature of the organisms which mostly rely on the nutritional and environmental conditions (Chen et al. 2013). The environmental factors and nutrients influencing carbohydrate content in cyanobacteria are temperature, pH, intensity of light, salinity, and the content of carbon, nitrogen, iron, phosphorus, sulfur, respectively (Markou et al. 2012). The microalgae such as *Chlorella*, *Scenedesmus* sp., *Chlorophyta* division, and the cyanobacteria such as *Synechococcus* sp. and *Synechocystis* sp. are primarily considered for the bioethanol production. Among the various factors, light intensity greatly influences the carbohydrate content. The increase of light intensity higher than the normal level increases the carbohydrate (Vitova et al. 2015). In general, increasing light intensity up to a particular level has a favorable effect on the increase of carbohydrate and lipid contents. Increasing light intensity is beneficial up to the photosynthesis saturation point of each organism. Nutrient concentration and availability in the environment indirectly or directly influence the photosynthesis rate of cyanobacteria. Hence, in addition to the light intensity, optimal nutrient components such as carbon, nitrogen, iron, sulfur, and phosphorus can be an efficient strategy to improve the carbohydrate content (Vitova et al. 2015). Carbon content, which is the most important nutrient for the cyanobacteria, determines the biomass content and carbohydrate content. Under nitrogen limitation conditions, cyanobacteria utilizing the supplied carbon or CO_2 aided by light showed a high amount of carbohydrate accumulation (Chen et al. 2013). Addition of nitrogen increased protein synthesis, DNA, and pigments which indirectly favors the carbohydrate accumulation by upregulating the enzymes involved in the carbohydrate synthesis pathways (Markou et al. 2012). Limiting iron content in the cells affects the photosynthesis-derived electron transport, nitrogen fixation, sulfate reduction, and ROS content (Sunda and Huntaman 1997). On the other hand, sulfur limitation affects the polysaccharides, protein, sulfolipids contents, and electron transport chain. Limiting sulfur content decreases the cell division and affects the biomass content, whereby a high concentration of sulfur decreases the assimilation of photosynthesis which in turn reduces the carbohydrates and carbon-rich compounds (Markou et al. 2012).

The *Scenedesmus obliquus* CNW-N produced 50% of the carbohydrate content of its dry weight after nitrogen was utilized from the normal growth medium (Ho et al. 2013c). After the nitrogen was utilized from the growth medium, protein content was decreased in the organism which favors the accumulation of more carbohydrate content. The carbohydrate content of *Synechococcus* sp. PCC 7002 was elevated to about 60% of its dry weight after the nitrogen was depleted from the medium (Mollers et al. 2014). On the other hand, the limitation of phosphorus in the growth medium did not affect the carbohydrate content of the *Tetraselmis subcodiformis*. However, cellular productivity of *Tetraselmis subcodiformis* was significantly affected and showed lower growth than that under the sulfur or nitrogen limitation (Yao et al. 2012, 2013a). In addition, the increasing salinity of the *Tetraselmis subcodiformis* growth medium increased the carbohydrate content 30–40% (dry weight) (Yao et al. 2013b). Nitrogen depletion conditions increased the *Chlorella vulgaris* FSP-E and ESP-6 carbohydrate content from 15% to 54%. In the same study, *Chlamydomonas* Tai-04 carbohydrate was increased from 34% to 47% under nitrogen depletion conditions (Ho et al. 2013b). In another study, the carbohydrate content of *C. vulgaris* Beijerinck, strain CCALA924, was significantly improved under sulfur-limiting condition showing higher content than that under nitrogen and phosphorus depletion (Brányiková et al. 2011). The content of starch in *Chlorella sorokiniana* which is the major carbohydrate was improved in a short time of nitrogen starvation conditions (Li et al. 2015). Although the cellular carbohydrate content of cyanobacteria was increased by various nutrient limitations, the biomass content was compromised under these conditions. Hence, it is necessary to optimize the nutritional limitation which favors both aspects, that is, an increase in both biomass and carbohydrate contents.

19.4.2 Bioethanol by Hydrolysis and Fermentation

The starch, glycogen, and cellulose are primary carbohydrates available in cyanobacteria and microalgae from which the derived sugars are suitable for fermentation to yield bioethanol. Starch available in the microalgae has a high impact on serving as a feedstock for bioethanol production. However, cellulose is also suitable for bioethanol production from microalgae (Ho et al. 2012). The general microorganisms used for the fermentation to ferment the cyanobacterial and microalgal carbohydrates are *Saccharomyces cerevisiae* and *Zymomonas mobilis*. The glycogen present in the cyanobacteria is the form of energy storage. Glycogen structure and its characteristics are like starch, that is, having polymeric glucose structure with higher solubility. Besides, cyanobacterial glycogen is an easily hydrolysable and readily available substrate in fermentation for bioethanol production (Mollers et al. 2014). Although the sugar extraction is important, mild conditions are required for sugar hydrolysis in both acid and enzymatic hydrolysis. The treatment with 2–3 N sulfuric acid at 120 °C for 30 min is required to hydrolyze 71–97% of carbohydrate (in which 65% is made up of glucose) in *S. obliquus* (Miranda et al. 2012). The obtained *Scenedesmus bijugatus* after lipid extraction was used for the sugar extraction using sulfuric acid concentration (0.36–1.08 N), at 130 °C and 45 min yielding 84% of fermentable sugar. The obtained hydrolysate was further converted into ethanol by fermentation and 70% of theoretical ethanol yield was achieved (Ashokkumar et al. 2015). Enzymatic hydrolysis was performed for the *Chlamydomonas reinhardtii* sugar extraction and the results showed that 94% of carbohydrates were extracted upon hydrolysis by 0.005% of amylase from *Bacillus licheniformis* and 0.2% of glucoamylase from *Aspergillus niger*. The extracted sugar was further converted into ethanol by *Saccharomyces cerevisiae* S288C in fermentation and 60% of yield was obtained (Choi et al. 2010). Another study reported that the mild conditions of ultrasonic treatment, that is, 20 kHz, 30 W, and 40 min, with glutase from *A. niger* showed 98% of sugar recovery. Further fermentation by *S. cerevisiae* AM12 converts 80% of extracted sugars into ethanol (Asada et al. 2012). Ho et al. (2013b) reported that acid hydrolysis is more efficient than the enzymatic treatment to extract sugars from the *C. vulgaris* FSP-E. Under acidic conditions (0.036–1.8 NH_2SO_4, 20 min and at 121 °C), 95% sugar extraction was achieved, and the fermentation of hydrolysate by *Zymomonas mobilis* showed 90% theoretical ethanol yield (Ho et al. 2013a). Another study compared the different cell disruption methods and enzyme treatment on sugar extraction, and the results showed that the bead beating cell disruption method and pectinase enzyme showed high sugar extraction efficiency (Kim et al. 2014). Further fermentation by *S. cerevisiae* KCTC 7906 showed 89% of ethanol yield after 12 h. In addition, the study also observed the presence of a significant amount of pectin on the *C. vulgaris* cell wall. Acidic treatment by hydrochloric acid (HCl) with the concentration of 0.3 N, 15 min and at 121 °C showed 90% sugar extraction yield from the *Chlorella* sp. KR-1 and conversion of sugars into ethanol fermented by *S. cerevisiae* showed 80% of ethanol yield (Lee et al. 2015). When comparing the HCl and H_2SO_4 acidic and enzymatic hydrolysis on *Dunaliella tertiolecta*, the highest sugar extraction of 80% was observed with the combination of HCl (0.5 N) and amyloglucosidase treatment with 82% of theoretical ethanol yield obtained after the *S.cerevisiae* YPH500 fermentation (Lee et al. 2013).

Harun et al. (2011) investigated the effect of acid hydrolysis on ethanol production in *Chlorococcum humicola*. Maximum sugar recovery was observed between 0.36 and 3.6 N H_2SO_4 concentration treatment and 7.2 g/L of ethanol yield was obtained from the *S. cerevisiae* fermentation. When using enzymatic hydrolysis, *C. humicola* biomass was efficiently hydrolyzed by *Trichoderma reesei* ATCC 26921 cellulases, and the extracted sugars showed high efficiency in terms of ethanol production (Harun and Danquah 2011). Under nitrogen depletion conditions, the cyanobacterium *Synechococcus* sp. PCC 7002 produced 60% of carbohydrate content (Mollers et al. 2014). The accumulated carbohydrate was efficiently extracted by lysozyme and α-glucanases (enzyme treatment). An 80% of sugar yield was obtained after the enzymatic hydrolysis and 86% of theoretical ethanol yield was obtained after the fermentation by *S.cerevisiae*. The combination of H_2SO_4 and HNO_3 showed 80% saccharification with the concentration of 0.25–2.5 N and 0.5 N, respectively (Markou et al. 2013). The extracted sugars produced 55% of ethanol yield by fermentation using *S. cerevisiae* MV 92081. Wang et al. (2014) studied the hydrolysis and fermentation of *Tribonema* sp. At optimized hydrolysis conditions, 80% of sugars were recovered and subjected to fermentation by *S. cerevisiae* which showed 70% of theoretical ethanol yield.

According to the previous reports, the diverse nature of microalgae and cyanobacteria requires different hydrolysis conditions. Amylases, pectinases, and cellulases are generally used for enzymatic hydrolysis. However, amylase is prominently preferred for efficient hydrolysis. In acid hydrolysis, sulfuric acid treatment is most efficient for sugar extraction with the temperature ranging from 120 to 140 °C and 15–30 min. On the other hand, care should be taken in using a minimal concentration of chemicals for hydrolysis because the presence of chemicals in the hydrolysate is inhibitory to fermentation process. Nevertheless, glucose is the prime sugar compound found in the hydrolysates of microalgae and cyanobacteria which has a high potential for ethanol production by fermentation.

19.4.3 Bioethanol by Dark Fermentation

In general, the term dark fermentation is used in biohydrogen production from organic substrates. Some microalgae and cyanobacteria convert organic polymers into monomers, which are further converted into ethanol or acetic acids or organic acids (Ueno et al. 1998). During the absence of light, some microalgae and cyanobacteria produce ethanol from organic sugar polymers and excrete through the cell wall. The photosynthetic organisms, *Chalmydomonas moewusii*, *C. reinhardtii*, *Oscillatoria limnetica*, *C. vulgaris*, *Oscillatoria limosa*, *Cyanothece sp.*, *Gleocapsa alpicola*, *Spirulina sp.*, *Synechoccoccus sp.*, and *Chloroccoum littorale*, are able to produce ethanol during dark conditions (Ueno et al. 1998; Deng and Coleman 1999). However, ethanol production during dark conditions by cyanobacteria affects the hydrogen production yield (Ueno et al. 1998). During stress conditions, a high amount of carbohydrates is accumulated in cyanobacteria through photosynthesis. The excess carbohydrate accumulation in cells induces fermentative metabolism to produce ethanol during dark conditions (Beer et al. 2009; Abo-Hashesh et al. 2011). Although cyanobacteria can produce ethanol through dark fermentation, it is not an efficient process due to the low ethanol production yield.

19.4.4 Bioethanol by Photofermentation

The ethanol can be produced directly through photofermentation from engineered cyanobacteria (ABO 2014; Algenol et al. 2015). Induction of fermentative metabolic pathway in cyanobacteria can produce ethanol. However, the process is rather universal because glycolysis-based fermentation can produce multiple products other than ethanol. Hence, suitable and specific genetic engineering strategies in a fermentative metabolic pathway are important for ethanol production (Angermayr et al. 2009). Mostly, genetic engineering strategies of photosynthetic microorganisms were preferably done in *Synechocystis* sp. PCC 6803, *Synechococcus elongatus* sp. PCC 7992, *Anabena* sp. PCC 7120, and *Syenchococcus* sp. PCC 7002. These organisms were primarily used as a model for genetic engineering to produce ethanol directly by altering fermentative metabolic pathways (Rosgaard et al. 2012). The genome sequences of these organisms are readily available which makes genetic engineering strategies easy and simple. The fermentative metabolic pathway of cyanobacteria is presented below.

$$CO_2 \xrightarrow{\text{Calvin cycle}} \text{3-phosphoglycerate} \rightarrow\rightarrow\rightarrow \text{pyruvate} \overset{\text{PDC}}{\rightarrow} \text{acetaldehyde} \overset{\text{ADH}}{\rightarrow} \text{ethanol}$$

At the end of the Calvin cycle, 3-phosphoglycerate produced is further converted into pyruvate and pyruvate into ethanol by PDC (pyruvate decarboxylase) and ADH (alcohol dehydrogenase), respectively. Hence, photosynthesis and fermentation are two steps involved in photofermentative ethanol production. However, the pathway is not specific in producing ethanol and hence genetically modified photosynthetic organisms are preferred to produce ethanol by photofermentation.

19.5 Cyanobacteria for Bio-Oil Production

Cyanobacteria can produce various complex organic molecules through photosynthesis by utilizing inorganic components. When compared to other terrestrial crops, microalgae have the highest growth rate, high photosynthetic efficiency, and high biomass rich in organic compounds (Sivaramakrishnan and Incharoensakdi 2018b). In addition, cyanobacteria can sequestrate the carbon dioxide from the atmosphere, those carbon molecules are later converted into organic compounds such as lipids or other bioproducts (Li et al. 2017a, b). Cyanobacterial lipids are considered for biodiesel production in which lipids are converted into biodiesel by conventional transesterification methods. The triglycerides in microalgae or diglycerides in cyanobacteria are primarily considered for biodiesel applications.

Transesterification can be done by using acid or alkali catalysts, such as potassium hydroxide, sodium hydroxide, sulfuric, phosphoric, sulfonic, and hydrochloric acids. However, alkali catalysts are preferable choice for efficient biodiesel production. Using chemical catalysts in transesterification affects the downstream processing. Hence, enzyme-catalyzed transesterification was widely preferred for biodiesel production which makes the recovery process easy. However, both enzyme and chemical catalysts show limited immiscibility of the reactants, hence the process requires adequate agitation to maintain mass transfer efficiency. In addition, transesterification comprises 40% of the overall energy consumption of biodiesel production (Sivaramakrishnan and Incharoensakdi 2018b).

Cyanobacterial lipid content can be improved by altering various parameters such as pH, temperature, light intensity, mode of culture, and medium compositions. Genetic engineering and other mutation induction strategies are also considered for cyanobacterial lipid improvement (Sitther et al. 2020). Altering light intensity and a slight increase in pH of the medium increased the lipid content in microalgae (Moheimani 2013). High light intensity improved the neutral lipid production which is the primary source for biodiesel production (He et al. 2015). A study revealed that the mixture of wastewaters from the municipal and industrial has a great potential to improve the lipid content in the microalgae (Gentili 2014). Altering microalgal culture modes, such as heterotrophic and mixotrophic, improved the lipid content upon the addition of glucose in batch mode and air supply during dark cycles (Praveenkumar et al. 2014). Increasing sodium carbonate concentration increased the lipid content, biomass, and total lipid production in three different microalgae (Sivaramakrishnan and Incharoensakdi 2017). Sivaramakrishnan and Incharoensakdi (2018a) overexpressed the glycerol kinase gene in *Synechocystis* sp. PCC 6803 which resulted in considerably increased lipid content (Sivaramakrishnan and Incharoensakdi 2018a). Hence it is clear that the cyanobacteria have a considerable amount of lipid content that can be considered for bio-oil production.

In general, bio-oils are generated from the lignocellulosic biomass, which is acidic, viscous, unstable, and contain solid residues and oxygenated compounds (Kan et al. 2016). There has been a study comparing the properties of lignocellulosic and microalgal bio-oils. The results indicated that the microalgal bio-oils have higher calorific value and stability than the lignocellulosic biomass with low oxygen content (Chagas et al. 2016). Apart from the microalgal lipids, other components such as proteins, aliphatic, and aromatic hydrocarbons are also present in the bio-oils with increased quality when compared to the lignocellulosic biomass (Chagas et al. 2016). Proteinaceous microalgae have the potential to adapt under different culture conditions and can be grown in wastewater. The schematic representation of cyanobacterial bio-oil is presented in Fig. 19.3.

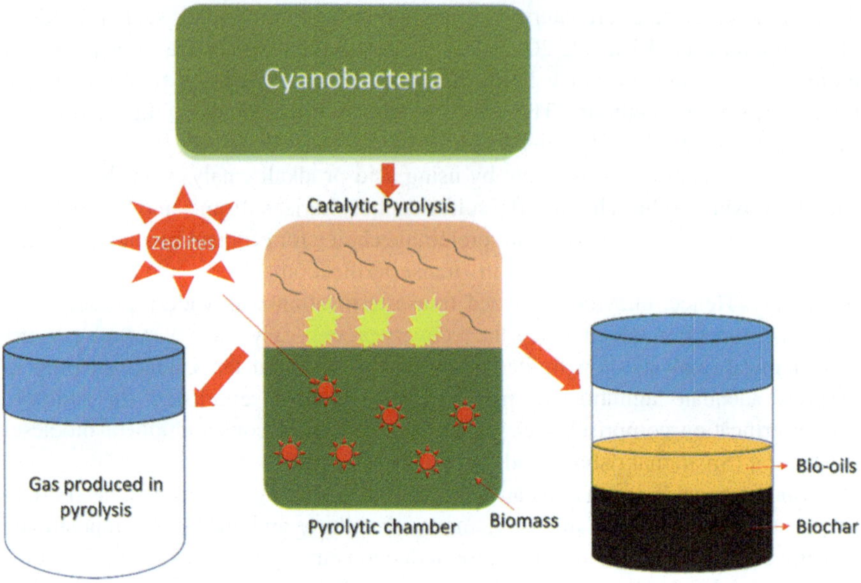

Fig. 19.3 Schematic representation of cyanobacterial bio-oil production

19.5.1 Pyrolysis

Pyrolysis is nothing but a thermochemical process in which oil or biomass is decomposed to produce volatile or non-condensable gases, biochar, and bio-oils (viscous fluids) in the absence of oxygen (Zainan et al. 2015). The thermochemical process converts cyanobacterial biomass to solid fuels and the residues are biochar which produces gases by gasification. Complex processes such as decarboxylation, dehydration, polymerization, and fragmentation occur during pyrolysis. The end products of pyrolysis are bio-oils and dark brown color viscous fluids. The bio-oils produced from biomass pyrolysis contain 300 various compounds, and the major important products are alcohols, sugars, hydrocarbons, acids, indoles, furans, polyaromatics, and carbonyls. In general, biomass stuffed with carbohydrates, proteins, and lipids is suitable for the pyrolysis for bio-oil production. The decomposition of carbohydrates starts first, followed by proteins in the range of 250–300 °C, and finally lipids in the range of 400–500 °C (Li et al. 2017a, b). The prominent product arising from the pyrolysis is bio-oils which have high industrial applications such as generation of power, heat, and fuel (Rago et al. 2018). In addition to bio-oils, other gases like CH_4, H_2, C_2H_2, CO, and C_2H_6 are also generated during pyrolysis. Cyanobacterial pyrolysis can be categorized into slow and fast pyrolysis which is dependent on the speed of temperature increase. Comparatively, fast pyrolysis showed high efficiency on bio-oil yields due to the short residence time which blocks the immediate secondary reaction that can reduce the yield of biochar (Bridgwater 2012). On the other hand, the long residence time of slow pyrolysis promotes the generation of a high amount of biochar (Lamers et al. 2012). Hence, fast pyrolysis is suitable to achieve high bio-oil yields and it also helps in the bio-oil recovery process which makes the process efficient (Dickerson 2013). Miao et al. (2004) reported that the bio-oils produced from fast pyrolysis of *Chlorella protothecoides* in a fluidized bed reactor showed low viscosity and high heating values when compared to the slow pyrolysis. The yield of bio-oils with high heating values and low viscosity can be increased by heterotrophic cultivation of cyanobacteria. About 3.4 times of bio-oil yield was achieved when the *Chlorella* sp. was cultured under the heterotrophic condition compared to that under the autotrophic condition (Miao and Wu 2004). In addition, cyanobacterial or microalgal biomass produced bio-oils with higher heating values than bio-oils produced from other woody biomass sources (Li et al. 2012). *Chlorella vulgaris* biomass can be utilized for the fast pyrolysis to produce bio-oils in a fluidized-bed reactor (Wang et al. 2013a, b). The high bio-oil yields of 49.2 and 55.4% could be obtained from the pyrolysis of low lipid-containing *Chlorella vulgaris* and *Dunaliella salina*, respectively (Gong et al. 2014). Hu et al. (2013) studied the pyrolysis temperature and particle size of cyanobacteria in a fluidized bed reactor. The gas yield was increased from 16.25% to 41.33% when the temperature was increased. On the other hand, an increase in particle size decreased the bio-oil yield from 54.97% to 42.86% and biochar yield from 57.09% to 20.39%, this is due to the heat and mass transfer limitations of large particle size biomass. Campanella and Harold (2012) studied the fast pyrolysis and its operating parameters of various microalgae, cyanobacteria, and duckweed in a falling solid reactor. The results showed that the high bio-oil yield was achieved beyond the 500 °C. At high temperatures, gas and vapors formed in the reactor swept from the biomass surface, and vapors crack immediately which prevents the secondary reactions causing an increase of the bio-oil yields and a decrease of the bio-char formation. The study also reported that the yield of bio-oils achieved in falling solid reaction is high when compared to the fixed-bed reactor. Apart from the reactor type, temperature also influences the bio-oil yield. Hence, it is clear that the fast pyrolysis is suitable to achieve high bio-oil yields and other factors such as heating rate, gas flow, temperature, and particle size also influence the bio-oil yields.

Bio-oil and biochar products can also be produced by microwave-assisted pyrolysis of algal biomass. Microwave is electromagnetic radiation that can produce radio frequency of 0.3–300 GHz with 400–800 °C temperatures. Due to the higher heating capacity of microwaves, it can produce a high yield of bio-oil and bio-syngas than other conventional methods. However, it requires an absorber to enhance the microwave-absorbing capacity by biomass which makes the process cost expensive (Ellison et al. 2020).

Hydropyrolysis is a novel method of pyrolysis that uses nitrogen as a carrier gas, and hydrogen is used for the thermal decomposition at high atmospheric pressure. The hydrocarbon yield of this method is higher with structural stability. The optimized conditions of hydropyrolysis to achieve maximum bio-oil and gas are 3 MPa, 60 min, and 310 °C. The important products obtained from the hydropyrolysis are CH_4, CO_2, CO, and unreacted H_2. However, controlling operating parameters is challenging for sustainable bio-oil production (Gamliel et al. 2018).

19.5.2 Pyrolyzed Bio-Oil Characteristics

Hydrocarbons, the compounds containing nitrogen group and oxygen, are the three main products obtained from the pyrolysis of cyanobacteria or microalgae. Straight and short-chain hydrocarbons (C_{10}) obtained from the pyrolysis can be referred to as petroleum products and considered for fuel purpose. The nitrogen- and oxygen-containing compounds evolved in the pyrolysis process are mainly from the cyanobacterial carbohydrate and protein decomposition (Li et al. 2017a, b). These nitrogen- and oxygen-containing compounds present in the bio-oils are prominently considered for fuel purposes. The presence of carboxylic acid in the bio-oils induces acidity which results in the oil polymerization as a secondary response. The secondary response increases the viscosity of the bio-oil which affects the fuel flow in engines and makes the fuel quality worse. Cyanobacterial bio-oils also contain nitrogen which affects the fuel quality. The presence of nitrogen in the bio-oils also causes the evolution of the nitrogen during combustion which has a negative environmental impact. Hence, it is necessary to eliminate the nitrogen and oxygen from the bio-oils produced from pyrolysis to retain the fuel quality. The aromatic hydrocarbons from *Chlorella vulgaris* can be produced by the catalytic pyrolysis method. The addition of a high amount of catalysts with high temperature increases the aromatic hydrocarbon yields by 24% in bio-oils (Thangalazhy-Gopakumar et al. 2012). Aromatic hydrocarbons of catalytic pyrolysis also increase the stability and octane number of bio-oils which favor the fuel quality. Hence, the addition of catalysts in the cyanobacterial pyrolysis process significantly reduces the oxygen and nitrogen content of bio-oils and improves the quality of the fuel.

19.5.3 Catalysts for Bio-Oil

To increase the quality of cyanobacterial bio-oils, catalytic pyrolysis is necessary. However, the selection of suitable catalysts is a challenging process to protect and improve fuel quality (Hazrat et al. 2015). Some catalysts may create an adverse effect on fuel quality. Hence, it is necessary to choose the catalyst that can eliminate the nitrogen and oxygen from the bio-oils. In addition, catalysts reduce the reaction time which in turn decrease the power consumption (Biffis et al. 2018).

19.5.3.1 Zeolite Catalysts

Zeolite catalysts are widely preferred for pyrolysis due to their porous nature. The most common zeolite used in pyrolysis is H-ZSM5, the carbonium ion mechanism is the specific characteristic of this catalyst that drives the elimination of carboxylation, oxygenation, and carbonylation of the bio-oils (Li et al. 2017a, b). The acidic nature of the H-ZSM5 catalysts increases the carbon content of bio-oil in the form of aromatic hydrocarbons. In addition, the active site of H-ZSM5 catalysts converts the complex molecules into simple molecules such as short hydrocarbons which can serve as an efficient fuel. During zeolite-catalyzed cyanobacterial pyrolysis, it produces some valuable compounds in the bio-oils other than the direct fuel-based compounds, such as anthracene, xylene, naphthalene, toluene, and benzene. Besides, Zeolite catalyst also reduces the exogenous compounds in bio-oils such as furans, aldehydes, and phenol which reduce the fuel quality (Thangalazhy-Gopakumar et al. 2012; Pan et al. 2010). The H-ZSM5-catalyzed *Chlorella vulgaris* pyrolysis improved the yield of aromatic hydrocarbons (Thangalazhy-Gopakumar et al. 2012). The investigation on the *Chlorella* sp. catalytic pyrolysis using three different versions of zeolite catalysts showed that the H-ZSM5 can produce high hydrocarbon yields (Du et al. 2013). Further hydrocarbon yields in bio-oils were improved by the impregnation of Cu in H-ZSM5 which also showed low nitrogen and oxygen content (Hanifzadeh et al. 2012).

19.5.3.2 Metal-Loaded Catalysts

ZSM5, another zeolite catalyst widely used in the pyrolysis process, allows the addition of other active metals in its porous structure that favors the pyrolysis process. The metal-loaded catalysts have a high beneficial advantage on denitrogenation and deoxygenation of bio-oils which improve the fuel quality. Besides, metal addition in the catalysts also increases the hydrocarbon production yield and decreases the coke formation when compared to the non-metal loaded catalysts (French and Czernik 2010). The study also reported the usage of catalyst ratios, and it was observed that the high catalyst ratio increased the hydrocarbon yield when pyrolysis was performed in a semicontinuous catalytic cracking reactor (French and Czernik 2010). However, metal-loaded multifunctional catalyst H-ZSM5 can help upgrade bio-oil characteristics (Gong et al. 2014). Therefore, more studies such as selection of catalyst, stability, preparative methods, and metal loading abilities are required to understand bio-oil upgradation. Apart from the metal-loaded catalysts, other catalysts like metal organic frameworks, metal organic frameworks with SO_3H groups, silica supported nickel phosphide catalysts, γ-Al_2O_3 catalysts, and nanoporous catalysts are also used in the pyrolysis process to produce bio-oils.

19.5.4 Catalytic Processing Methods

Ex situ and in situ are the two major routes involved in cyanobacterial pyrolysis. The vapors generated from cyanobacterial pyrolysis are swept through the catalyst in ex situ method, whereas in situ pyrolysis involves the mixing of both catalyst and cyanobacteria together (Dong et al. 2013). However, comparative studies of ex situ and in situ have been performed, and the results showed that the yield of olefin matters is higher in ex situ processes than that of the in situ process (Wang and Brown 2013). The use of an ex situ fixed-bed reactor for *Chlorella pyrenoidosa* pyrolysis showed 31.9% of olefin yields (Xinglong et al. 2013). The catalytic pyrolysis of *Chlorella pyrenoidosa* using ZSM-5 catalysts with SiO_2/Al_2O_3 group in ex situ method was performed in the steam reaction atmosphere and the results showed the higher olefin yield than the yield obtained from in situ pyrolysis method (Xinglong et al. 2013). The carbon and hydrocarbon yield of ex situ catalytic pyrolysis is higher than the yield obtained from the in situ pyrolysis method. However, the bio-oil yield from pyrolysis is dependent on the cyanobacteria used in the process, and a different reactor is required for the desired product requirements. In this aspect, multifunctional catalyst can overcome the setback acquired in the single compound catalysts and the multiple products can be produced in a single process (Wang et al. 2013a, b). The type of reactor also influences the bio-oil yields and hence it is necessary to design the suitable reactor to eliminate multiple steps performed in the current reactor designs. More investigations are required in both ex situ and in situ catalytic pyrolysis methods to eliminate multiple steps and to reduce nitrogen and oxygen content for bio-oil upgrading.

19.5.5 Deoxygenation and Denitrogenation of Bio-Oils

The cyanobacterial bio-oil contains oxygen and nitrogen-related compounds. Oxygen-related compounds affect the fuel quality by decreasing the flowability and reduce the heating value of bio-oils. On the other hand, nitrogen-related compounds generate nitrogen oxide during fuel combustion which is eliminated in the atmosphere causing a negative environmental impact. To utilize bio-oils for fuel purposes, it is necessary to increase the hydrocarbon content in bio-oil by eliminating oxygen and nitrogen-related compounds. Oxygen and nitrogen present in the catalytic pyrolysis bio-oil can be removed by dehydration, decarboxylation, hydrodeoxygenation, decarbonylation, and hydrodenitrogenation which improve the fuel quality in terms of viscosity, stability, higher heating value, and high hydrocarbon content (Ahmed et al. 2013).

19.6 Conclusion

The present chapter describes the cyanobacterial potentials on biofuel production such as biohydrogen, bioethanol, and bio-oil. Technical advancement is encouraged to improve cyanobacterial biohydrogen production for fuel purpose. The major setback involved in cyanobacterial biohydrogen production is that the viability of the process, power energy consumption-related issues, cost of production, storage, and delivery. In the case of cyanobacterial ethanol production, it is necessary to improve the carbohydrate content of the cells, and traditional fermentation is a promising method to produce bioethanol. Like biohydrogen production, the production costs need to be reduced by employing genetic engineering strategies to improve carbohydrate content, and process intensification of hydrolysis and fermentation can help cut the production cost further. To produce bio-oils, catalytic pyrolysis is a promising one. The catalysts such as zeolites, metal-loaded zeolites, aluminosilicate, metal-organic frameworks, and mesoporous silica have been extensively considered for both bio-oil production and bio-oil upgradation. However, proper understanding is still required in reactor designing to produce efficient bio-oil production and scale-up. The major difficulty that arises during catalytic pyrolysis is that the formation of oxygen and nitrogen-related compounds, and it should be eliminated to improve the bio-oil yield and ensure fuel quality. Although various setbacks in the cyanobacterial biofuel production still remains, biofuels can be produced in higher volumes due to the high biomass production capacity. By successfully addressing the technical drawbacks, the cyanobacterial biomass can become the potential candidate for biofuel production.

Acknowledgments R. Sivaramakrishnan is thankful to the Graduate School and Faculty of Science, Chulalongkorn University (CU), for senior post-doctoral fellowship from Ratchadaphiseksomphot Endowment Fund. A.I. acknowledges the research grant from CU on the Frontier Research Energy Cluster (CU-59-048-EN) and from the Thailand Research Fund (IRG 5780008).

References

ABO (2014) Photosyntetic production of biofuels from CO_2 by cianobactéria using Algenol's direct to ethanol®process-strain development aspects. AlgalBiomass Summit

Abo-Hashesh M, Wang R, Hallenbeck PC (2011) Metabolic engineering in dark fermentative hydrogen production; theory and practice. Bioresour Technol 102(18):8414–8422

Ahmed I, Hasan Z, Khan NA, Jhung SH (2013) Adsorptive denitrogenation of model fuels with porous metal-organic frameworks (MOFs): effect of acidity and basicity of MOFs. Appl Catal B Environ 129:123–129

Algenol IP, Friedrich A, Duhring U, Baier K, Inaba M, Shi T, Wang K, Enke H, Kramer D (2015) Cyanobacterium sp. host cell and vector for production of chemical compounds in cyanobacterial cultures. Patent, US 8846369 B2.15.01.2015

Anahas AMP, Muralitharan G (2018) Characterization of heterocystous cyanobacterial strains for biodiesel production based on fatty acid content analysis and hydrocarbon production. Energy Convers Manag 157:423–437

Angermayr SA, Hellingwerf KJ, Lindblad P, Teixeira de Mattos MJ (2009) Energy biotechnology with cyanobacteria. Curr Opin Biotechnol 20(3):257–263

Asada C, Doi K, Sasaki C, Nakamura Y (2012) Efficient extraction of starch frommicroalgae using ultrasonic homogenizer and its conversion into ethanol bysimultaneous saccharification and fermentation. Nat Resources 3:175–179

Ashokkumar V, Salam Z, Tiwari ON, Chinnasamy S, Mohammed S, Ani FN (2015) An integrated approach for biodiesel and bioethanol production from *Scenedesmus bijugatus* cultivated in a vertical tubular photobioreactor. Energy Convers Manag 101:778–786

Bamberger ES, King D, Erbes DL, Gibbs M (1982) H_2 and CO_2 evolution by anaerobically adapted *Chlamydomonas reinhardtii* F-60. Plant Physiol 69(6):1268–1273

Basak N, Das D (2007) The Prospect of purple non-sulfur (PNS) photosynthetic bacteria for hydrogen production: the present state of the art. World J Microbiol Biotechnol 23(1):31–42

Basri RS, Abd Rahman RN, Kamarudin NH, Ali MS (2020) Cyanobacterial aldehyde deformylating oxygenase: structure, function, and potential in biofuels production. Int J Biol Macromol 164:3155–3162

Bavandi R, Emtyazjoo M, Saravi HN, Yazdian F, Sheikhpour M (2019) Study of capability of nanostructured zero-valent iron and graphene oxide for bioremoval of trinitrophenol from wastewater in a bubble column bioreactor. Electron J Biotechnol 39:8–14

Beer LL, Boyd ES, Peters JW, Posewitz MC (2009) Engineering algae for biohydrogen and biofuel production. Curr Opin Biotechnol 20(3):264–271

Bennoun P (2001) Chlororespiration and the process of carotenoid biosynthesis. Biochim Biophys Acta (BBA) – Bioenerget 1506(2):133–142

Biffis A, Centomo P, Del Zotto A, Zecca M (2018) Pd metal catalysts for cross-couplings and related reactions in the 21st century: a critical review. Chem Rev 118(4):2249–2295

Brányiková I, Maršálková B, Doucha J, Brányik T, Bišová K, Zachleder V, Vítová M (2011) Microalgae – novel highly efficient starch producers. Biotechnol Bioeng 108(4):766–776

Bridgwater AV (2012) Review of fast pyrolysis of biomass and product upgrading. Biomass Bioenergy 38:68–94

Campanella A, Harold MP (2012) Fast pyrolysis of microalgae in a falling solids reactor: effects of process variables and zeolite catalysts. Biomass Bioenergy 46:218–232

Chagas BME, Dorado C, Serapiglia MJ, Mullen CA, Boateng AA, Melo MAF, Ataíde CH (2016) Catalytic pyrolysis-GC/MS of *spirulina*: evaluation of a highly proteinaceous biomass source for production of fuels and chemicals. Fuel 179:124–134

Chen C-Y, Zhao X-Q, Yen H-W, Ho S-H, Cheng C-L, Lee D-J, Bai F-W, Chang J-S (2013) Microalgae-based carbohydrates for biofuel production. Biochem Eng J 78:1–10

Choi SP, Nguyen MT, Sim SJ (2010) Enzymatic pretreatment of *Chlamydomonas reinhardtii* biomass for ethanol production. Bioresour Technol 101(14):5330–5336

de Farias Silva CE, Bertucco A (2016) Bioethanol from microalgae and cyanobacteria: a review and technological outlook. Process Biochem 51(11):1833–1842

Demirbas A (2008) Biofuels sources, biofuel policy, biofuel economy and global biofuel projections. Energy Convers Manag 49(8):2106–2116

Demirbas A, Fatih Demirbas M (2011) Importance of algae oil as a source of biodiesel. Energy Convers Manag 52(1):163–170

Deng M-D, Coleman JR (1999) Ethanol synthesis by genetic engineering in cyanobacteria. Appl Environ Microbiol 65(2):523

Dickerson TSJ (2013) Catalytic fast pyrolysis: a review. Energies 6(1):514–538

Dong T, Wang J, Miao C, Zheng Y, Chen S (2013) Two-step in situ biodiesel production from microalgae with high free fatty acid content. Bioresour Technol 136:8–15

Du Z, Ma X, Li Y, Chen P, Liu Y, Lin X, Lei H, Ruan R (2013) Production of aromatic hydrocarbons by catalytic pyrolysis of microalgae with zeolites: catalyst screening in a pyroprobe. Bioresour Technol 139:397–401

Ellison CR, Hoff R, Mărculescu C, Boldor D (2020) Investigation of microwave-assisted pyrolysis of biomass with char in a rectangular waveguide applicator with built-in phase-shifting. Appl Energy 259:114217

Florin L, Tsokoglou A, Happe T (2001) A novel type of iron hydrogenase in the green alga *Scenedesmus obliquus* is linked to the photosynthetic electron transport chain. J Biol Chem 276(9):6125–6132

French R, Czernik S (2010) Catalytic pyrolysis of biomass for biofuels production. Fuel Process Technol 91(1):25–32

Gamliel DP, Bollas GM, Valla JA (2018) Two-stage catalytic fast hydropyrolysis of biomass for the production of drop-in biofuel. Fuel 216:160–170

Gentili FG (2014) Microalgal biomass and lipid production in mixed municipal, dairy, pulp and paper wastewater together with added flue gases. Bioresour Technol 169:27–32

Ghosh S, Dutta B, Banerjee A, Roy S (2018) Bio-H_2 production using de-oiled cake as cheap nitrogen source and subsequent electricity generation by hybrid system. Bioresource Technol Rep 4:1–8

Gong X, Zhang B, Zhang Y, Huang Y, Xu M (2014) Investigation on pyrolysis of low lipid microalgae *Chlorella vulgaris* and *Dunaliella salina*. Energy Fuel 28(1):95–103

Greenbaum E, Guillard RRL, Sunda WG (1983) Hydrogen and oxygen photoproduction by marine algae. Photochem Photobiol 37(6):649–655

Hanifzadeh MM, Sarrafzadeh MH, Tavakoli O (2012) Carbon dioxide biofixation and biomass production from flue gas of power plant using microalgae. Renewable energy and distributed generation (ICREDG), second Iranian conference (2012), pp. 61–64

Hankamer B, Lehr F, Rupprecht J, Mussgnug JH, Posten C, Kruse O (2007) Photosynthetic biomass and H_2 production by green algae: from bioengineering to bioreactor scale-up. Physiol Plant 131(1):10–21

Harun R, Danquah MK (2011) Enzymatic hydrolysis of microalgal biomass for bioethanol production. Chem Eng J 168(3):1079–1084

Hazrat MA, Rasul MG, Khan MMK (2015) A study on thermo-catalytic degradation for production of clean transport fuel and reducing plastic wastes. Procedia Eng 105:865–876

He Q, Yang H, Wu L, Hu C (2015) Effect of light intensity on physiological changes, carbon allocation and neutral lipid accumulation in oleaginous microalgae. Bioresour Technol 191:219–228

Ho S-H, Chen C-Y, Chang J-S (2012) Effect of light intensity and nitrogen starvation on CO_2 fixation and lipid/carbohydrate production of an indigenous microalga *Scenedesmus obliquus* CNW-N. Bioresour Technol 113:244–252

Ho S-H, Huang S-W, Chen C-Y, Hasunuma T, Kondo A, Chang J-S (2013a) Bioethanol production using carbohydrate-rich microalgae biomass as feedstock. Bioresour Technol 135:191–198

Ho S-H, Huang S-W, Chen C-Y, Hasunuma T, Kondo A, Chang J-S (2013b) Characterization and optimization of carbohydrate production from an indigenous microalga *Chlorella vulgaris* FSP-E. Bioresour Technol 135:157–165

Ho S-H, Kondo A, Hasunuma T, Chang J-S (2013c) Engineering strategies for improving the CO_2 fixation and carbohydrate productivity of *Scenedesmus obliquus* CNW-N used for bioethanol fermentation. Bioresour Technol 143:163–171

Hosseini F, Kasaeian A, Pourfayaz F, Sheikhpour M, Wen D (2018) Novel ZnO-ag/MWCNT nanocomposite for the photocatalytic degradation of phenol. Mater Sci Semicond Process 83:175–185

Hu Z, Zheng Y, Yan F, Xiao B, Liu S (2013) Bio-oil production through pyrolysis of blue-green algae blooms (BGAB): product distribution and bio-oil characterization. Energy 52:119–125

Kan T, Strezov V, Evans TJ (2016) Lignocellulosic biomass pyrolysis: a review of product properties and effects of pyrolysis parameters. Renew Sust Energ Rev 57:1126–1140

Khetkorn W, Rastogi RP, Incharoensakdi A, Lindblad P, Madamwar D, Pandey A, Larroche C (2017) Microalgal hydrogen production – a review. Bioresour Technol 243: 1194–1206

Kim KH, Choi IS, Kim HM, Wi SG, Bae H-J (2014) Bioethanol production from the nutrient stress-induced microalga *Chlorella vulgaris* by enzymatic hydrolysis and immobilized yeast fermentation. Bioresour Technol 153:47–54

Ko JK, Lee JH, Jung JH, Lee S-M (2020) Recent advances and future directions in plant and yeast engineering to improve lignocellulosic biofuel production. Renew Sust Energ Rev 134:110390

Kumar G, Mathimani T, Sivaramakrishnan R, Shanmugam S, Bhatia SK, Pugazhendhi A (2020a) Application of molecular techniques in biohydrogen production as a clean fuel. Sci Total Environ 722:137795

Kumar M, Sun Y, Rathour R, Pandey A, Thakur IS, Tsang DCW (2020b) Algae as potential feedstock for the production of biofuels and value-added products: opportunities and challenges. Sci Total Environ 716:137116

Lakaniemi AM, Hulatt CJ, Thomas DN, Tuovinen OH, Puhakka JA (2011) Biogenic hydrogen and methane production from Chlorella vulgaris and Dunaliella tertiolecta biomass. Biotechnol Biofuels 4:34

Lamers P, Junginger M, Hamelinck C, Faaij A (2012) Developments in international solid biofuel trade: an analysis of volumes, policies, and market factors. Renew Sust Energ Rev 16 (5):3176–3199

Lee OK, Kim AL, Seong DH, Lee CG, Jung YT, Lee JW, Lee EY (2013) Chemo-enzymatic saccharification and bioethanol fermentation of lipid-extracted residual biomass of the microalga, *Dunaliella tertiolecta*. Bioresour Technol 132:197–201

Lee OK, Oh Y-K, Lee EY (2015) Bioethanol production from carbohydrate-enriched residual biomass obtained after lipid extraction of *chlorella* sp. KR-1. Bioresource Technol 196:22–27

Li H, Liao JC (2013) Biological conversion of carbon dioxide to photosynthetic fuels and electrofuels. Energy Environ Sci 6(10):2892–2899

Li R, Zhong Z, Jin B, Zheng A (2012) Selection of temperature for bio-oil production from pyrolysis of algae from lake blooms. Energy Fuel 26(5):2996–3002

Li T, Gargouri M, Feng J, Park J-J, Gao D, Miao C, Dong T, Gang DR, Chen S (2015) Regulation of starch and lipid accumulation in a microalga *Chlorella sorokiniana*. Bioresour Technol 180:250–257

Li F, Srivatsa SC, Batchelor W, Bhattacharya S (2017a) A study on growth and pyrolysis characteristics of microalgae using thermogravimetric analysis-infrared spectroscopy and synchrotron Fourier transform infrared spectroscopy. Bioresour Technol 229:1–10

Li P, Chen X, Wang X, Shao J, Lin G, Yang H, Yang Q, Chen H (2017b) Catalytic upgrading of fast pyrolysis products with Fe-, Zr-, and co-modified zeolites based on pyrolyzer–GC/MS analysis. Energy Fuel 31(4):3979–3986

Lindblad P, Lindberg P, Oliveira P, Stensjö K, Heidorn T (2012) Design, engineering, and construction of photosynthetic microbial cell factories for renewable solar fuel production. Ambio 41(Suppl. 2):163–168

Manish S, Banerjee R (2008) Comparison of biohydrogen production processes. Int J Hydrog Energy 33(1):279–286

Markou G, Angelidaki I, Georgakakis D (2012) Microalgal carbohydrates: an overview of the factors influencing carbohydrates production, and of main bioconversion technologies for production of biofuels. Appl Microbiol Biotechnol 96(3):631–645

Markou G, Angelidaki I, Nerantzis E, Georgakakis D (2013) Bioethanol production by carbohydrate-enriched biomass of *Arthrospira* (*spirulina*) *platensis*. Energies 6(8):3937–3950

Markov S, Thomas A, Bazin M, Hall D (1997) Photoproduction of hydrogen by cyanobacteria under partial vacuum in batch culture or in a photobioreactor. Int J Hydrog Energy 22 (5):521–524

Marone A, Ayala-Campos OR, Trably E, Carmona-Martínez AA, Moscoviz R, Latrille E, Steyer JP, Alcaraz-Gonzalez V, Bernet N (2017) Coupling dark fermentation and microbial electrolysis to enhance bio-hydrogen production from agro-industrial wastewaters and by-products in a biorefinery framework. Int J Hydrog Energy 42(3):1609–1621

Mazard S, Penesyan A, Ostrowski M, Paulsen IT, Egan S (2016) Tiny microbes with a big impact: the role of cyanobacteria and their metabolites in shaping our future. Mar Drugs 14(5):97

Miao X, Wu Q (2004) High yield bio-oil production from fast pyrolysis by metabolic controlling of *Chlorella protothecoides*. J Biotechnol 110(1):85–93

Miao X, Wu Q, Yang C (2004) Fast pyrolysis of microalgae to produce renewable fuels. J Anal Appl Pyrolysis 71(2):855–863

Miranda JR, Passarinho PC, Gouveia L (2012) Pre-treatment optimization of *Scenedesmus obliquus* microalga for bioethanol production. Bioresour Technol 104:342–348

Moheimani NR (2013) Inorganic carbon and pH effect on growth and lipid productivity of *Tetraselmis suecica* and *chlorella* sp (Chlorophyta) grown outdoors in bag photobioreactors. J Appl Phycol 25(2):387–398

Mollers KB, Canella D, Jorgensen H, Frigaard N (2014) Cyanobacterial biomass as carbohydrate and nutrient feedstock for bioethanol production by yeast fermentation. Biotechnol Biofuels 7 (64):1–11

Monir MU, Abd Aziz A, Kristanti RA, Yousuf A (2018) Co-gasification of empty fruit bunch in a downdraft reactor: a pilot scale approach. Bioresource Technol Rep 1:39–49

Oliver NJ, Rabinovitch-Deere CA, Carroll AL, Nozzi NE, Case AE, Atsumi S (2016) Cyanobacterial metabolic engineering for biofuel and chemical production. Curr Opin Chem Biol 35:43–50

Pan P, Hu C, Yang W, Li Y, Dong L, Zhu L, Tong D, Qing R, Fan Y (2010) The direct pyrolysis and catalytic pyrolysis of *Nannochloropsis* sp. residue for renewable bio-oils. Bioresour Technol 101(12):4593–4599

Parmar A, Singh NK, Pandey A, Gnansounou E, Madamwar D (2011) Cyanobacteria and microalgae: a positive prospect for biofuels. Bioresour Technol 102(22):10163–10172

Patel A, Matsakas L, Rova U, Christakopoulos P (2019) A perspective on biotechnological applications of thermophilic microalgae and cyanobacteria. Bioresour Technol 278:424–434

Perera F (2018) Pollution from fossil-fuel combustion is the leading environmental threat to global pediatric health and equity: solutions exist. Int J Environ Res Public Health 15:16

Praveenkumar R, Kim B, Choi E, Lee K, Park J-Y, Lee J-S, Lee Y-C, Oh Y-K (2014) Improved biomass and lipid production in a mixotrophic culture of *chlorella* sp. KR-1 with addition of coal-fired flue-gas. Bioresour Technol 171:500–505

Quintana N, Van der Kooy F, Van de Rhee MD, Voshol GP, Verpoorte R (2011) Renewable energy from cyanobacteria: energy production optimization by metabolic pathway engineering. Appl Microbiol Biotechnol 91(3):471–490

Rago YP, Mohee R, Surroop D (2018) A review of thermochemical technologies for the conversion of waste biomass to biofuel and energy in developing countries. In: The nexus: energy, environment and climate change 2018. Springer, Cham, pp 127–143

Rajneesh SSP, Pathak J, Sinha RP (2017) Cyanobacterial factories for the production of green energy and value-added products: an integrated approach for economic viability. Renew Sust Energ Rev 69:578–595

Rastogi RP, Sinha RP (2009) Biotechnological and industrial significance of cyanobacterial secondary metabolites. Biotechnol Adv 27:521–539

Rastogi RP, Madamwar D, Pandey A (eds) (2017) Algal green chemistry – recent progress in biotechnology. Elsevier, Amsterdam, The Netherlands, pp 1–336

Rastogi RP, Pandey A, Larroche C, Madamwar D (2018) Algal green energy – R & D and technological perspectives for biodiesel production. Renew Sust Energ Rev 82(3):2946–2969

Robertson DE, Jacobson SA, Morgan F, Berry D, Church GM, Afeyan NB (2011) A new dawn for industrial photosynthesis. Photosynth Res 107(3):269–277

Rosgaard L, de Porcellinis AJ, Jacobsen JH, Frigaard N-U, Sakuragi Y (2012) Bioengineering of carbon fixation, biofuels, and biochemicals in cyanobacteria and plants. J Biotechnol 162 (1):134–147

Sarsekeyeva F, Zayadan BK, Usserbaeva A, Bedbenov VS, Sinetova MA, Los DA (2015) Cyanofuels: biofuels from cyanobacteria. Reality and perspectives. Photosynth Res 125 (1):329–340

Sharma NK, Tiwari SP, Tripathi K, Rai AK (2011) Sustainability and cyanobacteria (blue-green algae): facts and challenges. J Appl Phycol 23(6):1059–1081

Show K-Y, Lee D-J, Chang J-S (2011) Bioreactor and process design for biohydrogen production. Bioresour Technol 102(18):8524–8533

Show K-Y, Yan Y, Ling M, Ye G, Li T, Lee D-J (2018) Hydrogen production from algal biomass – advances, challenges and prospects. Bioresour Technol 257:290–300

Sitther V, Tabatabai B, Fathabad SG, Gichuki S, Chen H, Arumanayagam ACS (2020) Cyanobacteria as a biofuel source: advances and applications. In: Advances in cyanobacterial biology. Academic Press, pp 269–289

Sivaramakrishnan R, Incharoensakdi A (2017) Enhancement of total lipid yield by nitrogen, carbon, and iron supplementation in isolated microalgae. J Phycol 53(4):855–868

Sivaramakrishnan R, Incharoensakdi A (2018a) Enhancement of lipid production in *Synechocystis* sp. PCC 6803 overexpressing glycerol kinase under oxidative stress with glycerol supplementation. Bioresour Technol 267:532–540

Sivaramakrishnan R, Incharoensakdi A (2018b) Microalgae as feedstock for biodiesel production under ultrasound treatment – a review. Bioresour Technol 250:877–887

Sivaramakrishnan R, Incharoensakdi A (2020) Plant hormone induced enrichment of *Chlorella* sp. omega-3 fatty acids. Biotechnol Biofuels 13(1):7

Sunda WG, Huntaman SA (1997) Interrelated influence of iron, light and cell size on marine phytoplankton growth. Nature 390(6658):389–392

Thangalazhy-Gopakumar S, Adhikari S, Chattanathan SA, Gupta RB (2012) Catalytic pyrolysis of green algae for hydrocarbon production using H+ZSM-5 catalyst. Bioresour Technol 118:150–157

Troshina O, Serebryakova L, Sheremetieva M, Lindblad P (2002) Production of H_2 by the unicellular cyanobacterium *Gloeocapsa alpicola* CALU 743 during fermentation. Int J Hydrog Energy 27(11):1283–1289

Turner J, Sverdrup G, Mann MK, Maness P-C, Kroposki B, Ghirardi M, Evans RJ, Blake D (2008) Renewable hydrogen production. Int J Energy Res 32(5):379–407

Ueno Y, Kurano N, Miyachi S (1998) Ethanol production by dark fermentation in the marine green alga, *Chlorococcum littorale*. J Ferment Bioeng 86(1):38–43

Vitova M, Bisova K, Kawano S, Zachleder V (2015) Accumulation of energy reserves in algae: from cell cycles to biotechnological applications. Biotechnol Adv 33(6):1204–1218

Wang K, Brown RC (2013) Catalytic pyrolysis of microalgae for production of aromatics and ammonia. Green Chem 15(3):675–681

Wang H, Ji C, Bi S, Zhou P, Chen L, Liu T (2014) Joint production of biodiesel and bioethanol from filamentous oleaginous microalgae Tribonema sp. Bioresour Technol 172:169–173

Wang K, Brown RC, Homsy S, Martinez L, Sidhu SS (2013a) Fast pyrolysis of microalgae remnants in a fluidized bed reactor for bio-oil and biochar production. Bioresour Technol 127:494–499

Wang L, Li Y, Sommerfeld M, Hu Q (2013b) A flexible culture process for production of the green microalga *Scenedesmus dimorphus* rich in protein, carbohydrate or lipid. Bioresour Technol 129:289–295

Xinglong D, Zhaoan C, Song X, JInling Z, Jiannan Z, Yapeng L, Yunpeng X, Zhongmin L (2013) Catalytic pyrolysis of microalga *chlorella pyrennoidosa* for production of ethylene, propylene and butene

Yao C, Ai J, Cao X, Xue S, Zhang W (2012) Enhancing starch production of a marine green microalga Tetraselmis subcordiformis through nutrient limitation. Bioresour Technol 118:438–444

Yao C-H, Ai J-N, Cao X-P, Xue S (2013a) Characterization of cell growth and starch production in the marine green microalga *Tetraselmis subcordiformis* under extracellular phosphorus-deprived and sequentially phosphorus-replete conditions. Appl Microbiol Biotechnol 97 (13):6099–6110

Yao C-H, Ai J-N, Cao X-P, Xue S (2013b) Salinity manipulation as an effective method for enhanced starch production in the marine microalga *Tetraselmis subcordiformis*. Bioresour Technol 146:663–671

Zainan NH, Srivatsa SC, Bhattacharya S (2015) Catalytic pyrolysis of microalgae *Tetraselmis suecica* and characterization study using in situ synchrotron-based infrared microscopy. Fuel 161:345–354

Zhu LD, Hiltunen E, Antila E, Zhong JJ, Yuan ZH, Wang ZM (2014) Microalgal biofuels: flexible bioenergies for sustainable development. Renew Sust Energ Rev 30:1035–1046

Cyanobacteria as a Competing Source of Bioenergy: Metabolic Engineering and Modeling Approach for Medium Optimization

20

Alexander Dimitrov Kroumov, Fabiano Bisinella Scheufele, Maya Margaritova Zaharieva, Reneta Gevrenova, and Hristo Najdenski

Abstract

Biofuels produced by cyanobacteria prove to be advantageous in a global sense including environment safety. This chapter will focus on the cell-to-fuel process and biotechnological value of cyanobacteria which exhibit high photosynthetic efficiency. It is imperative to discuss only cyanobacterial strains with desirable fatty acid composition and other precursors for high and pure quality of fuels. Special attention will be given to metabolic engineering as a tool for strain design. The siderophores responsible for metal uptake into the cell will be discussed, as well. Further, mathematical analysis of subsystems of the biorefinery concept will be made toward the modeling approach of medium optimization. The complex approach of nutrients' calculation is a base for optimization of target metabolites: hydrogen, bioethanol, biodiesel, and other products. Process development of cyanobacteria will be analyzed from the view of system analysis theory and

A. D. Kroumov (✉)
Department of Biotechnology—Laboratory of Bioconversion and Biosynthesis of Microbial Metabolites, The Stephan Angeloff Institute of Microbiology—Bulgarian Academy of Sciences, Sofia, Bulgaria

F. B. Scheufele
Graduation Program of Biotechnology and Bioprocess Engineering, Federal University of Technology—Paraná—UTFPR, Toledo, Paraná, Brazil
e-mail: fabianob@utfpr.edu.br

M. M. Zaharieva · H. Najdenski
Department of Infectious Microbiology, The Stephan Angeloff Institute of Microbiology—Bulgarian Academy of Sciences, Sofia, Bulgaria

R. Gevrenova
Department of Pharmacognosy, Faculty of Pharmacy, Medical University of Sofia, Sofia, Bulgaria
e-mail: rgevrenova@pharmfac.mu-sofia.bg

R. P. Rastogi (ed.), *Ecophysiology and Biochemistry of Cyanobacteria*,
https://doi.org/10.1007/978-981-16-4873-1_20

principles of decomposition, which are in tremendous help of realization of the integral biorefinery concept.

Keywords

Cyanobacteria · Cell-to-fuel process · Mathematical analysis · Bioenergy · Engineered strains · Medium optimization · Linear programming procedure · System analysis theory

20.1 Introduction

In the last century, birth control in many big countries has been lowered which led to an enormous increasing in the human population worldwide. This change demonstrated that exploration of untapped land and untouched resources is over. Cultivation of microscopic organisms such as cyanobacteria (prokaryotic blue-green algae) as a renewable energy source is an attractive alternative for biotechnology today especially for healthcare, overcoming global warming, food supply, and other humanitarian issues. One of the promising resources is cyanobacteria/microalgae. Cyanobacteria grow in a simple nutrient medium and they have small genome sequences. The most challenging feature of cyanobacteria, in terms of biorefinery concept, is their ability to produce many high values products and biologically active compounds (BAC). Hence, their potential for applications in many fields of bio-industry is visible (Farrokh et al. 2019).

Nevertheless, this alternative requires a complex approach (Kroumov et al. 2017; Hinterholz et al. 2019; Scheufele et al. 2019) in order to meet the great expectations for fuel production. Our previous works about this subject highlighted in detail the problems and challenges (Kroumov et al. 2016) which bioengineers and biochemist algologists faced cultivating microalgae (Hinterholz et al. 2019; Schuelter et al. 2019; Gonçalves et al. 2019a, b). A complex approach for microalgae that can be successfully applied to cyanobacteria taking into account the specificity of their physiology, morphology, and cell behavior can be taken as a principle of analogy from our previous data (Kroumov et al. 2016; Schuelter et al. 2019). Starch and cellulose from cyanobacteria can be broken down and further to form dextrins, glucose, and other sugars which can be transformed by yeasts/bacteria to ethanol. The process requires a high energy supply (Sanderson 2011). Cyanobacteria/microalgae nowadays are most probably cost-effective alternatives to biofuel production (Machado and Atsumi 2012). This is because they may utilize CO_2 from waste gases (substrate with no cost) and can grow faster than plants being able to synthesize cell compounds, which are target as bio-energy sources (Dismukes et al. 2008). Various cyanobacteria species are highly tolerant to the high concentrations of CO_2 supplied by gas flow (Takano et al. 1992; Sheehan 1998). This particular cyanobacteria potential allow CO_2 purification of emissions from waste industrial sources (Kroumov et al. 2016). A recent achievement in culturing cyanobacteria demonstrated their ability to adapt to environmental changes and rich high-density growth. On the other hand, cyanobacteria are good objects for molecular biologists. Hence, in bioenergetics, efforts have been focused to make genetically engineered constructs in order to produce several different biofuels (Oliver et al. 2016; Khan et al. 2019).

Compared to heterotrophic microorganisms, cyanobacteria possess a low CO_2 fixation rate which requires many efforts to go from cyanochemicals to cyanofactories. It must be noticed, that cyanobacteria have a remarkable growth rate in comparison to other microalgae. Hence, they could be highly competitive as producers of BAC intended for application in medicine, agriculture, bio-energy, etc.

Let us start from bioethanol as a simple chemical. Briefly, details of scenario about microalgae ethanol production (Kroumov et al. 2017), can be presented as follows:

Carbon dioxide from industrial gases (for example, flue gas, and waste gas from ethanol fermentation) is an excellent carbon source for cyanobacteria growth and carbohydrate production. This alternative is very attractive and cost-effective when the task is bio-ethanol production. Simple stoichiometry shows (Eq. 1) that from 1 g of glucose cells could be produced 0.51 g of ethanol and 0.49 g of CO_2 is emitted.

$$C_6H_{12}O_6 \rightarrow 2C_2H_5OH + 2CO_2 \quad (1)$$

Biorefineries are flexible and they allow a significant reduction of the cost of all generations (first, second, and third) biofuels production. The applied technologies may incorporate existing infrastructure of transport as well as to integrate the water, energy, feedstocks, and by-products resources. The carbon cycle environmental benefits must not be neglected, as well. An example for the last decade showed that the classical technologies of CO_2 utilization remained costly (22–36 € per ton of Carbon) in terms of energy consumption (Van Den Hende et al. 2012). A very complex techno-economic analysis with a modeling procedure investigating several CO_2 capture technologies was recently published (Sukor et al. 2020).

This chapter will focus on the mathematical analysis of sub-systems of the biorefinery concept. Special attention will be paid to modeling approach of medium optimization. Nutrients present a high part of the overall costs of process development and realization. The complex approach of nutrients' calculation is a base for optimization of target metabolites: hydrogen, bioethanol, biodiesel, and other products. Hence, the process development of cyanobacteria will be analyzed by using complex approaches such as system analysis theory. Its principles of decomposition could be successfully applied for medium optimization aiming to minimize the costs and research efforts. In fact, this is the main objective of the zero-waste policy of biorefinery concept.

20.2 Role of Metabolic Engineering to Achieve Effective Technology from Cyanobacteria

In the context of the economic constraints of the process of bioethanol/biofuels by cyanobacteria, one crucial point remains—the metabolic engineering of targeted strain to enhance ethanol/carbohydrates of cyanobacteria. Analysis of many metabolic engineering tools adapted to cyanobacteria is discussed in the review of Carroll et al. (2018).

The authors outlined the cyanobacteria as a target for metabolic manipulation. They analyzed the metabolic pathways in which energy valuable metabolites can be synthesized. Cyanobacterial chemical production was systematized for *Synechococcus elongatus* PCC 7942, *Synechococcus* sp. PCC 7002, *Synechocystis* sp. PCC 6803 strains, and recently for *Synechococcus elongatus* UTEX 2973. The authors logically compared the physiology of studied strains and their potential and limitations to produce targeted compounds.

Conclusions from this work were that efforts on cyanobacteria metabolic engineering gave a very positive message for cyanobacteria as a host organism. The new technology usage of CRISPR-cas9 and GSMs (Carroll et al. 2018) is a novel way and definitely will increase the activity of CO_2 utilizing strains which is very perspective as well as the overall cell behavior under these changes.

Therefore, many scientists are searching for the most cost-effective methodology and approach in metabolic engineering aiming to overcome the low productivity of cyanobacteria by manipulating their pathways before industrial biofuel production can be fully realized.

Classification of research efforts on crucially important topics in the field is organized as follows (Zahra et al. 2020):

- Engineering the genetic structures of cyanobacteria.
- Improving the physiological effectiveness.
- Optimizing various parameters of the targeted pathways.
- Optimization of the pathway flux by using different approaches.
- Improving the carbon fixation (Gustavsson and Lee 2016).

The role of synthetic biology of cyanobacteria has been increased substantially and relevant details can be found elsewhere (Weiwen Zhang 2018; Knoot et al. 2018).

20.3 Biofuel Production Perspectives

Recently, scientific efforts resulted in the creation of efficient cyanobacterial strains and were reported in publications referring to the production of various chemical compounds—ethanol, isobutanol, 2,3-butanediol, etc. Nevertheless, it should be noticed, the synthesis of most chemicals in cyanobacteria is still at an early stage. Low metabolic output of desired metabolites still limits the wide applicability of cyanobacteria as their producers in the industry. Innovative methodologies have to be developed in order to solve these multilayer problems. Interdisciplinary studies incorporating researchers in the field of algology, metabolic, and process engineering definitely will provide robust future achievements. Progress in structural and synthetic biology will guarantee more efficient cyanobacterial strains for the market in the near future. The expectations are connected with the improvement of their photosynthetic efficiency.

Though in the synthesis of lipids as a source of bioenergy the photoautotrophic cyanobacteria have obvious advantages over heterotrophic microorganisms and plants, several known drawbacks should be overcome.

1. Solving the scale-up problems when culturing cyanobacteria for the production of bio-energetic metabolites via CO_2 recycling from industrial waste gases is crucial. Autotrophic mode of cultivation has the disadvantage of having a daily light/dark cycle which decreased synthetic productivity.
2. The mixotrophic and heterotrophic mode of cultivation of cyanobacteria can be studied as cost-effective alternatives.
3. How to prevent contamination in open pond systems (which are the solely cost-effective large-scale ones) becomes a technological problem.
4. Downstream unit operations such as separation and extraction techniques are costly and should be considered in the working capital together with cultivation ones in the algal biofuel industry. It is obvious that they need further steps of overall optimization. A good solution is to use self-flocculating strains in which the excretion of desired metabolites in the medium is satisfactory. This is a very vast topic and cannot be discussed in detail in this book chapter.

Further, we are going to focus on molecular biology achievements, siderophores, modern robust methods of medium optimization, and principles of photobioreactors modeling and design.

Let us follow this step-by-step procedure by showing some achievements of researchers in the field. They can be summarized as follows:

Responses of cyanobacteria to various stress factors such as salt among multiple organisms (Stanley and Bandara 2010; Foo et al. 2014). The application of the sRNA tools for genetic manipulation, the genome-wide regulation of target genes, and the knockdown of essential genes have become a promising approach in the field of synthetic biology (Na et al. 2013; Gaida et al. 2013). The influence of light as a harmful component to photosynthetic organisms and their metabolism is in direct link with efficiency optimization of metabolic machinery in cyanobacteria (Li et al. 2009). Production of high-value chemicals from cyanobacteria was accelerated and well documented (Klemenčič et al. 2017). Metabolic engineering boost the creation of new cyanobacteria and their potential to synthesize BAC was enriched by internal metabolic machinery modification (Englund et al. 2016).

Therefore, metabolic methods offer an innovative design of cyanobacteria/microalgae cells linked with the control of metabolic system response to targeted valuable BAC. This is because the current state of art in the field showed low productivity which does not meet the requirements for transferring the process in large-scale facilities. This can be demonstrated by analyzing the results from culturing of transgenic strains and their different behavior under the conditions of scaling. This is the so-called Effect of scale-up which has the highest priority in technology transfer to industrial scale. Further, the modeling tools helping to solve this task will be discussed substantially in this review. They are the most reliable methods for any process development including the application of cyanobacteria as green biorefineries.

The latest achievements and knowledge database about all aspects of research on cyanobacteria are summarized in many books (Lüttge et al. 2012; Weiwen Zhang 2018; Luan et al. 2020; Singh et al. 2020).

The reader may find in these books the molecular methods to examine cyanobacterial diversity such as PCR-based DNA profiling methods and PCR-independent/genome-based DNA profiling methods as well as database resources for cyanobacterial research—Cyanobacterial KnowledgeBase; TAU-MAC Culture Collection; CyanoBase, etc.

20.4 Nutrient Medium Effects over Cyanobacteria Performance at the Cellular Level

20.4.1 Siderophores as Key Factors in Metal Transport

Knowledge on siderophores' role in cyanobacteria cultivation in controlled conditions is the key to optimize medium components, which is crucial to the overall process development. For this reason, here we are going to provide information about the robust achievements in this field.

Cyanobacteria, Actinobacteria, Firmicutes, and Proteobacteria contain comparable numbers of coding genes for trace metals such as metalloproteins (Zerkle 2005). For example, many metal cofactor-containing proteins have existed before Gram-negative bacteria became independent branches (Hug et al. 2016).

20.4.2 Cynobacteria and their Siderophores

Schizokinen is a known siderophore synthesized by some eubacteria (Gross et al. 1985; Singh et al. 2020). In *Synechococcus* sp. PCC 7002 was found the synechobactin (Řezanka et al. 2018). The structure of this siderophore is analyzed in detail by the authors (Ito and Butler 2005). Works on genes identification involved in aerobactin biosynthesis in *Anabaena* sp. and other cyanobacteria were published by authors (Jeanjean et al. 2008; Nicolaisen et al. 2008; Hopkinson and Morel 2009; Singh et al. 2020).

The researchers believed that the key enzymes involved in the anachelin synthesis can be classified as a salicylate synthase, a salicylate-specific loading module and a tyrosine hydroxylase.

The information found in bioinformatics (Johnson et al. 2008), and the bioinformatic tool BLAST directed research efforts to experimental verification of the putative anachelin gene cluster. Complete understanding of its function most probably will lead to findings of new BAC in cyanobacteria.

20.4.3 Current Studies on Iron Uptake by Cyanobacteria

Hypotheses of iron uptake by cyanobacteria are published (Sonier et al. 2012). For *Anabaena* sp., forms of iron (Fe^{2+} and Fe^{3+}) complexed with siderophores are suggested (Rudolf et al. 2016). The most common conditions of iron uptake are connected with starvation or access to the environment. The interaction between the environmental iron and cell requirements for it were studied by the authors (Sharma and Johri 2003). Such iron utilization can be an important criterion for the adequacy of photoautotrophs under iron-depleted conditions. A study shows that under iron access conditions, the siderophores of *A. variabilis* performed allelopathic activity (Matz et al. 2004). Further, an interesting observation is that the anachelin of *A. cylindrica PCC 7122* inhibits the metabolic activities of *Kirchneriella contorta* and *Chlamydomonas reinhardtii* (De Sarkar et al. 2016). Nevertheless, this mode of allelopathic interactions in cyanobacteria needs further investigation. It is observed that some cyanobacterial strains synthesize various siderophores (Wilhelm and Trick 1994). Hence, most likely the cyanobacterial siderophores perform various functions besides iron transport.

Several studies reported the uptake of other metals by cyanobacteria. For example, the important function of Mn in photosystem II has been studied (Nelson and Junge 2015). In presence of other metals, Mn has shown competitive function for the active site of enzymes (Lynch and St.Clair 2004). Hence, the transport and storage of Mn have to be monitored and controlled. Details of Mn transport mechanisms in cyanobacteria can be found elsewhere (Tottey et al. 2008; Sharon et al. 2014; Brandenburg et al. 2017; Singh et al. 2020). Also, details about the copper and zinc transport across membranes can be found in many works (Thelwell et al. 1998; Grass et al. 2005; Singh et al. 2020). Furthermore, studies of uptake of nickel and cobalt by cyanobacteria are essential because these metals are biologically fundamental transition metals (Huertas et al. 2014; Singh et al. 2020).

The regulation of metal transport is studied extensively and specific studies can be found elsewhere (Giedroc and Arunkumar 2007; Giner-Lamia et al. 2014; Foster et al. 2014; Sharon et al. 2014).

20.4.4 Siderophores and Future Perspectives in the Area

The understanding of the molecular and biological activities of some compounds of Fe uptake in cyanobacteria has increased, but this is not enough to fully clarify other metal ion's transport. Characterization of the transcriptional regulators in cyanobacteria was done. It helped to identify the main systems involved in iron uptake and those connected with the nitrogen metabolism and the photosynthesis chain (Hernández et al. 2004a, b; López-Gomollón et al. 2007a, b; González et al. 2010). Studies demonstrated a direct interconnection between the major metabolic pathways and their corresponding key regulators (Singh et al. 2020).

Therefore, the main goal is to obtain successfully modified and metabolically active cyanobacteria, and after that to enrich our knowledge and data on the important stages linked to process development on a big scale. Hence, the improvement of engineered designs of cyanobacteria will highlight a wide range of possibilities and will boost the biorefinery concept success.

20.5 Process Development at Macro Population Level

Intimate cellular mechanisms of metal transport have a direct link with cyanobacterial behavior on the population level. This knowledge must be taken into consideration in the medium nutrients' optimization as a subsystem of process development and scaling up. Hence, the next step includes the analysis on a macro population level, where the researcher may use a modern and robust method for optimization of macro- and micro components of the nutrient media. The industrial application and technical–economic competitiveness of any cyanobacterial/microalgal strain are impossible if its culturing conditions (substrates supply—macro and micro components for cellular growth) are not optimal. Therefore, special attention has to be given to this subject.

20.5.1 A Complex Theoretical Approach for Cyanobacteria/Microalgae Nutrient Medium Optimization

In our group, recently, a robust algorithm for optimization of microalgal nutrient media was developed (Kroumov et al. 2015). The algorithm uses all the information about algae chemical elements and those of inorganics in the flue gas. The precise values of nutrient components in the medium can be calculated with linear programming procedure (LPP) through non-equality equations wherein the right-hand side appeared macro- and micro-components contained in the cell. This methodology is coded in MAPLE software and has been under continuous examination and assessment in our research (in Bulgaria and Brazil) on different algae systems (Hinterholz et al. 2019; Schuelter et al. 2019; Scheufele et al. 2019); Hence, application of this approach for optimization of macro- and micronutrients of the medium for cultivation of cyanobacteria will be of a great help.

20.5.2 Description of the Algorithm

Traditionally, the optimal values of nutrients in the medium for the production of biomass and high-value products (HVPs) are performed by applying the simple trial and error procedure or by applying statistical experimental design methods. (Kathiresan et al. 2007; Yang et al. 2014; Kanaga et al. 2016).

By considering that the CO_2 stands out as a major portion of the operational costs in mass production of microalgae (Kadam 1997; Kroumov et al. 2016), as well as for cyanobacteria, a flue gas (among other waste gases from industries) can be considered as an excellent economical alternative. Hence, carbon bio-sequestration of flue gas from fossil-fuel power plants (Benemann 1993; Maeda et al. 1995; Pandit et al. 2012), industrial heater (Chae et al. 2006), natural gas-fired boiler (Doucha and Lívanský 2006), and biogas (Bose et al. 2019), was performed in an attempt to make the process competitive in industrial scale.

It must be highlighted that, besides the major importance of this aspect over the cultivation systems, there is very scarce literature about the understanding on interactions between flue gas composition, water chemistry, and algal physiology.

Hence, by applying the new algorithm for optimization of nutrients in the medium using:

(i) the principles of System Analysis Theory (Kaffarov et al. 1979, 1985),
(ii) suitable mathematical methods,
(iii) and available chemical equilibrium software for calculation of species of main gases of flue gas in water.

one may clarify the understanding of their multiple interactions. The algorithm was used to design cultivation processes with *Scenedesmus* and *Chlorella* species. The aim was the maximum fixation of CO_2 from flue gas emissions. The results were presented in several meetings of the Society of American Engineers in Massachusetts and elsewhere in the United States (Crofcheck et al. 2009a, b, 2010, 2012a). The experimental verification of this approach is published in detail (Crofcheck et al. 2012b). Furthermore, the algorithm was applied in our research in Brazil (Hinterholz et al. 2019; Schuelter et al. 2019), reaching substantial success.

Shortly in this book chapter, highlights on this methodology will be given and the benefits from it when used as a subsystem for complex photobioreactor (PBR) model development.

Firstly, the most commonly used medium recipes for culturing freshwater *Chlorella* species were analyzed by applying the knowledge on the elemental composition of algae biomass in calculation with LPP. Secondly, the macro and micro-elements composition were changed in order to design an optimal medium for outdoor cultivation studies. Finally, we succeed to simulate the $SO_{2(aq)}$ concentration and to find out the algae tolerance to it. Moreover, the knowledge about SO_2(aq) interactions with algae for a given pH was enriched.

Principles of analogy can be excellently adapted by applying this algorithm for the cyanobacterial nutrients optimization procedure.

Hence, Fig. 20.1 presents the mutual influence of flue gas composition over the aqueous medium and, consequently, on the elemental composition of the cyanobacteria. It must be highlighted that in order to reach maximum desirable responses (e.g., biomass, protein, sugars, lipids, or other HVPs contents or even the

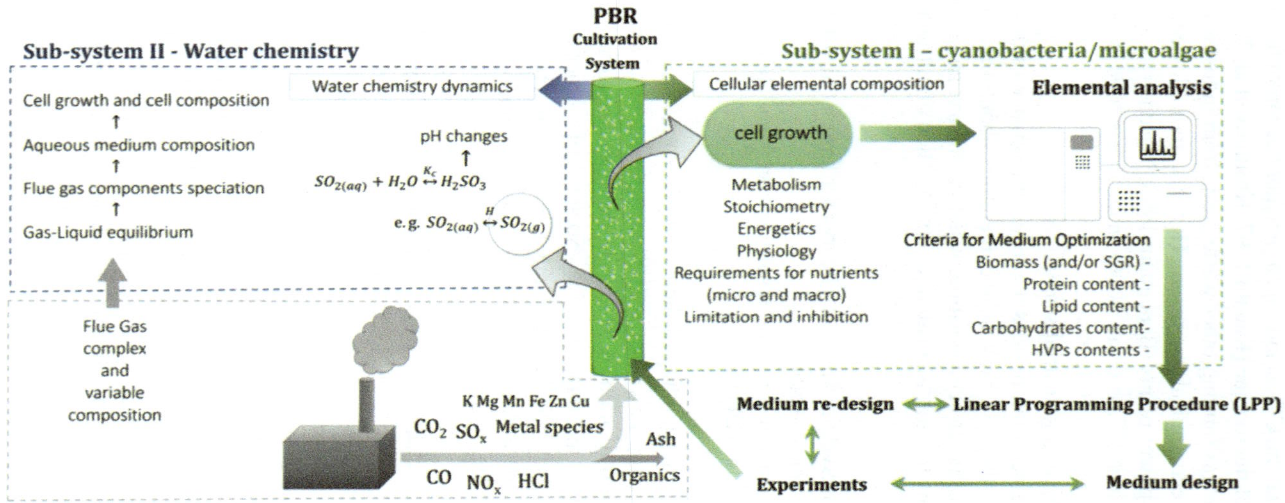

Fig. 20.1 Graphical presentation of complex theoretical approach for medium optimization

treatment of a specific compound—i.e., wastewater treatment), cyanobacteria metabolism, and physiology requirements for nutrients (macro- and micro-) must be meticulously considered. Therefore, by applying adequate analysis techniques to determine the elemental composition, along with linear programming, an optimal medium can be achieved for one specific strain and operational conditions, aiming at one or multiple maximization criteria.

Case study: The modeling of the water chemistry of flue gas absorption proved that pH decreased and reached a value of pH = 2.0. It is obligatory to control the pH by adding suitable growth buffers. In this case, the application of sodium bicarbonate was the best choice. The bicarbonate ions are utilized as a main source of carbon by cyanobacteria/microalgae under high pH (Mokashi et al. 2016; Kroumov et al. 2016; Li et al. 2018).

Recent research in the area showed an innovative approach with the application of two steps of cultivation of *Chlorella vulgaris* UTEX 395 for optimization of growth and lipid synthesis by the strain under low and high concentrations of $NaHCO_3$ (Lohman et al. 2015). The results were impressive/an increasing of specific growth rate by 69%, and total fatty acid methyl esters from 53.3 (control) up to 61% under optimal conditions/ and most importantly, such technique can be applied for other target metabolites in the contest of integral biorefinery concept. In the liquid phase, the sodium bicarbonate is considered to be a depot for CO_2 capture (Chi et al. 2011) and an excellent option, which is a key for improving the microalgal behavior in any PBR design and decreasing the cost for delivery and storage of CO_2.

20.5.3 Theoretical Basis for Algorithm Development

The existence of a deep understanding of microalgal systems including cyanobacteria gives opportunities to calculate precisely the components of macro- and microelements for optimal algae growth. Hence, the design and application of cost-effective medium are essential for process optimization. The procedure must rely on robust practical approaches, considering important relationships between the components in the broth (including dissociation and speciation of flue gases in water, i.e. water chemistry, which strongly influences pH, strain physiology, and metabolism).

Overall, the study and description of such complex systems cannot be performed by a simple analysis. Application of innovative mathematical approaches for the analysis, monitoring, and synthesis of medium is required:

- For the design of optimal medium, the application of modern direct and numerical and optimization methods (linear programming) coded in MAPLE®, MATLAB®, etc. is necessary (e.g. Jacobi, LU, Gauss-Seidel).
- Chemical reactions in water can be calculated on the basis of MINEQL+4.6 software. States such as chemical equilibrium aqueous speciation of gases, solid-phase saturation, and precipitation–dissolution are included in this package.

Other speciation software is also available, such as, MEDUSA—Make Equilibrium Diagrams Using Sophisticated Algorithms software (Puigdomenech 2004). Figure 20.1 is shown the developed algorithm for optimal medium design.

The stages could be separated conditionally as follows:

- Subsystem I—cyanobacteria/microalgae.
- Subsystem II—flue gas.
- Subsystem III—water chemistry. Knowledge about stages I, II, and III would help to formulate the optimization criterion (objective function) and by using the "procedure of decision-making" to choose the option medium design/ experiments/medium redesign.

20.5.4 Subsystem I—Cyanobacteria/Microalgae Physiology

Algae physiology is a key point for process development and it is localized as the first subsystem. The estimation of freshwater medium for culturing of *Chlorella* strains is a topic. The stoichiometry of chosen microalgae allows to evaluate physiological requirements of the strain for nutrients. It has to be noticed that the calculation procedure is not a single act and requires loop estimates where the medium selection for industrial application is a final step.

20.5.4.1 Cyanobacteria/Microalgae Biomass Elemental Composition

Aiming to investigate the element requirements for each particular cyanobacterial/ microalgae strain, elemental analysis can be performed on inorganic medium components which are present in the biomass composition. Several analytical techniques can be used for such purpose (e.g., ICP-OES—inductively coupled plasma–optical emission spectrometry; TXRF—total reflection X-ray spectrometry (TXRF); LIBS—Laser-induced breakdown spectroscopy, etc.) (Pořízka et al. 2012; Espinoza-Quiñones et al. 2015). Biomass nitrogen content was used as a reference element. Also, for simplicity, the elemental composition of the biomass can be assumed constant (i.e., the dynamic changes are not considered in this procedure).

The first step in the calculations of medium components is to evaluate the most common medium recipes (such as N-8 and M-8 (Mandalam and Palsson 1998), BG-11 (Borowitzka and Borowitzka 1988), M4N medium (Watanabe and Saiki 1997), and Watanabe Medium (Scragg et al. 2002), used extensively for the cultivation of freshwater microalgae. Moreover, any nutrient medium which can be applied for industrial culturing of cyanobacteria/microalgae or target strain must be evaluated qualitatively and quantitatively in this step. The standard of optimal growth was achieved by precise calculation of the chemical components of the medium. It must be noticed, that any medium for cyanobacteria culturing for industrial application can be evaluated precisely by this algorithm. Furthermore, we are going to show the details and achievements for designing the optimal medium for culturing *Chlorella*. The algorithm can be used straightforward for cyanobacteria.

Hence, the linear programming technique (coded in Maple software) was applied to obtain the necessary amounts of chemical components. They were compared with the values used in the recipes.

Note: Values of macro- and microelements (Min, Max) for N, P, K, Mg, S, Fe, Zn, Cu, Mn can be found in (Oh-Hama and Miyachi 1988).

20.5.4.2 Linear Programming Procedure

Nine chemical elements were chosen to appear in the highest concentrations. The trace elements are not the problem because they are comprised enough as impurities in the used chemicals. Therefore, through this approach, the nutrients can be calculated to meet the growth requirements for optimal biomass production. To calculate the optimal value of nutrients the LPP was used. Details can be found elsewhere (Kroumov et al. 2015).

There are different approaches to design the medium composition including statistical (factorial design) and recently developed innovative search methods of global extremum.

20.5.4.3 Requirements for Nutrients in Algology

The algal physiology of cyanobacteria/microalgae as a function of nutrients and working conditions are presented in (Borowitzka and Borowitzka 1988; Richmond 2004). It is important to study sources of cheap nutrients for example nitrogen to achieve a competitive process for optimal algae growth. If the source of microelements such as V, Mo, Co, N is a flue gas, adding growth factors in some applications has to be considered. From a financial point of view, it is challenging to evaluate the cyanobacteria growth response on cheap fertilizers as a medium. The utilization of nutrients from wastewaters is a plus. In this sense, the balance of water flows in the cultivation step has to be carefully calculated. This will allow to avoid some inhibition effects from algae products available in the cultivation broth. During autotrophic metabolism under noncontrolled conditions, the dynamics of cell growth can be measured indirectly by monitoring the pH changes. Following this logic, siderophores are important (Naito et al. 2008) in the scheme of the metal uptake (Fe, Ca, Mg, amongst other ions) where the precipitation processes of metal ions take place and their bio-availability is lowered.

20.5.4.4 Future Perspective: Medium Optimization for Cyanobacteria/Microalgae

Empirical approaches for the calculation of nutrients are of low efficiancy and require sets of experiments to be performed. Certainly, achievements in the understanding of metabolic pathways helped to design better nutrient media for culturing of microalgae/cyanobacteria. Balances based on the metabolic engineering approach supported by stoichiometric estimates most likely will be the preferable way for determination of the medium composition. Nevertheless, applied numerical methods and techniques have to be verified for adequacy in any particular case. The search for optimal medium always has to be done under physiological and financial constraints. State of the art requires further studies in the field.

20.5.5 Subsystem II—Flue Gas

According to the developed algorithm of the medium design (Fig. 20.1), two steps have to be highlighted:

1. Gas–liquid equilibrium and flue gas components speciation:
   - Dynamics of pH changes linked with the absorption of flue gas, solubilization, and speciation of its components.
2. Flue Gas Components Effect on the Cell Growth Would Be Studied Namely and Clarify
   - How the flue gas components (and their soluble species for a given pH condition) in the aqueous medium will affect the cyanobacteria/microalgae growth (i.e., nutrient requirements, limitation, and inhibition). The need to give a priority to study in detail regarding the tolerance of cyanobacteria/ microalgae to waste gases impurities such as SO_x, NO_x, HCl, etc. (flue gas, in our case) and the cell response to different flue gas compositions from the various coal combustion plants or other flue gas sources must be highlighted.
   - The modeling of flue gas absorption in alkaline solutions is well researched (Desch et al. 2006; Gómez et al. 2007; Marocco and Inzoli 2009). The recommendation from these studies is to buffer the nutrient broth in order to keep the setup pH value. Usually adding NaOH ($NaHCO_3$) solved the problem from a chemical (Chang and Rochelle 1985; Wylock et al. 2008) and microalgal physiology point of view (Hsueh et al. 2007).

The capture of CO_2 from flue gas by a chemical reaction to produce bicarbonates (e.g., $NaHCO_3$) and the use of the latter as the carbon source for microalgal cultivation is an attractive perspective. However, for some microalgae species, a combination of SO_x and NO_x has some toxic influence on the microalgae (Lee et al. 2002) and cyanobacteria (Lee et al. 2002; Bhola et al. 2014; Singh et al. 2016) growth. Hence, special attention has to be given to the capture of flue gas (actual or simulated) by microalgae/cyanobacteria. Inhibition effects related to microalgae tolerance to high CO_2 concentration and the presence of SO_x and NO_x impurities were extensively studied (Negoro et al. 1991, 1993; Maeda et al. 1995; Yanagi et al. 1995; Lee et al. 2002; Jeong et al. 2003; Yen et al. 2015).

An interesting study on the adaptation of green microalgae to unfiltered flue gas can be found elsewhere (Aslam et al. 2017). The authors claimed that a slow adaptation period for chosen green alga and increasing of flue gas supply by small doses can achieve success after several months of adaptation. The study once again pointed out that any selected strain for industrial application should be tested carefully in order to achieve 100% adaptation to untreated flue gas.

Flue gas compositions vary from one to another flue gas source (thermoelectric power plants, cement industry, etc.). If studies are performed under specific flue gas content (e.g., CO_2 up to 15% and SO_2 up to 700 ppm) it is possible to build a proper and optimal medium.

20.5.6 Subsystem III—Water Chemistry

The choice of gases absorption unit passed through calculation of chemical species in the water environment. Hence, analysis of flue gas composition showed that the dynamics and instantaneous equilibrium reactions of six dissolved species are important: $CO_{2(aq)}$, HCO_3^-; CO_3^{2-}, $SO_{2(aq)}$, HSO_3^-, and SO_3^{2-}. Details can be found elsewhere (Kroumov et al. 2016, 2017; Scheufele et al. 2019). In the case of flue gas absorption, the bicarbonate/carbonate system plays a crucial role (Ebrahimi et al. 2003).

20.5.7 Procedure of Decision-Making

In the step "Medium design/Experiments/Medium re-design" calculations used all available information from other subsystems under the chosen criterion of optimization. It must be noted that in preliminary studies the water chemistry, flue gas absorption and speciation, and microalgae growth can be simulated step by step without the need for experimental verification. Undoubtedly, subject to inspection are the interactions between levels/subsystems as well as the competing hypotheses for the chosen environmental conditions and microalgae strains.

20.5.8 Choice of Criterion for Medium Design

It must be kept in mind that the criterion for nutrient medium design in research and practice is different. In experimental conditions, it is important to maximize the microalgae/cyanobacteria growth in the broth, where the influence of sources of growth factors and vitamins supporting maximum growth are under evaluation. On a big scale, the financial reasons dominate and the process of CO_2 fixation requires minimization of costs. Therefore, the medium design has to be calculated on the basis of techno-economical criterion where optimal functioning of microalgal cells meets the cheapest possible sources of nutrients. Hybrid strategies can be applied where inorganic components of wastewaters can be used as well. The obtained information can be applied to build manuals of nutrient media. Further technology transfer on different scales will be facilitated. The constraints of nutrient medium design in lab scale determine the real intervals of research and development aiming to obtain desired optimal medium. The boundary conditions of an industrial nutrient medium design are availability of raw materials throughout the year; variation of quality of nutrients; time for mixing in order to reach homogeneity in the broth;

transport costs of all nutrient medium components; price fluctuations of chemicals; contamination problems; water recovery, etc.

The algorithm was verified for two different microalgae species and was discussed in detail (Crofcheck et al. 2012b).

Note: Practical usefulness of linear programming procedure (LPP) was verified in calculations of nutrient media applied in industrial scale and conditions (e.g., BG-11 medium; Zarrouk's Medium; Bold's Basal Medium-BBM; Ogbonna-Tanaka Medium; Modified Ogbonna-Tanaka Medium). A manual containing media recipes specified by calculation with LPP is available and some of the recipes were checked out under the Brazilian project "Science without borders," 2014–2018 (CNPq #313737/2014–4).

20.6 Modeling Procedure and Photobioreactors (PBRs) Design

The algal technology has several milestones of crucial importance. One of them is PBR design (Chandra et al. 2020), which answers the question on biomass maximization when a highly effective cyanobacterial/microalgal strain is selected (Taleb et al. 2016). Without a coherent theory for understanding the phenomena of parallel processes occurring in the bioreactor, its design will not be effective (Cui et al. 2020). Our group established a very robust approach for analysis of PBR as a sophisticated system by using system analysis theory (Kroumov et al. 2016, 2017; Hinterholz et al. 2019; Scheufele et al. 2019). The evaluation of PBRs performance was discussed in detail elsewhere (Hinterholz et al. 2019; Scheufele et al. 2019). The development of microalgae kinetics and the use of computational fluid dynamics (CFD) are milestones for the description of key subsystems which tremendously help for overall optimal PBR design. The group of Pruvost J. published many excellent studies on the combination of microalgal kinetics, light irradiation and distribution in PBRs, and their connections with hydrodynamics and mass transfer phenomena (Pruvost et al. 2006, 2008, 2015, 2016; Loubiere et al. 2011; Lee et al. 2014). This can be the basis for novel PBRs configurations with benefits for overall process development and technology transfer to industrial application in bioenergy and high-value products production fields.

20.7 Complex Biorefinery Concept for Cyanobacteria Biomass Use

The theoretical basis of the integral biorefinery concept is not new. Nevertheless, it is worth noting that it should be applied for cyanobacteria/microalgae as a visible and highly effective economic alternative to obtain HVPs and proteins, lipids, and complex sugars (González Delgado and Kafarov 2011; ParraSaldivar 2014; de Farias Silva et al. 2019; Vernès et al. 2019; Chandra et al. 2019; Bhattacharya and Goswami 2020). It must be noticed that the biorefinery concept applied for wastewater purification could be a unique perspective with definitive financial and environmental benefits (García-galán et al. 2020; Arias et al. 2020).

Hence, despite the great potential of cyanobacteria/microalgae to produce BACs, especially the multiproduct refineries (Kareya et al. 2020), the economic feasibility must meet the requirements to reduce production costs for given facilities and product quality. Because of the high competition with fuel technologies, the cyanobacteria/microalgae-based biofuels must be highly cost-effective.

20.8 Conclusions

The analysis of the tools for overall process development in innovative PBR construction was an object of this work. Starting with strain selection and metabolic engineering work on the construction of targeted strains, siderophores in cyanobacteria were also analyzed. Special attention was given to the mathematical approach for nutrient media optimization as a milestone of overall process development. The main advantage of this algorithm is minimizing the scientific efforts of multidisciplinary teams and avoiding excessive research experiments. The most important task is to identify the proper strains and nutrients in a loop procedure. Afterward, robustly can be executed experiments on a bigger scale in order to evaluate optimal schemes of microalgae functioning. The biofuels from cyanobacteria could be a feasible technology if waste gases from coal-fired plants (or other flue gas) are used. The combined use of wastewaters as a cheap supply of carbon, nitrogen, phosphorus, etc. is especially beneficial. This complexity is discussed when PBR design and integral biorefinery concept form the cost effectivity of the overall technology.

20.9 Future Perspectives

The biomass of cyanobacteria offers a reliable platform for the extension of biofuel production which is a serious step for diversification of energy sources for human society. Any innovations resulting in the creation of novel closed PBRs, modification of microalgae strains with the high ability for overproduction of target metabolites, cost-effective downstream processes under the integrated concept of biorefinery will contribute to improve the state of the art in the near future.

Acknowledgments The author gratefully acknowledged the financial support of the Brazilian project of CNPq #313737/2014-4, "Science without borders," 2014–2018 and scholarship for "Special Visiting Researcher" of the author. This program gives the opportunity to apply and verify the LPP for many algae cultivation processes. Special parts of this work were done under the Bulgarian Grant (КП-06-Н37/12), as well. The authors gratefully acknowledge the work of Assist. Prof. Yana Ilieva, Ph.D for revision and proofreading of the manuscript.

References

Arias DM, Uggetti E, García J (2020) Assessing the potential of soil cyanobacteria for simultaneous wastewater treatment and carbohydrate-enriched biomass production. Algal Res 51:102042. https://doi.org/10.1016/j.algal.2020.102042

Aslam A, Thomas-Hall SR, Mughal TA, Schenk PM (2017) Selection and adaptation of microalgae to growth in 100% unfiltered coal-fired flue gas. Bioresour Technol 233:271–283. https://doi.org/10.1016/j.biortech.2017.02.111

Benemann JR (1993) Utilization of carbon dioxide from fossil fuel-burning power plants with biological systems. Energy Convers Manag 34:999–1004. https://doi.org/10.1016/0196-8904(93)90047-E

Bhattacharya M, Goswami S (2020) Microalgae – a green multi-product biorefinery for future industrial prospects. Biocatal Agric Biotechnol 25:101580. https://doi.org/10.1016/j.bcab.2020.101580

Bhola V, Swalaha F, Ranjith Kumar R et al (2014) Overview of the potential of microalgae for CO_2 sequestration. Int J Environ Sci Technol 11:2103–2118. https://doi.org/10.1007/s13762-013-0487-6

Borowitzka M, Borowitzka L (1988) Micro-algal biotechnology. Cambridge University Press, Cambridge

Bose A, Lin R, Rajendran K et al (2019) How to optimise photosynthetic biogas upgrading: a perspective on system design and microalgae selection. Biotechnol Adv 37:107444. https://doi.org/10.1016/j.biotechadv.2019.107444

Brandenburg F, Schoffman H, Kurz S et al (2017) The Synechocystis manganese exporter Mnx is essential for manganese homeostasis in cyanobacteria. Plant Physiol 173:1798–1810. https://doi.org/10.1104/pp.16.01895

Carroll AL, Case AE, Zhang A, Atsumi S (2018) Metabolic engineering tools in model cyanobacteria. Metab Eng 50:47–56. https://doi.org/10.1016/j.ymben.2018.03.014

Chae SR, Hwang EJ, Shin HS (2006) Single cell protein production of *Euglena gracilis* and carbon dioxide fixation in an innovative photo-bioreactor. Bioresour Technol 97:322–329. https://doi.org/10.1016/j.biortech.2005.02.037

Chandra R, Iqbal HMN, Vishal G et al (2019) Algal biorefinery: a sustainable approach to valorize algal-based biomass towards multiple product recovery. Bioresour Technol 278:346–359. https://doi.org/10.1016/j.biortech.2019.01.104

Chandra R, Vishal G, Sánchez CEG, Uribe JAG (2020) Bioreactor for algae cultivation and biodiesel production. In: Singh L, Yousuf A, Mahapatra DMBT-B (eds) . Elsevier, Bioreactors, pp 289–307

Chang CS, Rochelle GT (1985) Sulfur dioxide absorption into sodium hydroxide and sodium sulfite aqueous solutions. Ind Eng Chem Fundam 24:7–11. https://doi.org/10.1021/i100017a002

Chi Z, O'Fallon JV, Chen S (2011) Bicarbonate produced from carbon capture for algae culture. Trends Biotechnol 29:537–541. https://doi.org/10.1016/j.tibtech.2011.06.006

Crofcheck C, Montross M, Kroumov A, et al (2009a) Selection of microalgal strains and medium optimization for CO_2 mitigation from flue gas. In: Annual Institute of Biological Engineering Meeting in Santa Clara. Santa Clara, CA

Crofcheck C, Montross M, Shea A, et al (2009b) Selection of microalgal strains and medium optimization for CO_2 mitigation from flue gas. In: ASABE Annual International Meeting in Reno. Reno, NV

Crofcheck C, Montross M, Cassidy K, et al (2010) Medium optimization for CO_2 mitigation from flue gas. In: Annual Institute of Biological Engineering Meeting in Cambridge. Cambridge, MA

Crofcheck C, Monstross M, Xinyi E, et al (2012a) Influence of media composition on the growth rate of *Chlorella vulgaris* and *Scenedesmus acutus* utilized for CO_2 mitigation. In: ASABE Annual International Meeting. Dallas, TX

Crofcheck C, Shea A, Monstross M et al (2012b) Influence of media composition on the growth rate of *Chlorella vulgaris* and *Scenedesmus acutus* utilized for CO_2 mitigation. J Biochem Technol 4:589–594

Cui X, Yang J, Feng Y, Zhang W (2020) Simulation of a novel tubular microalgae photobioreactor with aerated tangent inner tubes: improvements in mixing performance and flashing-light effects. Archaea 2020:1–16. https://doi.org/10.1155/2020/8815263

de Farias Silva CE, Barbera E, Bertucco A (2019) Biorefinery as a promising approach to promote ethanol industry from microalgae and cyanobacteria. In: Ray RC, Ramachandran SBT-BP from FC (eds) Bioethanol production from food crops. Elsevier, pp 343–359

De Sarkar S, Blom JF, Bethuel Y et al (2016) Allelopathic activity of the iron chelator anachelin - a molecular hybrid with a dual mode of action. Helv Chim Acta 99:760–773. https://doi.org/10.1002/hlca.201600123

Desch W, Horn K, Propst G (2006) Computation of equilibria in models of flue gas washer plants. Comput Chem Eng 30:1169–1177. https://doi.org/10.1016/j.compchemeng.2006.02.016

Dismukes GC, Carrieri D, Bennette N et al (2008) Aquatic phototrophs: efficient alternatives to land-based crops for biofuels. Curr Opin Biotechnol 19:235–240. https://doi.org/10.1016/j.copbio.2008.05.007

Doucha J, Lívanský K (2006) Productivity, CO_2/O_2 exchange and hydraulics in outdoor open high density microalgal (*chlorella* sp.) photobioreactors operated in a middle and southern European climate. J Appl Phycol 18:811–826. https://doi.org/10.1007/s10811-006-9100-4

Ebrahimi S, Picioreanu C, Kleerebezem R et al (2003) Rate-based modelling of SO_2 absorption into aqueous $NaHCO_3/Na_2CO_3$ solutions accompanied by the desorption of CO_2. Chem Eng Sci 58:3589–3600. https://doi.org/10.1016/S0009-2509(03)00231-8

Englund E, Liang F, Lindberg P (2016) Evaluation of promoters and ribosome binding sites for biotechnological applications in the unicellular cyanobacterium *Synechocystis* sp. PCC 6803. Sci Rep 6:36640. https://doi.org/10.1038/srep36640

Espinoza-Quiñones FR, Módenes AN, de Pauli AR, Palácio SM (2015) Analysis of trace elements in groundwater using ICP-OES and TXRF techniques and its compliance with Brazilian protection standards. Water Air Soil Pollut 226:32. https://doi.org/10.1007/s11270-015-2315-8

Farrokh P, Sheikhpour M, Kasaeian A, Asadi H, Bavandi R (2019) Cyanobacteria as an eco-friendly resource for biofuel production: a critical review. Biotechnol Prog 35(5):e2835

Foo JL, Jensen HM, Dahl RH et al (2014) Improving microbial biogasoline production in *Escherichia coli* using tolerance engineering. MBio 5:1–9. https://doi.org/10.1128/mBio.01932-14

Foster AW, Osman D, Robinson NJ (2014) Metal preferences and metallation. J Biol Chem 289:28095–28103. https://doi.org/10.1074/jbc.R114.588145

Gaida SM, Al-Hinai MA, Indurthi DC et al (2013) Synthetic tolerance: three noncoding small RNAs, DsrA, ArcZ and RprA, acting supra-additively against acid stress. Nucleic Acids Res 41:8726–8737. https://doi.org/10.1093/nar/gkt651

García-galán MJ, Arashiro L, Santos LHMLM et al (2020) Fate of priority pharmaceuticals and their main metabolites and transformation products in microalgae-based wastewater treatment systems. J Hazard Mater 390:121771. https://doi.org/10.1016/j.jhazmat.2019.121771

Giedroc DP, Arunkumar AI (2007) Metal sensor proteins: nature's metalloregulated allosteric switches. Dalt Trans:3107–3120. https://doi.org/10.1039/b706769k

Giner-Lamia J, López-Maury L, Florencio FJ (2014) Global transcriptional profiles of the copper responses in the cyanobacterium *Synechocystis* sp. PCC 6803. PLoS One 9:e108912. https://doi.org/10.1371/journal.pone.0108912

Gómez A, Fueyo N, Tomás A (2007) Detailed modelling of a flue-gas desulfurisation plant. Comput Chem Eng 31:1419–1431. https://doi.org/10.1016/j.compchemeng.2006.12.004

Gonçalves VD, Fagundes-Klen MR, Trigueros DEG et al (2019a) Statistical and optimization strategies to carotenoids production by *Tetradesmus acuminatus* (LC192133.1) cultivated in photobioreactors. Biochem Eng J 152:107351. https://doi.org/10.1016/j.bej.2019.107351

Gonçalves VD, Fagundes-Klen MR, Trigueros DEG et al (2019b) Combination of light emitting diodes (LEDs) for photostimulation of carotenoids and chlorophylls synthesis in *Tetradesmus* sp. Algal Res 43:101649. https://doi.org/10.1016/j.algal.2019.101649

González Delgado ÁD, Kafarov V (2011) Microalgae based biorefinery: issues to consider. CT&F - Ciencia, Tecnol y Futur 4:05–21. https://doi.org/10.29047/01225383.225

González A, Bes MT, Barja F et al (2010) Overexpression of FurA in *anabaena* sp. PCC 7120 reveals new targets for this regulator involved in photosynthesis, iron uptake and cellular morphology. Plant Cell Physiol 51:1900–1914. https://doi.org/10.1093/pcp/pcq148

Grass G, Franke S, Taudte N et al (2005) The metal permease ZupT from *Escherichia coli* is a transporter with a broad substrate spectrum. J Bacteriol 187:1604–1611. https://doi.org/10.1128/JB.187.5.1604-1611.2005

Gross R, Engelbrecht F, Braun V (1985) Identification of the genes and their polypeptide products responsible for aerobactin synthesis by pColV plasmids. Mol Gen Genet MGG 201:204–212. https://doi.org/10.1007/BF00425661

Gustavsson M, Lee SY (2016) Prospects of microbial cell factories developed through systems metabolic engineering. Microb Biotechnol 9:610–617. https://doi.org/10.1111/1751-7915.12385

Hernández JA, López-Gomollón S, Bes MT et al (2004a) Three fur homologues from *anabaena* sp. PCC7120: exploring reciprocal protein-promoter recognition. FEMS Microbiol Lett 236:275–282. https://doi.org/10.1111/j.1574-6968.2004.tb09658.x

Hernández JA, Peleato ML, Fillat MF, Bes MT (2004b) Heme binds to and inhibits the DNA-binding activity of the global regulator FurA from *anabaena* sp. PCC 7120. FEBS Lett 577:35–41. https://doi.org/10.1016/j.febslet.2004.09.060

Hinterholz CL, Trigueros DEG, Módenes AN et al (2019) Computational fluid dynamics applied for the improvement of a flat-plate photobioreactor towards high-density microalgae cultures. Biochem Eng J 151:107257. https://doi.org/10.1016/j.bej.2019.107257

Hopkinson BM, Morel FMM (2009) The role of siderophores in iron acquisition by photosynthetic marine microorganisms. Biometals 22:659–669. https://doi.org/10.1007/s10534-009-9235-2

Hsueh HT, Chu H, Yu ST (2007) A batch study on the bio-fixation of carbon dioxide in the absorbed solution from a chemical wet scrubber by hot spring and marine algae. Chemosphere 66:878–886. https://doi.org/10.1016/j.chemosphere.2006.06.022

Huertas M, López-Maury L, Giner-Lamia J et al (2014) Metals in cyanobacteria: analysis of the copper, nickel, cobalt and arsenic homeostasis mechanisms. Life 4:865–886. https://doi.org/10.3390/life4040865

Hug LA, Baker BJ, Anantharaman K et al (2016) A new view of the tree of life. Nat Microbiol 1:16048. https://doi.org/10.1038/nmicrobiol.2016.48

Ito Y, Butler A (2005) Structure of synechobactins, new siderophores of the marine cyanobacterium *Synechococcus* sp. PCC 7002. Limnol Oceanogr 50:1918–1923. https://doi.org/10.4319/lo.2005.50.6.1918

Jeanjean R, Talla E, Latifi A et al (2008) A large gene cluster encoding peptide synthetases and polyketide synthases is involved in production of siderophores and oxidative stress response in the cyanobacterium *anabaena* sp. strain PCC 7120. Environ Microbiol 10:2574–2585. https://doi.org/10.1111/j.1462-2920.2008.01680.x

Jeong ML, Gillis JM, Hwang JY (2003) Carbon dioxide mitigation by microalgal photosynthesis. Bull Kor Chem Soc 24:1763–1766. https://doi.org/10.5012/bkcs.2003.24.12.1763

Johnson M, Zaretskaya I, Raytselis Y et al (2008) NCBI BLAST: a better web interface. Nucleic Acids Res 36:W5–W9. https://doi.org/10.1093/nar/gkn201

Kadam KL (1997) Power plant flue gas as a source of CO_2 for microalgae cultivation: economic impact of different process options. Energy Convers Manag 38:S505–S510. https://doi.org/10.1016/S0196-8904(96)00318-4

Kaffarov V, Vinarov A, Gordeev L (1979) Modelling biochemical reactors (in Russian). Forrest Industry, Moscow

Kaffarov V, Vinarov A, Gordeev L (1985) Modelling and system analysis of biochemical industrial production (in Russian). Forrest Industry, Moscow

Kanaga K, Pandey A, Kumar S, Geetanjali (2016) Multi-objective optimization of media nutrients for enhanced production of algae biomass and fatty acid biosynthesis from *Chlorella pyrenoidosa* NCIM 2738. Bioresour Technol 200:940–950. https://doi.org/10.1016/j.biortech.2015.11.017

Kareya MS, Mariam I, Arumugam Nesamma A, Jutur PP (2020) CO2 sequestration by hybrid integrative photosynthesis (CO_2-SHIP): a green initiative for multi-product biorefineries. Mater Sci Energy Technol 3:420–428. https://doi.org/10.1016/j.mset.2020.03.002

Kathiresan S, Sarada R, Bhattacharya S, Ravishankar GA (2007) Culture media optimization for growth and phycoerythrin production from *Porphyridium purpureum*. Biotechnol Bioeng 96:456–463. https://doi.org/10.1002/bit.21138

Khan AZ, Bilal M, Mehmood S et al (2019) State-of-the-art genetic modalities to engineer cyanobacteria for sustainable biosynthesis of biofuel and fine-chemicals to meet bio–economy challenges. Life 9:1–22. https://doi.org/10.3390/life9030054

Klemenčič M, Nielsen AZ, Sakuragi Y et al (2017) 13 – Synthetic biology of cyanobacteria for production of biofuels and high-value products. In: Gonzalez-Fernandez C, Muñoz RBT-M-BB (eds) Woodhead publishing series in energy. Woodhead Publishing, pp 305–325

Knoot CJ, Ungerer J, Wangikar PP, Pakrasi HB (2018) Cyanobacteria: promising biocatalysts for sustainable chemical production. J Biol Chem 293:5044–5052. https://doi.org/10.1074/jbc.R117.815886

Kroumov AD, Módenes AN, Trigueros DEG (2015) A complex theoretical approach for algal medium optimization for CO_2 fixation from flue gas. Acta Microbiol Bulg 31:61–70

Kroumov AD, Módenes AN, Trigueros DEG et al (2016) A systems approach for CO_2 fixation from flue gas by microalgae—theory review. Process Biochem 51:1817–1832. https://doi.org/10.1016/j.procbio.2016.05.019

Kroumov AD, Scheufele FB, Trigueros DEG et al (2017) Modeling and technoeconomic analysis of algae for bioenergy and coproducts. In: Algal green chemistry. Elsevier, pp 201–241

Lee J-S, Kim D-K, Lee J-P et al (2002) Effects of SO_2 and NO on growth of *chlorella* sp. KR-1. Bioresour Technol 82:1–4. https://doi.org/10.1016/S0960-8524(01)00158-4

Lee E, Pruvost J, He X et al (2014) Design tool and guidelines for outdoor photobioreactors. Chem Eng Sci 106:18–29. https://doi.org/10.1016/j.ces.2013.11.014

Li Z, Wakao S, Fischer BB, Niyogi KK (2009) Sensing and responding to excess light. Annu Rev Plant Biol 60:239–260. https://doi.org/10.1146/annurev.arplant.58.032806.103844

Li J, Li C, Lan CQ, Liao D (2018) Effects of sodium bicarbonate on cell growth, lipid accumulation, and morphology of *Chlorella vulgaris*. Microb Cell Factories 17:111. https://doi.org/10.1186/s12934-018-0953-4

Lohman EJ, Gardner RD, Pedersen T et al (2015) Optimized inorganic carbon regime for enhanced growth and lipid accumulation in *Chlorella vulgaris*. Biotechnol Biofuels 8:82. https://doi.org/10.1186/s13068-015-0265-4

López-Gomollón S, Hernández JA, Pellicer S et al (2007a) Cross-talk between iron and nitrogen regulatory networks in *anabaena* (*Nostoc*) sp. PCC 7120: identification of overlapping genes in FurA and NtcA regulons. J Mol Biol 374:267–281. https://doi.org/10.1016/j.jmb.2007.09.010

López-Gomollón S, Hernández JA, Wolk CP et al (2007b) Expression of furA is modulated by NtcA and strongly enhanced in heterocysts of *anabaena* sp. PCC 7120. Microbiology 153:42–50. https://doi.org/10.1099/mic.0.2006/000091-0

Loubiere K, Pruvost J, Aloui F, Legrand J (2011) Investigations in an external-loop airlift photobioreactor with annular light chambers and swirling flow. Chem Eng Res Des 89:164–171. https://doi.org/10.1016/j.cherd.2010.06.001

Luan G, Zhang S, Lu X (2020) Engineering cyanobacteria chassis cells toward more efficient photosynthesis. Curr Opin Biotechnol 62:1–6. https://doi.org/10.1016/j.copbio.2019.07.004

Lüttge U, Beyschlag W, Büdel B, Francis D (eds) (2012) Progress in botany, vol 73. Springer, Berlin, Heidelberg

Lynch JP, St.Clair SB (2004) Mineral stress: the missing link in understanding how global climate change will affect plants in real world soils. F Crop Res 90:101–115. https://doi.org/10.1016/j.fcr.2004.07.008

Machado IMP, Atsumi S (2012) Cyanobacterial biofuel production. J Biotechnol 162:50–56. https://doi.org/10.1016/j.jbiotec.2012.03.005

Maeda K, Owada M, Kimura N et al (1995) CO_2 fixation from the flue gas on coal-fired thermal power plant by microalgae. Energy Convers Manag 36:717–720. https://doi.org/10.1016/0196-8904(95)00105-M

Mandalam RK, Palsson B (1998) Elemental balancing of biomass and medium composition enhances growth capacity in high-density *Chlorella vulgaris* cultures. Biotechnol Bioeng 59:605–611. https://doi.org/10.1002/(SICI)1097-0290(19980905)59:5<605::AID-BIT11>3.0.CO;2-8

Marocco L, Inzoli F (2009) Multiphase Euler–Lagrange CFD simulation applied to wet flue gas desulphurisation technology. Int J Multiph Flow 35:185–194. https://doi.org/10.1016/j.ijmultiphaseflow.2008.09.005

Matz CJ, Christensen MR, Bone AD et al (2004) Only iron-limited cells of the cyanobacterium *Anabaena flos-aquae* inhibit growth of the green alga *Chlamydomonas reinhardtii*. Can J Bot 82:436–442. https://doi.org/10.1139/b04-022

Mokashi K, Shetty V, George SA, Sibi G (2016) Sodium bicarbonate as inorganic carbon source for higher biomass and lipid production integrated carbon capture in *Chlorella vulgaris*. Achiev Life Sci 10:111–117. https://doi.org/10.1016/j.als.2016.05.011

Na D, Yoo SM, Chung H et al (2013) Metabolic engineering of *Escherichia coli* using synthetic small regulatory RNAs. Nat Biotechnol 31:170–174. https://doi.org/10.1038/nbt.2461

Naito K, Imai I, Nakahara H (2008) Complexation of iron by microbial siderophores and effects of iron chelates on the growth of marine microalgae causing red tides. Phycol Res 56:58–67. https://doi.org/10.1111/j.1440-1835.2008.00485.x

Negoro M, Shioji N, Miyamoto K, Micira Y (1991) Growth of microalgae in high CO_2 gas and effects of SO_X and NO_X. Appl Biochem Biotechnol 28–29:877–886. https://doi.org/10.1007/BF02922657

Negoro M, Hamasaki A, Ikuta Y et al (1993) Carbon dioxide fixation by microalgae photosynthesis using actual flue gas discharged from a boiler. Appl Biochem Biotechnol 39–40:643–653. https://doi.org/10.1007/BF02919025

Nelson N, Junge W (2015) Structure and energy transfer in photosystems of oxygenic photosynthesis. Annu Rev Biochem 84:659–683. https://doi.org/10.1146/annurev-biochem-092914-041942

Nicolaisen K, Moslavac S, Samborski A et al (2008) Alr0397 is an outer membrane transporter for the siderophore schizokinen in *anabaena* sp. strain PCC 7120. J Bacteriol 190:7500–7507. https://doi.org/10.1128/JB.01062-08

Oh-Hama T, Miyachi S (1988) Chlorella. In: Michael A, Borowitzka LJ (eds) . Cambridge University Press, Micro-algal biotechnology, pp 3–26

Oliver NJ, Rabinovitch-Deere CA, Carroll AL et al (2016) Cyanobacterial metabolic engineering for biofuel and chemical production. Curr Opin Chem Biol

Pandit S, Nayak BK, Das D (2012) Microbial carbon capture cell using cyanobacteria for simultaneous power generation, carbon dioxide sequestration and wastewater treatment. Bioresour Technol 107:97–102. https://doi.org/10.1016/j.biortech.2011.12.067

ParraSaldivar R (2014) Algae biofuels production processes, carbon dioxide fixation and biorefinery concept. J Pet Environ Biotechnol 05(4):184. https://doi.org/10.4172/2157-7463.1000185

Pořízka P, Prochazka D, Pilát Z et al (2012) Application of laser-induced breakdown spectroscopy to the analysis of algal biomass for industrial biotechnology. Spectrochim Acta Part B At Spectrosc 74–75:169–176. https://doi.org/10.1016/j.sab.2012.06.014

Pruvost J, Pottier L, Legrand J (2006) Numerical investigation of hydrodynamic and mixing conditions in a torus photobioreactor. Chem Eng Sci 61:4476–4489. https://doi.org/10.1016/j.ces.2006.02.027

Pruvost J, Cornet JF, Legrand J (2008) Hydrodynamics influence on light conversion in photobioreactors: an energetically consistent analysis. Chem Eng Sci 63:3679–3694. https://doi.org/10.1016/j.ces.2008.04.026

Pruvost J, Cornet JF, Le Borgne F et al (2015) Theoretical investigation of microalgae culture in the light changing conditions of solar photobioreactor production and comparison with cyanobacteria. Algal Res 10:87–99. https://doi.org/10.1016/j.algal.2015.04.005

Pruvost J, Le Gouic B, Lepine O et al (2016) Microalgae culture in building-integrated photobioreactors: biomass production modelling and energetic analysis. Chem Eng J 284:850–861. https://doi.org/10.1016/j.cej.2015.08.118

Puigdomenech I (2004) HYDRA: Hydrochemical equilibrium-constant database software

Řezanka T, Palyzová A, Sigler K (2018) Isolation and identification of siderophores produced by cyanobacteria. Folia Microbiol (Praha) 63:569–579. https://doi.org/10.1007/s12223-018-0626-z

Richmond A (2004) Handbook of microalgal culture: biotechnology and applied phycology. John Wiley & Sons, p 472. https://doi.org/10.1002/9780470995280

Rudolf M, Stevanovic M, Kranzler C et al (2016) Multiplicity and specificity of siderophore uptake in the cyanobacterium *anabaena* sp. PCC 7120. Plant Mol Biol 92:57–69. https://doi.org/10.1007/s11103-016-0495-2

Sanderson K (2011) Lignocellulose: a chewy problem. Nature. https://doi.org/10.1038/474S012a

Scheufele FB, Hinterholz CL, Zaharieva MM et al (2019) Complex mathematical analysis of photobioreactor system. Eng Life Sci 19:844–859. https://doi.org/10.1002/elsc.201800044

Schuelter AR, Kroumov AD, Hinterholz CL et al (2019) Isolation and identification of new microalgae strains with antibacterial activity on food-borne pathogens. Engineering approach to optimize synthesis of desired metabolites. Biochem Eng J 144:28–39. https://doi.org/10.1016/j.bej.2019.01.007

Scragg A, Illman A, Carden A, Shales S (2002) Growth of microalgae with increased calorific values in a tubular bioreactor. Biomass Bioenergy 23:67–73. https://doi.org/10.1016/S0961-9534(02)00028-4

Sharma A, Johri BN (2003) Growth promoting influence of siderophore-producing *Pseudomonas* strains GRP3A and PRS 9 in maize (Zea mays L.) under iron limiting conditions. Microbiol Res 158:243–248

Sharon S, Salomon E, Kranzler C et al (2014) The hierarchy of transition metal homeostasis: iron controls manganese accumulation in a unicellular cyanobacterium. Biochim Biophys Acta Bioenerg 1837:1990–1997. https://doi.org/10.1016/j.bbabio.2014.09.007

Sheehan J (1998) A look Back at the U.S. Department of Energy's aquatic species program—biodiesel from algae. Program

Singh JS, Kumar A, Rai AN, Singh DP (2016) Cyanobacteria: a precious bio-resource in agriculture, ecosystem, and environmental sustainability. Front Microbiol 7:1–19. https://doi.org/10.3389/fmicb.2016.00529

Singh PK, Kumar A, Singh VK, Shrivastava AKBT-A (eds) (2020) Advances in cyanobacterial biology. Elsevier

Sonier MB, Contreras DA, Treble RG, Weger HG (2012) Two distinct pathways for iron acquisition by ironlimited cyanobacterial cells: evidence from experiments using siderophores and synthetic chelators. Botany 90:181–190. https://doi.org/10.1139/B11-099

Stanley D, Bandara A, Fraser S et al (2010) The ethanol stress response and ethanol tolerance of *Saccharomyces cerevisiae*. J Appl Microbiol 109:13–24. https://doi.org/10.1111/j.1365-2672.2009.04657.x

Sukor NR, Shamsuddin AH, Mahlia TMI, Mat Isa MF (2020) Techno-economic analysis of CO_2 capture technologies in offshore natural gas field: implications to carbon capture and storage in Malaysia. PRO 8:350. https://doi.org/10.3390/pr8030350

Takano H, Takeyama H, Nakamura N et al (1992) CO_2 removal by high-density culture of a marine cyanobacterium *Synechococcus* sp. using an improved photobioreactor employing light-diffusing optical fibers. Appl Biochem Biotechnol 34–35:449–458. https://doi.org/10.1007/BF02920568

Taleb A, Kandilian R, Touchard R et al (2016) Screening of freshwater and seawater microalgae strains in fully controlled photobioreactors for biodiesel production. Bioresour Technol 218:480–490. https://doi.org/10.1016/j.biortech.2016.06.086

Thelwell C, Robinson NJ, Turner-Cavet JS (1998) An SmtB-like repressor from *Synechocystis* PCC 6803 regulates a zinc exporter. Proc Natl Acad Sci 95:10728–10733. https://doi.org/10.1073/pnas.95.18.10728

Tottey S, Waldron KJ, Firbank SJ et al (2008) Protein-folding location can regulate manganese-binding versus copper- or zinc-binding. Nature 455:1138–1142. https://doi.org/10.1038/nature07340

Van Den Hende S, Vervaeren H, Boon N (2012) Flue gas compounds and microalgae: (bio-) chemical interactions leading to biotechnological opportunities. Biotechnol Adv 30:1405–1424. https://doi.org/10.1016/j.biotechadv.2012.02.015

Vernès L, Li Y, Chemat F, Abert-Vian M (2019) Biorefinery concept as a key for sustainable future to green chemistry—the case of microalgae. In: Li Y, Chemat F (eds) Green chemistry and sustainable technology. Springer Singapore, Singapore, pp 15–50

Watanabe Y, Saiki H (1997) Development of a photobioreactor incorporating *chlorella* sp. for removal of CO_2 in stack gas. Energy Convers Manag 38:S499–S503. https://doi.org/10.1016/S0196-8904(96)00317-2

Weiwen Zhang XS (2018) Synthetic biology of cyanobacteria

Wilhelm SW, Trick CG (1994) Iron limited growth of cyanobacteria: multiple siderophore production is a common response. Limnol Oceanogr 39:1979–1984. https://doi.org/10.4319/lo.1994.39.8.1979

Wylock CE, Colinet P, Cartage T, Haut B (2008) Coupling between mass transfer and chemical reactions during the absorption of CO_2 in a $NaHCO_3$-Na_2CO_3 brine : experimental and theoretical study. Int J Chem React Eng 6. https://doi.org/10.2202/1542-6580.1502

Yanagi M, Watanabe Y, Saiki H (1995) CO_2 fixation by *chlorella* sp. HA-1 and its utilization. Energy Convers Manag 36:713–716. https://doi.org/10.1016/0196-8904(95)00104-L

Yang F, Long L, Sun X et al (2014) Optimization of medium using response surface methodology for lipid production by *Scenedesmus* sp. Mar Drugs 12:1245–1257. https://doi.org/10.3390/md12031245

Yen H-W, Ho S-H, Chen C-Y, Chang J-S (2015) CO_2, NO_x and SO_x removal from flue gas via microalgae cultivation: a critical review. Biotechnol J 10:829–839. https://doi.org/10.1002/biot.201400707

Zahra Z, Choo DH, Lee H, Parveen A (2020) Cyanobacteria: review of current potentials and applications. Environments 7:13. https://doi.org/10.3390/environments7020013

Zerkle AL (2005) Biogeochemical signatures through time as inferred from whole microbial genomes. Am J Sci 305:467–502. https://doi.org/10.2475/ajs.305.6-8.467

The manufacturer's authorised representative in the EU is Springer Nature Customer Service Centre GmbH, Europaplatz 3, 69115 Heidelberg, Germany. If you have any concerns regarding our products, please contact ProductSafety@springernature.com

Printed and bound by CPI Group (UK) Ltd, Croydon, CR0 4YY
30/09/2026
02225332-0014